An Introduction to MRI
for Medical Physicists and Engineers

An Introduction to MRI for Medical Physicists and Engineers

Anthony Wolbarst
and Nathan Yanasak

Medical Physics Publishing
Madison, Wisconsin

26 25 24 23 22 21 20 19   1 2 3 4 5

First printing May 2019

Library of Congress Control Number: 2018950587

ISBN hardcover: 978-1-930524-20-0
ISBN eBook: 978-1-930524-58-3

Published by:
Medical Physics Publishing
4555 Helgesen Drive
Madison, WI 53718
(608) 224-4508 or 1-800-442-5778
mpp@medicalphysics.org
medicalphysics.org

Printed in the United States of America

Contents

Dedication

We joyfully dedicate this effort, and our love,
to our most splendid partners,
Ling and Wendy

"At the still point of the turning world."
–T.S. Eliot

On education, attributed to Einstein:

"Everything should be made as simple as possible, but no simpler."

Preface

This is an introduction to the science and technology of magnetic resonance imaging, written at the beginning graduate level for professional medical physicists and engineers in training.

MRI is perhaps the most celebrated and versatile of the imaging technologies now found in the clinic. It can obtain a range of types of diagnostic information, in part because of its capability to exploit a variety of unique and distinct soft-tissue contrast mechanisms at the sub-millimeter level of resolution. While employed primarily to detect and assess potential pathologies, MRI can also make possible the appraisal of tissue and organ function, and it can help to guide the direction of interventions and treatment follow-ups. It can even fuse with other modalities (e.g., PET, SPECT) to generate more specific diagnostic information. And as the literature clearly shows, the rate of MR innovation continues to grow.

The first medical MR images were published nearly a half century ago, and currently upwards of 25,000 MR devices over the globe are performing tens of millions of studies annually. Only Japan carries out more MRI studies *per capita* than the United States. Despite the costs, which are declining but still high, these numbers continue to rise.

Based on the phenomenon of *nuclear magnetic resonance* (NMR), which was devised during the Second World War, the theory and technology of MRI are generally considered to be relatively challenging for medical physicists, engineers, and technically oriented physicians to master. This introductory book is intended to help people get going in this exciting and critically important field—whether as a general clinical physicist or engineer, or a beginning MR specialist.

It is assumed that the reader has had at least two years of college general physics and calculus, including a semester of modern physics, and one of calculus. Some familiarity with computers, general imaging science, and physiology would be helpful, but not essential. Several more advanced topics, like Fourier analysis, ***k***-space, and statistical distributions, will be introduced where they are needed.

Most of the classical and modern physics terminology and symbols adopted will be those of a few widely used undergraduate texts [Eisberg et al. 1985; Feynman et al. 1964; Purcell et al. 2013; and Sears et al. 2015, 14th Ed.]. Several texts provide discussions of general medical imaging [Bushberg et al. 2012; Webb 2012; and Wolbarst 2005]. The NMR notation is primarily that of Slichter [1990], and some of the arguments are derived from those in advanced texts [Abragam 1961; Bernstein et al. 2004; Brown et al. 2014; Chen et al. 2009; Dixon et al. 1985a & 1985b; and Liang and Lauterbur 2000]. Smith et al. [1989] and Elster et al. [2001] provide non-mathematical but nontrivial and useful introductions to the field.

0.1 Welcome to MRI

MRI is a noninvasive, thin-slice or volume imaging technology that reveals the structural, anatomic details of soft tissues, often much better than do *computed tomography* (CT) or the other imaging methods (Figure 0.1). Moreover,

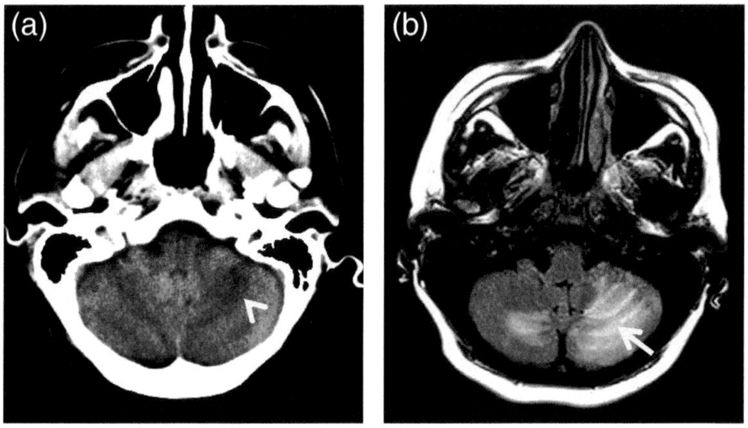

Figure 0.1 CT and MRI transverse slices at the level of the posterior fossa, displaying the cerebellar hemispheres. The patient suffers from posterior reversible encephalopathy syndrome (PRES), which presents with edematous changes of the white matter. (a) In the CT axial (transverse) reconstruction, the slight linear artifacts that cross the cerebellum in an arc are associated with the overlying skullcap; there is also a slight, ill-defined hypo-density in the white matter of the left cerebellar hemisphere (arrowhead). (b) With one particular MRI technique known as FLAIR (FLuid Attenuation Inversion Recovery), the hyper-intensity of the cerebellar white matter at the same level indicates edema (arrow) and PRES. (*Courtesy of Patrizio Capasso, MD, DSc.*)

Table 0.1 Characteristics that distinguish MRI from other imaging approaches. All of these will be discussed further on in the book.

Magnetic Resonance Imaging

Extraordinary soft-tissue contrast
- Clinical maps of both anatomy and physiology
- Multiple forms of contrast created through and reflecting different aspects of the tissue biophysics: T1, T2, blood flow, water diffusion, oxygen consumption, etc.
- Assesses rotations, flow, etc., of water and lipid molecules

Fundamental trade-off : SNR, resolution, acquisition time
- Optimizable (with RF/gradient pulse sequences, contrast agents)
- Sub-millimeter resolution
- Cardiac speed

No ionizing radiation, but...
- Relatively small risks from magnetic fields and RF
- Costly main and gradient magnets
- Abstract physics, complex pulse sequences

Table 0.2 Mechanisms of the major categories of clinical imaging modalities. ***X-ray-based:*** Computed Radiography (CR); Digital Radiography (DR) Digital Mammography (DM); Digital Tomography (DT); Digital Fluoroscopy (DF), including Digital Subtraction Angiography (DSA); Computed Tomography (CT). ***Radionuclide-based:*** Single Photon Emission Computed Tomography (SPECT) and Positron Emission Tomography (PET). ***Ultrasound-based:*** B-Mode; Doppler. ***MRI-based:*** PD-*weighted;* T1-*w;* T2-*w;* MR Angiography (MRA): MR Spectroscopy (MRS); functional MRI (*f*MRI); Diffusion Tensor Imaging (DTI), etc.

X-ray – Radiography, CR, DR, DM, DT, DF, DSA, CT

Differential attenuation of x-ray photons by tissues of differing thickness, density, chemical makeup; and affected by photon energy.

Primarily involves photoelectric absorption and Compton interactions; Compton scatter radiation reduces contrast.

Nuclear Medicine – Planar NM, SPECT, PET

Differential uptake and concentration of radiopharmaceuticals that target specific tissues/biological compartments.

Subsequent emission of detectable high-energy photon radiation.

Ultrasound – B-Mode, Doppler

Differential reflection of high-frequency mechanical vibrations at boundaries between tissues of *different elasticity* or *density*.

MRI – T1-*w*, T2-*w*, PD-*w*, MRA, MRS, fMRI, DTI, etc.

Differences among tissues in hydrogen nucleus (proton) density, spin relaxation times T1 & T2, blood flow, chemical shift, water diffusion, deoxy- vs. oxyhemoglobin concentration in blood, etc.

Arise primarily from interactions of protons in tissue water and lipids with local molecular environments.

unlike CT but similar to *nuclear medicine* —including positron emission tomography (PET)—it can also provide invaluable information on the physiology and pathology of organs, muscles, nerve trunks, and other tissues.

MRI is extraordinarily flexible in the ways in which it can generate forms of anatomic contrast among radiologically and mechanically similar tissues (Table 0.1). It can provide 2D, 3D, or 4D images for visualization of human physiology and biochemistry, with in-plane resolution of better than 1 mm. Like PET, MRI can often distinguish between healthy and damaged regions within an organ. And as with diagnostic *ultrasound* (US), MRI carries out all its good deeds with virtually no hazard to the patient or staff from ionizing radiation since no high-energy photons or radioactive nuclei are involved. Risks from MRI do exist, but they are largely manageable through patient screening and selection.

The various MRI imaging modalities utilize a diversity of physical *probes* to examine the body, as summarized in Tables 0.2 and 0.3. These probes interact with the tissues through mechanisms that can be sensitive not only to the specific characteristics of the tissues, but also to the nature of the probes themselves. Most importantly, they draw upon dissimilar and unique physical processes to create and display different kinds of contrast in the tissues that are being observed. The images they create can, therefore, provide visually and clinically different types of information on the patient's condition, making one or another of the several technologies better suited for following up on a particular set of symptoms. MRI is among the most multi-talented of the modalities in this regard, and also the most expensive and complex.

X-ray radiography, on the other hand, is the oldest form of clinical imaging, the easiest to describe, and globally the most widely used. Bundles of x-rays originating from the focal spot of an x-ray tube transit the body along their separate individual geometric paths, and they may or may not undergo photoelectric or Compton or classical absorption, or scattering by different amounts (Figure 0.2). The degree to which this happens is determined by the thicknesses of the tissues traversed, their physical density, and their chemical makeup (effective atomic number), and also by the energy spectrum of the photon probes. This process imprints an x-ray intensity pattern, or *primary x-ray image*, into the beam which, upon exiting the patient, can be captured by a sheet of film in a *film cassette.* The film is subsequently developed, fixed, dried, and hung on a *film box* for inspection. In the developed countries, film is rarely used now; it has been replaced with a digital, solid state *image receptor* (IR) and display *monitor*. The same physical interactions and processes also underlie digital radiography (DR), computed radiography (CR), full-field digital mammography (FFDM), digital tomosynthesis (DT), digital fluoroscopy (DF), digital subtraction angiography (DSA), computed tomography (CT), etc. (Figure 0.3a). While an x-ray image can be sufficiently informative for many clinical tasks, there are still only three gross anatomical factors (tissue dimension, density, and atomic number) that can affect the beam so as to generate the visual contrast. MRI, on the other hand, is highly sensitive to a number of elusive differences in the chemical or biological makeup and environment at the molecular level, allowing it to create tissue contrast in a wide range of ways.

Table 0.3 The major categories of imaging technologies, including the probe or signal involved in generating an image, the image receptor or detector, and the factors involved in creating contrast.

Modality	Probe / Signal	Detector	Source of *Contrast*: differences in...
Planar x-ray, CT	X-rays thru body	AMFPI; II+CCD; GdO, etc., array	$\int \mu(\rho, Z, kVp)\, ds$
Nuc. med; SPECT; PET	Gamma-rays from body	NaI crystal, PMTs; LSO array	Radiopharm. uptake, concentration; emission
Ultrasound	MHz sound	Piezoelectric transducer	ρ, κ, μ_{US}
MRI	Magnet, RF	RF radio receiver	PD, T1, T2, blood flow, chemical shift, [O], water diffusion, etc.

As opposed to the anatomical x-ray projections of radiography and CT, *nuclear medicine* informs primarily on physiological or molecular processes. It indicates where a tissue-specific radiopharmaceutical agent is taken up and concentrates. Irregularities in apparent gamma- or annihilation-ray emission from the targeted biological compartment may indicate excessive or diminished uptake of the nuclide, or a normal organ being obscured by overlying tis-

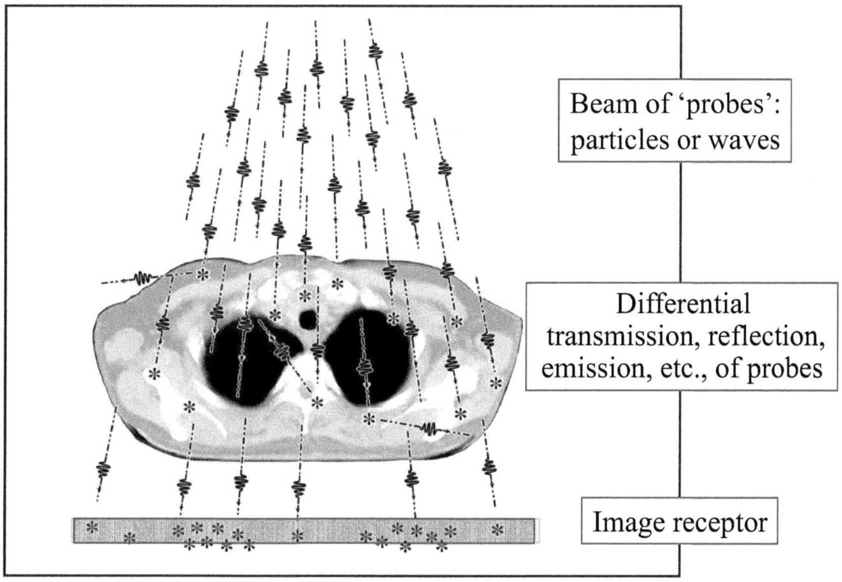

Figure 0.2 Radiography is the medical imaging technology most widely used throughout the world. An x-ray vacuum tube supplies a brief pulse of high-energy photons (typically up to 150 keV) that enter the patient and are differentially attenuated by tissues having a range of thicknesses, densities, and chemical makeup (effective atomic numbers), producing contrast in the primary x-ray image emerging from the body. Those photons that pass through are detected and mapped by the screen-film or solid-state/digital image receptor. Nuclear medicine and ultrasound work in analogous fashions, but MRI is a good deal more subtle.

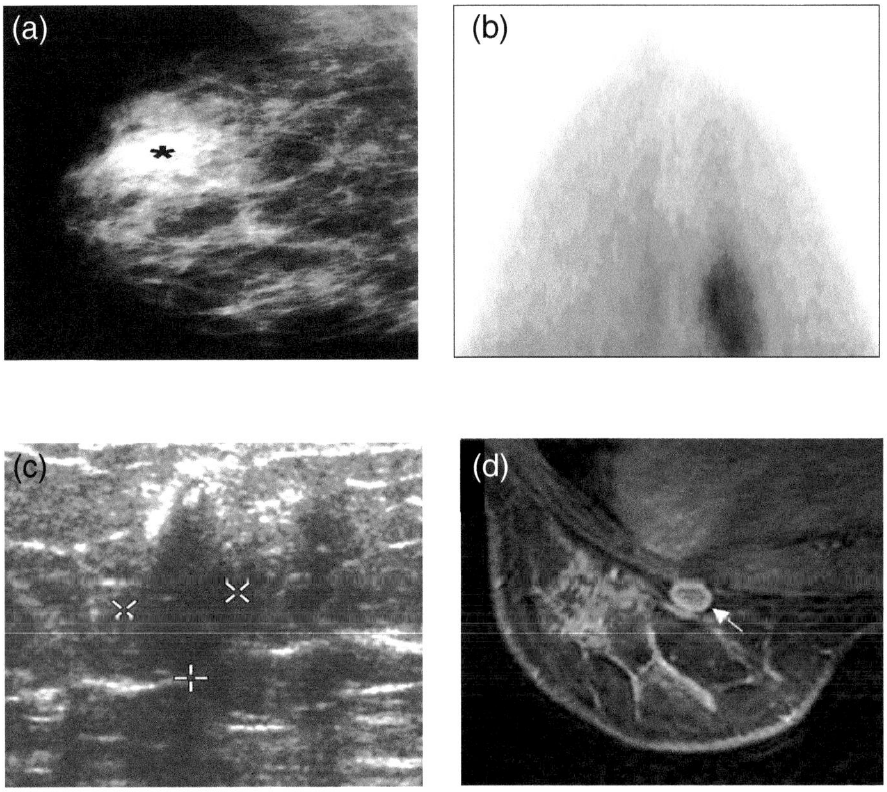

Figure 0.3 Breast imaging of post-40-year-old patients with biopsy-proven lobular carcinomas. The four modalities illustrated—mammography, PET, US, and MRI—produce visible contrast through radically different biophysical mechanisms, and provide complementary information. (a) Mammography must have both high soft-tissue radiographic contrast to distinguish neoplasms from normal breast tissue and fine resolution to examine microcalcifications. At the same time, the deposition of radiation dose must be very low. An x-ray-based digital (in this case) mammogram indicates regions that differ in thickness, density, or chemical makeup from those of the healthy breast tissue surrounding it. (b) PET, a 3D form of nuclear medicine, is highly sensitive to tissues that, like many tumors, take up and concentrate an excess amount of a glucose-analog radiopharmaceutical such as fluoro de-oxy glucose (^{18}FDG). The F-18 nucleus ejects a positron, which immediately collides with an electron, thereby creating a pair of 511 keV annihilation photons that can be detected outside the body. (c) B-mode ultrasound is often able to distinguish quickly, reliably, and inexpensively between a solid tumor (such as the one here noticed earlier with mammography) and a fluid-filled cyst, the interior of which produces no echoes and appears uniformly black. (d) Because it can provide information on high-spatial-resolution morphology without radiation, MRI is often a screening tool of choice for those younger patients who happen to be at higher risk for breast cancer.

sue—a problem that can be largely eliminated for tomographic approaches such as *Single Photon Emission CT* (SPECT) and *Positron Emission Tomography* (PET) (Figure 0.3b).

Ultrasound (US) differs considerably in that it does not utilize high-energy, ionizing photons as probes, but rather packets of high-frequency, short-wavelength mechanical waves. As these propagate through the body, a fraction of their energy is reflected back from sufficiently flat and smooth interfaces between separate tissues that differ in density or elasticity (or, conversely, compressibility). Such dissimilarity in material properties comes primarily from the macromolecules governing the tissue structure, and it helps define the anatomical boundaries that provide ultrasound's distinct form of contrast (Figure 0.3c). Additionally, ultrasound can display blood flow and slight tissue motions, such as those of a heart valve (Doppler), and also the elastic properties of tissue (elastography). Ultrasound images can be acquired without ionizing radiation and at rapid frame rates, making it an inexpensive, transportable, and powerful method for characterizing tissues or for guidance of diagnostic or therapeutic procedures.

Each of these modalities calls upon, in effect, a single physical mechanism to bring forth contrast. Clinical *magnetic resonance imaging*, on the other hand, exploits a still growing number of radically distinct, abstruse, and unique approaches to probe the molecular environs of tissue protons. It harnesses several magnetic fields functioning in unison to provide information on the molecular surroundings of the nuclei of the ordinary hydrogen atoms occurring naturally in water and lipid molecules within and around cells. (Two thirds of the atoms in our bodies, incidentally, are hydrogens.) And that, in turn, leads to a variety of one-of-a-kind forms of soft-tissue contrast that are exquisitely sensitive to and reflect upon tissue biophysics (Figure 0.3d).

The protons that produce clinical MRI signals nearly all belong to intra- and inter-cellular water or lipid molecules, and hereafter we shall refer to them simply as *water* or *tissue protons*. Clinical MRI may be defined as the art and science of creating and interpreting spatial maps of the *biochemical/physical environments* of tissue protons. Observable distinctions in these environments can be related clinically to differences in the anatomic and physiological properties and conditions of the tissues which, in turn, can reveal pathologies that may be present. This approach can thereby provide separate and unique kinds of information on tissue anatomy and physiology to which CT and the other modalities are oblivious.

0.2 Organization and Style of the Book

Some comments about the organization and style of the book: MRI is a multi-faceted and multi-layered technology. There are several major, nearly disjointed, but nonetheless interwoven groups of ideas that must be introduced and combined in explaining it, and a major question is how to put them in the most reasonable order for the beginning student. For example:

- The simplified, non-rigorous quasi-quantum (QM) depiction of protons in spin-up and spin-down states, aligned along and against the strong external principal field, $\boldsymbol{B}_0$, is relatively easy to grasp and pedagogically useful. Should this picture of NMR come before or after (or not at all) the generally more powerful classical (pulsed energy) view? To help develop physical intuition, we choose to do both, with the quasi-spin-up/spin-down picture coming first.
- Likewise, spin-lattice (T1) relaxation is explicable with the above quasi-quantum picture alone, while T2 makes sense only in the more rigorous classical treatment (which happens to drop out intact from a full QM treatment!). Should we introduce T1 early on to impart some color and relevance at the beginning of the story? Or would it be more effective to present both mechanisms together later on? To avoid muddying the waters more than necessary early on in the story, we choose the latter path.
- A number of MRI phenomena involve, in essence, only NMR in a voxel (e.g., the NMR process itself, including proton spin relaxation). These descriptions can readily be extended so as to encompass a single row of voxels comprising a 1D patient/phantom. But should we bypass this 1D process altogether and defer any discussion of imaging until it is possible to discuss MRI of 2D slices and in 3D? This is possible, building in significantly greater rigor and power, but at the price of more complexity of notation and mathematics. Our decision is to do as much as possible in a single voxel, then move on to 1D MRI with a row of voxels, and wait until chapter 12 to consider the more sophisticated encoding considerations needed for 2D, 3D, and 4D imaging.
- Then there is the continually expanding zoo of radiofrequency (RF) and magnetic field gradient pulse sequences, becoming ever more sophisticated, complex, and daunting. How many of these should we examine, and in how much detail? It is clearly important to deconstruct T1-weighted and T2-*w* MR imaging, along with FLAIR (FLuid Attenuation Inversion Recovery), but what about FLASH (Fast Low Angle SHot, marketed by Siemens); CE-T2 FFE (Contrast-Enhanced T2 Fast Field Echo, from Philips); General Electric's GRASS (Gradient-Recalled Acquisition in the Steady State), and the many, many others? Which ones from this alphabet soup should be covered here to illustrate the basic sequences used in clinical MRI, and in how much depth? Our objective is to give you the tools you'll need to understand these and the new sequences that you'll run into, but not to turn this into an encyclopedic handbook for specialists.

So how do we select and order the most basic and critical topics, and design the mode of presentation for best telling a unified, comprehensive story of MRI at a beginning graduate level, while also making it as clear, direct, and intelligible as possible? In other words, what pedagogic approach will promote an intuitive understanding that best helps the reader relate to the physical and mathematical descriptions introduced here and elsewhere?

Our three guiding principles have been, to the extent achievable, to:

- explain the NMR and MRI phenomena and technical underpinnings in a fashion sufficiently rigorous to satisfy the needs of a beginning graduate student of medical imaging;
- optimize the smoothness and continuity of flow of the fundamental ideas; and
- adhere to the wonderful maxim on education attributed to Einstein, "*Everything should be made as simple as possible, but no simpler.*"

We shall attempt to present the ideas underlying MRI in an integrated fashion, but avoid the vast amounts of specialist detail one picks up later while working in the field. Having thought long, over many semesters, about how to achieve this in our teaching, your authors have chosen the sequence of topics in the table of contents. Hopefully this will work for you.

This is a serious and nontrivial book written for professionals in training. But the style of writing and the presentation may seem a little unusual. We have tried to make the tone informal, even conversational, more or less how you might explain the subject to a colleague who happens to be unfamiliar with MRI. It starts off slowly and gently, with the depth, rigor, and density increasing as you move along. The text contains a little repetition of important ideas, especially in the earlier chapters. This is done intentionally, because expressing an idea in several different ways, and even just repeating it a few times, can help the student learn a convoluted and wide-ranging new subject. In any case, fresh information, which might have been less appropriate earlier on, is often added when a topic is revisited.

While some readers might prefer a more rapid, bare-bones, and axiomatic approach from the outset—as with typical upper-level physics and math texts—the present one reflects our teaching styles. In particular, we try to *explain* as much as possible about MRI, rather than merely to *describe* the steps of what is going on by listing the basic facts and equations. We believe that nearly all equations tell stories, and it can be illuminating to treat them as storytellers. We hope that the book will leave the reader with an intuitive feel for the underlying physics and the ability to put into words, equations, and diagrams the essence of this complex and demanding area. We have found that this approach appears to work well for clever and interested students starting out in so challenging an undertaking.

Finally, about the *exercises*. Many of them ask you to give some preliminary thought to a topic upcoming shortly. Others involve your working through ideas, rather than our stating them explicitly. As such, they should be thought of as *part of the main text*, albeit in denser form. Some you will figure out the exercises in seconds, while others may require a bit more cogitation. Please do consider each of them, where it occurs, before you proceed. Similarly, the figures and their captions may assist in following the text.

We have done our best to make this a fruitful and enjoyable experience. But we are well aware that there are alternative ways to attack the issues that you might come to prefer, so we really welcome your suggestions for other approaches.

Anthony Wolbarst and
Nathan Yanasak

Acknowledgments

This is a textbook for upper-level students new to MRI. It introduces no original physics or research results; rather, we have built upon what others before us have found and created, and we express our profound appreciation to those many. What we have done here is to organize and present a number of ideas differently—hopefully that will prove to be of value.

Some colleagues have offered advice and comments on the manuscript. Kevin King of GE, co-author of the indispensable *Handbook of MRI Pulse Sequences*, and Jason Stafford at MD Anderson helped clarify our thinking on a number of subtle points. Wendell Lutz, a good friend of ABW at Harvard, read the first part of the book carefully and caught some basic physics mistakes that the authors should not have made. Raymond Benton Pahlka of MD Anderson read parts of the draft and found a few others. ABW's understanding of the field was deepened by interactions with Charles Smith, director of the University of Kentucky's Magnetic Resonance Imaging and Spectroscopy Center (MRISC), and by Dave Powell and Peter Hardy there. David Clark, an Australian geophysicist, told us the tale of Earth's magnetism, and Robert Zamenhof provided the ART joke, oh, those so many moons ago.

Finally, the hard work and perseverance of Bobette Shaub and Todd Hanson at Medical Physics Publishing made what could have been a painful ordeal into a good experience.

Our thanks to you all.

Your authors have been collecting NMR and MRI teaching images for a combined total of more than 40 years. Despite our efforts, we are not sure of the provenance of a few included here. If you see something that is not credited properly, please let us know, and we shall happily fix that oversight in the next printing.

Finally, we ask any reader to contact us, by way of the publisher, if you find errors or think of better ways to present ideas. We hope this first edition will be well received, but that the next one will be even better.

CHAPTER 1

Introduction to MRI

The story of Magnetic Resonance Imaging (MRI) begins with the nuclear magnetic resonance phenomenon, discovered just before and during the Second World War. The technique was refined over the ensuing decades, and a multitude of experimental and theoretical studies were devised to explore the factors that affected the precise magnetic resonance frequencies at which nuclei resonate in a strong external field. A key revelation was that even slight differences in the molecular environments of the nuclei cause detectable shifts in their behaviors. Researchers realized that such studies, especially when in combination with x-ray diffraction results, could be invaluable in unraveling molecular structures. Also of great importance, it became possible to explain the associated nuclear-spin relaxation (T1, T2) mechanisms, which would come to play so central a clinical role in MRI.

In the mid-1970s, scientists designed methods to carry out NMR voxel-by-voxel within a thin slice of tissue, thereby enabling the delineation of two-dimensional MRI maps of T1 and of T2 throughout the body. This led to further approaches—such as *MR spectroscopy* (MRS), *functional MRI* (fMRI), *diffusion tensor imaging* (DTI), and *MR angiography* (MRA)—that portray the spatial dependences of other physiological and metabolic tissue properties, too. All of this has established MRI as a powerful and extraordinarily versatile diagnostic modality with a number of unique imaging capabilities. Were it not for their still significantly greater cost, MRI machines might well drive CT largely out of business.

This chapter will say a bit about NMR and MRI in general, then illustrate some of the remarkable capabilities of MRI with a real case study.

There follows a sketch of the rich history of the field, and an important caveat: there are potentially lethal but preventable hazards associated with strong and rapidly switching magnetic fields and with the intense pulses of radiofrequency (RF) energy from an MR device. Be careful for yourself, your patients, and all others.

1.1 A Stratospheric, Warp-Speed View of NMR and MRI

This brief section zip lines through much of the physics of NMR and MRI. We shall provide only cursory descriptions now, but you can expect the explanations and fuller descriptions to follow in the rest of the book. Many of the whys and wherefores may not be apparent here, but they will come later.

The nuclear magnetic moment is proportional to the nuclear spin

The normal chemical properties of virtually any atom are determined solely by the arrangement of its electrons, hence ultimately by the number of protons in the nucleus, the atomic number, Z. Gases, liquids, solids, and living cells behave the way they do because of, and almost only because of, the configurations of the Z electrons of their constituent atoms.

The properties of a nucleus itself, however, depend on the number of neutrons, N, as well. The values of both Z and N together govern whether or not a specific isotope is radioactive, of importance in nuclear medicine, and in the cleanup of nuclear power reactors and weapons facilities. The makeup of a nucleus also determines its magnetic properties, which are fundamental to MRI. It can be fruitful to think of a nucleus as being somewhat like a spinning ball of positive charge [Eisberg et al. 1985]. A proton, a neutron, or a composite nucleus has a quantum mechanical (QM) attribute that is called, by analogy with the corresponding vector property of classical physics, *nuclear spin angular momentum*, ***I***, or just spin, as in Figure 1.1.

The only nucleus currently of major interest in clinical MRI is the simplest of them all, namely that of an atom of ordinary hydrogen, i.e., a solitary proton. A nuclear physicist views a proton as a subatomic particle made up of two up-quarks and one down-quark held together by the strong nuclear force. Just as interactions among charged particles are brought about by the virtual photons of quantum electrodynamics, likewise quarks are bound by the virtual gluons of chromodynamics, within a sea of virtual quarks that are constantly appearing and vanishing. Mother Nature has

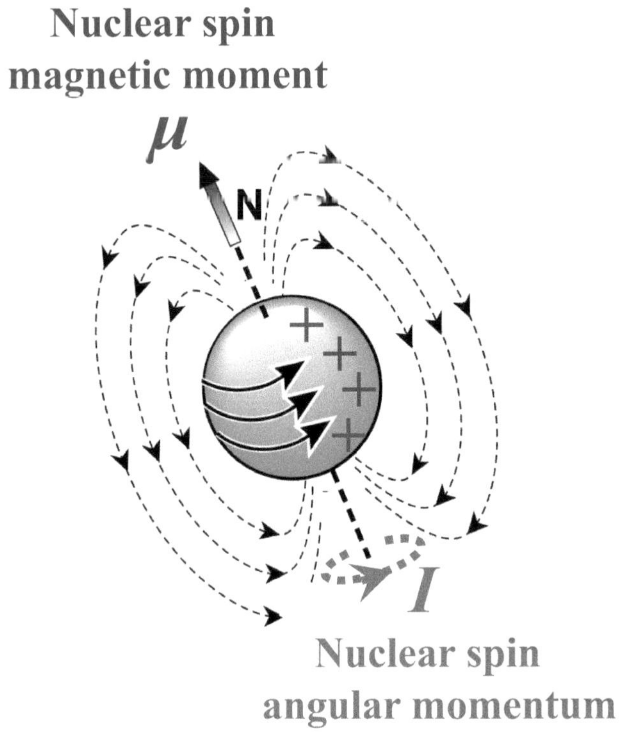

Figure 1.1 A free nucleus has a quantum mechanical attribute analogous to classical spin and is said to possess *spin angular momentum, S*. As a 'spinning' charged body, it produces its own weak magnetic field, like that of a compass needle or small bar magnet. The strength of this nuclear magnet is parameterized by its nuclear spin magnetic moment, represented as the vector quantity μ_n, of magnitude $\mu_n \equiv |\mu_n|$.

decreed that, despite the problems theorists are having in getting the spins of its quarks and gluons to add up right (a situation known as the 'proton spin crisis'), a lone proton is one of many nuclei considered to be a 'spin-½' particle, therefore a fermion.

Every nucleus comprised of an odd number of protons or of neutrons or both possesses non-zero spin. In any nucleus with even numbers of both protons and neutrons, at the other extreme, each proton pairs up and aligns anti-parallel to another, for reasons that can be explained with QM, and their two dipole magnetic fields cancel—they have zero net spin. The same is also true of the neutrons. As a result, carbon-12, oxygen-16 (which, together with hydrogen, make up ¾ of the mass of soft tissues), and calcium-40 play no active role in imaging.

As with any other charged and moving entity, a 'spinning' nucleus produces its own magnetic *dipole* field, its *nuclear magnetic moment*, a vector entity that is represented by a lower-case italic Greek mu, $\boldsymbol{\mu}$. The magnitude or strength of the inherent nuclear moment of the nth isotope, designated $\mu_n \equiv |\boldsymbol{\mu}_n|$, is a principal measure of its 'magnetness,' the strength of the field it itself produces.

The *gyromagnetic ratio*, γ_n, is a slightly different measure of the same thing, the strength of the intrinsic nuclear magnetic field of the nucleus. Encountered more commonly as $(\gamma_n/2\pi)$, it relates the nuclear magnetic moment to the nuclear spin angular momentum vector, $\boldsymbol{I}_n$,

$$\mu_n = \gamma_n \boldsymbol{I}_n \tag{1.1a}$$

At times it is more convenient to use $\gamma_n/2\pi$ rather than μ_n in discussion, or vice versa, but otherwise the two are equivalent, apart from a constant.

The subscripts on I_n, μ_n, and γ_n label specific <u>n</u>uclear isotopes by their atomic weight,

$$n \equiv (Z_n + N_n). \tag{1.1b}$$

Clinical MRI deals almost exclusively with the nuclei of regular hydrogen atoms, $^1H_1$ or H-1 in the standard notation, so there is no further need for the subscript, and the magnitudes of the two vectors can be related as

$$\mu = \gamma \boldsymbol{I} \tag{1.1c}$$

The protons that are imaged are found almost entirely in cellular or extra-cellular water or lipids and, again, we shall refer to them generally as *tissue protons*. So hereafter, γ and μ will appear unadorned with an 'n,' and they will refer only to hydrogen nuclei in tissues.

From here, traditional introductions to NMR and MRI physics take one of two pedagogic paths, the quasi-quantum 'spin-up/spin down' and the 'classical.' We shall, however, be guided by the incontrovertible wisdom of Yogi Berra, philosopher and former catcher for the resplendent New York Yankees: "If you come to a fork in the road, take it." Both routes happened to pass by his house, and such is the case here, too.

Some Notes on Vectors

Before proceeding with our story, we should clarify the notation for vectors. A magnetic field, ***B***, which has both magnitude and direction, is a *vector* entity and is represented with a **bolded** symbol. Its strength, *B*, a scalar, is not in bold. ***B*(*r*)** is a *vector field* that has magnitude or direction that may vary according to the value of the position vector ***r***. We shall deal frequently with the particular one-dimensional situation of a 1D patient reposing along the *x-axis* and in an external field whose strength varies with *x—but that always points along the z-axis*! This might be written ***B***(*x*), but when it is necessary to emphasize the fixed direction of the external field, it will be presented as $\boldsymbol{B_z}(x)$.

Individual components of a vector will *not* be denoted in bold: for example, *x* is the first spatial component of the three-space position vector $\boldsymbol{r} \equiv \{x,y,z\} = x\mathbf{i} + y\mathbf{j} + z\mathbf{k}$, where **i**, **j**, and **k** are the orthonormal unit vectors of 'real-' or 'three-' space. This may also be expressed as $\boldsymbol{r} \equiv \{x_1, x_2, x_3\} = x_1\mathbf{i} + x_2\mathbf{j} + x_3\mathbf{k}$.

In mathematics and computer science, on the other hand, some quantities with multiple, discrete values can be treated as vectors or arrays, such as with the vector $\boldsymbol{x} \equiv \{x_1, x_2, x_3, \ldots, x_n\}$ that specifies the positions of voxels 1, 2, ..., *n* along the x-axis of an image. We shall *not* denote the individual members of this set as vectors with bold font.

Similarly, when discussing the rather enigmatic construct known as **k**-space, $\boldsymbol{k} \equiv \{k_x, k_y, k_z\}$ (which is totally, absolutely *not* related to the unit **k**-vector of real-space) will represent a single vector location in that space.

Every concept introduced briefly in this chapter, by the way, will be pursued more fully later on.

First picture of NMR: Quasi-quantum mechanical (spin-up/spin-down along B_z and the z-axis)

The easier way to begin addressing proton NMR is by way of a great simplification of the full quantum mechanical (QM) treatment.

Imagine that in a magnetically shielded room, a tray of compasses is being mechanically rattled gently. With no external magnetic field, their needles are jostled about and randomly point in all directions because of the effects of the 'noise' energy being inputted. When a strong external field, B_z, is present, however, then the needles will tend to align relative to it, each with its own north pole heading toward the magnet's

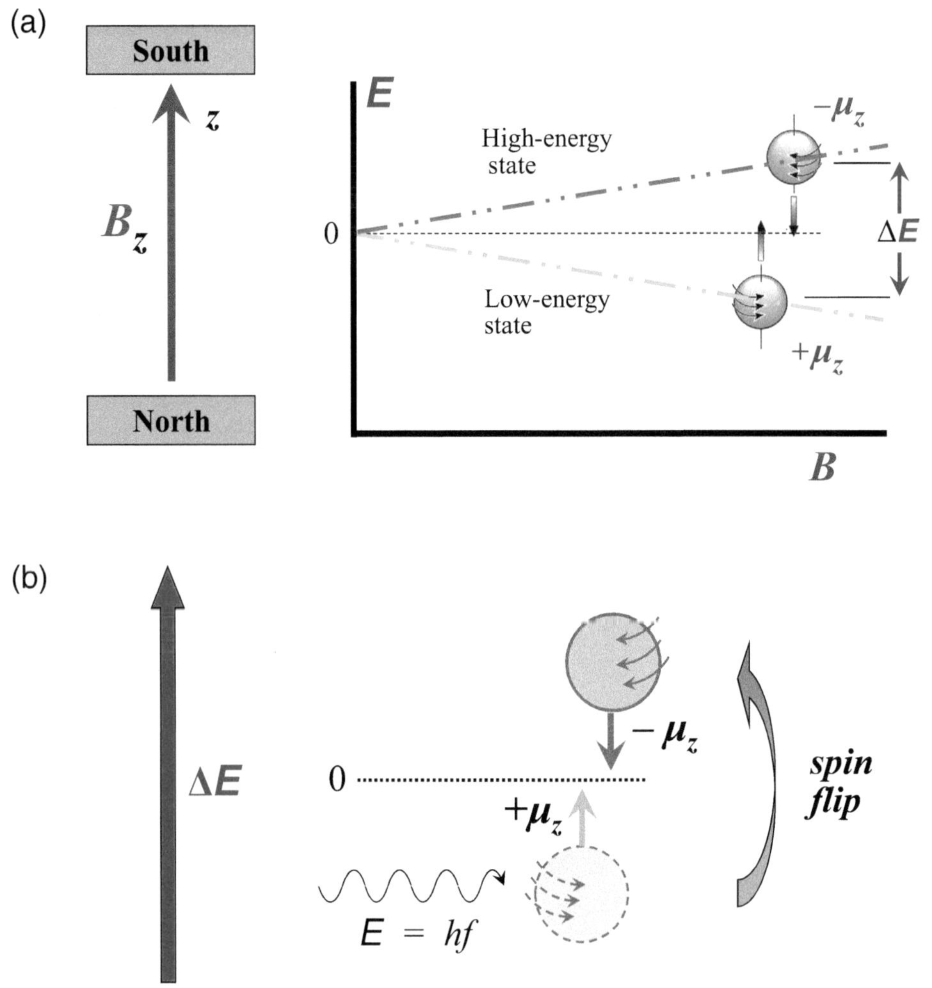

Figure 1.2 There are two pedagogical pictures of NMR, the *quasi-QM* and the *classical*, that are commonly drawn upon in introducing MRI physics. The two are mutually incompatible and inconsistent, and they cannot be combined together, because of the quite different assumptions adopted in constructing each of them out of the fully correct, rigorous quantum theory. But both are useful in different ways. Here we discuss only the greatly simplified *quasi-QM* picture. (a) A single proton behaves like a tiny compass needle, but it can align in an externally magnetic field, $\mathbf{B_z}$, *only along* or *against* it and the **z**-axis. The energy levels for the two quasi-QM proton spin states diverge by the amount of the Zeeman energy, which is linear in the strength of the total local magnetic field. Nearly all of this $\mathbf{B_z}$ comes from the main magnet, and is constant. Much of the rest is from the intermittently activated *gradient coils*, and from the brief pulses from the RF (radiofrequency) *transmitter*. Finally, the amount of random RF 'noise' generated by fluctuations in molecular motions is tiny, but it greatly influences proton spin relaxation processes. A spin-½ nucleus such as a proton is allowed to inhabit either one of the two levels: in the lower-energy, 'spin-up' state, the projection of $\boldsymbol{\mu}$ onto the **z**-axis is aligned parallel to $\mathbf{B_z}$. Alternatively, in the 'down,' high-energy state, it points against $\mathbf{B_z}$. Also, *un*like a compass needle, a proton can remain for long periods of time in a quasi-stable, higher-energy spin-orientation state, pointing the 'wrong' direction. (b) The two spin states differ by the Zeeman energy, $\Delta E = 2\mu_z \mathbf{B_z}$. A photon (or phonon) of the right energy can cause a transition up or down between the two. The horizontal dashed line in the middle indicates the energy the nucleus has at $\mathbf{B_z} = 0$.

south pole. But by grabbing and twisting a needle you can make it point in any direction you choose, until you let it go.

Protons behave only somewhat like that. When a bunch of them are inserted into the uniform field $\boldsymbol{B}_z$, they will be *polarized*, but—as it happens, perhaps somewhat mysteriously—with only half of them aligned *along* it and the other half *against*. This is a wonderful example of what is affectionately known as *quantum weirdness*. The spin magnetic moment of a proton is *spatially quantized*, and its *z*-component can align only 'up' along the direction of the external field or 'down' against it. According to this simplified, quasi-QM picture, the nucleus can exist in either of these two possible spin states in the external field—and, *un*like a compass needle, with no other alignments (Figure 1.2a). In other words, the *z-components* of a proton's nuclear magnetic moment can assume only the values

$$\mu_z \;=\; \pm\tfrac{1}{2}(h/2\pi)\gamma \qquad (1.1\text{d})$$

If the field produced by the nucleus itself aligns *along* an external field, as with a compass needle, then the nucleus is said to inhabit the *lower-energy*, or *spin-up*, state.

It simplifies many of the illustrations in this book to align the main external field vertically, somewhat like the energy scale. While the fields of superconducting magnets are mostly horizontal, those of permanent and electromagnets are vertical, so there's nothing really to unlearn to assume, for now, that they point upward.

The classical *energy of alignment* of a magnetic dipole moment in a *z*-orientated external field is given by the vector *dot* (or *scalar*) product

$$\mathsf{E} \;=\; \boldsymbol{\mu}\cdot\boldsymbol{B}_z, \qquad (1.2\text{a})$$

and this general form remains valid quantum mechanically. To avoid confusion with the electric field $\boldsymbol{E}$, the E intended to represent the non-vector *energy* is shown in Arial typeface. From Equations (1.1) and (1.2a), the QM *Zeeman splitting*, $\Delta\mathsf{E}_{\text{Zeeman}}$, between the two energy levels of a spin-½ nucleus, is proportional both to its innate magnetness (μ or γ) and to the strength of the local magnetic field there, $\boldsymbol{B}_z$:

$$\Delta\mathsf{E}_{Zeeman} \;=\; 2\mu_z B_z \;=\; h(\gamma/2\pi)B_z. \qquad (1.2\text{b})$$

EXERCISE 1.1 What are the units of μ?

At the heart of MRI is the critical fact that a proton in its lower-energy spin state can be elevated to a high-energy state by interaction with an applied radiofrequency photon that carries exactly the right energy. Indeed, one easy way to determine that the *nuclear magnetic resonance* process is occurring in a sample of water, say, is to turn on an external magnetic field $\boldsymbol{B}$, pass a beam of RF photons through it, and detect the rate at which it absorbs the energy as a function of the frequency, ν. Bearing in mind the ubiquitous Planck-Einstein relationship between the energy of a photon and its frequency, $\mathsf{E} = h\nu$, then the energy required to flip over a proton in the field $\boldsymbol{B}_z$ is

$$h\nu \;=\; \Delta\mathsf{E}_{Zeeman} \;=\; h(\gamma/2\pi)B_z,$$

or (1.2c)

$$\nu_{Larmor}(B_z) \;=\; (\gamma/2\pi)B_z.$$

This is the central *Larmor equation*, which relates the frequency at which a cohort of protons gathered together in a magnetic field of local magnitude B_z will undergo the phenomenon of nuclear magnetic resonance. NMR is one of the two essential concepts of MRI (the other being *spin-relaxation*), and here you are at only page 5! Imagine how much more you'll pick up in the rest of the book.

EXERCISE 1.2 What is the Larmor frequency, ν_{Larmor}*, at which the NMR process takes place for a free proton?*

EXERCISE 1.3 What is the conceptual difference between the two forms of Equation (1.2c)*?*

The essential Larmor equation for isolated protons is

$$\nu_{Larmor}[\text{MHz}] \;=\; [42.58\,(\text{MHz/tesla})] \times B_z[\text{T}] \quad (\textit{free protons}), \qquad (1.3)$$

describes the condition under which RF photons from a transmitter will be absorbed by tissue protons and induce nuclear spin transitions (both up and down). Like the Bohr atom, the notion is conceptually and mathematically uncomplicated, yet it is still solid, even if not derived here in a quantum mechanically rigorous manner. In any case, the up-down space quantization constraint is powerful enough for the approach to lead to this fundamental result.

EXERCISE 1.4 What is the Larmor frequency for protons in a 1.5 T main magnetic field? 3 T? 7 T?

MRI of a 1D phantom

We can advance from NMR of tissue in a single voxel, at a single point in space, to MRI throughout a region with this spin-up/spin-down model. Consider here a 1D phantom consisting of a single row of voxels lying along the $\boldsymbol{x}$-axis and containing different amounts of water (Figure 1.3a). The trick is just to establish a gradient magnetic field, $\boldsymbol{G_x}$, that varies along the $\boldsymbol{x}$-direction,

$$G_x \equiv \partial B_z / \partial x. \tag{1.4a}$$

The *direction* of the gradient magnetic field itself always points 100% *upward* everywhere, but its magnitude is designed to increase linearly with x-position,

$$B_z(x) = B_0 + G_x x, \tag{1.4b}$$

where $\boldsymbol{B_0}$ is the very strong (1.5 or 3 tesla), uniform field from the MRI instrument's main magnet.

*EXERCISE 1.5 Find an expression for the frequency at which the protons in the voxel at position **x** of a linear phantom resonate.*

This elementary generalization of the Larmor equation provides the critically important prescription for connecting a measured reading of $\nu_{Larmor}(x)$ to the corresponding value of $\boldsymbol{B_z}(x)$. And once the precise magnitude of $\boldsymbol{G_x}$ has been established and is routinely checked, knowledge of the local $\boldsymbol{B_z}(x)$, in turn, gives up voxel position, x. Now, with the location of each voxel and a measurement of the RF signal strength from it, it

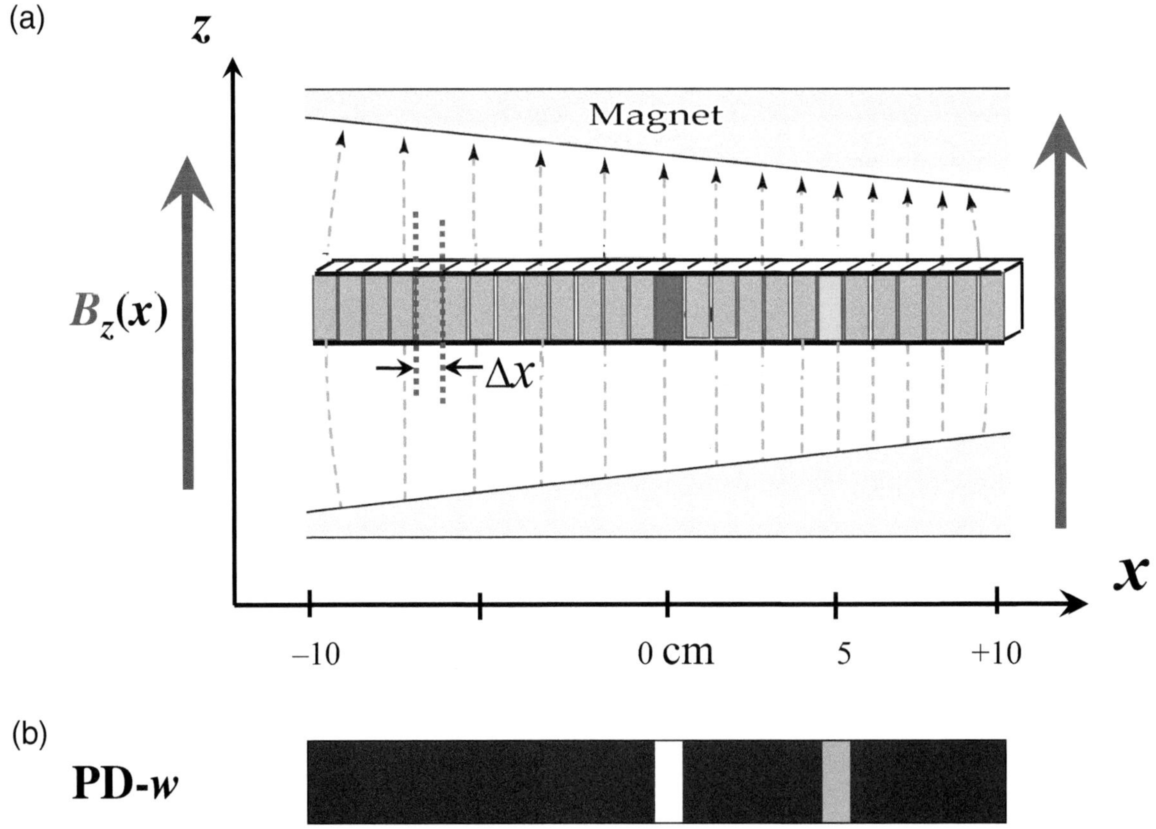

Figure 1.3 This 1D phantom consists of a row of small, hollow chambers, all of width Δx. (a) The one at x = 0 cm contains twice as much water as that at 5 cm, and the others are empty. The main magnetic field points upward along the z-axis at each chamber, but the x-gradient component causes the strength of the *net local* field at each voxel, $\mathbf{B_z}(x)$, to increase linearly with position x. Field strength increases as the dashed field lines come closer together. Each of the two water samples undergoes NMR at a different, but precisely measurable frequency. This is an encoding technique that allows determination of voxel position along the x-axis (b), which makes possible the creation of a MR image of proton density (PD) along the phantom. The greater the PD in a voxel, the brighter the pixel shines on the display.

is but a hop, skip, and jump to determining the spatial distribution of *proton density* (PD) within a one-dimensional patient lying along the x-axis—which amounts to creating an honest-to-goodness 1D PD MR image (Figure 1.3b). By convention, the greater the PD in a voxel, the brighter the pixel is made to be on the display.

Second picture of NMR: Classical (magnetization precessing in x-y plane)

The other general didactic technique for introducing the NMR phenomenon, the so-called *classical* picture, can actually be derived fully and rigorously from QM theory. It focuses on the composite *nuclear magnetization*, $\boldsymbol{m}(\boldsymbol{r}_j, t)$, the overall magnetic field produced by the tissue protons in the jth voxel at position $\boldsymbol{r}_j$ in 2D or 3D real space:

$$\boldsymbol{m}(\boldsymbol{r}_j, t)\text{: at position } \boldsymbol{r}_j \text{ in 2D- or 3D-space}$$

And despite its quantum mechanical roots, the QM *expectation value* of the time-dependent magnetization, $<\boldsymbol{m}(\boldsymbol{r}_j, t)>$, ends up behaving entirely *classically*, governed by the *Bloch equations*, in complete agreement with Newtonian dynamics.

When in a strong magnetic field, the component of the magnetization in the jth voxel that inhabits (at least temporarily) the $\boldsymbol{x}$-$\boldsymbol{y}$ plane normal to $\boldsymbol{B}_z$, namely $\boldsymbol{m}_{xy}(\boldsymbol{r}_j, t)$, will *precess* about it, quite like a toy top or a gyroscope in a gravitational field (Figure 1.4a). (For both proton and gyroscope, the rate of precession increases linearly with field strength.) This rotation of $\boldsymbol{m}_{xy}(\boldsymbol{r}_j, t)$ about $\boldsymbol{B}_z$ happens at the same Larmor frequency as was obtained from the spin-up/spin-down model (Exercise 1.2). Meanwhile, the component of the voxel magnetization that lies along $\boldsymbol{B}_z$ and the z-axis, $\boldsymbol{m}_z(\boldsymbol{r}_j, t)$, does not change (unless outside factors such as RF pulses or *spin-relaxation* processes are at work conspire to disrupt the situation). Hereafter we will adopt the $\boldsymbol{r}$ notation or the j, but not both, unless necessary for book-keeping purposes.

The classical picture plays an essential role in discussing the T2 relaxation process, and also the various sequences of RF pulses and gradient magnetic fields that generate signal and contrast in clinical MRI, such as those of the *saturation-recovery*, *spin-echo*, and *gradient-echo* techniques.

The classical approach is conceptually very different from the quasi-QM, and the two are mutually incompatible, as will be demonstrated later. If you try to combine them, you inevitably end up in trouble. A Sunfish sailboat and a Corvette convertible can each get you from here to the other side of the lake, but you won't get far if you try to fuse the two. The quasi-QM picture has the spin axes of protons pointing up or down in it, while the classical picture has them lying in the $\boldsymbol{x}$-$\boldsymbol{y}$ plane and precessing about the z-axis—both obviously cannot hold at the same time, an impossibility. To maintain your sanity, don't try to figure out how both can be going on simultaneously, just accept the mystery of it as just being so. More later.

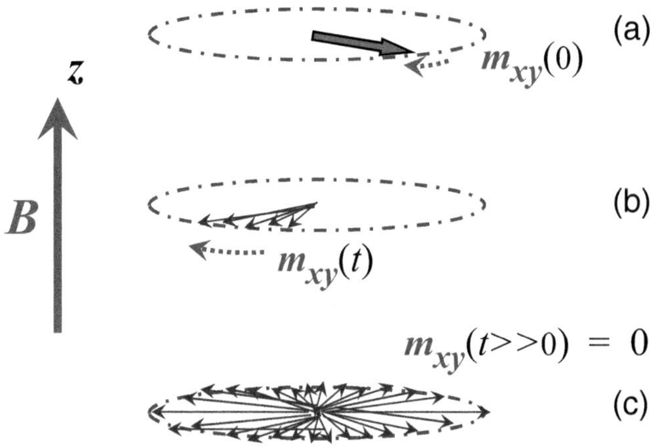

Figure 1.4 The 'classical' view is clearly incompatible with the 'quasi-QM'. (a) The magnetization of a cohort of protons, $\boldsymbol{m}_{xy}(t)$, starts out precessing like a single, narrow vector in the plane transverse to $\boldsymbol{B}_z$. (b) Over time, the protons precess at slightly different rates because of minute differences in their local environments, and they begin to lose phase coherence. The *transverse magnetization* spreads out and diminishes at the rate 1/T2, as $e^{-t/T2}$. (c) After a period several times the tissue's *transverse* or *spin-spin relaxation time*, T2, the protons will be pointing every which way, the sum of their spin vectors will be zero, and $\boldsymbol{m}_{xy}(t \gg T2)$ will vanish.

But while seemingly inconsistent, the two models are complementary to one another. In fact, each can be valuable in the right context. But avoid trying to think about NMR with both pictures at the same time. Doing so leads only to weeping and wailing and gnashing of teeth. So just don't try it!

Proton spin relaxation

As the theory and practice of NMR evolved, physicists undertook a second main line of inquiry into the interactions of a nucleus with its environs, namely the *spin-relaxation* mechanisms. The two most important of these are nuclear *longitudinal spin relaxation* (parameterized by the relaxation time T1) and nuclear *transverse spin relaxation* (T2). These, along with the proton density of the tissue contributing to the signal, play fundamental roles in creating the exquisite soft-tissue contrast associated with clinical MRI.

EXERCISE 1.6 What are some other systems, mechanical or otherwise, that exhibit near-exponential fall-off behavior over time, distance, radiation dose, etc.?

Perhaps the best way to get into classical resonance is by way of a familiar analogy. After an initial push, a child's swing resonates at its inherent normal-mode frequency (Figure 1.5), as with many other disturbed physical systems. But any of these will dissipate its energy over time, because of frictional or similar effects, and the amplitude of its oscillation will undergo near-exponential relaxation (decay) with a characteristic *relaxation* or *damping time* T.

Longitudinal proton spin relaxation time, T1

Imagine a bunch of identical compass needles fidgeting about on a gently vibrating table, each trying to align along the Earth's magnetic field. In your mind's eye, give the table a hard whack, immediately after which the needles will point randomly in all directions. Now leave it alone again, and the needles will settle down and begin to re-align mostly northward again. The average amount of time this recovery process takes for the ensemble of compasses is the system's relaxation time, $T_{needles}$. It is determined by the strength of the interaction between the field and a needle, by its mass and shape, and by the nature of the dissipative forces present, such as those occurring at the mechanical pivot points where the needle is supported.

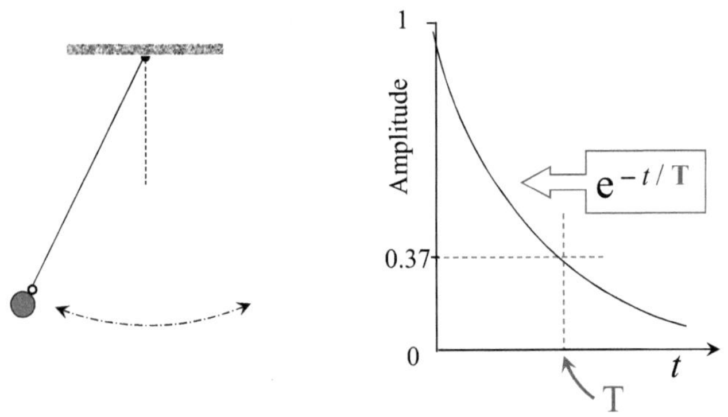

Figure 1.5 A pendulum serves as simple analogue (but not at all a direct model!) for proton resonance and T1 relaxation. The oscillatory motion is affected by friction-like forces that cause it to fall off exponentially over time with a characteristic time 'T.' Analogous phenomena are responsible for the decline of a NMR signal at the rate 1/T1 or 1/T2.

A proton is a fully quantum mechanical entity, of course, but in some ways a cluster of them in a voxel of tissue do behave rather like a bunch of tiny compass needles. They differ, however, in that when a patient lies in the strong magnetic field of an MRI device, a very little bit more than half of them will settle into the lower energy-state. (Here, of course, we're thinking only in terms of the *quasi-quantum mechanical picture*, in which proton spin axes can point only up or down relative to the strong external field!) They don't all end up in the more comfortable and 'natural,' lower-energy alignment along the field, like real compass needles, because magnetic noise from random molecular motions will continuously be knocking their spins both up and down. In a standard 1.5 tesla field, the system achieves a dynamic thermal equilibrium with about five in a million more protons in the lower-energy state, as predicted by Boltzmann statistics. (That may not seem like much, but how many of them are there in a 1 mm^3 voxel?)

You can, with RF excitation at the Larmor frequency, flip some of the lower-energy protons up, and tickle some of the others down, and thereby disturb the equilibrium condition. Do this effectively, and the system eventually has about the same numbers in the two energy levels, a situation known as *spin saturation*. Turn off the RF, and the spins will begin to undergo a kind of spontaneous, thermally induced relaxation process over time analogous to (but in actuality quite different from) that of the compass needles, approximately as

$$[1 - e^{-t/T1}], \qquad (1.5a)$$

with the exponential decay rate of 1/T1. You might assume that it is the photons emitted during this relaxation process that are detected and used to create T1-type MRI images, but that is too rudimentary an explanation. A big issue, indeed, is how to follow all this in time as it happens, a topic to be explored on several occasions along the way.

The time it takes for a voxel-sized cohort of protons to return 63% of the way back toward their equilibrium condition is designated T1, known as the proton spin-lattice or *longitudinal spin-relaxation time.* At t = T1, $[1 - e^{-t/T1}] = [1 - e^{-1}] = 0.63$. But be forewarned —T1 relaxation is only part of what's going on. This argument, by the way, applies to a number of other phenomena that diminish exponentially, such as radionuclides popping off over time, x-ray photons passing deeper into tissue, cells exposed to various amounts of high-LET (linear attenuation coefficient) radiation dose, etc.

Transverse proton spin relaxation time, T2

At the heart of the classical picture of NMR is the notion of the magnetization, $\boldsymbol{m}(t)$, of a voxel of protons precessing in the $\boldsymbol{x}$-$\boldsymbol{y}$ plane. This commonly starts out with all the protons aligned together at $t = 0$, so that $\boldsymbol{m}_{xy}(t)$ can be represented as a rotating narrow vector, the arrow back in Figure 1.4a. The individual protons in the voxel, however, sit in magnetic fields that differ slightly from place to place, so that they do not all rotate at exactly the same frequency. For this and other reasons, the orientations of individual spin packets begin to spread out over time as

$$e^{-t/T2}, \qquad (1.5b)$$

and they lose their initial phase coherence (Figure 1.4b). The time it takes to fall 37% of the way to this situation is called T2, the spin-spin or *transverse relaxation time.* After several more periods of duration T2, the spin alignments are so dispersed that their vector sum has dropped to near zero (Figure 1.4c) and $\boldsymbol{m}_{xy}(t)$ vanishes.

T1 and T2 clinical images

T1 and T2 are measures of different but overlapping physical processes, such as the rotations of water molecules near a large biomolecule. These times involve the interplay of protons with their biochemical environments in and between the cells of a tissue. T1 and T2 in a voxel are exquisitely sensitive to the detailed nature of friction-like and other interactions of the water and lipid protons with one another and with nearby ions, biomolecules, and membranes. The concentrations and biophysical characteristics of these ions and biomolecules are affected by the type and physiologic status of the cells they constitute—which, in turn, depend on the biological makeup and pathological condition of the tissues. MRI may, therefore, be able to distinguish not only between histologically different tissues, but even between normal and unhealthy regions of the same tissue.

T1 and T2 can be assessed point by point throughout a portion of the patient, in effect, and utilized to create two related but separate kinds of medical images. Through selection of the operating parameters of the MRI device, it is a straightforward matter to display spatial variations in T1, T2, and a range of other clinically revealing tissue characteristics. Put another way, the numerous types of MRI *contrast*, arising from and reflecting separate physical and physiological processes, are obtained with distinct operational settings of the device (sequences of RF and gradient-field pulses, echo time TE, repetition time TR, etc.)

With the widely employed *spin-echo pulse sequences*, the proton MRI signal strength, $s(x,t)$, from the voxel at position x commonly decays through spin-relaxation mechanisms approximately as

$$\begin{aligned} s(x,t) \;\propto\; & PD(x)\times[e^{-2\pi i\nu_L t}] \\ & \times[1-e^{-TR/T1(x)}]\times[e^{-TE/T2(x)}] \end{aligned} \qquad (1.6a)$$

PD($\boldsymbol{x}$) is the proton density for the voxel at x, $\nu_L(x)$ the Larmor frequency there, and T1(x) and T2(x) are the two principal spin-relaxation times.

If given the value of $s(x_j, t)$ over time for the voxel at each position x_j, it would be a simple matter to extract and map PD(x), T1(x), and T2(x) throughout the patient. Regrettably, an MR device does not detect individual signals from all the voxels separately, but rather only their superimposed, summed value:

$$S(t) \;=\; \Sigma_j\, s(x_j,t)\,. \qquad (1.6b)$$

Much of this book will explore the physical origin, meaning, and analysis of this equation, and of the three clinically important tissue parameters—PD, T1, and T2—and related matters. The mathematical process of separating $S(t)$ into its (thousands of) constituent parts $\{s(x_j, t)\}$ is known as *MR image reconstruction.* A very simple example of this appeared in connection with Figure 1.3

The above has been a rapid sketch of some central aspects of NMR and MRI. The following chapters will explain and discuss in detail what has been noted so briefly here, starting over at the beginning.

1.2 Uniqueness of MRI

MRI can do much of what CT does, but it is much better at defining the anatomy and finding pathologies in soft tissues. Similarly, it can reveal much of what can be found with nuclear medicine, ultrasound, and other modalities. But there are also valuable studies that only MRI can carry out.

MRI reflects primarily upon the interactions of water and lipids with the biomolecules that happen to be present within and between the cells of tissues. T1- and T2-imaging are powerful because they provide forms of image contrast, based on those interactions, that are altogether different from those of other modalities—ones that are sensitive, in particular, to differences in soft tissue types and that correlate with specific pathologies. As medical physicists, computer scientists, and others continue to provide technical improvements at a rapid clip, the transfer of these evolving technologies to the clinic will open new areas of medical research and, thereafter, of widespread application.

One area of significant clinical growth potential in MRI is in the prediction and monitoring of response to treatment, whether radiation therapy, chemotherapy, local heat ablation with US, gene therapy, or whatever. In cancer therapy, for example, ionizing radiation and certain tumoricidal drugs are relatively ineffective on tumor cells that are poorly oxygenated. Indeed, the most radio-resistant cells are commonly those inhabiting a barely viable hypoxic rind between active tumor, on the outside, and necrotic tumor cells within that have died from anoxia. MRI may well be able to reveal whether a pharmaceutical intended to promote angiogenesis in that rind, or to facilitate the perfusion or diffusion of water or oxygen into it (which may render it more radiation-sensitive), will be successful on a specific patient. Likewise, following the administration of a drug or dose of radiation, MRI and perhaps MR spectroscopy can assess the degree to which the opening or proliferation of blood vessels has increased the oxygen level in hypoxic regions and, therefore, the likelihood of destroying the tumor. These are but two out of countless examples of image-guided therapy (involving MRI alone or in collaboration with other modalities, e.g., PET) that either already exist or are forthcoming.

1.3 A Real MRI Case Study

A female associate professor of genetics experienced erratic diffuse headaches that responded to an Advil or Tylenol. The headaches were mild but still a little disruptive, so after several months she turned to her general practitioner. She presented as a healthy 52-year-old in no apparent distress apart from mild hypertension that was controlled by medication. She followed a good diet and exercised moderately several times a week. Other aspects of her physical examination were unremarkable and, in particular, a neurological examination revealed no deficits. She claimed to be happily married to her best friend and reported no major stresses or anxieties, apart from those arising from rearing two independent and adventurous teenage daughters.

Computed tomography (CT)

Uncomfortable with the duration of the problem, her internist ordered a CT scan of the brain at the local clinic. The results were suggestive of an irregularity in the right posterior temporo-occipital region, partially replacing some of the occipital horn of right lateral ventricle. He felt that an MRI head study could provide better information.

The next day she and her husband drove a half hour to one of her university's teaching hospitals. In her haste at home, she misplaced the DVD of the CT study. After her check-in at the medical center, however, all of her subsequent images, along with lab results and other medical records, were stored safely and kept available for immediate retrieval on the Department of Radiology's *Picture Archiving and Communications System* (PACS).

MRI: T1-w, T2-w, and FLAIR imaging

Although not available, the CT had already performed an invaluable service by drawing attention to the possible irregularity. In any case, the MRI studies would be more revealing of soft-tissue problems. The patient underwent a number of standard MR studies, including the gathering of T1-weighted and so-called FLAIR images (Figure 1.6).

The T1-*w* image supported the CT report, and also noted the extension of a well-circumscribed $1.7 \times 1.1\ cm^2$ lesion to the lining of the occipital horn of the

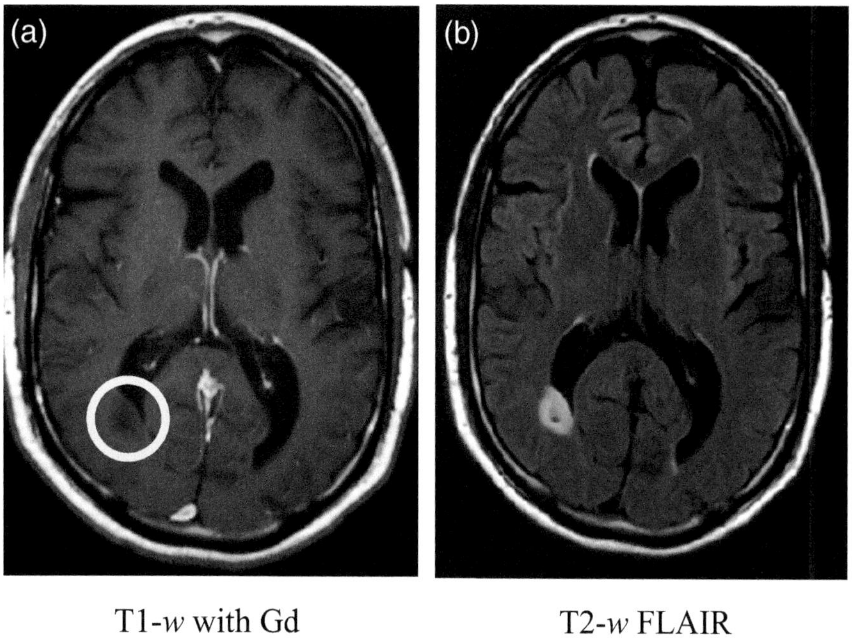

Figure 1.6 Two MR images of the same thin (1 mm), transverse slice. As with CT, a transverse MRI scan looks upward from the feet. (a) The T1-weighted study reveals a right posterior temporo-occipital lesion adjacent to occipital horn of the right lateral ventricle. Injection of a gadolinium-based MRI contrast agent, as shown here, made no difference visually from an earlier scan obtained without the contrast agent, suggestive that the blood-brain barrier had not (yet) been breached by a growing tumor. (b) The FLAIR (*FLuid Attenuation Inversion Recovery*) study provided different but complementary clinical information. Here signals from *cerebralspinal fluid* (CSF) and other fluids are suppressed, yielding a somewhat different type of MRI contrast, one that better demonstrates the lesion, which is now hyperintense. [*Courtesy of Charles Smith, Peter Hardy, David Powell, University of Kentucky MR Imaging and Spectroscopy Center (MRISC)*].

right lateral ventricle. The absence of bilateral symmetry of the brain was striking, but there was no midline shift or hydrocephalus. The T1-*w* scan was then repeated following the injection of 20 ml of gadolinium (Gd) contrast agent, and little change occurred. Higher-grade neoplasms in the brain are more likely to disrupt the blood-brain barrier, and are thus associated with more avid contrast enhancement. The failure of the region to noticeably take up the agent implied that the lesion was probably not a high-grade neoplasm, a hopeful sign.

A FLAIR image (*FLuid Attenuation Inversion Recovery*) is similar, but suppresses signals from aqueous fluids. In the brain, for example, it eliminates much of the appearance of *cerebrospinal fluid* (CSF), facilitating a better evaluation of the surrounding edema, as well as enhancing lesion conspicuity (Figure 1.6b). The reading radiologist felt that the image indicated a lower-grade glioma protruding into the ventricle, or possibly (although less likely) a tumor-like demyelinating lesion. (A glioma is a common type of tumor that occurs in glial cells, most frequently of the brain; the primary job of these cells is to nurture and maintain neurons, but researches have recently found that they also play a role in neuron functionality.) He also reported several small point-like T2 signal abnormalities scattered throughout the subcortical white matter of the cerebral hemispheres, possibly representing chronic small-vessel blockages or conceivably further areas of neural demyelation.

Magnetic resonance spectroscopy (MRS)

The gold standard in identifying a tumor is a biopsy (physically sampling and microscopic analysis of the tissue). But MRI can provide, in addition to images, quick and non-invasive preliminary information regarding tissue pathology by means of an advanced procedure known as *MR spectroscopy* (MRS).

The Larmor frequency of a proton is determined precisely by the exact magnetic field it experiences. The electrons circulating throughout a biomolecule themselves create a weak magnetic field, and this will affect the magnetic field seen by its tissue protons at various locations. This results in parts-per-million *chemical shifts* in the resonance frequencies of the protons in different molecular surroundings. Then Equation (1.3a) should be modified, if the protons are in a uniform external magnetic field B_0, as

$$\nu_{Larmor} = (\gamma / 2\pi) \times B_0 (1 + \sigma), \qquad (1.7)$$

where the *chemical shift coefficient*, σ, for a particular proton is strongly dependent on the details of its unique molecular environment. This may seem a little like the proton density imaging of Figure 1.3, but here the Larmor frequency is being shifted by the field from orbit-

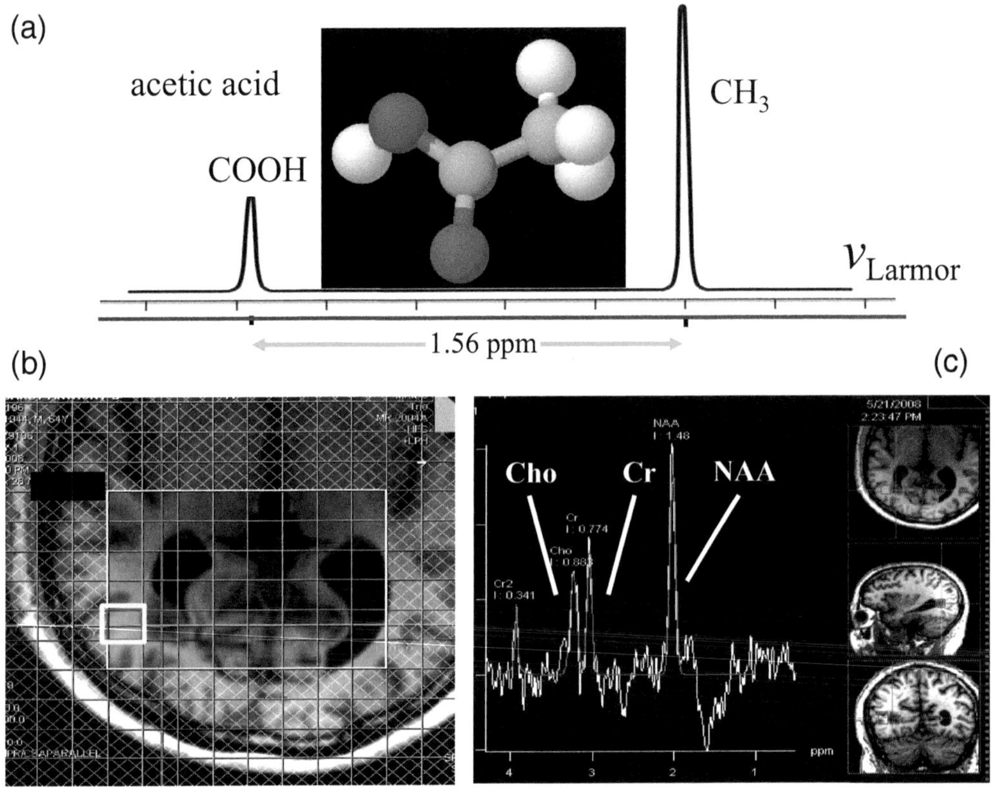

Figure 1.7 An MRS spectrum (a) reflects the uneven circulation of electrons within a molecule (acetic acid) that give rise to small variations in *local* magnetic field, hence in parts per million (ppm) *chemical shifts* in proton Larmor frequency. (b) Tissue just outside the patient's lesion ( ) yields a normal MRS spectrum. The spectrum from the lesion itself appears nearly identical, (c) but its higher Cr/NAA and Cho/NAA ratios are characteristic of a glioma.

ing molecular electrons, not by an externally applied gradient field.

The proton nuclear magnetic resonance spectrum of the simple acetic acid (CH_3COOH) molecule, for example, appears as two distinct MR peaks, both very close in resonance frequency to that of a free proton. They are separated by a 1.56 parts per million (ppm) chemical shift in the Larmor frequency, indicating the two slightly different local magnetic fields that the protons can feel (Figure 1.7a). One of the resonance lines is three times the other in amplitude: the three protons within the methyl (CH_3–) group all experience the same swirl of electrons and identical surroundings. The electron flow within the acid (–OOH) group, on the other hand, is different, and its single proton resides in a slightly lower local field.

Protons in molecules imposing a chemical shift are usually too few in number, and with too small a shift, to be seen in standard MRI images, other than lipids. But an MRS instrument combines exceptional homogeneity of the external field with the excellent RF frequency-resolution of an NMR spectroscope, so that protons in different molecular structures can often be distinguished. This allows *in vivo* examination of small, well-localizable volumes of soft tissues, and the spectra obtained may lead to the identification of certain irregularities.

MRS is capable of performing nuclear resonance spectroscopy on small, multi-voxel blocks of tissues at specific locations within the body. For tissues away from the lesion (Figure 1.7b,c) the proton spectra for N-acetylaspartate (NAA), creatine and phospho-creatine (Cr), and choline (Cho) appeared normal. The lesion itself, however, reveals a statistically significant irregularity in the concentrations (areas under the peaks) of Cr and NAA, relative to that of Cho. The spectral signatures for numerous normal and abnormal tissues have been studied, and the pattern seen here, when combined with the imaging information, is indicative of a glioma, but of indeterminate stage.

Functional MRI (fMRI)

Treatment of a tumor depends on its type, anatomical location, grade (degree of abnormality and growth rate seen under a microscope), and stage (size and degree of spread). In the present case, the lesion's position is such that surgery or radiation therapy might well result in the loss of one of the patient's two fields of vision (to the left or right). While she could accept that, she made it unambiguously clear that she would not agree to any action that would seriously jeopardize her ability to read; she would prefer to leave the disease untreated and take her chances.

Before proceeding to a needle biopsy—which itself could impose some risk to reading vision—she underwent two noninvasive MRI-based studies that would help to determine the closeness of the apparent tumor to optically active regions. The first was *functional MRI* (f MRI) (Figure 1.8a).

In an f MRI study, a subject is asked to undertake a mental process of some sort, such as repeatedly viewing a visual stimulus or tapping a finger. In the process, associated brain tissues become unusually active and consume extra oxygen, transforming oxyhemoglobin into deoxyhemoglobin there more rapidly than when at rest. These two molecules differ magnetically, and deviations in their balance may produce detectable tissue contrast where a region of the brain is being triggered. You might suspect that the relevant neurons are simply burning more oxygen; that indeed happens but, as will be seen later, the brain is far cleverer than that, and the process is actually more subtle and interesting.

Functional MRI data complement those of PET, which lights up those parts of the brain that happen to be burning excess radio-labeled glucose. Other modalities, like electroencephalography (EEG) and magnetoencephalography (MEG), are also being developed for mapping brain function. Establishing correlations among their results and those of f MRI and PET will greatly enhance the medical value of all of them.

The patient underwent f MRI with two separate sets of periodic stimuli, self-directed finger tapping and visual images. Temporal variations (which is what is of interest here) in the MRI signal are much smaller than the average value of the signal voltage itself, and effective noise-rejection and statistical information-processing programs must be invoked. The finger-tapping task demonstrated robust activation within the associated region of the motor cortex of the cerebral hemispheres. The patient's response to an intermittent visual stimulus is shown here in a thin sagittal slice that passes through the circled lesion; it indicates that one of the optically active regions of the brain (in green) lies adjacent to it, and possibly within it. Damage to it could cause catastrophic sight loss.

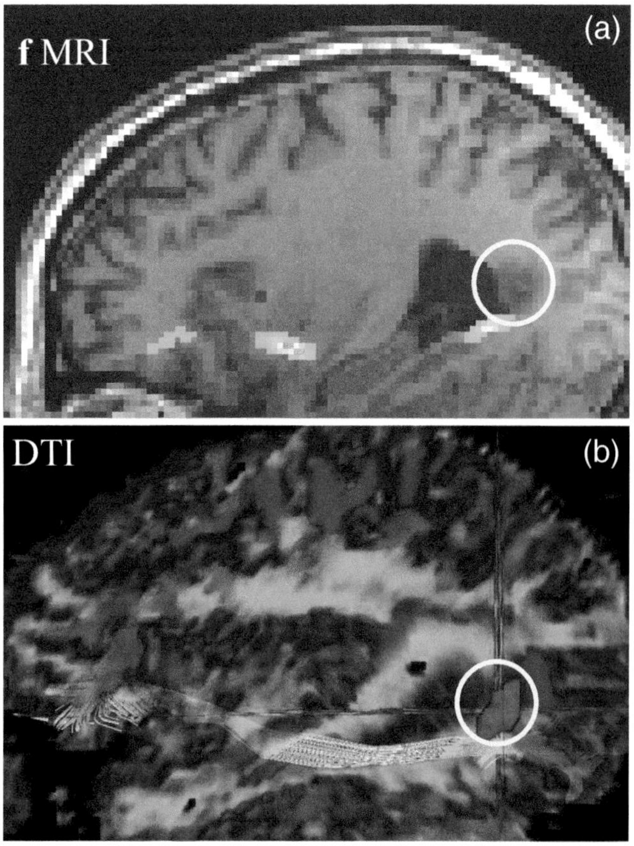

Figure 1.8 MRI for pre-surgical planning. (a) A thin (1 mm) slice sagittal T1-w image with a functional MRI (fMRI) overlay highlights small regions of the brain (yellow-green), near the circled lesion, that respond to a flashing light stimulus. (b) A diffusion tensor image (DTI) further indicates the close proximity of the patient's nerve trunks, the 'optic radiation' (white horizontal strands) to the glioma (red solid). [With thanks to Charles Smith and David K. Powell, University of Kentucky MRISC].

Diffusion tensor MR imaging (DTI)

In view of the rather discouraging f MRI results, it was felt to be worthwhile to carry out a *diffusion tensor image* study, as well. With DTI, contrast arises from the unidirectional *diffusion* of water molecules along the long axons of a nerve trunk, against a background of other water molecules that are diffusing isotropically, thereby bringing the nerve trunks into view. Figure 1.8b. A sagittal, thin-slice DTI corroborated the earlier f MRI finding that the patient's probable glioma lies directly adjacent to, and possibly infiltrating, superior portions of a nerve trunk, the *optic radiation* (the pale horizontal ribbon).

MRI-guided fine-needle biopsy

After viewing all the evidence, and particularly the DTI results, a neurosurgeon felt that with MRI guidance, she could very probably obtain a fine-needle tissue biopsy sample with low chance of damaging the optic radiations. After consulting with her personal physician, the patient agreed to the three-step invasive process.

First, in the operating room and under local anesthesia, a rigid, nonmagnetic frame was screwed firmly into the skull, providing the platform for establishing a fixed Cartesian coordinate system within which to localize the tumor, and later to reach it with a fine biopsy needle.

A separate assembly comprising a box with four flat, transparent walls embedded with weakly magnetic orthogonal positional markers and lines was attached. This provides an orthogonal frame of reference that is visible to the MRI device. Any point within the brain could now be expressed as a set of x-, y-, and z-coordinates relative to the frame, to within one millimeter (Figure 1.9a). MR images were now taken, and from these the neurosurgeon decided where in the lesion she would obtain the biopsy sample, and the path she would follow in getting there.

Then, with the frame still attached rigidly to the skull, the first (MRI-imaging) assembly is removed from it and the second, a stereotactic biopsy needle guidance assembly, is fixed to it (Figure 1.9b). A medical physicist experienced in the use of the equipment made the necessary calculations, based on the images just obtained, and he and the neurosurgeon set the required angles and distance limiters on the mechanical needle-guidance assembly. Soon thereafter, after drilling a small bore-hole in the skull, the surgeon advanced

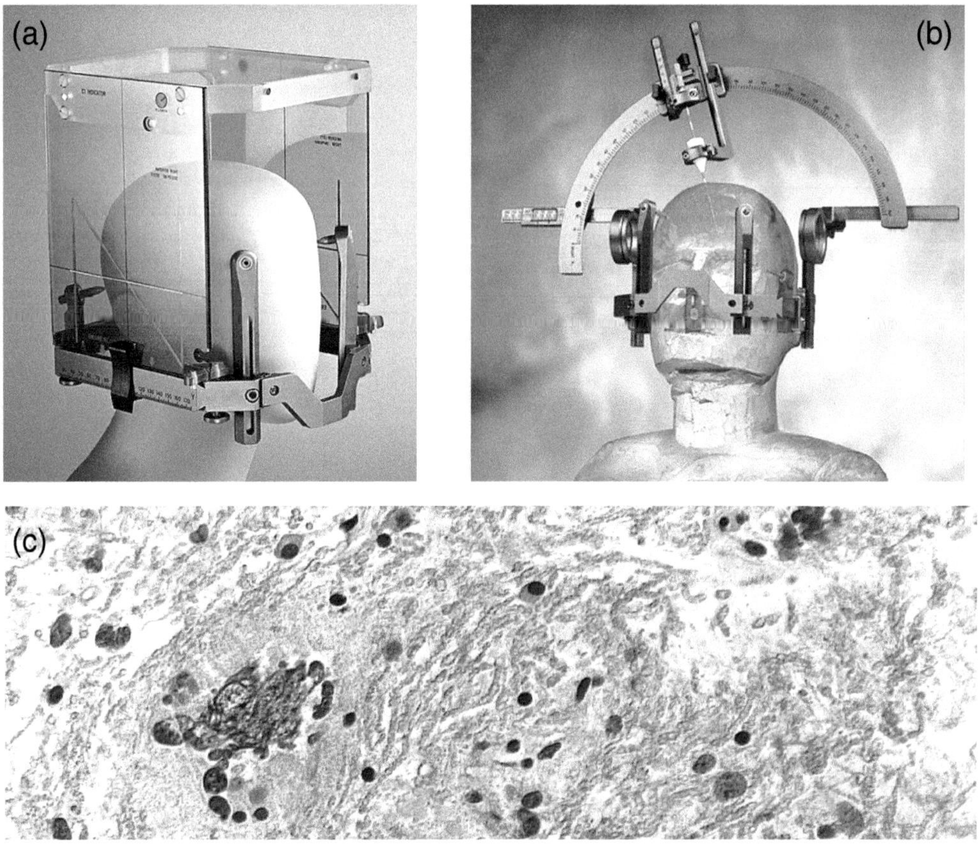

Figure 1.9 A stereotactic fine-needle biopsy device consists of an immobile frame screwed firmly into the scull in the operating room, along with two separate attachments used in sequence. (a) The first of these consists of an array of paramagnetic fiducial markers embedded in plastic planes; these can be read precisely by the MRI device during a clinical study. These make possible accurate localization of a target point within the lesion. (b) Then, with the frame still fastened rigidly to the skull, the first assembly is removed and a second attached. This one can guide insertion of a hollow sampling needle along any direction chosen by the neurosurgeon, such that its tip will end up within 1 mm of the target point. [*Courtesy of Elekta*]. (c) Photomicrograph from one of the biopsy slides. The pathologist reported that the sample collected from this lesion displays scattered and atypical cellular nuclei, suggestive of an astrocytoma, a form of glioma, but of undeterminable grade.

the hollow needle exactly the right distance and obtained the sample without incident.

It was expected that the samples obtained would be definitive, but they were not (Figure 1.9c). The surgeon phoned the patient the evening of the biopsy with a very unoptimistic report from the pathologist: the tissues obtained were equivocal, but suggestive of an advanced, fast-growing high-grade glioma. The patient reacted to the news calmly, her greatest concern being the impact this would have on her husband and daughters, to whom she was devoted. Further examination of the slides a few days later by several other pathologists, remarkably, led to the far less awful prognosis that the glioma may be only of stage-1 or -2, which either can be watched for changes or treated with drugs and radiotherapy with a reasonable chance of control.

Should we try positron emission tomography?

A final diagnostic test was considered, but rejected. Tumors commonly oxidize glucose at a faster rate than do healthy tissues of the same type, and positron emission tomography (PET) is a nuclear medicine modality highly sensitive at detecting excessive cellular uptake and use of it, Figure 0.4b. Sugar molecules are labeled with radioactive fluorine-18, and the injected 18-fluorodeoxyglucose (^{18}FDG) concentrates preferentially in fast-metabolizing tissue, such as neoplasms. The fluorine nucleus decays with the emission of a positively charged *positron*, which travels 1 mm or so and immediately collides with its anti-particle, an electron. The two particles mutually annihilate, giving birth to a pair of 511 keV *annihilation photons*, which fly away from the site of the interaction in almost exactly (to within ¼°) antiparallel directions. If the two are detected simultaneously on the opposite sides of a PET imager, within a tiny time window, they will contribute to the formation of a PET image of the region that had taken up the radiopharmaceutical. Thus, PET is another source of tissue image contrast, but one that exploits the metabolism of radioactive glucose, rather than of normal oxygen, like f MRI.

PET findings might be helpful, such as in finding other lesions in the brain, but it would probably have little or no effect on the patient's treatment, and she and her physicians decided against pursuing it.

Treatment guidance and follow-up

The patient and her physicians, hopeful that the glioma might turn out to be slow-growing or even quiescent, decided that the best course of action was to do nothing. An MRI examination every six months for three years has revealed that, in fact, her tumor was not growing appreciably, and she's still alive and well. In an ironic twist, upon returning home she learned that other residents of her condo were also experiencing headaches. The epidemic ended, as did her own headaches, when the large construction project next door, which frequently produced unpleasant fumes, terminated. So her glioma may well have had nothing at all to do with her symptoms, and was identified by accident.

One area of significant clinical growth potential in MRI is in the prediction and monitoring of response to treatment, whether radiation therapy, chemotherapy, local heat ablation with ultrasound, gene therapy, etc. For example, MRI continued to play a role in treatment of our patient, in part by monitoring any change in the size, shape, or any other aspect in the appearance of the lesion. But some developments in her status might call for additional or alternative treatments.

The purpose of nearly any curative (non-palliative) surgical or other therapy, of course, is to eliminate the problem without causing unacceptable damage to healthy (especially critical) tissues. A procedure similar to that for the fine needle aspiration biopsy (Figure 1.9) can be used to direct a narrow but intense beam of x-rays at the lesion. Linear accelerators (linacs) have been modified to perform microradiosurgery on brain tumors with very small fields and, guided by CT or MRI, with very high precision. Brachytherapy with suitable and meticulously located radioactive seeds might also be appropriate. And for some kinds of surgery, a growing number of institutions have MRI in the operating room in order to aggressively, and more safely, resect the disease. Similarly, chemotherapy drugs can be infused under guided delivery alone or in conjunction with radiation or surgery. How might the general approach described here be applied in other medical procedures [Jolesz 2008]?

In the long run, perhaps the most extraordinary (from our current perspective) clinical advances will arise from ongoing efforts to establish strong linkages among the various forms of physiologic imaging—especially MRI, nuclear medicine, and optical imaging—with genomics, proteomics, molecular biology, and nanotechnology. These fields are just beginning to overlap and unify, yet already they have created novel families of biologically derived *imaging biomarkers* as measures of biological processes. It is easy to suppose that this is still just the lowest hanging fruit, and that eventual results could be far beyond our current imaginings.

1.4 A Brief History of MRI: Bloch, Purcell, Damadian, Lauterbur, et al.

Wilhelm Conrad Röentgen flung open the doors of modern medicine on the chilly evening of November 8, 1895. Within weeks he was able to demonstrate that his totally unanticipated x-rays can generate astounding images of broken bones and bullets within the body. Magnetic resonance imaging (MRI) entered the scene in the mid-70s, also with fanfare, but only after much essential groundwork had been laid down (for other reasons) over the previous quarter century. MRI grew out of nuclear magnetic resonance (NMR), which itself has been around since before mid-century [Mattson et al. 1996].

Probably as good a place as any to start the story of MRI is in the year 1913, when Niels Bohr unveiled his model of the hydrogen atom. The Bohr atom was largely a classical planetary system but, as with Planck's 1901 derivation of the blackbody radiation spectrum, he imposed a severe quantum mechanical constraint on it: he demanded that the angular momentum of the electron orbiting its nucleus be quantized. His derivation indicated that the electron could travel only in distinct orbitals, and had to inhabit discrete energy levels. Despite its rather ad hoc construction, the model agreed astoundingly well with the results of earlier optical and infrared spectroscopy. When Heisenberg and Schrödinger separately unveiled the two standard versions of modern QM in the mid-1920s, one of their first great successes was a rigorous application to the hydrogen model.

A critically important milestone in the early days of quantum mechanics was the widely debated demonstration in 1922 by Otto Stern and Walther Gerlach, physicists at the University of Frankfurt, that a beam of silver atoms in a vacuum would be deflected in an unexpected way by a magnetic field gradient. It was not for another five years that the results of the experiment were interpreted correctly: in addition to any orbital angular momentum it may enjoy, an atomic electron also acts somewhat like a spinning ball of charge that (like a proton) creates a weak magnetic field of its own. It is said to possess a purely quantum mechanical attribute called, by analogy to that of a rotating ball, *electron spin angular momentum*, and the magnetic field it generates is its *electron spin magnetic moment*. It was found, moreover, that unlike a compass needle, the spin moment is subject to *spatial quantization*; that is, the measured component of spin along the external magnetic field could take on only certain specific values. The field was said to split the electron's quantum state into a pair of Zeeman levels, first detected in optical spectra in 1896. (Zeeman himself, incidentally, had earlier been fired from his job for working on this project against the explicit orders of his supervisor; he was vindicated six years later with the awarding of the Nobel Prize.) Stern, a Jew, came to Pittsburgh's Carnegie Institute of Technology in 1933 when Hitler became Chancellor of Germany, and he received the Nobel Prize in physics a decade later. While not a Nazi, Gerlach did contribute to the German war effort, which is presumably a reason he was not invited to join his colleague on the podium in Stockholm.

The following year, the Nobel Prize went to Isidor Isaac Rabi for a different series of studies involving atomic and molecular beams [Breit 1931]. After suggesting that the Stern-Gerlach experiment could be modified so as to explore the properties not only of electrons but of nuclei, too, Rabi for the first time demonstrated the phenomenon of *nuclear magnetic resonance*. He sent beams of hydrogen, lithium, sodium, potassium, and other atoms speeding through a magnetic field. Then, as radiofrequency energy of fixed frequency was being pumped in, the strength of the magnetic field was swept slowly, and the current of the beam underwent a dip at the NM resonance condition, Equation (1.2). Among much else, Rabi's group at Columbia found that an isolated proton, like an electron, has an inherent spin angular momentum of ½, because of which it produces its own nuclear magnetic moment, μ. And it, too, is spatially quantized, allowed to align only up or down in an external magnetic field. It is largely because of this property, and clever and subtle manipulations of the NMR phenomenon, that MRI is possible.

NMR spectroscopy in solids and liquids was developed independently toward the end of World War II by Felix Bloch [Bloch 1946] at Stanford University (protons in paraffin) and Edward Purcell [Purcell et al. 1946] at Harvard (water) and their colleagues. Their approaches and equipment were somewhat different, but it soon became clear that they were observing essentially the same phenomenon. At first, NMR was of concern primarily to physicists, who rapidly brought about significant advances, in particular high-resolution NMR spectroscopy. This, in turn, proved an invaluable tool, allowing chemists to unravel the molecular structures of complex organic and other compounds. The importance of the work of Bloch and Purcell was recognized immediately, and the two shared the 1952 Nobel Prize in physics for it. A few years later, Nicolaas Bloembergen (also a Harvard physics Nobel laureate, and the father of the field of nonlinear optics), and Purcell and Robert Pound published a theory of nuclear spin relaxation of the kind that still underlies the universal biophysical parameters T1 and T2 [Bloembergen 1947].

Many others carried out further NMR work that later proved essential to the invention of MRI. Among these, Erwin Hahn discovered spin-echoes in NMR in 1950 [Hahn 1950]. And after high-speed computers became available, Richard Ernst [Ernst et al. 1966] pioneered the use of short pulses of RF (rather than the continuous-wave systems then being employed) and Fourier-Transform NMR (Nobel Prize in Chemistry, 1991).

In parallel with this, improvements in superconductor technology boosted the potency of NMR with the introduction of magnets an order of magnitude stronger than permanent magnets and electromagnets. Nearly all clinical MRI magnets being purchased today in developed countries are superconducting.

Hints of the possibility of imaging built upon NMR first appeared in the 1970s, not long after Godfrey Hounsfield and the company EMI brought CT to the market. In 1971, Raymond Damadian, an Armenian-American physician and entrepreneur, published in *Science* magazine his findings that tissues surgically removed from various organs in rats have different relaxation times T1 and T2 for potassium and sodium (rather than protons) [Damadian 1971]. Those from tumors, moreover, tend to have measurably longer values than do the surrounding healthy tissues from which they were excised. He suggested that a device be built to monitor relaxation times coming from different parts of the body, to search for cancer *in vivo*, but he did not describe a way to make spatial localization of NMR signals happen in practice. Nonetheless, Damadian's papers motivated other researchers to explore NMR as a basis of a medical imaging technology.

In his 1952 Ph.D. thesis, H. Y. Carr generated a one-dimensional MR image by means of a single gradient field that varied linearly along a 1D phantom, as in Figure 1.3 [Carr 1993]. This allowed signal strength to be linked to RF Larmor resonance frequency, thus to field strength, hence to spatial position, Equation (1.2b). While it was not yet clinically useful, Carr was the first to demonstrate a form of spatial localization for MRI.

Aware of Carr's work, Paul Lauterbur, a professor of chemistry at the State University of New York at Stony Brook, generalized the approach to 2D imaging and devised a way by means of which the NMR signals coming from different parts of the body could be distinguished from one another. In 1973 he published in *Nature* the first two-dimensional MRI image, that of a small clam his daughter had found on the beach [Lauterbur 1973; Dawson 2013]. Something to inspire hope in the rest of us: the first time Lauterbur submitted his paper, the editors of *Nature* rejected it! He later said, "You could write the entire history of science in the last 50 years in terms of papers rejected by *Science* or *Nature*."

Peter Mansfield of the University of Nottingham soon thereafter developed a mathematical method that made it possible to greatly reduce image reconstruction time and to produce clearer images [Garroway et al. 1974]. While Lauterbur had employed a projection-reconstruction technique analogous to that of CT, Mansfield's Fourier transforms between *k*- (reciprocal-) and real spaces provided a much faster method based on frequency and phase spatial encoding. This made it possible for NMR, a novel research tool primarily of interest only to chemists, to become an imaging modality that radically transformed much of clinical medicine.

On July 3, 1977, after seven years of effort, Damadian carried out the first whole-body scan on the thorax of his slender assistant, Dr. Larry Minkoff, in a procedure that took almost five hours (Figure 1.10a) [Damadian et al. 1977]. Damadian named the first whole-body MRI scanner 'Indomitable,' and it is now on permanent display at the Smithsonian Institution in Washington, D.C.

The 2003 Nobel Prize in Physiology or Medicine went to Lauterbur (Figure 1.10a) and Mansfield (Figure 1.10b) in recognition of their seminal work. Carr protested the failure of his being included in the award, and a number of others supported him in this; at least his name does live on in the important Carr-Purcell spin-echo pulse sequence. More publicly, Damadian's company, FONAR, took out full-page ads in *The New York Times* and *The Washington Post* protesting his omission, a "shameful wrong that must be righted." It has been suggested that his exclusion is attributable, in part, to his belief in new-earth creationism and a literal interpretation of Genesis. Needless to say, this is not the only decision that the Nobel Committee has made over the years that has stirred considerable controversy, as attested to by the stories of Lise Meitner (discovery and explanation of nuclear fission), Oswald Avery (isolation of DNA), and Rosalind Franklin (structure of DNA). Meitner, at least, had chemical element 109 named after her.

There were, of course, a number of other near misses. Before the realization of NMR, the Dutch physicist C. J. Gorter imagined that transitions among nuclear spin states might be discernible. His strenuous efforts to detect them around 1940 were unsuccessful, in part because of insufficiently sensitive equipment and his choice of sample materials with unfortunate relaxation times. It was a few months after his visit to Columbia and his critical conversations with Rabi that the latter succeeded in demonstrating NMR in a particle beam.

A second near miss was Y. K. Zavoisky, who performed electron paramagnetic resonance (EPR or ESR)

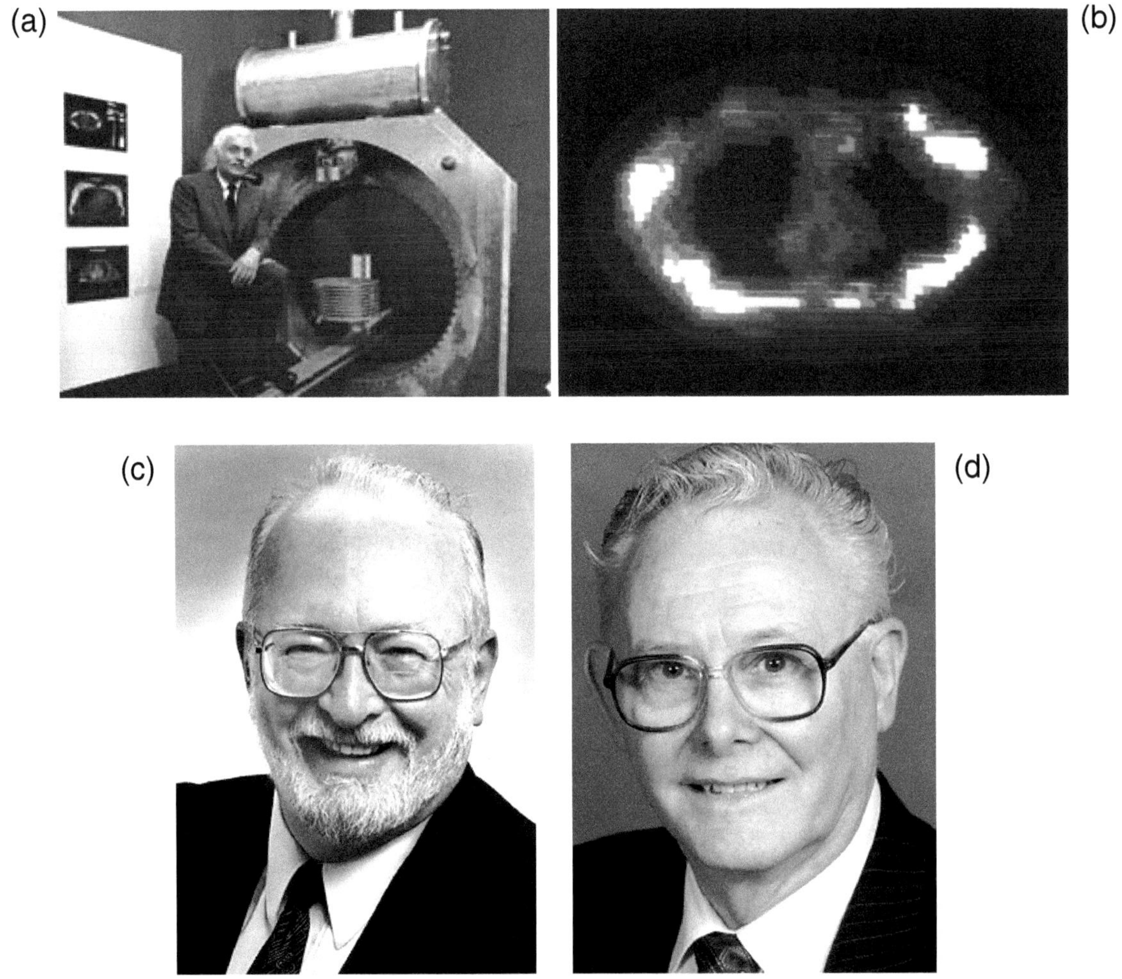

Figure 1.10 Pioneers of MRI. (a) Raymond Damadian and the first whole-body MRI scanner, built in 1977. Damadian himself served as the lead-off guinea pig, but was too large for the machine to produce a signal. (b) Indomitable's first image, of Lawrence Minkoff's chest, took nearly five hours to complete. *[Courtesy of the FONAR Corporation]*. (c) Paul Lauterbur and (d) Peter Mansfield were awarded the 2003 Nobel Prize in Physics for their contributions to the early development of MRI.

in war-ravaged Russia a year before Bloch and Purcell first carried out NMR. He published in *Journal of Physics USSR* in 1945, which the Nobel committee apparently overlooked, but at least he ended up with the Stalin and Lenin prizes. Next best thing.

The 1980s and later have seen the continuous maturation of research efforts by manufacturers and academicians and the development of the equipment and software that have led to continuous growth and improvement. By now, nearly all of the basic technology and theory have been well worked out, but efforts are ongoing to speed up data acquisition, enhance contrast-to-noise, and reduce costs, especially for the magnets. MRI will continue to evolve for the foreseeable future and, no doubt, provide us with nice surprises from time to time.

1.5 Critical Caveat

Just about everything good in life has its downsides, and MRI is no exception. Imaging times tend to be longish (minutes, sometimes) compared to other cross-sectional anatomical techniques such as CT, although, like cost, these keep coming down. There are more artifacts than with other modalities, and some of these are hard to detect and rectify. The learning curves for physicians, technologists, and physicists are steep and high. But perhaps the most pressing issue is the need for constant, proactive caution to prevent life-threatening injuries.

While MRI is widely touted for its inherent safety, MRI systems are not totally free of hazards—indeed, people have been severely injured and killed by them. Staff and physicians must remain ever vigilant for possible shrapnel and implanted metallic objects such as aneurysm clips, wires, and pacemakers within patients and anyone else who enters the imaging suite—and also for potentially lethal flying screw drivers, oxygen bottles, scalpels, hand-cuffs, and other things you may be less inclined to worry about (Figure 1.11). Chapter 15, which covers MR quality assurance and safety, will go into these problems in detail.

The effects on tissues of intense static magnetic fields, of rapidly switched gradient fields, and of radiofrequency power, have been studied extensively. Many new commercial imagers use a field strength of

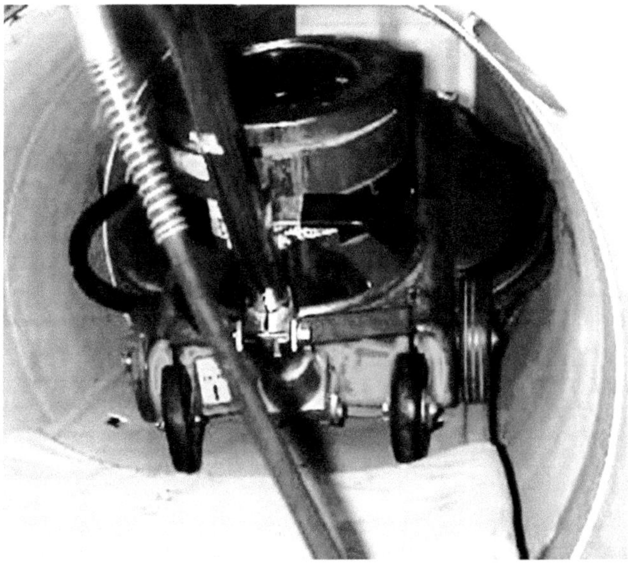

Figure 1.11 CAVEAT! How *not* to clean the insides of a superconducting MRI magnet. You should *always* assume, unless you know otherwise for certain, that a gun is loaded, and that there are strong fields both inside and outside the magnet which may well exert extreme, wrenching forces on any metallic objects within the patient or nearby. [*From www.simplyphysics.com.*]

three tesla (with 7 T systems nearing FDA clearance), which can lead to higher RF power deposition or the need for greater gradient field strength and switching speed. Studies of their long-term biologic effects will continue, but meanwhile, it clearly makes sense to exercise serious caution, especially in examining pregnant women, infants, and patients on life-support systems. In the United States, the Food and Drug Administration publishes recommendations for limiting the principal magnetic field strength, the rate of switching gradient fields, and especially levels of RF power (which can cause acute burns), and so on, which are thought to be protective. The majority of these guidelines come from the international and national advisory bodies that provide expert recommendations in these areas.

The bottom line, however, is this: no aspect of MR imaging in and of itself appears to pose serious risks under normal clinical conditions when staff follow currently accepted operational and safety guidelines. But constant vigilance is essential!

CHAPTER 2

Quasi-Quantum, Two-State Picture of Proton NMR in a Single Voxel

A quantum mechanical (QM) treatment of matter and electromagnetism (EM) is largely what the theory of MRI is all about. But interactions between the two can be pretty bazaar and hard to describe outright, so let's start out at the beginning, with a very brief review of the basics of classical EM physics. We then expand on the elementary ideas on nuclei that were mentioned in the previous chapter, and in particular, those relevant to the quasi QM (spin-up/spin down) picture NMR in a single voxel.

In the next chapter, we shall extend this to carry out a simple MRI *gedanken*-experiment (thought-experiment) with a one-dimensional patient.

2.1 A Brief Review of Electromagnetism

Humans have been aware of electricity since before we began depicting lightning on cave walls. About 600 years BC, along with bringing us a workable democratic political system, the Greeks discovered that rubbing wool on amber (petrified resin) would cause it to attract hair, feathers, and the like, and to produce lightening-like sparks. The earliest use of the phenomenon was probably when the 1st century Roman physician Scribonius Largus applied Mediterranean electric rays directly to the heads and feet of sufferers of severe headaches and gout, apparently with some success. A century later, the great Galen of Pergamon, perhaps the most influential medical researcher of antiquity, clarified that the fish had to be alive.

But it took the efforts of William Gilbert at the outset of the 17th century and those who followed him—like Charles-Augustin de Coulomb (force law, 1784), Luigi Galvani (who made a dead frog's leg twitch, 1780), Alessandro Volta (electric battery, 1799), Hans Christian Ørsted (electric current produces magnetism), André-Marie Ampere (a magnetic field circles the current in a wire, 1823), Georg Simon Ohm (resistance, 1827), Michael Faraday (magnetic induction; motors and dynamos, 1831) and Thomas Edison (incandescent light, 1878)—to start making sense of it and harnessing it.

Because of his numerous inventions, theoretical writings, and experiments—most notably the one with lightning and a kite in 1752 that somehow didn't kill him—the great American autodidact and polymath Benjamin Franklin became world renowned for his studies of electricity long before the American Revolutionary War. He was not only a leading figure in the creation of his country and author of *Poor Richard's Almanac*, but also the inventor of bifocal glasses, the Franklin stove, and the lightning rod, and originator of the words *charge*, *battery*, *conductor*, *positive*, and *negative*.

The discovery of magnetism, on the other hand, was intimately tied to the early use of iron. As long ago as 4000 BC, the Egyptians and Mesopotamians fashioned ornaments out of small amounts of iron from fallen meteorites. But while they may have recognized the metal's superiority over copper and bronze in tools and weapons, they did not have the ability to produce significant amounts of it from iron ore. The technology necessarily for smelting iron ore was developed by the Hittites in what is now Turkey, about 1500 BC. And within only a few more centuries, the Iron Age was in full swing throughout the Middle East and southeastern Europe, and spreading.

At some point—and estimates vary from 2500 BC in China to 800 BC in Greece—people noted that a certain gray-black, metallic mineral was naturally magnetic. Pieces of the iron ore magnetite (named for the Greek province of Magnesia, where it was mined) cling to unmagnetized iron. Moreover, an iron needle brought into contact with this sort of natural magnet could itself become a magnet, and even persist in that condition. Such a magnetizable material is said to be *ferromagnetic*.

Magnets were doubtless viewed as amusing diversions until some clever soul found the right application: a magnetized needle or a piece of magnetite, if suspended by a thread or supported by a cork on water, will twist about and always end up aligning north-south. The oldest extant written reference to compasses comes to us from 12th century England; indeed, the common term for magnetite, "lodestone," derives from the Middle English for "stone that leads." But by then, compasses may well already have been aiding in navigation for hundreds of years or more.

A major advance in the understanding of magnetism came in 1820 with Ørsted's discovery that a wire carrying an electric current creates a field that causes a compass needle to deflect. A current-carrying wire loop, moreover, itself behaves like a magnetic dipole, and itself tends to twist about in an external magnetic field—the basis for how an electric motor works. These findings demonstrated two related but separate mysterious and profound facts of life: first, moving charges, such as electrons flowing along a wire, create a magnetic field; and second, any small magnet that can move about freely, such as a compass needle or a current-bearing coil of wire, will tend to align along a magnetic field already present [Purcell et al. 2013].

A most basic, defining characteristic of any magnet, regardless of its shape, size, or material makeup, is that it always has two separated 'opposite' *poles,* with its north and south poles at its opposing ends. Two north or two south poles on separate magnets repel one another, while a north and a south attract. The *magnetic dipole field* of a single bar or needle magnet has both magnitude and direction, and so is a *vector* entity (Figure 2.1a). The nature of the field may vary from place to place throughout space and over time, as a *vector field*, but at any instant and location, the direction and length of an arrow can designate the local field's orientation and strength. By convention, the lines and arrows representing a dipole magnetic field are taken to run outside the magnet from its north pole to its south.

The Système International d'Unités (SI, or MKS) measure of magnetic field strength is the *tesla* (T); the traditional unit, the gauss, is 10^4 times smaller. The field strength at the surface of the Earth, viewed as a huge bar magnet, is about 0.00005 T (½ gauss), give or take, depending on where you are. The field from the small magnet that holds your favorite finger painting to the refrigerator door is 0.1 T or so; the fridge magnet may be small, but it is far from weak. The fields from the main superconducting magnets of the Large Hadron Collider at CERN, the largest and most expensive instrument ever built, produce fields greater than 8 tesla. And much smaller research magnets have been built that reach over 20 tesla. This doesn't amount to much, however, compared to the most potent magnet field known to exist, the one at the center of a type of x- and gamma-ray-emitting neutron star known as a *magnetar*, which is of the order of 10^8 to 10^{11} T.

Clinical MRI is carried out most commonly, by far, with a 1.5-tesla superconducting magnet, but about a quarter of new machines operate at 3.0 T, which produce a noticeably better signal-to-noise ratio. Any field must be stable over time and uniform, to within a few parts per million, within a volume large enough to accommodate a significant part of a body. The technology needed to produce such a field does not come from the local hardware store, nor is it cheap. Fewer than a hundred 7.0 T instruments have been constructed, primarily for clinical research but, because of their high sticker price, their numbers are not likely to swell. Permanent and room-temperature electromagnets cost considerably less, but max out at 0.3 T to 0.7 T, and they are rarely utilized in developed countries.

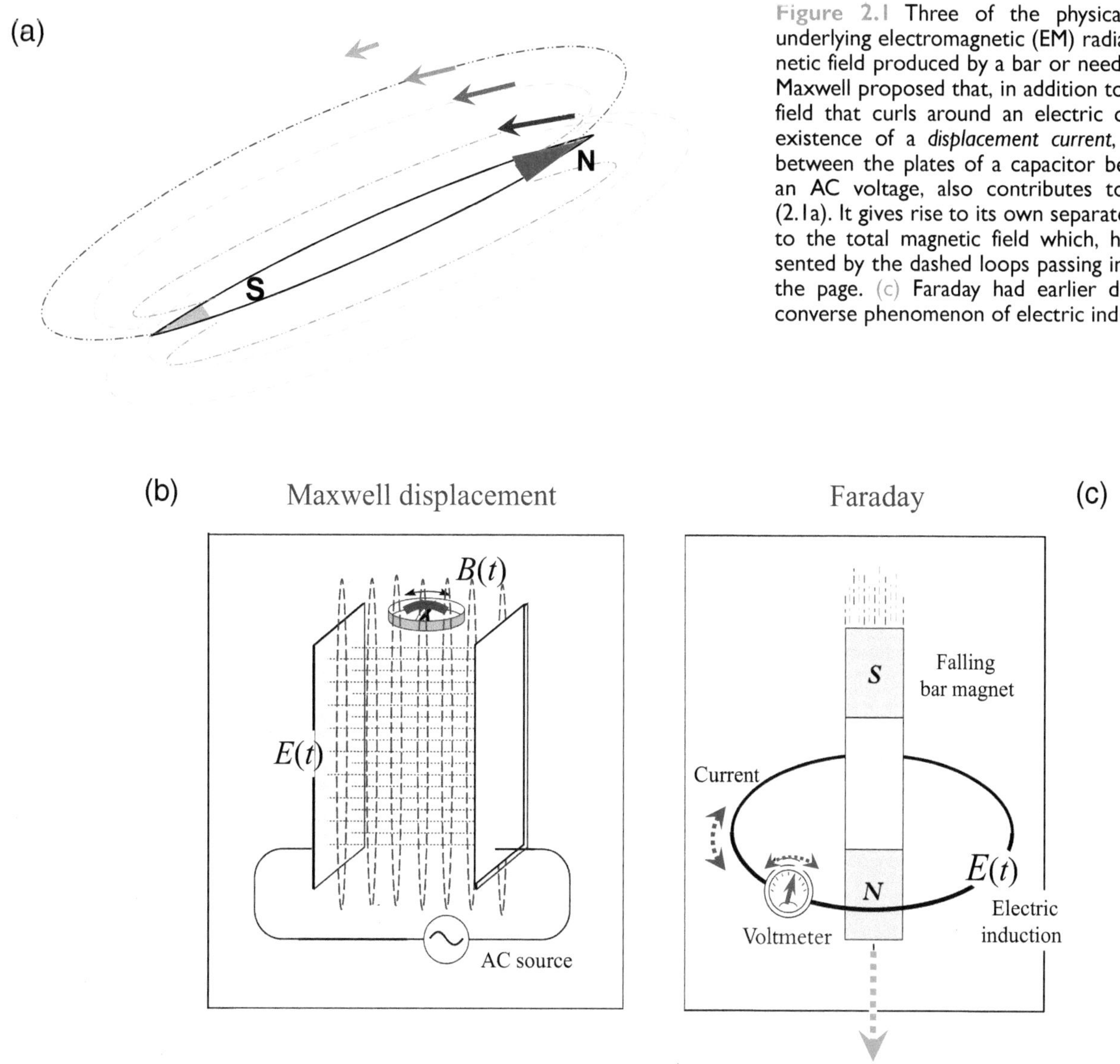

Figure 2.1 Three of the physical phenomena underlying electromagnetic (EM) radiation. (a) Magnetic field produced by a bar or needle magnet. (b) Maxwell proposed that, in addition to the magnetic field that curls around an electric current, J, the existence of a *displacement current*, such as that between the plates of a capacitor being driven by an AC voltage, also contributes to *B*, Equation (2.1a). It gives rise to its own separate contribution to the total magnetic field which, here, is represented by the dashed loops passing into and out of the page. (c) Faraday had earlier discovered the converse phenomenon of electric induction.

Nikola Tesla

The tesla (T), incidentally, is named after Nikola Tesla, who was born in Croatia of Serbian parents in 1856 and emigrated to America 28 years later. A dreamer and a man of remarkable and far-ranging genius, Tesla invented (along with much else) major elements of the technology that is used now in generating and harnessing electricity. He and George Westinghouse, who bought a number of Tesla's 700 patents, carried on an extended battle with Thomas Edison regarding the merits of alternating vs. direct current power. It was Tesla who got it right (although Edison's favored DC is used for transmitting very high voltage power).

Tesla took the occult seriously, and as he grew older, his idiosyncrasies became more pronounced, as described in this excerpt from a March 4, 1991, story about him in *New Yorker* magazine:

> "He couldn't tolerate the sight of pearl earrings, the smell of camphor, the act of shaking hands, or close exposure to the hair of other people. He strongly favored numbers that were divisible by three. He seldom ate or drank anything without first calculating its volume. He counted his steps. He washed his hands compulsively. He never married, but he told a friend he had once loved a particular pigeon 'as a man loves a woman.'"

So much for animal magnetism.

Fields, Maxwell's equations, and the wave equation

Nearly all phenomena of everyday life are governed by either gravitational or electromagnetic forces. (The weak and strong nuclear forces are largely hidden from view, as are the bases for mental processes.) While the first of these pins us to the Earth's surface, the latter hold us together, along with the molecules, liquids, and solids of which we are made.

Table 2.1 The four Maxwell equations and two electromagnetic (EM) wave equations in a vacuum, in SI units. The speed of light in a vacuum is $c = 2.9979 \times 10^8$ m/s. $\mu_0 = 8.854 \times 10^{-12}$ C^2s^2 kg m^3 is the vacuum *magnetic permeability*, or the *permeability* of free space; more generally, the permeability of a substance is a measure of the magnetic field within it that arises when it is subject to an applied magnetic field. The corresponding *electric permittivity* of free space is $\varepsilon_0 = 1.257 \times 10^{-12}$ kg m $/C^2$. All three of these are irreducible fundamental constants of nature. Things can get more complicated for EM radiation flowing through matter.

$\nabla \times \boldsymbol{B} = \mu_0(\mathbf{J} + \varepsilon_0\, \partial \boldsymbol{E}/\partial t)$	Maxwell-Ampere	(2.1a)
$\nabla \times \boldsymbol{E} = -\partial \boldsymbol{B}/\partial t$	Maxwell-Faraday	(2.1b)
$\nabla \cdot \boldsymbol{E} = \rho/\varepsilon_0$	Gauss	(2.1c)
$\nabla \cdot \boldsymbol{B} = 0$	Gauss, magnetism	(2.1d)
$(c^2\nabla^2 - \partial^2/\partial t^2)\,\boldsymbol{E} = 0$; $(c^2\nabla^2 - \partial^2/\partial t^2)\,\boldsymbol{B} = 0$	EM wave, vacuum	(2.2)

Newton described gravity in terms of the force acting at a distance between masses. To explore electricity and magnetism, Faraday and Maxwell extended the concept of force by visualizing and inventing fields. Instead of viewing objects as pushing and pulling directly on one another, as Newton did, they imagined an intermediary—a shadowy, ethereal electrical or magnetic presence that an electrical charge or magnetic pole generates and responds to in space—and it is the *field* from each that interacts with the other and causes it to react. The field enveloping the bar magnet of Figure 2.1a, for example, is illustrated with lines of force that would suggest the direction of force on a nearby pole. Are they real?

The stage was now set for what was one of the greatest leaps of scientific imagination since Newton's *Principia Mathematica* (1687). In the last year of the American Civil War, and four years before publication of *On the Origin of Species*, James Clerk Maxwell demonstrated in *A Dynamical Theory of the Electromagnetic Field* that the four equations bearing his name (Table 2.1) can be combined so as to give rise to an equation for the propagation of *electromagnetic* (EM) *waves* [Maxwell 1865]. The Maxwell equations are sketched here in modern notation and SI units. [Fleisch 2008].

Ampere's circuital law described how the magnetic field ***B*** 'circulating' or curling around an electric current, **J**, is proportional to that current. This can be expressed in differential (as opposed to the equivalent integral) form as

$$\nabla \times \boldsymbol{B} \;=\; \mu_0 \,\mathrm{J},$$

It contains a universal constant, the *magnetic permeability of vacuum*, which is commonly but regrettably represented by the symbol μ_0—an entity that has virtually nothing to do with the magnetic moment $\boldsymbol{\mu}$ of a proton or a compass needle. Seems our elders sometimes just ran out of symbols.

Maxwell argued that Ampere's law was incomplete in this form. He amended it to include a novel fundamental physical phenomenon that he proposed, the converse of magnetic induction, in which a *changing electric field* gives rise to a newly imagined transient *displacement electric current* which, in turn, creates an additional magnetic field. The full *Ampere-Maxwell* equation reads

$$\nabla \times \boldsymbol{B} \;=\; \mu_0 (\mathbf{J} + \varepsilon_0 \,\partial \boldsymbol{E}/\partial t), \qquad (2.1a)$$

ε_0, another fundamental constant, is known as the *electric permittivity* of empty space. When there are no real charges flowing nearby, $\mathbf{J} = 0$, and this equation reduces to

$$\nabla \times \boldsymbol{B} \;=\; \mu_0 \varepsilon_0 \,\partial \boldsymbol{E}/\partial t).$$

So even in the absence of moving charges, a changing electric field, such as between the plates of an open-air capacitor driven by AC, gives rise to a magnetic field that curls around the changing ***E***-field.

Conversely, by *Faraday's law*, a changing magnetic field, $\boldsymbol{B}(t)$, gives rise to a transient electric field, $\boldsymbol{E}(t)$, in the surrounding space (Figure 2.1c):

$$\nabla \times \boldsymbol{E} \;=\; -\,\partial \boldsymbol{B}/\partial t. \qquad (2.1b)$$

The voltage induced in a nearby loop of wire that surrounds *B*, say, is linear in the *rate of change* of that field.

The two other Maxwell equations bear Gauss's name. *Gauss's law* describes how an electric field emanates from a static electric charge. It states that the strength of the field, $\boldsymbol{E}(\boldsymbol{r})$, outside a region, is proportional to the net *charge density*, ρ, contained within it,

$$\nabla \cdot \boldsymbol{E} \;=\; \rho \,/\, \varepsilon_0 , \qquad (2.1c)$$

Equivalent to Coulomb's law, it describes the manner in which the electric field diverges from the enclosed charge. And in the absence of any charge, so that $\rho = 0$, this also simplifies in the obvious manner.

Finally, *Gauss's law for magnetism*,

$$\nabla \cdot \boldsymbol{B} \;=\; 0, \qquad (2.1d)$$

informs us that a magnetic field ***B*** has no divergence, i.e., that it is a 'solenoidal' vector field. It states explicitly that magnetic fields are illustrated with *closed* field lines—unlike the field lines spreading out from electric charges, no lines of magnetic field diverge forever outward. Magnetic field lines that start on a North pole must terminate on some South pole somewhere. This is to say that under normal conditions, and unlike the case for electric charges, magnetic monopoles do not exist. While magnetic 'monopoles' have been postulated by Paul Dirac and in superstring theory, they have never yet been observed.

EXERCISE 2.1 Is Equation (2.1d) equivalent to saying that Dirac's magnetic monopoles do not exist?

Maxwell combined these four equations to derived a *wave equation* that described the propagation of electromagnetic (EM) radiation through free space:

$$\left[(\mu_0 \varepsilon_0)^{-2} \nabla^2 - \partial^2/\partial^2 t)\right] \boldsymbol{E}(\boldsymbol{x}, t) \;=\; 0, \qquad (2.2a)$$

with a parallel formulation for the $\boldsymbol{B}(x,t)$ field.

With Maxwell's waves, a changing magnetic field at a point in space generates an adjacent changing electric field which, in turn, creates a nearby changing magnetic field, which itself produces a changing magnetic field a little further on, and so on and on. The wave just keeps on rolling along (or 'bootstrapping') (Figure 2.2). A wave of frequency ν and wavelength λ propagates in the direction perpendicular to those of the two fields and with the speed of wave propagation, c,

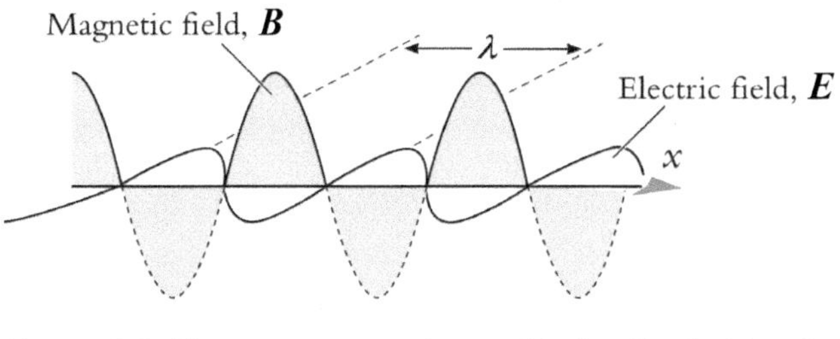

Figure 2.2 Electromagnetic radiation: $E(\boldsymbol{x},t) = E_{max}(\boldsymbol{x},t)$ [sin 2π $(vt - x/\lambda)$] for the 1D propagation of EM waves, where $v \times \lambda = c$.

$$\nu\lambda = c. \quad (2.2b)$$

Maxwell's theory predicts that the speed of electromagnetic radiation in a vacuum, c, can be related to the magnetic permeability and electric permittivity of vacuum as

$$c = (\mu_0 \varepsilon_0)^{-\frac{1}{2}}. \quad (2.2c)$$

The two parameters μ_0 and ε_0 had already been measured, and from them Maxwell calculated the velocity of EM radiation in a vacuum to be about $c \sim 3.00 \times 10^8$ m/s. Noting the closeness of this to the measured speed of light, he proposed that they were, in fact, one and the same. This additional triumph was the first time that c had been obtained not from direct measurements, but rather by combining other universal physical constants. The wave equation for $\boldsymbol{E}$ in free space became, and remains

$$(c^2\nabla^2 - \partial^2/\partial^2 t)\boldsymbol{E}(\boldsymbol{x},t) = 0. \quad (2.2d)$$

The simplest solution to this partial differential equation is, in 1D, just

$$\boldsymbol{E}(\boldsymbol{x},t) = \boldsymbol{E}_{\max}(x,t)\,\sin 2\pi(\nu t - x/\lambda). \quad (2.2e)$$

So the speed of light obtained from the Maxwell equations agreed closely with the value determined experimentally. Not only that, but the equation also implied the existence of the radiation at higher and lower frequencies, as well, that would be detected only later—x-rays, ultra-violet, infrared, microwave, radiofrequency, etc. (Table 2.2).

Photons

The story of electromagnetism reached a climax in the *anno mirabilis*, when the obscure Albert Einstein published three earth-rattling papers, any of which alone would have made his name. One of these proposed that the measured speed of light is an absolute constant of nature, unlike space and time, irrespective of the

Table 2.2 The electromagnetic (EM) spectrum, including the frequency, ν, and wavelength, λ, of its wave-like properties, and the energy, E, of its particle-like photons.

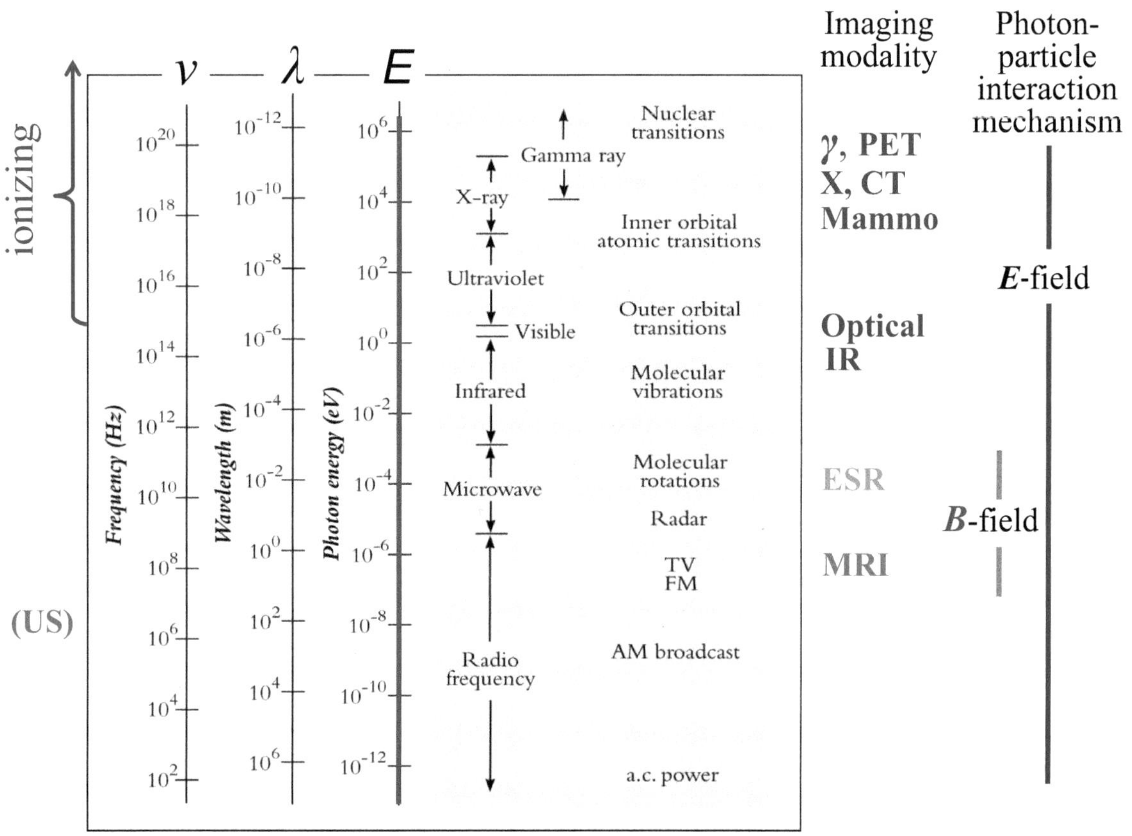

motion of an observer. This assertion is at the heart of his special theory of relativity.

A second manuscript, which dealt with Brownian motion, provided the first firm evidence that atoms are genuine entities, not merely imaginary constructs useful to chemists.

And in the same year, he argued that an individual photon is a *wave-particle* entity subject to the ubiquitous Planck-Einstein relationship

$$E = h\nu. \tag{2.3}$$

This became the cornerstone of quantum mechanics, and for it Einstein was awarded the Nobel Prize in Physics in 1921. E is the photon's *particle-like energy*, ν is its *wave-like frequency*, and h is the *Planck constant* (Table 2.2). It says that the higher the frequency of a wave of electromagnetic radiation (or of any particle), the greater the energy carried by each of the individual quanta that comprise it. Optical-range photons, for example, are a few electron-volts in energy. This is where many of the chemical reactions of everyday life occur, which is why so many of them emit light. The very high-frequency x- and gamma-ray photons can transport thousands or millions of eV each. Those of NMR and MRI, bearing *micro*-eV, are 12 orders of magnitude less energetic than x- and gamma-rays—yet MRI and CT produce similar-appearing results!

Also shown are the kinds of quantum transitions or technologies in which they are involved and the imaging modalities in which they are employed. In medical imaging, for most interactions between probes and matter, it is the electric field of a photon that interacts with an atomic electron at any energy. In NMR and electron spin/paramagnetic resonance (ESR/EPR), however, it is an RF photon's magnetic field that causes spin transitions of a nucleus with spin, or an unpaired atomic/molecular electron, respectively. The energy of an ultrasound (US) wave doesn't fit into this diagram, since its radiation is acoustic/mechanical in nature, not EM.

2.2 Magnetic Fields within Matter: Magnetic Susceptibility

Much of the above has concerned magnetic fields in free space. Yet those that arise within the body itself are also of critical importance. As must be apparent by now, there is a great deal going on within a tissue voxel

Table 2.3 There are magnetic fields with a number of separate functions at play within a voxel of tissue. Three of these—the main, gradient, and ν_{Larmor} RF fields—are applied by the MR device. Diamagnetism, paramagnetism, and ferromagnetism materials are properties of bulk matter, all attributable to the behavior of molecular orbital *electrons*. *Electron paramagnetism*, which can affect NMR and MRI, is to be distinguished from *nuclear paramagnetism*, which is the fundamental physical process that underlies NMR and MRI.

$\boldsymbol{B}_0$	1.5 T, 3 T (7 T)
$\boldsymbol{G}_{grad} = d\boldsymbol{B}_z/dx$	20–80 mT/m
$\boldsymbol{B}_1$	0–50 mT
$\boldsymbol{B}_{\text{proton-proton}}$ (T1,T2)	< 0.4 mT
$\boldsymbol{B}_{\text{electron diamagnetism}}$	–10 μT
e.g., chemical shift	~μT
$\boldsymbol{B}_{\text{electron paramagnetism}}$ (Gd)	+10 μT

during MRI. Some of this comes about because of the fields from the MR instrument's main and gradient magnets and RF antenna (Table 2.3). The values of the T1 and T2 proton spin relaxation times, on the other hand, are largely determined by the direct magnetic interactions between the pairs of partner protons on water molecules, or on lipids, as they tumble and move about in the cellular soup. Yet other forces arise from the electrons circulating around molecules, resulting in diamagnetism and the chemical shifts found in magnetic resonance spectroscopy. And the fields coming from gadolinium and other paramagnetic atoms can be harnessed as contrast agents to enhance the visibility of tumors and other lesions for MR scanning.

Applied main, gradient, and RF fields in voxel

With MR imaging, the *main field*, $\boldsymbol{B}_0$, is ideally uniform over space and time—but the field that a proton within a patient experiences is neither. Room-temperature coils just interior to the superconducting coils responsible for $\boldsymbol{B}_0$ are intermittently producing *gradients*, such as $\boldsymbol{G}_x = d\boldsymbol{B}_z/dx$. These are intentionally made to vary linearly by as much as 80 mT/m during the brief, millisecond periods that they are being activated. In situations where such a field gradient is present, we replace $\boldsymbol{B}$ with $\boldsymbol{B}_z(\boldsymbol{x},t)$, which includes $\boldsymbol{B}_0$ as only one of its components. The subscript z is a reminder that while the magnitude of $\boldsymbol{B}_z(\boldsymbol{x},t)$ may vary over space or time, it always points directly up, along the z-axis.

In addition, an RF antenna emits short pulses of Larmor-frequency EM energy that, under the right conditions, is absorbed by the population of protons. This process can be monitored indirectly by detecting and following over time the amplitude of the resulting magnetization vector, $\boldsymbol{m}_{xy}(t)$, precessing in the transverse plane in each voxel (in the classical picture of NMR).

Spontaneous magnetic fields that affect T1 and T2

The previous chapter sketched T1 and T2 proton spin relaxation and hinted at how they occur. The take-home message is this: the molecular magnetic environment of a proton consists primarily of the externally applied $\boldsymbol{B}_0$, $\boldsymbol{G}_{\text{grad}}$, and $\boldsymbol{B}_1$, but others that arise naturally within the patient also play essential roles.

A proton in a water molecule, for example, is well aware of its overlap with the field that its partner proton generates. The spin of each proton will stay aligned along or against $\boldsymbol{B}_0$. But the relationship between the two varies rapidly as the water tumbles, as does the field each one experiences due to the relative position of its comrade. Likewise, as the hydrogen atoms close to one another in a lipid molecule interact, they feel the quickly (if the lipid is light, small, and not attached too strongly to a membrane) or slowly (otherwise) varying magnetic fields.

In both cases, that component of $\boldsymbol{B}_{\text{proton-proton}}(t)$ that happens to be fluctuating at the Larmor frequency is capable of exciting protons in the lower Zeeman state into the higher, or of 'tickling' those in the higher state down into the lower. The system achieves and remains at thermal equilibrium, with equilibrium magnetization $\boldsymbol{m}_0$, when the rates at which these two processes, mediated by Larmor-frequency *phonons* (quanta of vibrations or sound) and EM, occur are the same. Moreover, after the spin system is disturbed, it will return toward equilibrium. This occurs at the rate of what is known as *longitudinal spin relaxation* up and down along the z-axis, and that *rate* is indicated 1/T1. During any brief excitation of the system with a Larmor-frequency RF, incidentally, spin-relaxation is ongoing, but it is far too small an effect to be noticed then.

EXERCISE 2.2 How is Figure 1.2 relevant in all this?

T1-relaxation is not the only consequence of perturbing a population of spins. Suppose some sort of disruption happens to leave the spins in a voxel all in the horizontal plane and precessing in phase about $\boldsymbol{B}_0$ (Figure 1.4a). (Is this possible in the quasi-QM picture?) If there were no relaxation or similar processes at play, then $\boldsymbol{m}_{xy}(t)$ would just keep on rotating like that indefinitely. But such is not the case, and Larmor-frequency T1 events, for example, will disrupt this idyllic state of affairs. They will kick some protons out of synchrony with the others, which has the effect of spreading out the separate spins that were initially moving coherently (Figure 1.4b).

But that's not all that happens. In addition, some waters and lipids will be bound at least briefly, to slowly turning macromolecules. Their proton spins remain pointing along or against $\boldsymbol{B}_0$, and their field strengths will change only *slowly* ($\nu \sim 0$), along with the macromolecules they are adhering to. This is nothing at all like oscillating at ν_{Larmor}. Because their fields overlap one another randomly, and stay that way briefly, $\boldsymbol{B}_{\text{proton-proton}}$ will be a bit stronger for some, because of which they will precess a little faster, at least for a while, than the *average* proton—and conversely for those momentarily in slightly weaker fields. This is a prescription, if there ever was one, for the individual spins that together create $\boldsymbol{m}_{xy}(t)$ to spread out (Figure 1.4b). These two effects collaborate to cause the magnitude of $\boldsymbol{m}_{xy}(t)$ to decay at the *transverse spin relaxation* rate 1/T2, where

$$1/\text{T2} = 1/\text{T1}(\nu \sim \nu_{\text{Larmor}}) + 1/\text{T}_{\text{proton-proton}}(\nu \sim 0),$$

which will be defined more rigorously later on.

Magnetic susceptibility: diamagnetism, paramagnetism, and ferromagnetism induced by B_0

Most commonly, the fixed, highly uniform main magnetic field, $\boldsymbol{B}_0$, is being applied from outside the patient by superconducting or electromagnet coils. Older treatments of NMR introduced the 'magnetic intensity,' $\boldsymbol{H}$, to indicate the field generated by external magnets. Indeed, in free space, $\boldsymbol{H}$ and $\boldsymbol{B}_0$ are essentially the same, even though they employ different units.

But the story is quite different within matter. The application of an external magnetic field perturbs the outer electron orbitals of its atoms and molecules, which leads to the creation of an additional *electronic magnetization* field, $\boldsymbol{M}$, which adds to or diminishes the local fields within the material. Inside a tissue, the field strength at point x is called $\boldsymbol{B}(\boldsymbol{x})$, and it may be somewhat different from the applied H. For the present discussion, it will be fine to simplify things, forgetting $\boldsymbol{H}$ and sticking with $\boldsymbol{B}_0$ and $\boldsymbol{B}(\boldsymbol{x})$.

The strength of the local field within tissue at a point inside it will differ from B_0 by a factor of χ, its *magnetic susceptibility*:

$$B(x) = B_0 + M(x) = [1 + \chi(x)]B_0$$

or (2.4)

$$M = \chi B_0 .$$

All materials display one of three general categories of bulk magnetic susceptibility (Table 2.4): *diamagnetism, paramagnetism, and ferromagnetism*. For each of these, again, the overall magnetic properties are determined by the behavior of its electrons, *not* by spinning nuclei.

For quantum mechanical reasons involving the Pauli exclusion principle, any material creates its own very weak magnetic field opposed to one that is externally applied. For some materials—metals, carbon, organic compounds such as wood, petroleum and plastics, oxyhemoglobin, many soft tissues, and water—that's the entire story. The clouds of electrons orbiting its atoms or molecules are spin-paired, in the sense that the magnetic field generated by one electron is effectively canceled out by that from another, on the same atom, that is oriented in the opposite direction. As a result, such an *electron diamagnetic* material reacts to an externally applied field by reducing the local field within itself by a tiny amount, and the susceptibility $\chi = M/B_0$ is typically less than 10^{-5} or so. If you draw magnetic lines of force that pass through a piece of diamagnetic substance, they will spread apart, as if the medium is forcing the field outward. A superconductor, strange in so many other ways as well, is strongly diamagnetic and repels an external field fully, apart from in a thin surface layer, in what is known as the *Meissner effect*.

Table 2.4 It is orbital electrons on individual atoms or molecules that give rise to diamagnetism and paramagnetism bulk media, like tissue. Ferromagnetism is a much more forceful in a phenomenon that arises in iron, nickel, etc. and their alloys as a result to the electrons' collective behavior.

$$M = \chi B$$

	Definition	Typical	
$\chi =$	<0	-10^{-5}	diamagnetic
	>0	$+10^{-5}$	paramagnetic
	>>0	10–1,000,000	ferromagnetic

We have already explored one example of diamagnetism in connection with Figure 1.7 and the *chemical shift*. Orbital electrons circulate around the separate atoms of a molecule in distinctly different ways, so the weak local fields they contribute at the various nuclei are not the same. As a result, each hydrogen nucleus may undergo NMR at a frequency that differs from that of a free proton by a minute factor known as the *chemical shift constant*, $\sigma \sim 10^{-5}$ to 10^{-6}.

EXERCISE 2.3 Does σ depend on the position of a hydrogen atom on a molecule?

EXERCISE 2.4 Would you expect to find a relationship analogous to Equation (2.3) applicable to the protons in water?

Atoms with one or more *un*paired electrons, conversely, may find it energetically favorable for those electrons to go against the grain and align parallel to an applied field. Such *electron paramagnetic* materials contribute to and enhance the strength of the external field, generally by a small amount, and for them χ is a little larger than 0. But as with diamagnetic matter, when the external field B is zero in Equation (2.4), then so also is M.

Some electron paramagnetic materials, like deoxyhemoglobin and methemoglobin (which is a product of blood degradation), occur naturally in the body, and others are introduced intentionally as *contrast agents.* The paramagnetic MRI contrast agent employed most commonly in the clinic contains chelated (held within a small molecular cage) ions of gadolinium (Gd), where the Gd ion is complexed with compounds such as diethelenetriamine-penta-acetic acid (Gd-DTPA) or its relatives. Gadolinium carries seven unpaired electrons; but unlike the case of a *free radical*, these are in *inner* (4f) orbitals and have little to do with the atom's chemical reactivity. (Free radicals, with unpaired electrons in the *outermost* orbital, tend to be highly reactive.) The magnetic moment (from both its spin and its angular motion) of an unpaired electron can be many hundreds or thousands of times greater than that of a proton.

Unpaired inner orbital electrons on an atom give rise to strong fluctuating magnetic fields at the nucleus, and these can substantially affect the local environment of nearby tissue protons. They can significantly affect, for example, the protons' T1 and T2 values. Likewise, it is MRI's sensitivity to the difference between diamagnetic oxyhemoglobin and paramagnetic deoxyhemoglobin that underlies functional MRI (fMRI).

Shortly after Øersted discovered the magnetism from an electric current, Ampere proposed that the magnetic field from a piece of *ferromagnetic* material is produced by vast numbers of microscopic loops of electric current; these attempt to align in an external magnetic field, as does an entire magnetized iron needle—an interpretation close to the modern one. This normally causes the fields inside or near a ferromagnet to be much greater than the external field present, and $\chi \gg +1$. Unlike a diamagnetic or paramagnetic substance, moreover, a ferromagnet may stay magnetized after the external field is turned off, so that Equation (2.4) is no longer applicable.

It is the flipping over of the magnetic poles of something quite a bit smaller than the Earth—namely a proton spinning rapidly on its axis (and thereby producing a magnetic field)—that underlies the workings of MRI.

All of the above, and in particular the discussion of diamagnetism and paramagnetism, has concerned magnetic fields caused or modified by molecular *orbital electrons*. The *protons* associated with lone hydrogen atoms in the water and in lipid molecules in a voxel of tissues also give rise to a *nuclear* paramagnetic field, or *nuclear magnetization*, when placed in an external field. Nuclei other than protons may or may not have nuclear magnetizations, but with MRI, we are concerned almost exclusively with solitary hydrogen nuclei. It is the behavior of these protons, and of the time-dependent magnetic field which they themselves generate in a voxel, that underlies clinical MRI.

The Earth's Magnetic Field

The Earth's own magnetic field, incidentally, tells a long and complex story. First of all, people have long called the end of a compass needle that swings toward the Earth's northern geographical pole in the Arctic the needle's 'North' pole. So by this convention, the Earth's internal magnet must actually have a south magnetic pole up there, where the geomagnetic field plunges vertically into the ice. Beneath our planet's surface and rock mantle lie an outer core of molten iron and other metals and then a solid inner core. Its magnetic field is produced by electric currents associated with the rotation and convection of these materials, which act as a giant fluid dynamo. Near the core-mantle boundary, the configuration of the field is complex, with strong quadrupole, octupole, et al., components. These fields fall off rapidly, so at the surface we see a predominantly dipolar field that wobbles around a bit over the millennia. Shearing and twisting instabilities in the fluid flow have caused variations in the geomagnetic field, including even pole reversals—from time to time, the magnetic north pole flops over and ends up near where the south pole had been. This happened most recently about three quarters of a million years ago, and another such transition may be due in perhaps 50,000 years.

Much about the sequence of these events has been inferred from study of the alignments of the magnetic fields in newer undersea rock, such as at the Mid-Atlantic Ridge, as it solidified from lava. As magnetic mineral grains (magnetite, etc.) cool below their curie temperatures, their small crystals become ferromagnetic (more precisely, ferrimagnetic), and their net magnetic moments tend to align with the prevailing magnetic field; with further cooling, the net magnetization of the assemblage of magnetic mineral grains is frozen in*.

*Private communication, with thanks to geophysicist Dr. Dave Clark, CSIRO, Australia.

2.3 Nuclear Magnetic Dipole Moment, μ

Magnetic fields arise near any moving electrically charged particles, including spinning nuclei.

In your study of electric motors, you came across the classical *magnetic moment*, $\boldsymbol{\mu}$, of a current-bearing circular loop of wire in a rotor, where the vector $\boldsymbol{\mu}$ lies normal to the plane of the loop. The *magnetic moment* of the loop, the strength of the magnetic field it produces, increases with both the current i and the area A or radius r as

$$\mu \equiv iA = \pi r^2 i, \tag{2.5a}$$

from Ampere's law. In an external field $\boldsymbol{B}'$, the *torque* on the loop, the force tending to twist it over, is described as the vector (cross) product $\tau = \mu \times \boldsymbol{B}'$. The power of a motor is proportional to $\boldsymbol{\mu}$, and the larger its value, the greater the horsepower.

Also from classical physics: the magnetic field produced at a distance r away from a long straight wire carrying electric current, i, is $\boldsymbol{B}(r,\boldsymbol{i}) = \mu_0 \boldsymbol{i} / 2\pi r$. Likewise, the field created by a 1-turn planar coil of radius r at its center is of strength $\boldsymbol{B}(r,\boldsymbol{i}) = \mu_0 \, \boldsymbol{i} / 2r$ where again, μ_0 is the permeability of vacuum, *not* a magnetic moment!

Like current in a coil, a charged, nonconducting hoop rotating about its axis of symmetry will create a dipole field normal to its plane, and so also will a stack of adjacent charged hoops covering or filling a sphere. Most atomic nuclei behave somewhat like spinning balls of positive charge, and create dipole vector magnetic fields that somewhat mimic a sub-microscopic bar magnet or compass needle. From the symmetry of the situation, the alignment of a proton's dipole field is associated with that of its spin axis—where else? Curl your right hand around a hydrogen nucleus, say, in the direction in which it is rotating: your thumb then defines 'the direction' of the spin. This is not arbitrary, but rather follows from the more basic 'right-hand rule' for current: When you point the thumb of your right hand along the direction of the flow of 'conventional' (positive) charge, then the direction in which your fingers curl defines the direction of the magnetic field produced around it.

The dipole magnetic moment of a nucleus of a given isotope, $\boldsymbol{\mu}$, points along its nuclear spin angular momentum, $\boldsymbol{I}$. The magnitude of the nuclear spin vector is $|\boldsymbol{I}| = (h/2\pi)\,[\mathrm{I}(\mathrm{I}+1)]^{1/2}$ where I is the scalar *nuclear spin quantum number* and, again, h is Planck's constant. The value of I is always a positive integer multiple of $\frac{1}{2}$: $\mathrm{I} = 0, \frac{1}{2}, 1, \frac{3}{2}, \ldots$ The spin quantum numbers of nuclei of interest in MRI appear in Table 2.5. It is especially relevant that Mother Nature has decreed that, despite the problems theorists are having in getting the spins of its constituent quarks and gluons to add up right (a situation known as the 'proton spin crisis'), the lone proton is to be a 'spin-½' particle, with I = ½.

EXERCISE 2.5 The value of the magnetic moment, μ, for any nucleus is comparable in magnitude to the so-called nuclear magneton, μ_N, defined as $\mu_N \equiv eh/4\pi m_p$, where e is the charge on the proton and m_p is its mass. Show that the value of μ for a spin-½ nucleus is related to its γ as $\mu = \gamma(3^{1/2}\,\mu_N\, e/m_p)$.

Every nucleus comprised of an odd number of protons or of neutrons possesses spin and a magnetic moment. In any with even numbers of both protons and neutrons, by contrast, each proton pairs up with and aligns anti-parallel to another, for reasons that can be explained with QM, so that their two dipole magnetic fields cancel; so also for the neutrons. As a result, the carbon-12 and oxygen-16 that make up most of soft tissues play no active role in imaging and, likewise, calcium-40 is notable for its absence in images, and the near-invisibility of bones.

Table 2.5 Some nuclear species, n, that can play a role in MRI. The nuclear spin magnetic quantum number, I, is central to determining the Larmor frequency, $\gamma_n / 2\pi$, at the field strength, $\mathbf{B}_0$, of the main magnet. Also listed are the isotope's natural abundance and the relative intensity of its NMR signal. Carbon-12, oxygen-16, and calcium-40 are even-even nuclei that have no nuclear spin and almost no impact in imaging.

Nucleus n	Spin I	$\gamma_n/2\pi$ MHz/T)	% Natural Abundance	Relative Sensitivity
H-1	**1/2**	**42.58**	**99.98**	**1**
C-13	1/2	10.71	1.1	0.02
F-19	1/2	40.06	100	0.83
Na-23	3/2	11.26	100	0.09
P-31	1/2	17.24	100	0.07
C-12	0	X	98.9	X
O-16	0		99.8	
Ca-40	0		96.9	

The magnitude and direction of a nucleus' dipole magnetic field vary throughout the space around it. But nuclear QM reveals that the magnitude of its inherent dipole moment is proportional to that of its spin angular momentum. As in Equation (1.1a),

$$\mu = \gamma |I|. \tag{2.5b}$$

where γ, like $\boldsymbol{\mu}$, is nuclide-specific. The scalar *gyromagnetic ratio*, γ, and the *magnetic dipole moment* vector are closely related measures of a nucleus' 'magnetness,' the strength of the field that it itself generates. (Both terms, $\boldsymbol{\mu}$ and γ, appear frequently in the NMR and MRI literature.) Each isotope of every element has its own, unique characteristic value of γ, and the stronger the field generated by a nucleus, the larger its γ. The magnetic field produced by a nucleus of common phosphorus, P-31, for example, is about 0.4 times smaller than that of a proton, and the length of its $\boldsymbol{\mu}$-vector would be correspondingly shorter. It is more convenient to deal with $\gamma/2\pi$ rather than with γ alone, and values of $\gamma/2\pi$ are listed in Table 2.5 for MRI-relevant isotopes. The magnitude of γ for a nucleus or, conveying the same information but in different units, of $\mu \equiv |\boldsymbol{\mu}|$, can be either found experimentally or be derived (with varying degrees of accuracy) by nuclear theorists. The measured value of $\gamma/2\pi$ of an isolated proton, in particular, is **42.58** MHz/T, Equation (1.1b).

Although nearly isolated, a proton in a molecule does feel the influence of its immediate electron cloud, Equation (1.6), and much more importantly, the presence of other magnetic entities nearby. This sensitivity is one of the factors that makes MRI so remarkably useful. Hereafter we shall focus only on tissue protons. As stated before, that leaves the field of clinical MRI open, at present, almost entirely to hydrogen nuclei alone.

Equation (2.5a) may be intuitively satisfying, but it's a bit of a red herring for our purposes. Both the nuclear spin and the nuclear magnetic moment are quantized. When a proton enters a strong magnetic field aligned along the **z**-axis, what are essential in the study of MRI are not $\boldsymbol{I}$ and $\boldsymbol{\mu}$, per se, but rather their *z*-components, I_z and μ_z,

$$\mu_z = \gamma I_z. \tag{2.5c}$$

Proton-proton dipole interaction

There are several ways that protons can talk to one another, but the most important of these is the nuclear dipole-dipole interaction. Imagine two small magnetic dipoles, such as a pair of water protons, of moments $\boldsymbol{m}_1$ and $\boldsymbol{m}_2$. Their magnitudes will be the same, of course, but not necessarily their orientations. If $\hat{\boldsymbol{r}}$ is a unit vector parallel to the line joining their centers, and $|r|$ is the distance between them, then the classical dipole energy of orientation of the system is

$$E = -(\mu_0 / 4\pi |r|^3)[3(m_1 \cdot \hat{r})(m_2 \cdot \hat{r}) - (m_1 - m_2)]. \tag{2.6}$$

With appropriate alteration of the notation, this is quantum mechanically correct, as well. This is the principal mechanism by means of which protons in a molecule interact with one another and effect proton spin relaxation.

2.4 Proton Spin-Up and Spin-Down States, |↑> and |↓>

As mentioned earlier, the easier of the two general approaches to NMR is to simplify the full quantum treatment by disregarding the phase factors of the quantum spin states (Table 2.6). This quasi-QM tactic is conceptually and mathematically less complicated than the real thing, but can still provide good heuristic arguments for important results, such as the Larmor equation and the mechanism of T1 relaxation (just as Bohr's quasi-QM model of the hydrogen atom seems to work pretty well).

In an external magnetic field $\boldsymbol{B}_z$ along the *z*-direction that is already in existence, a magnetic dipole will experience a torque, or twisting force stemming from the magnetic *Lorentz force*, that tries to align its moment along that field. You can hold the needle of a compass at any angle relative to the direction of the $\boldsymbol{B}_z$, which is aligned along the *z*-axis. When released, the needle tends to swing around to settle down into the orientation of lowest energy, such that its own north pole points up, toward the south pole of the external magnet. When pointing north, straight up along the positive *z*-axis, it is in its configuration of lowest potential energy. That is, of course, what enables the device to guide your path which, as those who enjoy sailing far from shore or hiking deep into the forests know, is very comforting. You naturally (and correctly) assume that in an external magnetic field aligned along the *z*-axis, the potential energy of orientation, E, of a compass needle should be proportional to the strength of that field. Likewise, it increases with the strength of the compass's own dipole's magnetic field. The energy of

Table 2.6 There are two non-rigorous pictures for introducing NMR, and both are useful in different ways. Each starts with the set of precisely correct QM spin-state functions, $\{|\psi>\}$. The first approach simplifies each $|\psi>$ so as to leave only the 'spin-up,' $|\uparrow>$, and 'spin-down,' $|\downarrow>$, states in their place. In this picture, transitions can take place only between $|\uparrow>$ and $|\downarrow>$, but this is enough to derive the Larmor frequency, ν_{Larmor}, and to discuss both the thermal equilibrium voxel net magnetization, $\boldsymbol{m}_0$, and T1 relaxation. The second introduction to NMR follows from a meticulous QM treatment. Quite remarkably, this leads directly to the Newtonian *Bloch equations* for the time dependence of the QM *expectation value* (which are of probabilistically predictive value) for the behavior of the nuclear magnetization $\boldsymbol{m}(t)$ in a voxel. These classical equations underlie virtually all aspects of NMR, such as the nature of the transverse relaxation time T2 and imaging in 2D $\boldsymbol{k}$-space, as will be introduced later.

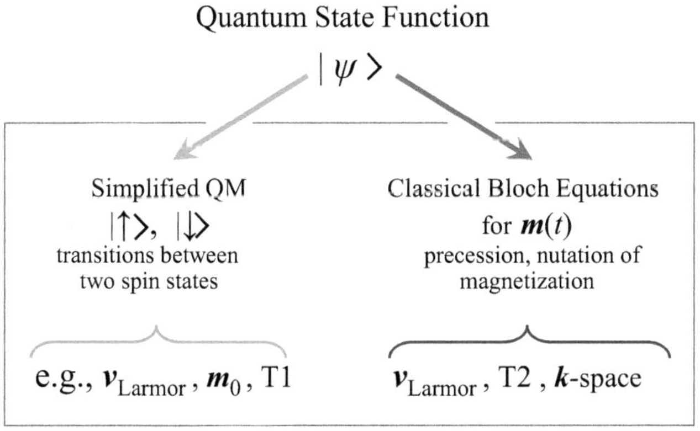

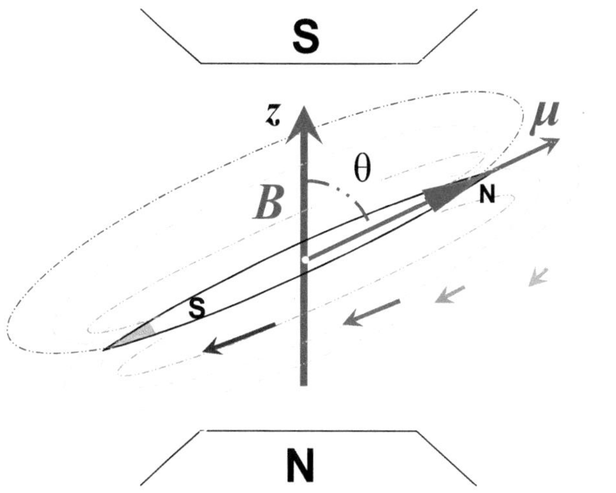

Figure 2.3 A compass needle, subject to Newton's laws, tends to align in an externally applied magnetic field, such as that produced by the main magnet of an MRI device. It prefers to have its north pole pointing toward the south pole of the external magnet, in which case it is in its configuration of lowest energy, if it is not moving. Because a proton is subject to the often counter-intuitive laws of quantum mechanics, instead, its behavior differs radically. It can remain for long periods of time in the higher-energy spin-orientation state, aligned in the 'wrong' direction. We are interested in the amount of energy required to twist a compass needle, or a proton, through 180°, so that it ends up pointing in this "wrong" direction. For both small bar magnets and protons, that energy, ΔE, is proportional both to the magnitude of the external field strength, B_z, and to the magnitude of its own "magnetness," as indicated by its value of μ.

orientation, E, is given classically by the scalar product, Equations (1.2) (Figure 2.3).

$$\begin{aligned} E &= -\mu \cdot B_z \\ &= -|\mu||B_z|\cos\theta = \mu B_z \cos\theta, \end{aligned} \tag{2.7}$$

If you twist a needle that is pointing north fully around to the south, its potential energy increases continuously until it reaches the amount $2\,\mu\,B_z \equiv 2\,|\boldsymbol{\mu}|\,|\boldsymbol{B}_z|$. The same is true for a bar magnet held between the poles of a magnet. The energy required to twist such an object over through 180° is separately proportional both to the 'magnetness' of the object and to the strength of the external field.

$$\boldsymbol{I}\cdot z = \boldsymbol{I}_z = \pm \tfrac{1}{2}(h/2\pi). \tag{2.8a}$$

Nature dictates that in an external field, a spin-½ particle such as a proton or electron, on the other hand, can reside quasi-stably (with our simplified, quasi-QM model) in one or the other of two, but only two, possible orientations: in marked contrast to a compass needle, the spin axis of a proton in $\boldsymbol{B}_z$ is *spatially quantized* and must lie either '*up*' along the external field and the z-axis or '*down*,' anti-parallel to it, but in no other possible direction in between. In QM terms, the *nuclear magnetic spin quantum number* of a proton (corresponding to its component along the z-axis) may assume only the values –½ or +½, and the z-component of the spin angular momentum will be of magnitude
It follows from Equations (1.2b) and (2.5) that

$$\mu_z = \gamma \boldsymbol{I}_z = \pm \tfrac{1}{2}(\gamma/2\pi)h. \tag{2.8b}$$

The z-subscript is appended as a mnemonic that μ_z always points in the upward or downward, $\pm z$, direction. The presence of Planck's constant, h, serves as a reminder that what is going on here is essentially quantum-mechanical in nature. Things get a bit more complicated for nuclei that are not of the spin-½ persuasion, but we shall have little to do with any of those here.

The QM energy operator, or *Hamiltonian*, $\hat{H}$, that appears in the time-independent Schrodinger equation for a system of noninteracting spins, happens to be of the same general form as the energy of Equations (2.7), and

$$\hat{H} = -\gamma\hat{I}\cdot B_z = -\mu_z B_z, \tag{2.9a}$$

where $\hat{I}$ is the QM nuclear spin operator.

The *lower-energy, spin-up* +½-spin state, commonly written |+½>, and the higher-energy, spin-down one, |–½>, are of energies

$$E_\pm = \pm\mu_z B_z = \tfrac{1}{2}h(\gamma/2\pi)B_z, \tag{2.9b}$$

respectively, and both vary linearly with the external field strength (Figure 1.2). This incorporates the quantum constraint that a proton can exist in the *up* configuration, or the *down*, but nowhere else.

EXERCISE 2.6 What is the energy required to twist a proton through 180° in a 1.0-T field? To what photon frequency does that correspond?

The difference in energy between the two levels,

$$\Delta E(B) = (E_+ - E_-) = 2\mu_z B_z = h(\gamma/2\pi)B_z, \tag{2.9c}$$

known as the *Zeeman* splitting, is the energy needed to elevate a proton from its lower spin-state to the higher (Figure 2.4). The solid blue line is straight, a reminder that the corresponding Larmor frequency photon required to bring about such a transition for hydrogen is directly proportional to B_z; the axis for the energy, in micro-eV, lies along the right-hand border of the graph. The dashed red line does the same for phosphorus-31, which has a nuclear moment only 40% of that of a proton.

The energy required to bring about a proton-NMR spin transition is typically in the range 0.1 to 0.6 μeV (10^{-6} eV). This is a half dozen orders of magnitude (a factor of 10^6) less than the energies ordinarily exchanged among molecules during a light-emitting chemical reaction (~1 eV), and 10^{12} times less than the energies involved in typical x-ray photon-electron (i.e., photoelectric absorption or Compton scatter) interactions or nuclear, gamma-ray, and positron annihilation events in imaging (10^5 to 10^6 eV).

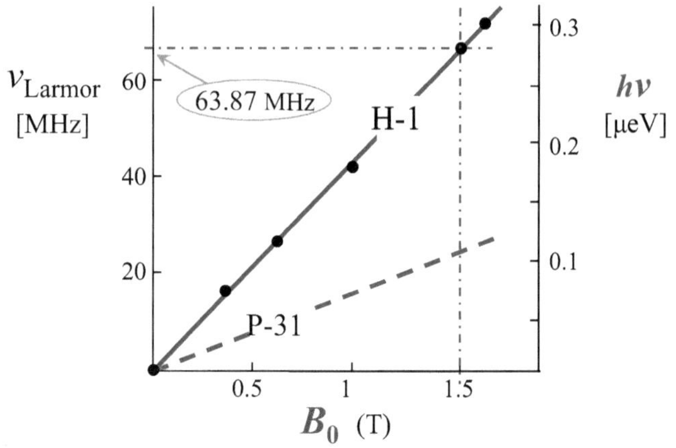

Figure 2.4 The proton NMR phenomenon involves inducing transitions between its two spin states through the input of a photon of exactly the correct energy and RF frequency. Flipping a proton over, or twisting it through 180°, requires the transfer of exactly $\Delta E = 2\mu_z\mathbf{B}_z$, as indicated in micro-eV on the right-hand vertical axis. $\Delta E(B)$, is plotted against B_z (tesla) as the solid blue line for normal hydrogen nuclei. The corresponding photon frequency (MHz), called its *Larmor frequency* (left axis), is also linear in B_z, $\nu_{\text{Larmor}} = 42.58\ B_z$. The Larmor frequency of a free proton in a 1-T field is 42.58 MHz, 63.87 MHz at 1.5 T, and 127.7 MHz at 3 T, all of which are in the radiofrequency (RF) range. The red dashed line describes the phosphorus-31 nucleus, whose nuclear magnetic moment is about 2½ times smaller than that of the proton, and which is of some research interest in MRI. It is the opposite process—in which protons in the upper spin state are induced to give off photons, phonons, etc., of that energy—that is responsible for T1 relaxation.

Our quasi-quantum picture assumes that only two spin states are possible, namely |+½> and |–½>. The full, complete quantum theory includes time-dependent phase factors on these, and also proposes that the correct QM spin configuration is a changing mixture, or *superpositioning*, of these. We shall not explore this, but you can rest confident that it explains much of the 'quantum weirdness' we encounter.

To summarize: in a strong magnetic field, any nucleus of MRI interest may be viewed, from the limited, quasi-QM picture, as a system that can inhabit either of two, but only two, spin states, with the *z*-component of the magnetic moment either parallel (ground state) or anti-parallel (high-energy state) to the external field. The Zeeman splitting apart of the two states, the difference in their energies, is proportional to the local B_z. A nucleus aligned in the low-energy orientation can be flipped over, and into the higher-energy state, by supplying this amount of energy. In that sense, it behaves rather like an ordinary compass needle.

2.5 NMR at the Larmor Resonance Frequency, $\nu_{Larmor} = (\gamma/2\pi)\, B_z$

We have talked of twisting over nuclear spins in a magnetic field, but have not said how actually to do the flipping.

From the simplified quasi-quantum perspective, imagine a proton sitting in a strong external magnetic field in its comfortable, low-energy "ground-state" spin-orientation, with its own north pole pointing toward the external magnet's south. The trick is somehow to grab hold of it and flip or twist it over through 180°. You do work on it to carry this out, so it is elevated in the higher-energy state, a transition that, by Equation (2.8c), requires $\Delta E = h\,(\gamma/2\pi)\,B_z$ of energy.

Perhaps the simplest method to excite a proton out of its lower-energy state is to enable it to absorb a photon of exactly the right energy and frequency, indicated by the squiggle in Figure 1.2b. The wave and particle characteristics of any sub-microscopic entity (whether more apparently wave-like or particle-like) are related through the Planck-Einstein relationship. In a field of strength B_z, photons of energy $E = 2\mu_z B_z$, and the *Larmor frequency* can provide the needed energy, where $h\nu_{Larmor} = 2\mu_z B_z = h(\gamma/2\pi)\,B_z$, or

$$\nu_{Larmor} = 2\mu_z B_z / h = (\gamma/2\pi) B_z, \qquad (2.10a)$$

which conveys the same information as Equation (1.3a). This *Larmor equation* forms the basis for NMR, hence for MRI. *For protons only*, and in the right units,

$$\begin{aligned}\nu_{Larmor}[\text{MHz}] &= (\gamma_{proton}/2\pi) B_z \\ &= 42.58[\text{MHz/T}] \times B_z[\text{T}]\end{aligned} \qquad (2.10b)$$

seen earlier as Equation (1.3b). It relates the Larmor frequency for a proton to the strength of the external magnetic field as in Figure 2.4, the left-hand side of the graph.

EXERCISE 2.7 How much more energy is needed to flip a proton than a P-31 nucleus in a 0.4-T field?

The Larmor frequency for protons in a 1.0-T field is **42.58** MHz. This is a useful number to remember, since it can be scaled up and down easily by multiplying by the actual field strength in tesla. In practice, MRI is performed most commonly with B_0 fixed at 1.5 T, for which the ν_{Larmor} is just under 63.87 MHz. In a 3.0-T field, $\nu_{Larmor}(B)$ is 127.7 MHz. Your National Public Radio station operates somewhere in the 88 to 108 MHz slot allotted to FM broadcasting.

EXERCISE 2.8 Demonstrate, using each of Equations (2.7) and (2.8) and Figure 2.4 that the proton Larmor frequencies at 1.5 T and 3 T—the fields employed most commonly in the clinic—are 64 MHz and 128 MHz.

EXERCISE 2.9 What is the photon energy, in electron volts, required to raise a P-31 nucleus from the lower into the higher energy state in a 3.0-T field?

So to induce protons to jump up into the higher-energy state, you can feed them photons of the appropriate Larmor frequency, produced by a radio transmitter and antenna. And that, in fact, points to a simple method to detect the NMR phenomenon, as seen in the next section.

EXERCISE 2.10 Frequency of rotation is commonly expressed as ν cycles/second. It can also appear as ω radians/second, where the angular frequency $\omega = 2\pi\nu$. Show that an equivalent form of the Larmor equation is $\omega_{Larmor} = \gamma B_z$.

While we're on the subject.... The NMR interaction between a nucleus sitting in a magnetic field and a Larmor-frequency photon is different, in a fundamental way, from the photon-electron interactions that occur at radio frequencies or at optical or x-ray energies. With photoelectric, Compton, and pair-production events between x-ray photons and bound electrons, it is the *electric* field of the photon that interacts with the *electric charge* of an inner atomic electron, resulting in a change in its quantum state. (What about optical transitions or RF?) With NMR, on the other hand, the rapidly oscillating *magnetic* component of the EM radiation interacts with the *magnetic moment* of the nucleus and brings about a change in nuclear spin quantum state (Figure 2.2 and Table 2.2). (It is quantum mechanically forbidden for the proton to absorb only some of a photon's energy and throw the rest away, a three-body process analogous to what *can* occur in electron Compton scattering, albeit at much higher energies.)

All of this suggests that *any* magnetic field (not only one from an RF photon) that varies at the Larmor frequency may be able to induce nuclear spin-state transitions, not only up but also down. Indeed, spontaneously occurring local fluctuations in the magnetic fields produced and felt by hydrogen nuclei are largely what determine the tissue-proton spin-relaxation time T1, and it also influences T2.

2.6 Preliminary Demonstration of the NMR Phenomenon in Water

This has been rather abstract so far, what with photon-induced transitions among quantum spin states and all. But what do you actually detect in an NMR (or MRI) study, and how do you carry out the measurements and interpret the results?

Let us design and perform (on paper, at least) the world's simplest NMR Gedanken-experiment (Figure 2.5). Since the water within tissues is a molecule of major clinical interest in MRI, we shall perform our NMR experiment on a sample of pure water, a satisfactory and far more convenient material to work with than livers and spinal cords. The magnet we have available happens to be fixed at 1.5 T.

EXERCISE 2.11 What is the Larmor frequency of hydrogen nuclei in a 1.5-T field?

We begin with a standard radio transmitter that pumps radiofrequency energy of a narrow, continuous band of frequencies, such as 10 MHz ±50 kHz, into an antenna. We arrange things so that initially, the antenna beams RF photons frequencies only within the band from 9.950 to 10.050 MHz into an empty plastic tank. On the far side of the tank is a receiver antenna, which is attached to an RF power meter, and the reading of the meter is recorded. We slowly but steadily step the central frequency, ν, of this narrow band of transmitted RF radiation upward, 100 kHz at a time, holding the

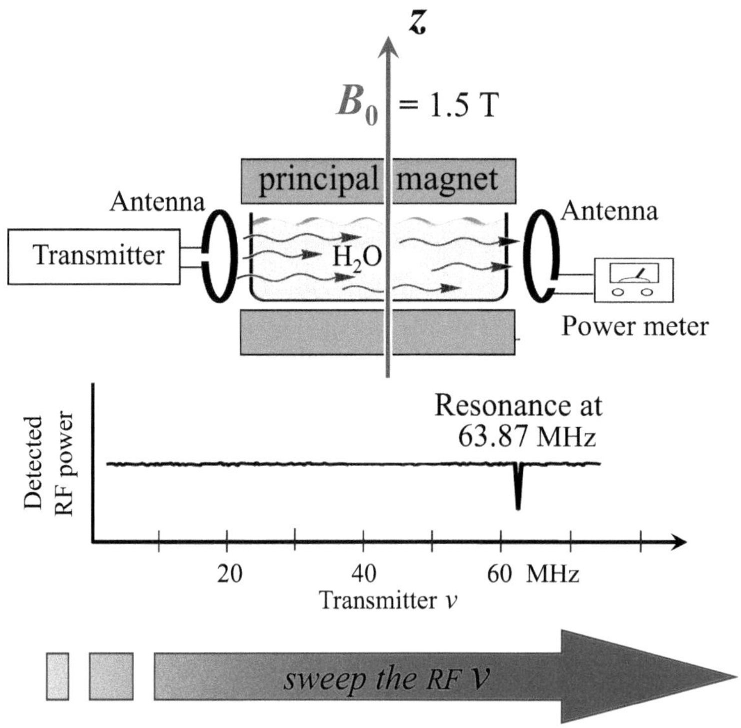

Figure 2.5 A rudimentary NMR experiment. A sample of water is sitting in a uniform magnetic field that is pointing upward (as with an 'open' magnet) along the **z**-axis, and held fixed at exactly B_0 = 1.5 tesla everywhere throughout the measurement. Radiofrequency (RF) power can be absorbed precisely at, but only at, the proton Larmor frequency. The equipment consists of a transmitter and antenna, along with a receive antenna that leads to a very sensitive power meter. The RF frequency is slowly and smoothly increased, and it is found that the level of detected RF power remains the same until the Larmor resonant frequency is reached (63.87 MHz for B_0 = 1.5 T). Abruptly here, over a very narrow band of frequencies, the water absorbs power, so that less energy reaches the detector. Immediately above the Larmor frequency, the meter jumps up again to its previous reading.

transmitted power level constant, and plot the *detected* power as a function of the frequency. Nothing interesting happens.

Let's fill the tank with pure water. Water is practically transparent to 10 MHz electromagnetic energy, just as it is to visible light, so the amount of RF power reaching the meter does not alter appreciably. Again, no surprises.

The electromagnet is turned on and adjusted so that its magnetic field within the water sample is uniform and fixed at a strength of 1.5 T. Again, nothing significant happens, and the level of power detected by the meter does not change. Still nothing very interesting!

After the water has had a chance to settle down in the external field, there is a slight relative excess of spins in the lower energy level, on the order of five parts per million. The extent of this *polarization* of the proton population is described in elementary statistical physics by the Boltzmann distribution (chapter 4). While 5×10^{-6} may seem like very, very little fraction of the protons, bear in mind how many of them there are in a cubic millimeter of water.

EXERCISE 2.12 In 1 mm^3 of water, about the size of a typical voxel, roughly how many protons are in the lower energy state in an MR field?

EXERCISE 2.13 What happens if too much ν_{Larmor} power is applied? What does 'too much' mean?

The application of Larmor-frequency RF could excite some of the spins in the lower energy level into the upper, with a slight amount of power absorbed from the beam, and this phenomenon is generally detectable by the power meter. This would indicate the occurrence of NMR!

So to carry out NMR in the water tank, we slowly but steadily sweep upward the central frequency, ν, of the narrow band of transmitted RF radiation, holding the *transmitted* power level constant, and plot the *detected* power as a function of the frequency. For a good while, the reading on the RF power gauge holds fast, but when the band of transmit frequencies reaches 63.90 ±0.05 MHz, the needle of the meter needle dips sharply. Here, and only here, some Larmor-frequency photons are being absorbed in the process of exciting water protons from the lower- into the higher-energy state—so less energy reaches the detector antenna. This is the indication we need that NMR is actually taking place, and at the frequency predicted by the Larmor equation. After that, the transmitter frequency continues to increase through and beyond the resonance condition, and the detected power returns to its previous flat level. We name the relative magnitude of the dip in detected power the MRI *signal strength*. And for consistency with what follows, we call the dip (negative, here) an *absorption peak*. (When we shift to the classical picture of NMR, however, resonance will *not* be detected by power absorption, as in Figure 2.5, but rather in quite different a manner: We shall listen for faint RF whispers coming from the patient's tissues.)

It's a little like plowing through a January blizzard in the Dakotas and slowly scanning the frequency dial of your FM radio. You hear nothing but static until suddenly the conditions are just right, and you hit Pink Floyd on an awesome soft-rock station!

What we have just gone through is not only a simple NMR study, but also an example of experimental *spectral analysis*. We pump RF energy into the physical system over a sequence of nearly monochromatic and adjacent frequency bands, and from the response determine how much power is absorbed (or emitted, or transmitted, etc.) in any of them.

If the whole experiment is repeated, but with several different fixed settings of the magnetic field, this same NMR phenomenon recurs, in each case at the Larmor frequency corresponding to the current value of the field, as with the distinct data points in Figure 2.4. In more realistic and complex situations, with multi-peak spectra, powerful mathematical techniques like *Fourier decomposition* are needed to untangle things.

EXERCISE 2.14 Would the same thing happen in a voxel of water as in a large tank full of it?

Nuclear magnetic resonance, just demonstrated experimentally, is the main phenomenon upon which magnetic resonance imaging is built. MRI, however, is carried out on a patient comprising hundreds of thousands of cubic tissue voxels. As you would expect, the greater the number of water or tissue protons in any given voxel, the stronger will be the signal from it. Indeed, the strength of NMR signal from the voxel at position x will always be linearly proportional to its *proton density* there, the number of water hydrogen nuclei per cubic millimeter of tissue:

$$s(x) \propto \text{PD}(x) \tag{2.11}$$

This and the above experiment suggest an easy method to map out tissue PD(x), at least in a one-dimensional phantom, coming up in the next chapter. And maps of the spatial distributions of T1 and T2 follow soon thereafter.

Finally, this picture is didactically OK, but it is different from what is normally done in a real laboratory NMR experiment or in a clinical MRI study, as we shall see. It is a fairly good description, however, of the workings of *electron spin resonance* (ESR), also known as electron paramagnetic resonance (EPR), a closely related and valuable research tool (Figure 2.2b). With ESR, it is a spinning, unpaired atomic electron attached to a molecule that can exist in either of two spin states that are Zeeman split apart by an applied magnetic field. The electron, too, aligns parallel or anti-parallel to a strong applied magnetic field. In part because an electron is nearly 2000 times lighter than a proton, its own magnetic field is much greater, and its resonance frequency is correspondingly higher, typically in the microwave, or gigahertz (GHz), range commonly used in radar.

ESR finds important application in the study of the free radicals created in cells by gamma- and x-rays, which are chemically very reactive and can attack DNA molecules. The 2015 Nobel Prize in Chemistry was awarded to three people who elucidated the ways in which cells can biochemically self-repair from such damage.

CHAPTER 3

Proton Density MR Study of a 1D, Multi-Voxel Patient and the Quality of the Image

The previous chapter discussed the performance of NMR in a water sample in a uniform external magnetic field, $\boldsymbol{B}_0$. An unsurprising and important finding was that the strength of an NMR signal associated with a voxel increases with the number of protons present or, equivalently, to the proton density (PD), Equation (2.11). If we had a way to determine what signals are coming separately from each of the voxels in a patient, we could map out the PD($\boldsymbol{r}$) everywhere, and with some additional cleverness, other attributes of clinical interest, such as T1($\boldsymbol{r}$), T2($\boldsymbol{r}$), flow of blood through veins and arteries, diffusion of water along axons, etc.

It is possible to produce a magnetic resonance image of a real, three-dimensional anatomic region by performing NMR experiments one point at a time throughout it [Hinshaw, 1976]. This point-by-point technique causes the three magnetic field gradients to oscillate vigorously, in such a manner that the overall field is constant over time at only one 'null' point, where conditions are stable enough for the NMR phenomenon to occur and be recorded. The gradient fields are then altered slightly, and the NMR experiment is repeated at another, nearby location, and so on. In theory one can obtain MRI-type information by examining each tissue voxel independently, with no need for reconstruction calculations.

EXERCISE 3.1 Can one do this for CT?

Such an approach is much too slow and inefficient to be used in practice, however, and sophisticated methods with efficient MRI reconstruction algorithms are always employed instead. These involve not only the systematic switching on and off of gradient magnetic fields, but also the production and detection of numerous carefully sculpted pulses of RF energy. We shall defer discussion of all this for a while, however, until the chapter on the classical picture of MRI.

For now we shall consider the simpler story of a one-dimensional phantom in the spin-up/spin-down picture in more depth, so that $\boldsymbol{r} = \boldsymbol{x}$ or just x. (In 1D, you can view it as either a scalar representing position or as a vector aligned along the $\boldsymbol{x}$-axis.) In essence, a gradient is applied intermittently in the $\boldsymbol{x}$-direction, $G_x \equiv dB_z/dx$, where the strength of the magnetic field, $\boldsymbol{B}_z(x)$ (which is aligned fully along the z-axis) increases linearly with position x. Therefore, the Larmor frequency, $\nu_{Larmor}(x)$, is a unique function of x, too, and an indicator of the $\boldsymbol{x}$-value. This allows creation of an MR image that displays proton density vs. position along the phantom, PD(x).

A centrally important question arises now, as it does every time a clinical examination is undertaken: is the outcome of a measurement faithful and reliable enough to be useful? To address this essential topic, the second half of this chapter turns to the issue of image quality. This concerns, in particular, quantifiable image attributes such as contrast, resolution, the modulation transfer function (the MTF, which covers both contrast and resolution at the same time), stochastic (random) noise, contrast-detail diagrams, detective quantum efficiency (DQE), the powerful Receiver Operator Characteristics (ROC) approach, systematic artifacts, and more.

3.1 The Three Distinct Applied Magnetic Fields in 1D MRI: B_0, $G_x \cdot x$, and B_1

In clinical MRI, three different categories of magnetic fields associated with the apparatus all play essentials roles. (Small magnetic fluctuation that arise within the patient are also of great importance, but are considered separately.) The patient is immersed in the very strong, constant, highly uniform field, $\boldsymbol{B}_0$, of the *main magnet* that is meant to *polarize* the spins. The magnitude of this field, $B_0 \equiv |\boldsymbol{B}_0|$, will be expressed in its International System (SI) unit, the *tesla* (T), where 1 T = 10,000 gauss.

In addition, an $\boldsymbol{x}$*-gradient* field is activated from time to time, as directed by the computer, as part of the procedure to localize individual voxels along the x-axis, a process known as *spatial encoding*.

And third, the *transmitter*, also under computer control, and its antenna radiate *pulsed RF power*. RF signals returning from the patient are then *detected* by these or other coils and the *receiver*, and are processed to form an image.

Ideally, the main magnetic field is intense and *homogeneous* (uniform) throughout the volume of the patient's body being imaged, and temporally *stable*, not changing appreciably over time. In clinical practice, it can be generated in either of two general spatial configurations. You are doubtless familiar with the *cylindrical superconducting magnets* that create a horizontal main field that runs through the hole in the donut (Figure 3.1a). The relatively small number of MRI systems that employ permanent or room-temperature electro-magnets, on the other hand, typically produce vertical fields (Figure 3.1b). They have the *open magnet* arrangement favored by patients who become uncomfortable in confining spaces. They provide less signal-to-noise, and are in use, primarily in less developed parts of the world because of their significantly lower cost.

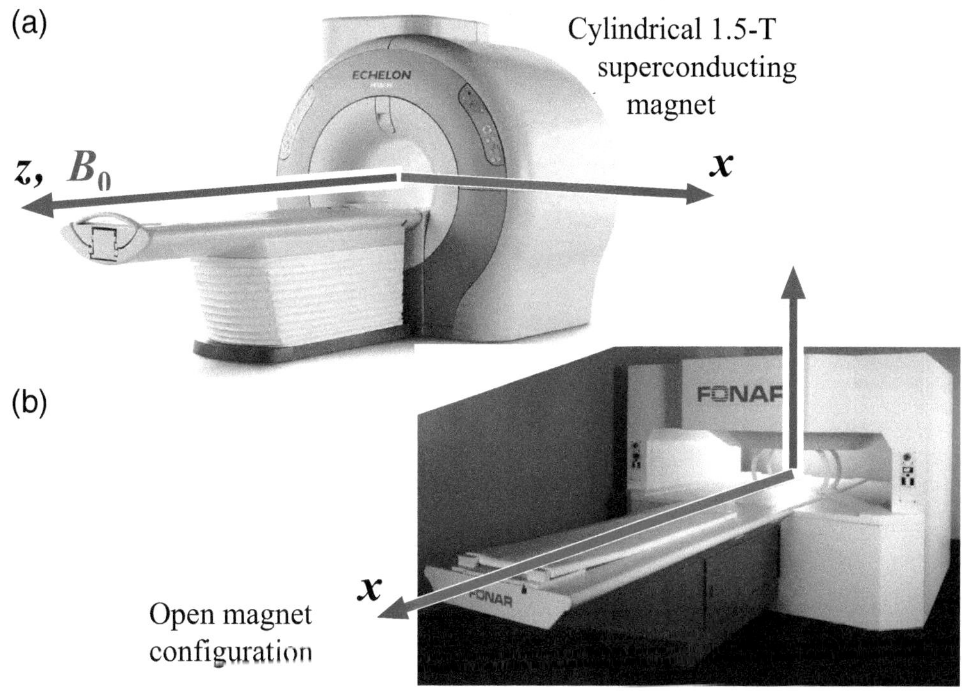

Figure 3.1 The two general geometries of MRI magnets, *closed* and *open*, normally result in horizontal and vertical main fields (purple), respectively. The three technologies that produce this field are *superconducting*, *electro-*, and *permanent* magnets, in order of decreasing strength. (a) The main field of a *closed* cylindrical solenoidal superconducting magnet is horizontal and aligned along what is defined as the z-axis. (b) For an *open* permanent or electromagnet, the main magnetic field, $\mathbf{B_0}$, is directed along a different **+z**-axis, in this case, pointing upward. Here the patient table lies along the **+x**-direction. This permanent magnet is comprised of heavy, adjacent C-shaped sections of magnetized ferromagnetic alloy. [Courtesy of the FONAR Corporation.]

The Cartesian *x*-, *y*-, *z*-coordinate axis system employed depends on the orientation of the main magnetic field. The *z*-axis, in any case, is chosen to lie along the direction of field of the main magnet (purple), $\boldsymbol{B}_0$ (Figure 3.2a).

The *gradient magnetic fields* (green) are normally pulsed on and off intermittently and briefly over time, and it is the mechanical movement of machine parts induced by the switching gradient fields that one hears as muffled booming during an MRI study. The *gradient* vector, $\boldsymbol{G}$, can be separated into *x*-, *y*-, and *z*-components that are defined as

$$\begin{aligned} G_x &\equiv \Delta B_z / \Delta x, \\ G_y &\equiv \Delta B_z / \Delta y, \\ G_z &\equiv \Delta B_z / \Delta z. \end{aligned} \tag{3.1a}$$

The field always points along the *z*-axis, as indicated in Figure 3.2b by the subscript, and so changes in

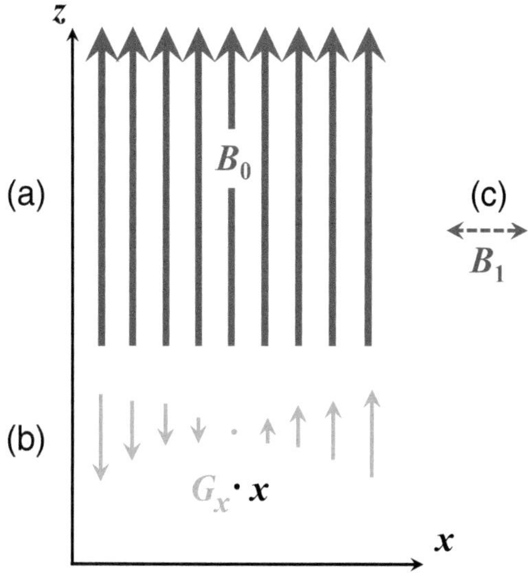

Figure 3.2 Magnetic fields for 1D MRI with an *open* main magnet. (a) The strong (usually 1.5 T; sometimes 3 T) unwavering and highly uniform main magnetic field, $\mathbf{B_0}$, is represented by the long purple vertical arrows. (b) The relatively weak ***x**-gradient*, $\mathbf{G_x}$ (green), is produced by the **x**-gradient coil, and its field varies in strength along the **x**-axis—but it is caused to ***align*** always along the **z**-axis! That is, unlike the main field, it is designed to be *non*uniform, varying in strength (but *not* in *orientation*) in a tightly controlled linear manner from place to place along the **x**-axis. $\mathbf{G_x} \equiv \Delta\mathbf{B_z}/\Delta x$ represents the relatively small change in the local ***z**-field*, ΔB_z, that occurs with a transverse shift in the ***x**-direction* by the distance $\Delta\mathbf{x}$. (Likewise, for 3D imaging, $\mathbf{G_y} \equiv \Delta\mathbf{B_z}/\Delta y$ and $\mathbf{G_z} \equiv \Delta\mathbf{B_z}/\Delta z$ generate gradient fields in the **y**- and **z**-directions. Again, for all three gradients, the *increments* in field strength ($\Delta\mathbf{B_z}$) lie only along the **z**-axis, parallel to $\mathbf{B_0}$.) One more time: while the strength of each gradient field depends on position within the patient, all give rise to changes in the strength of the net field which (in the absence of RF energy) always points only along the **z**-axis. (c) The weaker Larmor-frequency field (red), $\mathbf{B_1}$, oscillates at or near the Larmor frequency. QM theory informs us that to cause nuclear spin transitions, its magnetic field must be aligned *perpendicular* to $\mathbf{B_0}$.

the field, ΔB_z, also point along z—but its magnitude can vary as a function of position along any of the three orthogonal axes.

EXERCISE 3.2 What would the y-gradient, $\boldsymbol{G_y}$, look like in Figure 3.2? $\boldsymbol{G_z}$?

Once again, because this is so important a point: in 1D imaging,

> The main and gradient fields point only along the +z-axis, but the magnitude of the net field, $|\boldsymbol{B}_z(x)|$, can vary with $\boldsymbol{x}$-position.

Apart from the RF field, the *externally applied* magnetic field within the imaging region always and everywhere points along, and only along, the z-direction. But if there is a gradient present, then the *magnitude* of the z-component of the net field, $\boldsymbol{B}_z(\boldsymbol{x})$, varies from place to place within it.

For pedagogical purposes, we shall for a good while restrict discussion to that of a 1D phantom lying along the $\boldsymbol{x}$-axis, where all the action is. Working in 1D x-space simplifies the notation considerably. For the several 1D examples that follow,

$$\Delta B_z = G_x \Delta x. \quad (3.1b)$$

As indicated by this and Figure 3.2b, the $\boldsymbol{x}$-gradient field (but not the gradient!) vanishes at, but only at, $x = 0$, the origin. The total local field at the distance x from the center of the magnet (at $x = 0$) is aligned only in the z-direction, and it is of magnitude:

$$B_z(x) = B_0 + (\mathrm{d}B_z / \mathrm{d}x)x = B_0 + G_x x. \quad (3.1c)$$

Position x can have both positive and negative values. Put another way, the momentarily static field of strength $B_z(x)$ at the voxel at position $\boldsymbol{x}$, during the short interval (milliseconds) that the gradient is switched on, is just the sum of two entities, B_0 and $G_x x$, that both are aligned in the z-direction (just in case we haven't emphasized that point enough!). In the 3D situation, coming later in the book, $G_x x$ will be replaced with the vector *dot* (*scalar*) *product* $\boldsymbol{G}\cdot\boldsymbol{r}$, but in the meanwhile, the subscript x provides a little harmless redundancy.

$\boldsymbol{B}_1$ (in red in Figure 3.2b) is the magnetic component of a relatively weak (μT) pulsed *radiofrequency field*, oscillating tens of millions of times a second, generally at or very near the Larmor frequency. It is turned on for short (tens of microseconds), precisely timed periods. Produced by antenna coils situated close to the patient, $\boldsymbol{B}_1$ *always* points *perpendicular* to the alignment of $\boldsymbol{B}_0$. Other RF energy, subsequently re-emitted in modified form from the body, actually bears the signal information from which images are created. At this point in the story, we shall suggest that NMR events are detected through the absorption of RF energy at resonance. Such a process is plausible, and easily demonstrated in the related undertaking of *electron spin resonance* (ESR/EPR) but, as we shall see in a few chapters, this is not what actually happens in practice with clinical MRI.

The three kinds of magnetic fields discussed so far are all produced and detected by an MRI machine. The other half of the MRI story—the part that deals with relaxation times T1 and T2 and related matters—involves local, random magnetic *fields originating from within the body itself.* These are produced by tissue protons in various molecular environments (and to a generally much lesser extent by the flows of electrons within those molecules.) It is important to keep precise track of all of these contributions to the overall local magnetic environment of a tissue proton.

3.2 Encoding of Voxel x-Position with an x-Gradient, and the Local NMR Frequency, $\nu_{Larmor}(x)$

Consider a 1D phantom comprising a row of identical small three-dimensional *voxels* (volume elements) laid out horizontally over a 20-cm-long *field of view* (FOV_x) along the x-axis. These are seen from the side in Figure 3.3. There are N_x of them, and each is Δx millimeters wide, so the total FOV must be

$$\text{FOV}_x = N_x \times \Delta x. \quad (3.2a)$$

Such a perspective enables us to treat the phantom digitally. If either N_x or Δx is chosen to be smaller, the FOV_x will be correspondingly narrower, and vice versa.

In the presence of the gradient of G_x, the range of field strengths from one side of the FOV to the other would be ($G_x \times N_x \times \Delta x$). The corresponding *bandwidth* of RF frequencies, BW_{RF}, that the MR receiver (or power detector) must be able to input would then be

$$\begin{aligned} BW_{RF} &= (\gamma / 2\pi) G_x \times N_x \times \Delta x \\ &= (\gamma / 2\pi) G_x \times FOV, \end{aligned} \tag{3.2b}$$

typically in tens or hundreds of kilohertz (kHz).

To simplify this phantom further, let's assume that only two of the voxels contain water (Figure 3.4). One

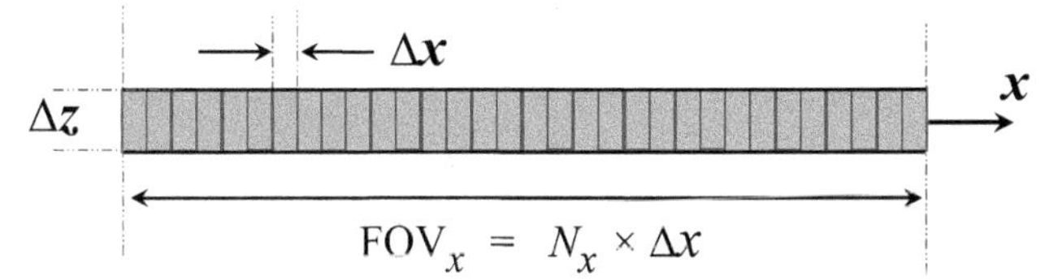

Figure 3.3 The phantom we shall frequently be referring to consists of a horizontal row of N_x small voxels lying along the *x*-axis, each Δx wide, containing various amounts of water or other materials. The overall field of view in the *x*-direction is $FOV_x = N \times \Delta x$ and, as discussed later, the RF receiver *bandwidth*, $BW_{reeceive} = (\gamma/2\pi)\ G_x \times N \times \Delta x$, required to detect and process all frequencies within this range. In practice, voxels are of the order of 0.5 mm across, and those illustrated here are drawn way out of proportion.

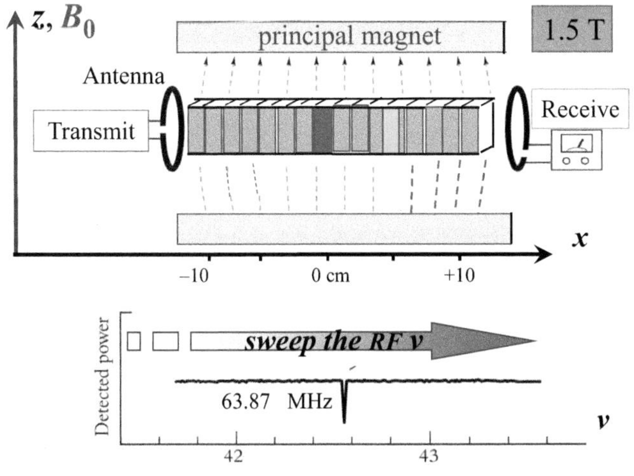

Figure 3.4 A 1D phantom in a *uniform* 1.5-T external field. The phantom is represented as a 20-cm-long set of very small 3D voxels, viewed here from the side. All of them are empty except for one compartment in its middle filled with water, and another one 5 cm toward its right-hand end that contains a 50%–50% mixture of water and Styrofoam powder. An obvious conclusion from this figure is that you cannot image without a magnetic field *gradient* or, rather, three of them for real 3D MR imaging.

EXERCISE 3.3 Explain why BW_{RF} does, or does not, depend on B_0? Does it double in switching from a 1.5 T to a 3.0 T system?

nonempty compartment, located at the center, is filled with pure water. The second, 5 cm to the right, contains a half-and-half mixture of water and inert filler, like Styrofoam, so that the average proton density is half that of water. Our objective will be to determine experimentally the water content of the phantom as a function of position along it, PD($\boldsymbol{x}$).

If the total magnetic field were uniform at 1.5 T from $x = +10$ cm at the right to $x = -10$ cm, as in Figure 3.4, however, then the water molecules in both compartments experience the same local fields. When the RF frequency is slowly swept upward, nothing would happen until 63.87 MHz, at which point the protons in both voxels would undergo resonance together, as in Figure 2.5. Not very informative.

An effective method to generate a real MRI map of proton density in 1D is based on *frequency-selective encoding* of voxel *x*-position, such as by means of an $\boldsymbol{x}$-gradient field (Figure 3.5a). The approach here is again to sweep the RF field slowly and keep track of the frequencies and signal amplitudes of any resonance peaks that may appear; but now we can associate the position of tissue with a particular resonant frequency using magnetic gradients—and vice versa!

It's just like playing a piano: the position of any key on the keyboard uniquely determines the frequency of the corresponding note you hear. Conversely, the frequency of a note is connected, in a 1-to-1 fashion, with the location of the associated key.

3.3 Proton Density (PD) MR of a Multi-Voxel, One-Dimensional Patient

In the 1D imaging of Figure 3.5, we adjust $\boldsymbol{G_x}$ so that the total local field strength is exactly 1.500 T at the center of everything, at $\boldsymbol{x} = 0$ (Figure 3.5b). The added $\boldsymbol{x}$-gradient field points only against the direction of $\boldsymbol{B_0}$ for $x < 0$, or along it for positive x (Figure 3.2). Suppose that the total local field strength, $|\boldsymbol{B_z}(\boldsymbol{x})|$, runs from 1.498 T at $x = -10$ cm, at the left-end voxel, to 1.502 T at the other end, $x = 10$ cm, as in Equations (3.1). The x-gradient is thus $G_x \equiv \Delta B_z / \Delta x = (0.004\ T / 20\ cm) =$

20 millitesla per meter (mT/m), a reasonable value in practice.

This indicates one way to go about determining the value of $B(x)$ so as to obtain x. The essential link is the Larmor equation itself, modified slightly to account for x-position in the gradient field. Measure

$$\nu_{Larmor}(x) = 42.58(B_0 + G_x x) \qquad (3.3a)$$

precisely (Figure 3.5c) and from this calculate the local field strength, $(B_0 + G_x\, x)$, equally precisely, at the x-coordinate of each water-containing voxel along the phantom. This, in turn, gives us the exact location of the voxel, $\boldsymbol{x}$, within the patient,

$$\begin{aligned} x &= \left(B_z(x) - B_0\right) / G_x \\ x &= \left(\nu_{Larmor}(x) - 42.58\, B_0\right) / 42.58\, G_x . \end{aligned} \qquad (3.3b)$$

In our example, this becomes: x [meters] = $(B_z(x) - 1.5)$ [T] / 20 [mT/m], so at the two water samples, the local fields are $B(x = 0) = 1.500$ T and $B(+5\text{ cm}) = 1.501$ T, respectively. To locate any slice of water, all we have to do is measure the frequency, $\nu_{\text{Larmor}}(x)$, of its resonance peak and plug it into Equation (3.3b). QED!

As we sweep the RF frequency in the example of Figure 3.5, the chamber at $x = 0$ will produce a peak at $\nu_{\text{Larmor}}(x = 0) = 42.58$ MHz/T × $(1.500 + G_x \times 0)$ T = 63.87 MHz. Similarly, when we detect the resonance frequency of the farther-right chamber as 63.91 MHz, its location is determined to be $x = 0.05$ m, from 63.91 MHz = (42.58 MHz/T) × [1.500 T + (20 mT/m × .05 m]. This finding, stemming from Equation (3.1c), is just a modification of the Larmor equation, but one that also provides an exclusive relationship that enables us to jump back and forth between voxel *x-position* within the patient and the *frequency* of the associated NMR signal.

Since the PD at the center is double that at x = 5 cm, the *amplitude of the NMR* peak at that frequency, $A(\nu)$, will be twice as great, also. By a reasonable convention, the brightness of the voxel at position x in the image is made to increase with the amplitude of the detected peak, hence with the PD there. There is now enough information to *reconstruct* an MRI map of water content for the phantom (Figure 3.5d).

To summarize: measuring NMR peak amplitude, $A(\nu)$, as a function of RF frequency, ν, allows us to generate a map of the proton density, PD(x), as a function of *position* within the phantom. This amounts to, in effect, obtaining the spectrum of the signal, $S(t)$, picked up by the receiver coil. The frequency of an NMR peak, which is easy to measure precisely, leads directly to the local magnetic field strength, through the Larmor equation. But field strength is known to increase in a controlled and calibrated linear manner along the length of the phantom. So we have an immediate, quantitative link between the NMR frequencies, and the positions of the only places in the body where the NMR is taking place and signals are being produced (Figure 3.5e).

We leave you now with a fundamental question: we have devised a way to correlate the frequencies of MR waves with positions along the x-axis—which will later also be referred to as the *frequency axis*, for obvious reasons. That's all that's needed in examining a 1D phantom. But most real patients exist in at least two dimensions, and it is necessary to embed information on y-position in the MR signals, as well. How are you going to do that?

3.4 Measures of Image Quality: Contrast, Resolution, and the MTF

We are highly pleased with ourselves for having created an MRI image, even if it's only of a one-dimensional patient. Now, however, we must turn immediately to the obvious and imperative next point: how good is it? Is a particular clinical study clear and uncluttered enough to lead to a correct diagnosis?

No single modality can perform all imaging tasks well, of course, so a physician first must understand what each type of medical examination can reveal about a patient's condition. Only then can she, guided by preliminary studies and her experience, choose an imaging technology with the specificity/selectivity and sensitivity most likely to provide the essential information needed for a given medical situation.

But what are the criteria by which *image quality* (and by extension, the effectiveness of an imaging device) be evaluated? Perhaps somewhat surprisingly, pretty much the same set of specifications have been found to gauge how well any imaging apparatus is performing. While we shall be focusing on MRI, ultimately the same basic criteria underlie the assessment of any medical imaging device: what is it that determines the detectability and identifiability of an anatomic or pathological irregularity?

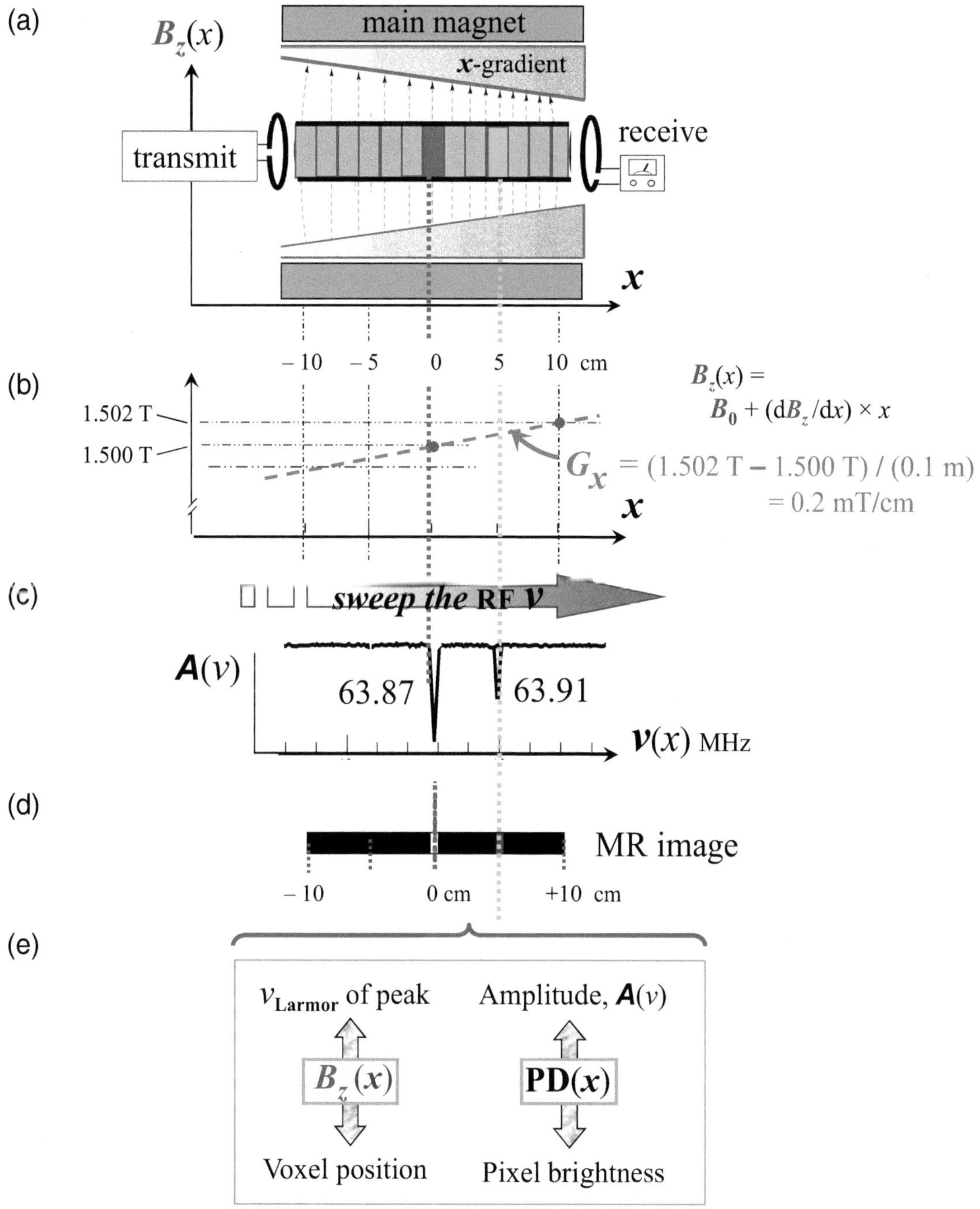

Figure 3.5 MRI study of the one-dimensional phantom in a *nonuniform* field comprised of the main field, of strength B_0, plus an **x**-gradient field. In such a gradient, there is a direct link from NMR signal frequency, through local field strength, to position in the phantom. (a) Here the magnet pole faces have been canted, superimposing a weak field gradient, G_x, onto the strong uniform vertical main field. The field is stronger where the field lines are closer together (their length doesn't matter). (b) The composite magnetic field still points only in the z-direction, but its strength increases with the **x**-coordinate as $G_x x$. By our design, the field is weaker (1.500 T) at the central compartment than at the one at x = 5 cm, where $B(x)$ = 1.501 T, so that G_x = (1.502 T – 1.500 T) / (0.10 m) = 20 millitesla/meter = 0.20 mT/cm. (c) The system slowly sweeps the RF frequency through the range of potential resonances and notes the frequency, ν_{Larmor}, and amplitude, $A(\nu)$, of any detected NMR signal, $S(\nu)$. One NMR signal appears at 63.87 MHz, associated with the voxel at x = 0, and another, half as strong, at 63.91 MHz, for x = 5 cm. The frequencies of the resonance peaks can be measured precisely and, through the Larmor equation, indicate the local magnetic field strength, $B_z(x)$ at any water chamber. But the field strength increases in a controlled, linear manner with x along the length of the phantom. (d) This rigid linkage provides the key that reveals where each chamber is located. Also, the amplitude, $A(\nu)$, of an NMR signal at any frequency is proportional to the proton density at the corresponding point in the phantom. One could thus relate proton density (from the NMR signal amplitude) to position along the phantom (from the NMR signal frequency), and thereby compute a map of water proton density throughout the phantom, PD(x)—that is, a 1D MR PD image of it! The display *grayscale* is set so that the brightness of a pixel increases with the proton density of the material within the corresponding voxel. (e) A summary of the above.

EXERCISE 3.3 Explain why BW_{RF} does, or does not, depend on B_0? Does it double in switching from a 1.5 T to a 3.0 T system?

An examining physician can provide an immediate response: either she can read it and make a valid determination of whether there is a medical problem, or not. Another doctor might have more or less training, experience, or better visual acuity and arrive at a different, or the same, conclusion. Would it improve matters to use another imaging instrument? And are there any ways to determine which combination of patient problem, physician, and equipment will perform best? (The same question applies when an *artificial intelligence* (AI) computer reads the results of a study, rather than a human.) Alternatively, the present device might just be in need of some fine adjustment, which might help immeasurably. But, again, how does one know when she has achieved optimal tuning?

Fortunately, there are several important image attributes that can be quantified, and this can assist greatly with assessing and improving the image quality from a diagnostic device. Foremost among these are tissue *contrast* and *resolution*, and the presence of distracting *stochastic noise* and preventable *artifacts*.

Any medical imaging system acts at least somewhat like a photographic camera (Figure 3.6a). As in Figure 0.3, it maps signals or probes transmitted through a three-dimensional object (x-ray, CT), emitted from it (gamma camera, SPECT, PET), or reflected (ultrasound) or re-emitted (light, MRI) from it onto a two- or three-dimensional display surface (Table 0.2).

The quality of a diagnostic system, and of the pictures it produces, is judged primarily by the goodness of the images. This, in turn, is commonly assessed in terms of the five general parameters mentioned above, namely the contrast and resolution, which are clinically desirable, and the random noise, distortion, and artifacts that detract from an image (Figure 3.7).

Contrast generally refers to an image's abilities to distinguish different physical attributes of the object, and to represent differences with shades of gray or colors. *Resolution* (also called *sharpness*, or absence of *blur*) describes its potential to capture fine detail, and to distinguish small nearby items as being separate. Both contrast and resolution may be diminished significantly by two general types of interfering *noise*. *Stochastic* (*random*) *noise* effects are most evident, of course, when a signal is weak. *Structured noise*, or *artifacts*, are nonrandom patterns of visual misinformation, analogous to the 60 Hz hum from a low-grade audio amplifier, that can either obscure or distort what is really there. Finally, *geometric distortions* can lead to catastrophic results during radiotherapy or surgical treatment planning.

Usually either contrast or resolution is of greatest interest, depending on the type of study. In a PET

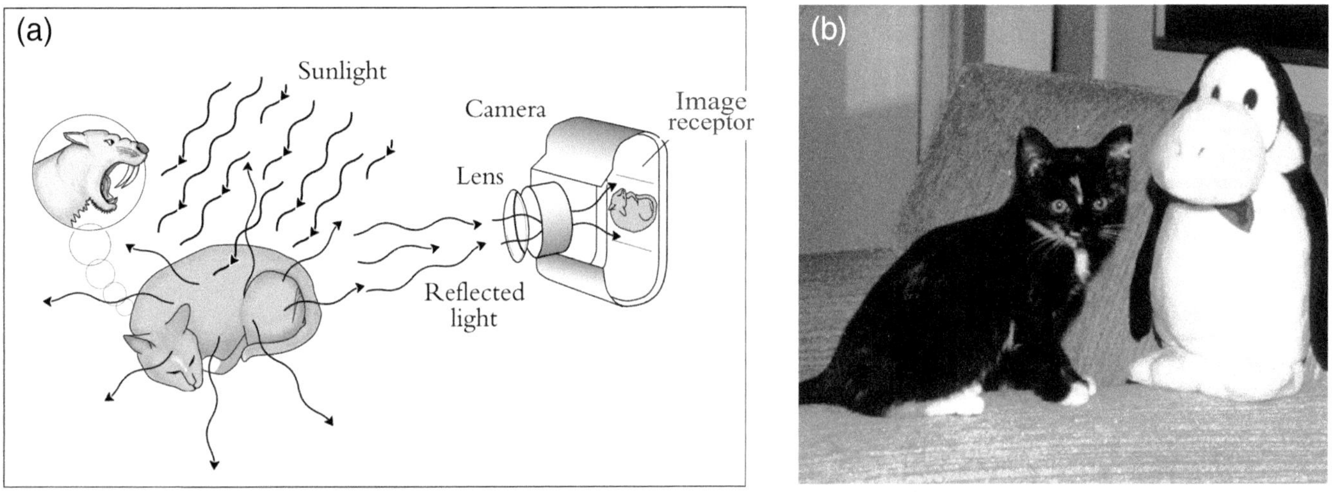

Figure 3.6 Imaging. (a) An optical camera creates an image by establishing a one-to-one correspondence between light absorbed by and re-emitted from a point on an object being photographed and a corresponding unique point on the film or digital image receptor. (b) A strong-contrast, high-resolution, low-noise rendering of a young Nadine Wolbarst with her special friend, Opus (don't ask, don't tell), demonstrating her normal glazed-over, catnip-deranged glare.

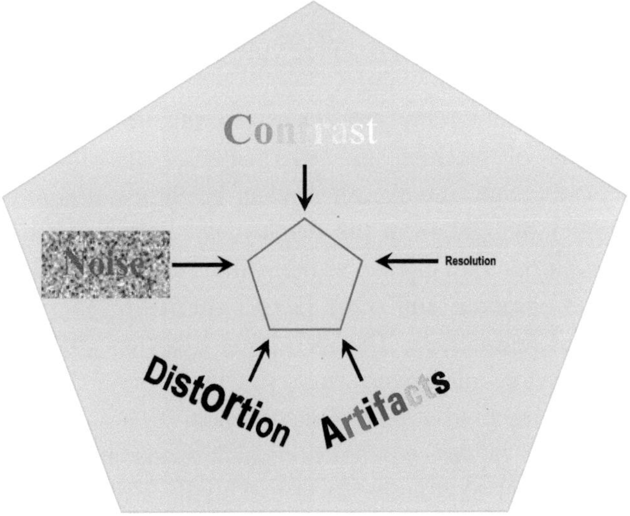

Figure 3.7 MR image quality. In any kind of imaging, not only medical, there are several fundamental standard measures that to a large extent determine and reflect the quality of the results. Visible *contrast* between the entity of interest and background is commonly of paramount concern, and a great virtue of MRI is that it allows creation of a number of radically different forms of contrast that reveal dissimilar, sometimes unique, forms of clinical information. *Resolution* refers to the ability to capture fine detail. *Random* or *stochastic noise* can significantly degrade both apparent contrast and resolution, even rendering them useless, if strong enough. Geometric *distortions* can be problematic when the imaging is being used for surgical or radiotherapy treatment planning. And nonrandom, structured noise, or *artifacts*, can also, like the 60 Hz hum you sometimes hear from inferior audio equipment, cause major problems; fortunately, the sources of these often can be tracked down and eradicated.

search for tumors, for example, contrast is all-important. For a digital x-ray examination of a hairline crack in bone, contrast is inherently good, and the dominant consideration is resolution. In addition, device designers and operators spend much of their time trying to eliminate, or at least minimize, random noise and artifacts. In doing so, and in other efforts to optimize performance, they may make use of a number tests and measures, such as the *Signal-to-Noise Ratio* (SNR), the *Modulation Transfer Function* (MTF), the *Detective Quantum Efficiency* (DQE), *receiver operating characteristics* (ROC), and others. These issues are a central focus for signal engineers and information scientists, and it all can get quite complex. But they are clearly relevant in an introduction to MRI, and some of them will be sketched here and later.

Contrast (C)

For many clinical MR purposes, image *Contrast*, *C*, is the image attribute of greatest concern. Contrast in a photograph, for example, commonly refers to differences in the level of light (signal strength) coming from anatomically or physiologically different parts of the body. If, for example, L_{obj} represents the intensity of *L*ight coming from an *obj*ect or region of interest, and L_{back} is that from a background or reference region, then one obvious measure of contrast might be

$$C \equiv (L_{obj} - L_{back}) / L_{back}. \tag{3.4}$$

This is just the difference in brightness between the object and the background region, taken relative to the background level.

Consider, again, Nadine's coat (Figure 3.6b). Most of it is comprised of 50 shades of gray, but the sclera of her eyes are much brighter. Suppose that when the photo is viewed in normal light, L_{sclera} = 50 units of light intensity reflect from the eyes, and L_{coat} = 10 units from the surrounding dark fur. Then with the contrast of the white of the eye against the background coat would be, with this first definition of contrast, $C = (50 - 10)/10 = 3.0$. A dark-colored object against a light background would yield a contrast numerically quite different—in fact, a negative value.

This definition of visual contrast is neither unique nor 'best,' however, and others might be more helpful in certain imaging situations. Instead of the relative difference between the two intensities, for example, one might consider the logarithm of their ratio:

$$C \equiv \log_{10}(L_{obj} / L_{back}), \tag{3.5}$$

which can be useful when levels of *L* differ by orders of magnitude. Although this may not be as self-evident a choice as Equation (3.5a), it is every bit as natural and legitimate a measure of the difference between two intensity levels. It is employed, in fact, with optical (optical density, OD) and ultrasound (decibels, dB) images, and elsewhere. But whatever the use, it is important to spell out clearly how you are going about defining and computing the *C*.

The perceived contrast for an image on a liquid crystal display (LCD) screen or on a film on a light box depends first and foremost on the way in which the imaging device transforms the characteristics of differ-

Table 3.1 Sources of contrast for the various major clinical modalities. The various planar x-ray modalities and CT all track differences of photon attenuation along different paths through the body that are due to nonequivalent combinations of tissue *density*, effective *atomic number*, and *path length*, with detectors such as an active matrix flat panel imager (AMFPI), image intensifier (II) plus charged-coupled device (CCD) camera, an array of gadolinium oxide (GdO) detectors, etc. For nuclear medicine, gamma (including SPECT) and PET cameras detect gamma rays and 511 keV annihilation photons from radiopharmaceuticals that have *concentrated* differentially in tissues, and then emit isotropically. Ultrasound follows brief pulses of high-frequency (>1 MHz) mechanical vibrations (not EM radiation) that propagate through the body at a nominally constant rate and reflect at boundaries between tissues of dissimilar *density* or *compressibility*. While each of the above technologies responds to one or a few physical attributes of tissue, MRI is sensitive to a number of its physiological characteristics, such as protons density, T1 and T1 proton spin relaxation times, the oxygen concentration in blood (f MRI), the flow and perfusion of blood, the diffusion of water along neuron axons (DTI), chemical shifts of ν_{Larmor} caused by variations in the electron currents within molecules, and a number of other phenomena.

Modality	Probe / Signal	Detector	Source of *Contrast*: Differences in...
Planar x-ray, CT, Fluoroscopy,...	X-rays transiting the body	AMFPI; II+CCD; GdO array, etc.	$\int_S \mu(\rho, Z, kVp)\, ds$
Nuclear Med: gamma camera, SPECT; PET	Gamma-rays emitted from body	NaI single crystal; multiple NaI; LSO array	Radiopharmaceutical uptake, concentration; Photon emission
US	MHz sound	Piezoelectric transducer	ρ, κ, μ_{US}
MRI	Magnet, RF	~AM radio receiver	T1, T2, PD, [O], blood flow, water diffusion, chemical shift, ...

ent tissues into shades of gray or color. It is often asserted that with ideal viewing conditions, contrast of a few percent can be discerned visually on a display device. But a number of factors affect the threshold of detectability of a pathology, in particular the amount and types of stochastic noise present in or around the relevant part of the image.

Also important are the sizes of the anatomic regions of interest, the sharpness of the borders between them, the overall level of brightness and uniformity of light from the display, the lighting conditions in the viewing room, the skills and state of mind of the observer, and other factors studied by psychophysics researchers. The increasing focus on artificial intelligence, moreover, suggests whole new dimensions in the field of pattern recognition.

Each of the standard clinical imaging modalities except MRI exploits what is essentially a single biophysical process, as indicated in Table 3.1 and in Table 0.2, to create clinical contrast. MRI uses more than a half dozen. In one of his lesser-known publications, Isaac Newton discussed how he collected and stored apples in a barrel on his back porch (Figure 3.8). He had a number of ways to cull out the few rotten ones, calling upon a set of radically different forms of contrast by means of which he could tell the good and the bad apart. Color and texture were easy giveaways, as were smell and taste, the presence of worm holes and soft spots, etc. Each of these types of contrast is *produced* by a unique biochemical process in the apple and, quite separately and independently, *detected* by way of a corresponding unique physiological sensory process of the viewer.

Similarly, as the rest of this book will show, MRI employs its amazing technologies to distinguish among a variety of biochemical and biophysical tissue processes, and to display the corresponding forms of contrast among tissues. These are based on variations

(a)

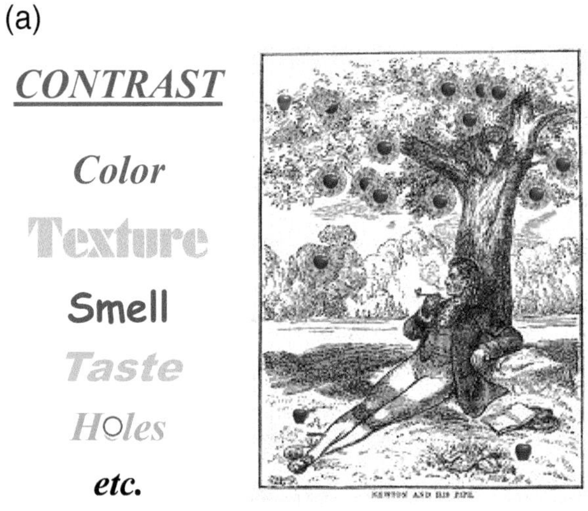

(b)

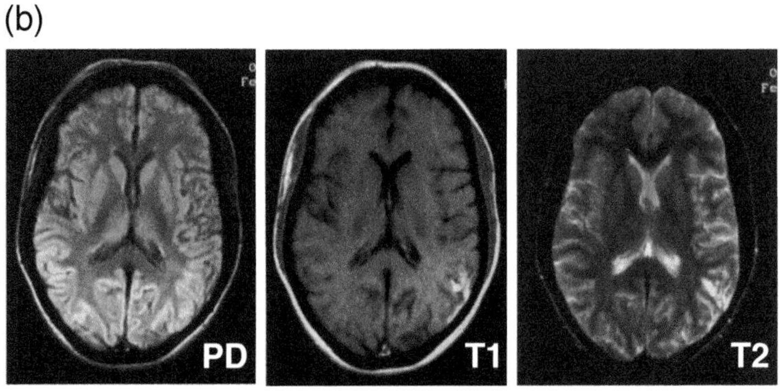

Figure 3.8 Forms of contrast. (a) Between good and rotten apples; it is even possible to grade apples through seismology [van Wijk et al. 2017]. (b) Three forms of MRI contrast created by totally different physical processes using three different imaging parameter settings.

among tissues of proton density, T1- and T2-relaxation rates, local relative concentrations of deoxyhemoglobin and oxyhemoglobin (f MRI), the flow of blood (MRA), the diffusion of water molecules within axons (DTI), and other factors. It is largely because of the diversity of such separate methods of soft-tissue contrast creation that makes MRI so extraordinarily versatile and powerful a diagnostic tool.

Resolution (R)

Resolution, R, is a second principle measure of image quality. It refers to the ability of an image or imaging system to capture fine detail. It may be reported as the minimum separation, ΔX, between small, high-contrast objects *within the patient's body* that can just barely be distinguished from one another *in the image* with a specified degree of certainty,

$$\text{Resolution, } R \equiv \Delta X \text{ in object (barely separate in image).} \tag{3.6}$$

If the objects were any nearer (Figure 3.9a), their images would blend together. In film radiography, resolution was traditionally expressed in terms of the closeness of adjacent thin, radio-opaque lines in a test phantom that can be barely made out in the film, and it can be much better than 10 line pairs per millimeter (10 lp/mm). We shall have more to say about resolution in chapter 5.

On modern digital systems, the amount of sharpness achievable is somewhat less than that on film. The sizes of the voxels for the imaging device, commonly less than a half millimeter on a side, impose inflexible limitations (Figures 3.3 and 3.9b). (The pixel dimensions of display monitors rarely pose problems these days.) The advantages of digital imaging, however, nearly always far outweigh those of the older analog technologies.

EXERCISE 3.4 Pixel size is too large in Figure 3.9b. What else might be degrading the image?

Glimpse of the modulation transfer function (MTF)

The *modulation transfer function* (MTF) of an imaging system is, in effect, the spatial counterpart to a stereo amplifier's frequency-response function.

The problem is one sadly familiar to all of you Bach aficionados. Of late, Glenn Gould's Goldberg Variations just haven't been sounding quite so sprightly on the old stereo. Not only that, but Pink Floyd seems a little off-color. So being physicists, we compare the output of the amplifier with a square-wave

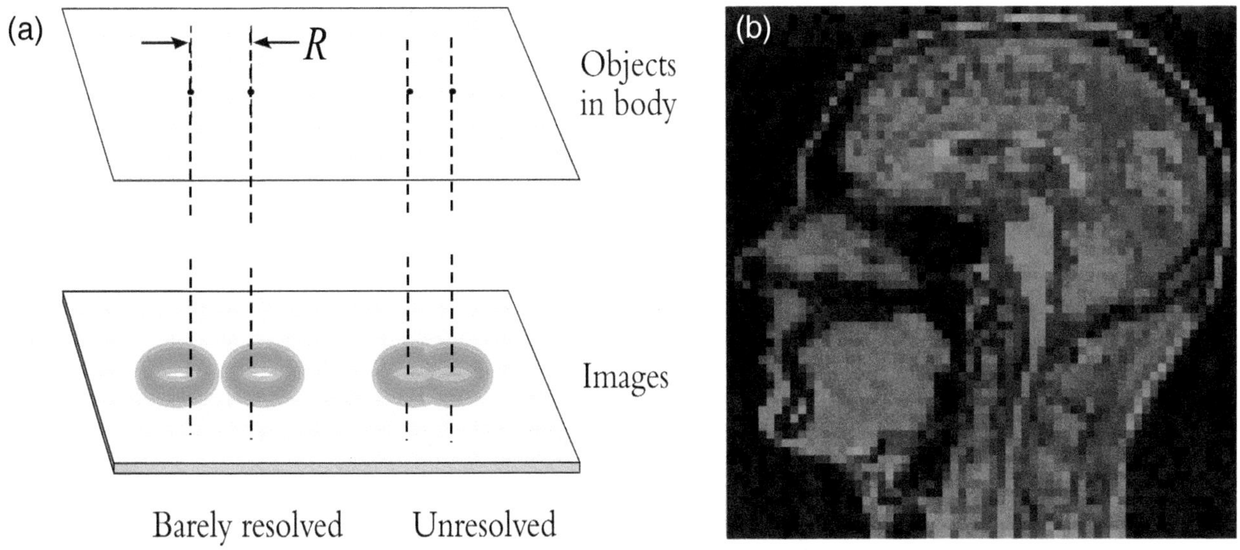

Figure 3.9 Resolution or sharpness. (a) Resolution in an image can be defined as the closeness, ΔX, of high-contrast point objects in the body that can barely be distinguished from one another in the image. (b) With digital imaging, resolution is ultimately limited by the dimensions of the smallest achievable voxels in tissue or pixels in the display. Here the voxels are so large that they give the impression of noise. What is another problem with this image?

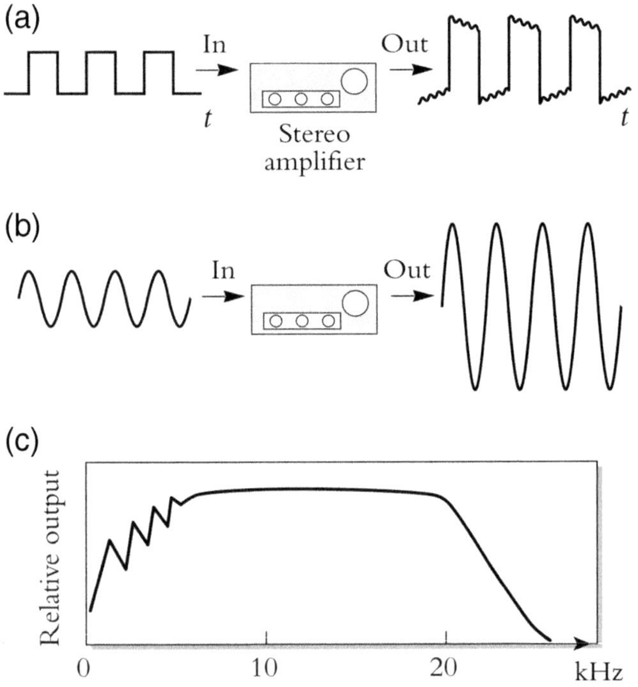

Figure 3.10 Distortions arising in an audio amplifier (a) arise because the response of the system is not strictly linear, the same at all frequencies. That is, it does not amplify the all components of the input signal by the same amount, and also it seems to shift the phases of some, which can be just as disruptive. (b) One way to check it out: Pump in monochromatic signals, one at a time, all of the same amplitude, and (c) plot the output amplitude (or phase shift) as a function of the input **v**.

input signal generated by a test device. Our worst suspicions are confirmed. The stereo system is doing something improperly, leading to horrible tonal distortion.

It is not apparent from Figure 3.10a, however, exactly where the trouble lies. It is much more revealing to examine the response of the system to a range of monochromatic sinusoidal input signals systematically (Figure 3.10b) and obtain the system's *frequency-response* curve (Figure 3.10c). This shows how the amplitude of the output sine wave varies with the frequency of the input sine wave (the amplitude of the input being held constant), and it should be flat over the range of audible frequencies. The frequency response actually measured may be able to tell a good deal about how well the amplifier is working and, in this case, it seems that something is grievously wrong at the low-frequency end. A separate measurement reveals that the phases down there are being shifted improperly. Since no one seems able to fix these things anymore, it may be time to give the machine the old heave-ho.

Figure 3.10 provides a nice analysis of a device that processes temporal signals. The same kind of thing can be done for *spatial* images, as well, which are made up of sine waves of the correct *frequencies*, *wavelengths*, *phases,* and *amplitudes*. For an electromagnetic or other time-varying wave, the notations for frequency, wave length, phase, speed of propagation, and amplitude are usually v or f, λ, φ, c, and $|\boldsymbol{E}|$ *or* $|\boldsymbol{B}|$, respectively, where $v \times \lambda = c$ (Figure 2.2). For spatial waves, on the other hand, the *spatial* frequency, also called the *wavenumber*, is commonly represented as k and quantified in units of *cycles per unit distance,* such as cycles/cm, or just mm^{-1}. When it indicates the direction of wave propagation, too, the *wave vector* assumes the mantle $\boldsymbol{k}$ and becomes a special kind of vector that exists in ***k****-space*, also known as *reciprocal space*. When wavelength is measured in SI units, the *spatial period*, $1/k = \lambda$, represents the distance separating the crests of adjacent waves in meters, generally called the *wavelength*.

The *modulation transfer function* (MTF) of an imaging system is, in effect, the spatial counterpart to the stereo amplifier's frequency-response function.

Equation (3.5) can be modified to provide an intuitively satisfying definition of the contrast of a single, localized object against a uniform background, where L_{obj} and L_{back} refer to the levels of light from the two (Figure 3.11a). If the intensity of an optical signal of interest, $L(x)$, happens to vary sinusoidally in space with wavelength λ, then its amplitude, L_{amp}, may be thought of as its peak value above background, $(L_{obj} - L_{bac})$, and a reasonable definition of contrast might be $C = L_{amp} / L_{bac}$. Since the intensity varies between a maximum value of $L_{max} = (L_{bac} + L_{amp})$ and a minimum of $L_{min} = (L_{bac} - L_{amp})$, we can represent the background level as $L_{bac} = ½(L_{max} + L_{min})$, and likewise for the signal amplitude, $L_{amp} = ½(L_{max} - L_{min})$. What we have been calling the contrast becomes:

$$C \equiv (L_{max} - L_{min}) / (L_{max}) + (L_{min}), \quad (3.7a)$$

(Figure 3.11b). This turns out to be easier to work with than $C = L_{amp} / L_{bac}$, even though the information content of the two expressions is the same.

If we replace the light wave $L(x)$ with a more general spatial signal $S(x)$, then we can define the *modulation*, M, (the first cousin of the contrast) of any spatial sine-wave signal, S, as

$$\mathtt{M}(k) \equiv (S_{max} - S_{min}) / (S_{max} + S_{min}). \quad (3.7b)$$

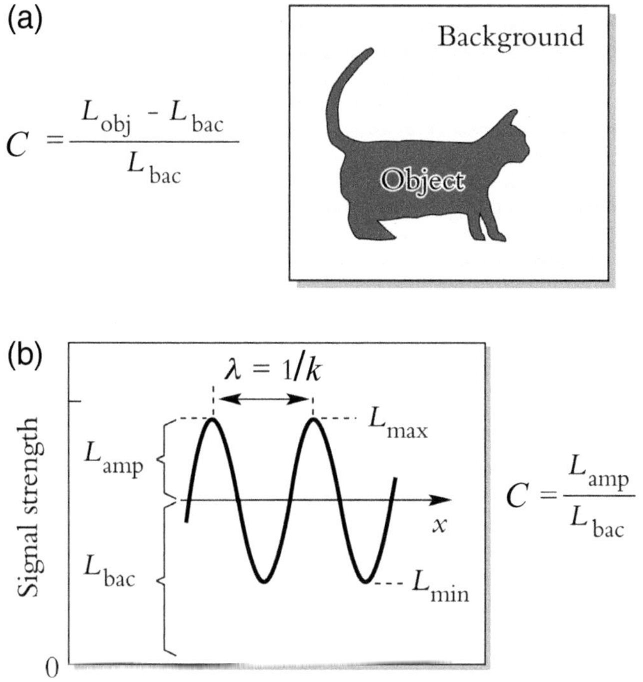

Figure 3.11 Contrast can be defined in terms of (a) object and background intensities or (b) signal maximum and minimum.

Since the amplitudes of the signals in Equations (3.7) are likely to be frequency-dependent, we should write the modulation as $M(\lambda)$ or $M(k)$.

Two RF carrier signals that are 50%- and 75%-modulated, for example, appear in Figure 3.12a, for example. Similarly, the input modulation for the situation of Figure 3.13a is $M_{in} = (3 - 1) / (3 + 1) = 0.5$, and $M_{out} = 0.4$. We are interested in the deleterious effects that, in carrying out its designated task, an image-processing apparatus might have on an input signal—and in particular, on its modulation.

EXERCISE 3.5 For situations in which the signal amplitude is small relative to background, (S_{max} – S_{min}) will be small. Does that mean that most of the energy of a signal with a modulation near 0 carries no information? Is a signal with a modulation of about 1 is being put to the most efficient possible use?

The standard method of assessing the impact of an operation on anything quantifiable is to compare the after with the before. Consider what a piece of equipment does to the three input waveforms of Figure 3.13a, each of which is of unit amplitude and $M_{in}(k) = 1$. It doubles the amplitude of the one with $k = 1$, but leaves its frequency and phase unchanged, as well as the modulation, too. So, it would seem, the apparatus has no deleterious effects on signals of this frequency. By contrast, the amplitude of the $k = 2$ wave is the same as the input's, but $M_{out}(2) = 0.5$, and this signal has been degraded. Finally, for the $k = 3$ signal, the amplitude of the output is halved and, more importantly, $M_{out}(3)$ has fallen to 0.33.

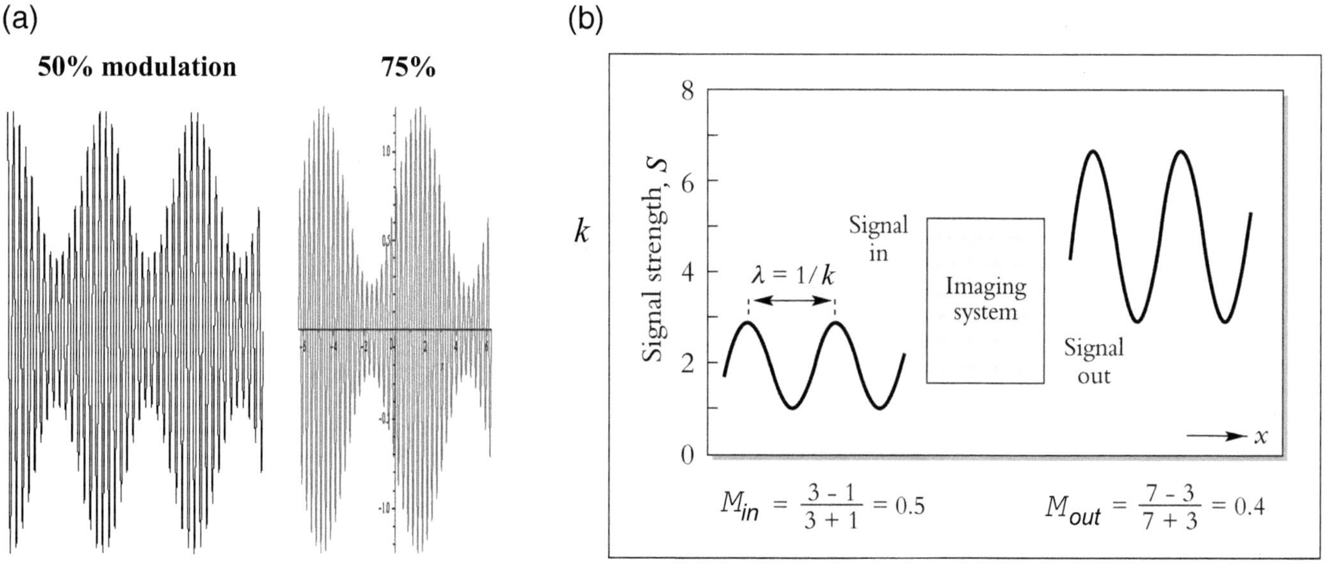

Figure 3.12 How the *modulation* of a sine wave can change in passing through a device. (a) Two RF carrier signals have been 50%- and 75%-modulated, respectively. (b) The modulation of the input signal is $M_{in}(k) = (3 - 1) / (3 + 1) = 0.5$, while that of the output is $M_{out} = (7 - 3) / (7 + 3) = 0.4$. That is, the apparatus reduced the modulation, and left it at only 80% of its former value. Is that significant?

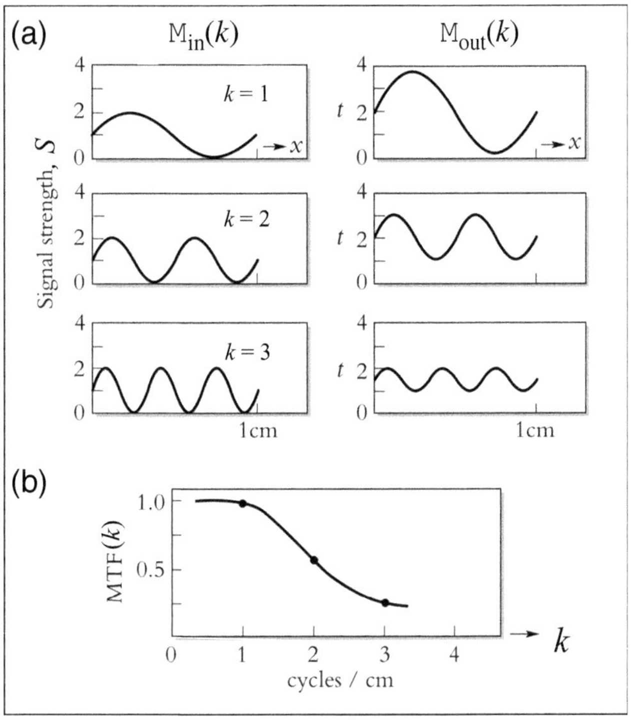

Figure 3.13 The Modulation Transfer Function, MTF(*k*), displays the effects of a device on input signals of unit modulation, but over a range of various spatial frequencies, in particular their output modulations, $M_{out}(k)$. (a) Input signals at three frequencies, all of unit amplitude and with $M_{in}(k) = 1$. The three output signals are frequency dependent: $M_{out}(1) = (4 - 0) / (4 + 0) = 1$, and likewise $M_{out}(2) = 0.5$ and $M_{out}(3) = 0.33$. (b) The MTF is defined as $\text{MTF}(k) \equiv M_{out}(k) / M_{in}(k)$. For the $k = 2$ signal, for example, $M_{out}(2) / M_{in}(2) = 0.5 / 1.0 = 0.5$, as indicated by the second data point in the graph of MTF(*k*) vs. *k*.

The general trend is summarized in Figure 3.13b, known as the *modulation transfer function* (MTF), which plots $M_{out}(k) / M_{in}(k)$ as a function of wavenumber:

$$\text{MTF}(k) \equiv M_{out}(k) / M_{in}(k), \tag{3.8}$$

An imaging device, like an audio amplifier, will be more successful at handling signal components of some frequencies than of others. The response is usually better at lower frequencies, so that the MTF normally falls off from a maximum value near $k = 0$. The MTF is commonly normalized to a value of 1 near $k \sim 0$,

$$\text{MTF}(\sim 0) \sim 1. \tag{3.9}$$

A system may not only reduce the modulation of a signal, but also introduce phase shifts into it. Phase shifts are critically important in some imaging situations—they carry essential information in 2D MRI, for example—and must be taken into account, but for simplicity we shall not consider them further yet.

The MTF(*k*) is thus a measure of the ability of the imaging system to handle contrast and modulation as a function of spatial frequency, hence of resolution. The greater the MTF(*k*) of a system at high spatial frequencies, the more adept it is at displaying fine detail as contrast. The roll-off region of the MTF(*k*) curve indicates the frequency range over which the system starts becoming incapable of reproducing the image components faithfully. In other words, the MTF provides a combined measure of contrast and resolving capacities together. Like, as we shall soon see, a *contrast-detail phantom*.

In practice, assessing an audio system by passing sinusoidal signals of various frequencies through it, one at a time, would work, but it is slow and laborious. There must be a better way! Indeed, there is.

Instead of using a set of sinusoidal test patterns of various frequencies as the input signal, we can employ a single straight, very narrow tube of high proton-density substance, lying along the ***y***-axis and immersed in a large tank of very-low-proton background material (Figure 3.14a). Looked at edgewise, the proton density has an extremely tall, narrow, Δx-wide peak, centered at $x = 0$, PD(***x***). If we perform an MRI scan of this subject on an absolutely perfect machine, which never happens in practice, the MR signal strength as a function of position, here called the *line-spread function*, LSF(*x*) will have exactly the same width and shape as the tube containing the sample of protons:

$$\text{LSF}_{perfect}(x) = \text{PD}(x). \tag{3.10}$$

But when we repeat the study on a machine that is in only OK condition instead, we obtain a somewhat blurry line described as $\text{LSF}_{OK}(x)$, which only somewhat resembles PD(*x*). A poor system yields an even more spread-out LSF. If we carried out this experiment in 3D and with a small point source of hydrogen, instead, we would record the *point spread function*, PSF(*r*), instead. The meaning of the third member of this family, the *edge-spread function* (ESF) is self-evident.

All this new stuff is interesting, but does it lead to anything useful? Time for a resounding YES! The LSF(*x*) and PSF(*r*) are relatively easy and quick to obtain by direct measurement. And as we shall see in chapter 5, the MTF is just the Fourier transform of the

LSF(x) or PSF(r), a computation that takes milliseconds on a digital computer (Figure 3.14b). So instead of having to employ a series of monochromatic spatially sinusoidal patterns of various frequencies, one at a time, as the input signals, we can choose a single, very narrow tube or sphere of proton-dense substance lying along the y-axis, at least in theory. As we shall see, the Fourier spectrum of this line-source contains all frequencies, and all at the same unit amplitude; this is to say that $M_{in}(k) = 1$ for all k, and they all will pass through an MR instrument at the same time.

One last MTF topic before moving on from contrast and resolution—we're saving the best for last! The MTF is useful not only in comparing various gradient field amplifiers with one another, or RF antennae, or other components, but also in determining how the overall performance of a complete scanner is influenced by the behavior of its separate building blocks. Suppose that the device consists of separable stages in series, where the output of each serves as the input to the next. The MTFs of the various stages, connected in series, are $MTF_1(k)$, $MTF_2(k)$, $MTF_3(k)$, etc., at spatial frequency of k cm^{-1}. We state without proof that the behavior of the whole system at that frequency is given by the product

$$MTF_{sys}(k) = MTF_1(k) \times MTF_2(k) \times MTF_3(k) \times \ldots \quad (3.11)$$

So if one particular component has a low MTF at some particular k, then the MTF for the entire system will be made low there. The imaging chain is truly only as strong as its weakest link.

3.5 Stochastic Noise, Signal-to-Noise Ratio, Contrast-Detail Curves, and DQE

As anyone knows who has ever had a teenager or been one, noise and chaos come in a remarkable variety of flavors. In all of its guises, however, noise always has the same practical effects: it obscures messages of importance by interjecting competing extraneous junk that carries no information, or quasi-data that is irrelevant but distracting, or (worst of all) information that seems valid but is wrong.

Visual noise refers to anything that interferes, to any extent, with access to the information content of an image. Such noise is caused in a number of ways and comes in diverse forms, but commonly it is of two general types. *Stochastic* (also called *random* or *statistical*) *noise* differs from study to study of the same object. In Figure 3.15, for example, a clean signal, S, is largely obscured by noise, N, when it appears as the much less helpful combination $S + N$.

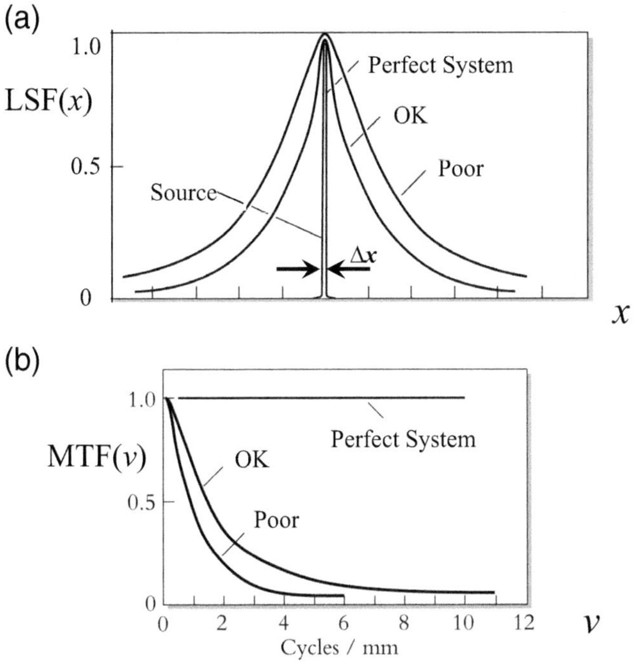

Figure 3.14 The LSFs and corresponding MTFs for three MR machines, with a very narrow, Δx-wide tube of material high in tissue protons displayed against a background nearly devoid of protons. (a) The first machine operates perfectly, unlike any real one, and its LSF(x) mimics the PD(x) (a Dirac delta function) exactly (see chapter 5). The second scanner, which works OK, generates a $LSP_{OK}(x)$ that is notably wider than the original PD(x) being imaged. The 'poor' system does even worse. (b) The MTF(k) functions are obtained from the corresponding LSFs by means of a Fourier Transform (FT), also discussed in chapter 5.

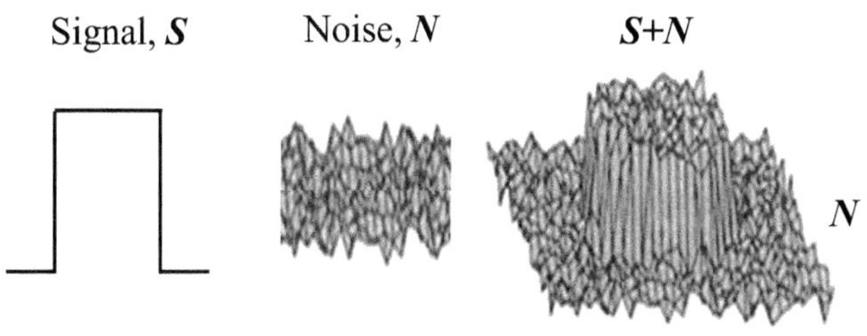

Figure 3.15 The pure, information-bearing signal power, S, and the background noise, N. An important quantifier of image quality is the *signal-to-noise ratio*, or SNR, defined as the ratio S/N. Also shown here is $S + N$.

With an *artifact* (*structural* or *deterministic noise*), on the other hand, a specific noise pattern (like 60 Hz hum in an audio system or, with CT, streak lines from a metal object) may recur among a number of images of the same or of separate objects.

With either kind of disruption, the concern is to extract as much useful signal as possible out of a background of noise—the amplitude of which may, at times, be considerably greater than that of the meaningful signal itself.

Stochastic noise

In the several types of film and digital radiography, a familiar source of stochastic noise is the x-ray Compton radiation scattered from within the patient. Scatter creates a haze, similar to that from silt stirred up near the bottom of a lake that obscures the sharp rocks and snapping turtles you'd prefer to avoid (Figure 3.16a). Fortunately, much of this kind of interference can be removed by way of an x-ray *anti-scatter grid* (Figure 3.16b). The problem in this case is that there are too many of the 'wrong' kind of photons, those that end up activating the image receptor where they shouldn't. Noise like this occurs in MRI, as well, but it tends to be harder to remove.

There is another form of apparent random noise that can arise for the converse reason—because insufficient numbers of the "right" kind of information-bearing "quanta" are involved in the creation of an image, as is the case with *quantum mottle* (Figure 3.16c). That is, if you are going to build an image out of vast numbers of minute dots (or of minute pixels of different degrees of brightness or color), you have to include a sufficient amount of them, and they have to be small enough for the image to look acceptably smooth (Figures 3.16d and 3.9b).

An important random noise occurring in MRI is thermal RF energy emitted from the patient, but MR system noise is comparable to it. Apart from that, if the radiofrequency shielding (e.g., Faraday cage) is inadequate, the presence of any extraneous RF fields—such as from a nearby AM radio transmitter, lightning storms, sparking of machinery, or fluorescent lamps—can induce unwanted electric jolts in the system. In addition, spontaneous thermally induced statistical fluctuations voltage/current in the RF receiver's electronic circuits (e.g., *Johnson-Nyquist noise, Schottky/shot noise*) can be disruptive. Even imperfections in a computer's digital reconstruction algorithms may lead to noise-like abnormalities.

Signal-to-noise ratio (SNR)

Much research has led to sophisticated, highly mathematical descriptions of stochastic noise and of methods

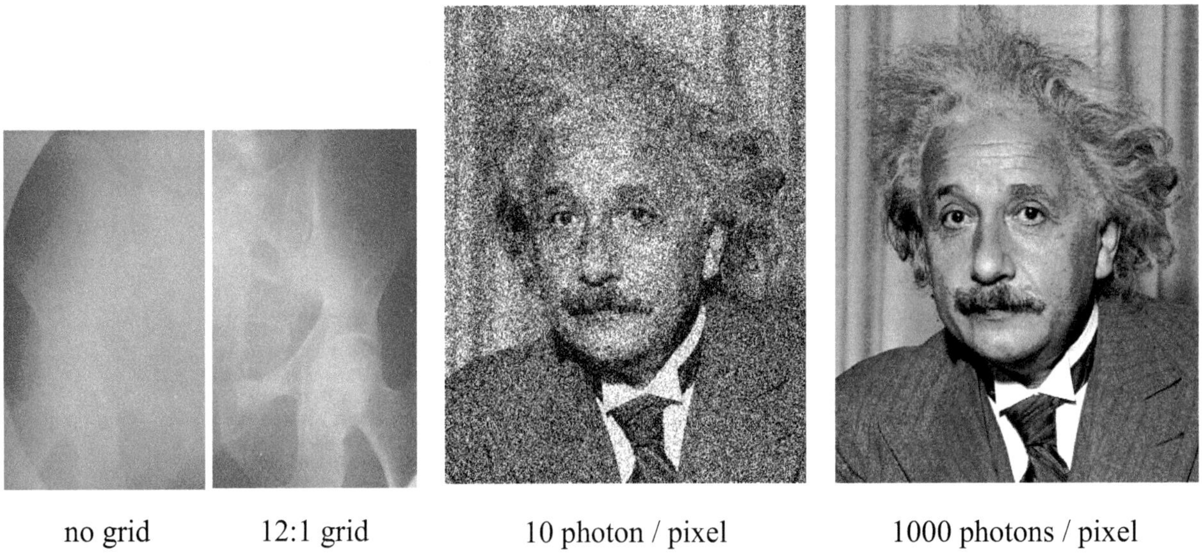

Figure 3.16 Two categories of noise. (a) In radiography, x-ray scatter radiation gives rise to a form of stochastic noise that can largely be removed with a grid. (b) To achieve the same level of brightness (average optical density) when the grid was added, the product of exposure time with current through the x-ray tube (mA-s) had to be increased by a factor of 10, which leads to 10 times the unwanted patient dose. (c) The mottled appearance in a photograph of the great man vanishes when (d) the number of light photons per pixel grows 1000-fold, improving the SNR by a factor of $\sqrt{1000}$.

to deal with it. Suppose that S is some gauge of the amount of clean, untainted information in a *S*ignal related, say, to the excess number or fraction of proton spins in the lower-energy state; and N is an indicator of the amount of background *N*oise energy intermixed with and obfuscating it, as in Figure 3.15. Then the signal-to-noise ratio, *SNR* or *S/N*, is a simple but valuable measure of the quality and usefulness of the information content of the mixture. It is defined as the ratio of pure signal power to noise power:

$$\text{SNR} = S / N. \tag{3.12}$$

S/N clearly improves as the signal alone increases or as the noise diminishes.

The SNR is a complex entity, the analysis of which requires the use of more advanced statistical analysis and information engineering. It is an important measure of the diagnostic utility of an image in its own right, and it also underlies other measures, such as the *contrast-to-noise ratio* (CNR) and the *detective quantum efficiency* (DQE), both of which play roles in modern imaging science.

According to one metric widely applied to *low*-contrast images, the *Rose Criterion*, the SNR must be at least 5 for an observer to be able to distinguish small features in an image unequivocally. We have here, as you doubtless noticed, just kicked the matter of quantifying S and N themselves down the road for now.

We shall return to the subject of the statistics of noise and the Rose Criterion in chapter 5.

Contrast-detail curves

Image *contrast*, *resolution*, and *stochastic noise* have been introduced and discussed separately here, in a reductionist fashion. But the contrast and resolving capabilities of a system can appear to influence one another, and stochastic noise can greatly diminish both. The ability to detect an entity in the body does not necessarily depend on contrast alone, or resolution, or on any of the other factors of image quality, but rather on a complex amalgam of all of them together. This is readily apparent from the detectability of the holes drilled in a radiographic *contrast-detail* QA test device, such as the one in Figure 3.17a. Hole diameter and depth (hence the resolution and contrast) vary systematically throughout in a metal plate. The obvious take-home message is that both the apparent resolution and contrast increase as the noise level diminishes and the SNR improves (Figure 3.17b).

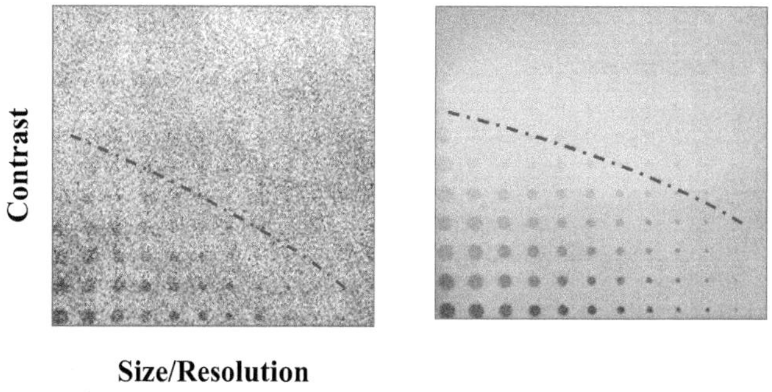

Figure 3.17 The impact of stochastic noise on apparent contrast and resolution. The radiographic *contrast-detail* QA tool for radiography has holes of ranges of diameters and of depths in a metal plate. The same kind of test can be carried out for the other imaging modalities. (a) The objective is to determine the combinations of contrast (as controlled primarily by hole depth) and resolution (diameter) for which the holes are just barely detectable in the image for a given noise level. It is easiest to make out the objects (holes) in the lower left-hand corner. (b) The test is repeated with less noise, and the line of detectability moves north-west.

EXERCISE 3.6 What does the motion of the line of detectability in Figure 3.17 tell you about the system?

Detective Quantum Efficiency (DQE)

A somewhat different approach to the same general end, and one that has long been used in imaging-system performance assessment, leads to the *detective quantum efficiency* (DQE). Nearly any component of an information-processing system will either degrade signal or introduce stochastic noise of its own, thereby diminishing the signal-to-noise ratio. The DQE reveals the frequency dependence of the extent to which this occurs:

$$\text{DQE}(k) = \left[\text{SNR}_{\text{out}}(k) / \text{SNR}_{\text{in}}(k)\right]^2, \tag{3.13a}$$

where the $\text{SNR}_{\text{out}}(k)$ and $\text{SNR}_{\text{in}}(k)$ are the signal-to-noise ratios for what goes into and for what comes out of the apparatus at spatial frequency k, respectively. You will want any component in an imaging chain to have $\text{DQE}(k) \sim 1$ over a wide range of spatial frequencies, as well as $\text{MTF}(k) \sim 1$.

From the definition of SNR, the DQE may be expressed also as

$$\begin{aligned} \mathrm{DQE} &= [S_{\mathrm{out}} / N_{\mathrm{out}}]^2 / [S_{\mathrm{in}} / N_{\mathrm{in}}]^2 \\ &= [S_{\mathrm{out}} / S_{\mathrm{in}}]^2 / [N_{\mathrm{out}} / N_{\mathrm{in}}]^2 . \end{aligned} \quad (3.13b)$$

Comparison with Equation (3.8a) suggests that the DQE is conceptually similar to MTF, as is discussed in more advanced texts. For example, DQE can be expressed as

$$\mathrm{DQE}(k) = \mathrm{g}[\mathrm{MTF}(k)]^2 / [\mathrm{NPS}(k)] , \quad (3.13c)$$

where g is a factor that is determined by signal intensity, the image-receptor contrast (also known as *gain*), and other items. NPS(k) is the *noise power spectrum*, which records the amplitude of the noise present as a function of its frequency. If this has a constant value over a band of frequencies, it is called *white noise*. Finding this spectrum, however, requires a Fourier transform of the *autocovariance* function, and other matters that are beyond the scope of this book [Barrett et al. 2014]. The DQE and NPS are mentioned here as a foretaste of what you may find in quality assurance directives.

3.6 Judging Performance with ROC Curves

The MTF, the DQE, and other quantitative assessment tools are useful for indicating weak links in an imaging chain. But that is not exactly the same as appraising the overall ability of a system to provide images that allow a physician to detect and classify tissue irregularities. The *receiver operating characteristic* (ROC) *curve* is a construct, borrowed from our radio engineering colleagues, by means of which one can quantify the all-around success with which an imaging system plus observer are together likely to generate clinically correct diagnoses. As with contrast-detail curves, the use of ROCs depends in part on observer perceptions and responses, rather than being derived entirely from physical measurements, like the SNR and DQE, but they are more widely applicable.

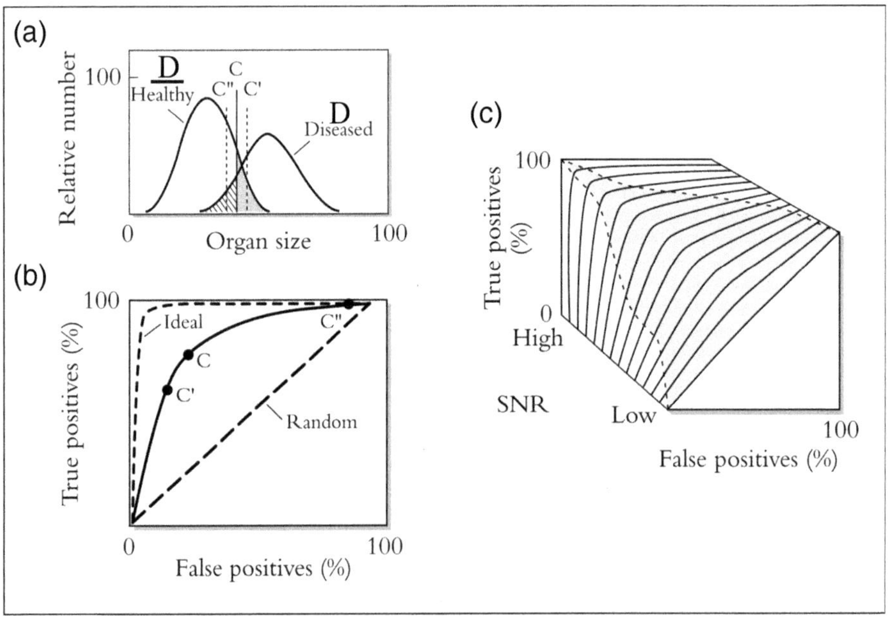

Figure 3.18 Construction of a *Receiver Operating Characteristics* (ROC) curve for a hypothetical clinical test based on detecting hepatomegaly with MRI. (a) Distribution of apparent liver sizes for a population of patients who have the disease, D, and for a group of healthy individuals who do not, D̲. Somewhat arbitrarily, the Principal Investigator for the study has chosen the curve marked C as a *clinical threshold*: Larger liver images are assumed to indicate the presence of the disease. (b) Point C indicates the percentage of false positives (FP) that are inadvertently included along with the true positives (TP) for this choice of clinical threshold liver size. C' and C" do the same for two other threshold choices. A number of points such as these define the ROC curve. An ideal curve is indicated by the upper dashed line, as opposed to the diagonal dashed one that comes from completely random data. If several different devices, or clinical techniques, or clinicians yield dissimilar ROC curves for the same set of test images, the system with the curve lying closest to the ideal is taken to provide the most reliable diagnostic information. (c) This ROC becomes more like the ideal as the signal-to-noise (SNR) improves.

True positives, false negatives, etc.

Suppose that some disease causes the liver to enlarge, and that in a particular MRI study of several hundred adults known to have the disease, marked D, the apparent sizes (however assessed) of the livers display a bell-shaped (*normal* or *Gaussian*) statistical distribution (Figure 3.18a). For healthy adults, $\underline{D}$, the imaged organ sizes are also normally distributed, but with a significantly smaller *mean* value. It is fairly easy to tell the two situations apart in most situations, but not for livers whose images are in the range of sizes where the distributions overlap. In that case, some healthy patients will appear to have the disease, and the images for some sick ones will look normal. Yet the clinician still must make decisions.

The principal investigator (PI) of a study proposes to use apparent liver size as the basis for diagnosis and picks, somewhat arbitrarily, the size marked C in Figure 3.18a to serve as the cutoff level: patients with a liver of image sizes larger than *threshold* C will be diagnosed as having the disease. That way, even though some with healthy livers will be picked incorrectly, it won't be too likely that many of those who actually have a problem will be missed. Those with image size larger than C, and thus a positive imaging finding, will be called diseased or *positive*. The others are labeled healthy, or *negative*.

A positive finding turns out to be a *true positive* (TP) if a biopsy, patient follow-up, or some other definitive test reveals unequivocally that the image-based diagnosis was correct. With a *false positive* (FP), on the other hand, the imaging study indicates that there is an abnormality, but the result of this study is wrong. So also for the negative studies: a negative diagnosis is a *true negative* (TN) only if there really is no abnormality present; an imaging study that misdiagnoses a diseased liver as being healthy is a *false negative* (FN). While FPs can be costly and terrifying, and have even led to extreme actions like suicide, some would consider the worst possible case for a serious but treatable disease to be the FN. When one does eventually learn the truth, it may be too late.

FP and FN are analogous to *Type 1* and *Type 2 errors* of statistical decision theory, incidentally, the probabilities for which are called α and β.

The four possibilities are laid out in Table 3.2a. The populations of false positive and false negative patients, for example, are indicated in Figure 3.18a

Table 3.2 A biopsy report can be either ***Positive*** (disease present) or ***Negative***, and that report can be either ***True*** (correct, T) or ***False*** (F). (a) True Positive (**TP**) and the other three possible combinations of image-based diagnosis and pathology lab confirmation/rejection. (b) The same kind of information, but expressed as the Probability (*P*) that the patient displays symptoms of the disease (+), given that he actually has it (D), *P*(+|D), *vs.* the probability *P*(−|$\underline{D}$) that a non-symptomatic patient does *not* have it.

(a)

		Pathology Report	
		Positive	*Negative*
Image-based Diagnosis	*Positive*	TP	FP (Type 1, α)
	Negative	FN (Type 2, β)	TN

(b)

		Pathology Report	
		Positive	*Negative*
Image-based Diagnosis	*Positive*	$P(+\|D)$	$P(+\|\underline{D})$
	Negative	$P(-\|D)$	$P(-\|\underline{D})$

with shading and cross-hatching, respectively. An individual patient with an imaged liver size just above C could easily turn out, of course, to be either true positive or false positive.

Sensitivity, selectivity, and accuracy

Naturally we are interested in knowing, and being able to discuss, the reliability with which you and the imaging system together will make diagnoses of the true positive, true negative variety. Several simple parameters have been devised for this purpose.

Suppose a study consists of a total of N_{total} cases, and that N_{TP} of them turn out to be TP, and so also for the other entries in Table 3.2a. First of all, as a check on the arithmetic,

$$N_{total} = N_{TP} + N_{TN} + N_{FP} + N_{FN}. \quad (3.14a)$$

Of them, the number $N_D = (N_{TP} + N_{FN})$ actually do have the Disease. That is, N_D is the number of cases in which disease really does exist, whether diagnosed correctly with imaging or not. The number who do <u>not</u> have the <u>D</u>isease is $N_{\underline{D}} = (N_{FP} + N_{TN})$, again regardless of the diagnosis.

The *sensitivity*, also called the *true-positive fraction* (TPf), is defined as the fraction of patients with *positive test results* who actually do *have the disease*:

$$\text{Sensitivity} = \text{TPf} = N_{TP} / N_D . \quad (3.14b)$$

A test is said to be sensitive when it can detect a disease that is, in fact, present. A sensitivity of 85% means that of 100 patients who do have the disease, 85 are diagnosed as positive. The *P*robability that, if a patient has the disease it will be detected, is indicated as $P(+|D)$ (Table 3.2b).

The *specificity*, or *true-negative fraction* (TNf), refers to the relative number of *healthy patients correctly diagnosed* as such:

$$\text{Specificity} = \text{TNf} = N_{TN} / N_D . \quad (3.14c)$$

somewhat like a kind of sensitivity for healthy patients. A specific test does not call healthy patients diseased.

The false-positive fraction (FPf),

$$\text{FPf} = N_{FP} / N_{\underline{D}} , \quad (3.14d)$$

on the other hand, describes those patients who actually are healthy but are misdiagnosed as having the disease. This correlates to a Type II error.

The *accuracy* of a clinical method is defined as the total fraction of cases that are *diagnosed correctly*:

$$\text{Accuracy} = (N_{TP} + N_{TN}) / N_{total} . \quad (3.14e)$$

A datum near the correct value is said to be accurate, but a group of measurements that differs from it in a systematic manner is *biased*. This is related to but not the same thing as *precision*: repeated measurements that are close to one another are *precise*, whether or not they are accurate. Bullet holes that are clustered in a target indicate precise marksmanship, but the holes are accurate only if they are in the bullseye.

All these parameters must be accepted with a bit of care, because they can sometimes be misleading. In mammographic screening, for example, breast disease is seen relatively rarely. If a clinician judged all films to be normal, even those with clear evidence of disease, the numerical accuracy still would be near 100%.

The parameters can nonetheless be helpful (and many medical tests are characterized by sensitivity and specificity alone), but they do depend on the choice of diagnostic cutoff level, C, of Figure 3.18a. If it were moved far to the left, for example, everything would be called positive, and the test would be clinically useless. So the big question is where to place the diagnostic cutoff level. The ROC method is much more powerful than the listing of a few parameters and can help significantly in finding a good answer.

EXERCISE 3.7 *In a set of 50 mammograms, four are of breasts with disease. A first-year medical student called all of them clear except for two normal (negative) films and three that he said show malignancies. Describe his performance quantitatively.*

EXERCISE 3.8 *Another student dealt with the same set of 50 cases by flipping a coin. What do you estimate his counts were for* N_{TP} *and the others?*

EXERCISE 3.9 *Finally, an experienced resident obtained these results for the 50 mammograms:* $N_{TP} = 5$, $N_{TN} = 38$, $N_{FP} = 2$, *and* $N_{FN} = 5$. *It was later generally agreed that the seven erroneous readings were due to inadequate image quality. Discuss.*

The Receiver Operating Characteristics (ROC) curve

Although it could have been designed differently, a receiver operating characteristic (ROC) curve plots the *true-positive* fraction against the *false-positive* fraction (Figure 3.18b) for various possible choices of the cutoff organ size, i.e., locations for C, C′, etc. [Obuchowski 2018]. The curve provides a valuable guide to help deal with a fundamental trade-off in clinical decision-making: if you decide you want to be sure to pick up a certain minimal fraction of true positives, how many false positives should be expected to be brought along inadvertently for the ride? How many wrong diagnoses must you tolerate as the price for a specified fraction of correct ones? While physicians rarely (if ever) use ROC curves to make clinical decisions, they do employ them to determine whether method X is generally "better" (i.e., more specific, sensitive, and accurate) than method Y.

In the study of enlarged livers, the areas under the curves of Figure 3.18a indicate that with size cutoff-level C, corresponding to a true-positive rate of about

90%, we must be willing to accept about a 15% false-positive rate, as indicated by Point C in Figure 3.18b. If you feel that no more than 5% false-positives is tolerable, you can increase the organ cutoff size to point C′ in Figure 3.18a, but you can find that the true positive rate falls to 70%. This yields a second point on the ROC curve. Conversely, if you are intent on catching nearly all of the true positives, point C″, then the number of false positives that sneak in also increases. By assuming a range of values of the threshold organ size, you can generate the entire ROC curve, the solid curve in Figure 3.18b.

Ideally, an ROC curve would look something like the left-most portion of the upper dashed line in Figure 3.18b, with all true positives and no false positives. But several factors conspire to prevent that. First, there are natural variations in liver size among healthy individuals, as for those with the disease; there obviously would be no control over the false positive or negative that would result. But also, the image of the liver has a blurred border, and the apparent size will depend on the contrast of the image signal relative to the background noise level—and on the experience of the radiologist. The percentage of correct diagnoses might, therefore, increase with improvements in the signal-to-noise ratio (Figure 3.18c) or perhaps in the abilities of the viewer. For a very bad test or reader, in which the results amount to random numbers, the ROC curve is like the straight dashed line of Figure 3.18b.

As the image quality or physician skill-level improves, the ROC curve should come closer to the ideal. If different imaging systems or techniques (which might lead to different values of the signal-to-noise ratio in Figure 3.15c) or different viewers yield dissimilar ROC curves, then the one with the curve closest to the ideal will most likely give correct results.

EXERCISE **3.10** *Why would random guessing lead to the diagonal line in Figure 3.18b?*

Bayes' theorem

Although The Rev. Thomas Bayes (1702–1761) spent most of his time attending to lost souls—the issue of transubstantiation vs. consubstantiation, and the like—he did set it aside long enough to invent the first methods of probabilistic inference, or using the frequencies with which a particular result has occurred in past trials to predict the probability of it (or something else) happening in the future.

The *conditional probability* of X *given* Y, $P(X|Y)$, also seen as $P_Y(X)$ and $P(X/Y)$, goes a step beyond the simple probability $P(X)$. It reveals $P(X)$, the probability of X occurring, given that another event, Y, is known to have taken place. If X and Y are unrelated, then $P(X|Y) = P(X)$.

An example of this would be to estimate the probability that a patient might have a Disease, P_{pat}D), from the *conditional probability* that when a certain diagnostic test is positive, +, then he truly has the affliction, $P(D|+)$, as confirmed in other ways. The overall *a priori* likelihood that the test turns out to be positive for the whole population, namely $P_{pop}(+)$, also gets involved. So, too, are the corresponding pair of terms for negative test results. Then the odds for the patient having thc disease are given by

$$P_{pat}(D) = \left[P(D|+)P_{pop}(+)\right] + \left[P(D|-)P_{pop}(-)\right]. \tag{3.15a}$$

This is self-evidently valid, but too simplistic to be of use—it doesn't even involve information on the test results for the individual patient.

The theorem that bears Bayes' name—and which is used widely in statistics, computer science, and medicine—is powerful and subtle. It relates to the basic question of how trustworthy a positive or negative test result really is, given a particular test. Tests are wrong far more often than we would wish for, so what should we trust if a particular study comes back with a specific result? *Bayes' theorem* may be helpful here, and it is applicable to situations describable with partially overlapping Gaussian probability distributions—exactly the condition summarized by Figure 3.18a. The theorem is commonly presented like as:

> Suppose you are concerned about the probability that a hypothesis or statement, X, is true. From the information currently available, you make a partially informed guestimate of its validity, $P(X)$. More evidence, Y, then arrives, and it is incorporated into the process, resulting in a revised assessment, $P(X|Y)$. Bayes' Theorem states that

$$P(X|Y)P(Y) = P(Y|X)P(X). \tag{3.15b}$$

Here we shall direct the theorem to a patient who, again, may or may not have the disease, and to a test whose results may or may not be positive. It tells us what we really want to know, namely $P(\mathrm{D}|+)$, the probability that if the test result is positive, then the patient truly is ill. Here the theorem would read

$$P(\mathrm{D}|+)P(+) = P(+|\mathrm{D})P(\mathrm{D}), \qquad (3.15c)$$

and lead to

$$P(\mathrm{D}|+) = P(+|\mathrm{D})P(\mathrm{D}) / \left[P(+|\mathrm{D})P(\mathrm{D}) + P(+|\underline{\mathrm{D}})P(\underline{\mathrm{D}})\right] \qquad (3.15d)$$

This computation involves only items that are already known from the literature:

- $P(+|\mathrm{D})$: The probability that if one has the disease, then the test will be positive. This is the *converse* of "the probability that with a positive test, the disease is, indeed, present," i.e., $P(\mathrm{D}|+)$, which appeared in Equation (3.13a) and which we are now trying to find.
- $P(+|\underline{\mathrm{D}})$: The probability of testing positive even though there is no Disease.
- $P(\mathrm{D})$: The overall rate of occurrence of the disease in the population.
- $P(\underline{\mathrm{D}}) =$ $[1 - P(\mathrm{D})]$, the overall likelihood that one does not have the Disease.

This time around, to assess the patient's status, you actually have to learn the results of the test for him, which seems sensible. (But beware: if the test is built on any unjustified assumptions that you neglected to account for, the results may appear more predictive than they really are.) A variant on this gives the other necessary piece of news, $P(\mathrm{D}|-)$.

EXERCISE 3.11 A certain disease afflicts one person in 1000. Some diagnostic test has been found to come back positive 99% of the time if a patient has it, but the probability of a false positive is 2%. What are the odds that a patient who tests positive actually has the disease?

EXERCISE 3.12 What is $P(+|\mathrm{D})\,P(\mathrm{D}) + P(+|\underline{\mathrm{D}})\,P(\underline{\mathrm{D}})$? When does $P(+|\mathrm{D}) = P(\mathrm{D}|+)$?

3.7 Some MR Artifacts

Now, on to the last image quality concerns, the possible presence of *distortions* and *artifacts*....

MRI is technologically and computationally the most flexible of the major imaging modalities, but also the most complex. To wit, a number of assumptions and approximations are made regarding the MR data acquisition and reconstruction processes, and some of them are not as good as perhaps we would like. Partly as a result of this, MRI gives rise to a variety of artifacts. These are spurious signal anomalies that do not actually represent real patient anatomy and physiology. They extend from the subtle and barely discernible to the overwhelming. One danger is that some of them can be mistaken for real anatomic or physiological effects, imparting an abnormal appearance to tissue that is actually healthy. More problematically, they may obscure signs of pathological irregularities that actually do exist. Each MR artifact is unique and must be addressed individually.

Artifacts fall into three somewhat overlapping general categories [Zhou et al. 2010]. These involve failures in MR data acquisition, with reconstruction computational algorithms, and with patient images that have been geometrically distorted, respectively (Table 3.3). We shall interject examples of common artifacts throughout the book.

Table 3.3 MRI artifacts range from the big, easily recognizable, and generally correctable to those that are rarely noticed, but all are potentially misleading and of danger to the patient [Zhou et al. 2010]. The three general types of MRI artifacts arise from faults (1) in the imaging system and in data acquisition, (2) in the image reconstruction procedure, and (3) in processes ongoing within the patient, respectively. Examples of the ones that are scattered throughout the book are underlined.

MRI Artifacts

Image Acquisition
- magnetic field inhomogeneity
- eddy currents
- chemical shift
- gradient-field distortion
- RF field inhomogeneity
- RF noise spike (herringbone)
- RF leakage (zipper)
- system stability (ghosts)

Image reconstruction
- aliasing
- blurring (i.e., exponential signal decay)
- parallel imaging
- Gibbs ringing

Physiology/patient
- blood flow pulsation (ghosts)
- respiratory motion (ghosts)
- susceptibility effects

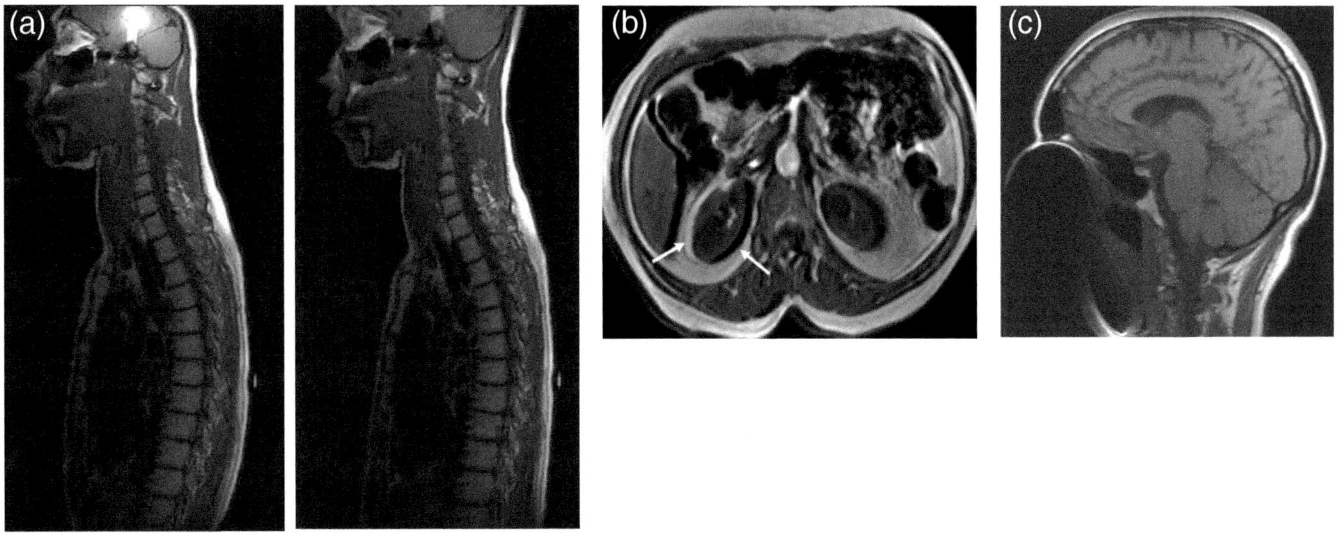

Figure 3.19 Three examples of common MRI artifacts. (a) One, related to imperfect data acquisition by the system, consists of a breakdown in the *linearity of a gradient* field, leading to geometric distortion of the large-FOV image of the spine, to the left. The problem vanished, to the right, with application of correctional software. (b) This chemical shift artifact is apparent as the white and black bands to the sides of each kidney. It arises from the 3.5 ppm difference between the resonant frequencies of water and lipid which, in a 1.5-tesla magnetic field, translates to a 220 Hz shift. (c) A susceptibility artifact from dental fillings containing magnetizable metals. (All three artifact images, and some others in the book, are courtesy of [Zhou et al. 2010].)

Nonlinear gradient-field artifact

An MRI artifact arises from a gradient field whose strength is not perfectly proportional to position—an instrumental effect. Gradient coils are effective in producing fields that normally change linearly with position near the center of the imaging field. Toward their edges, however, it is harder to maintain the linearity of the gradients, and they may tend to twist out of shape, giving rise to *gradient-induced distortion artifacts* (Figure 3.19a). Such deformation can arise also if the gradient coil has somehow been disturbed. This kind of problem is exacerbated on wide- or short-bore MRI scanners.

Software distortion-correction programs are routinely employed to help compensate for these artifacts. They may not be adequate, however, when extreme spatial precision is required to perform surgery, small-field radiotherapy, etc. (Figure 1.9).

Chemical shift artifact

The Larmor frequency of a proton in a molecule depends primarily on the field strength from the sum of the external fixed and gradient magnets. But it is exquisitely sensitive also to any other influences on the local magnetic field, such as that from the current flow of the molecule's own orbital electrons. The resulting *chemical shift* in the proton's ν_{Larmor} is thus determined by the details of the molecule's chemical structure (Figure 1.7).

The local field at a proton can be altered either by an inherent chemical shift or by an applied gradient magnetic field, and sometimes the system reconstruction software, in effect, manages to confuse the two [Servoss 2011]. Indeed, lipids and water in the same voxel may appear at slightly different spatial locations in an MR image for that reason, giving rise to a *chemical-shift artifact*, the white and black bands indicated by the two arrows near the patient's right kidney (Figure 3.19b). This is normally not apparent in MR images, but it can become so with MR systems of very good external field homogeneity and frequency resolution, such as those used for MRS. It is rather ironic that the artifact is more likely to appear with better systems!

Susceptibility artifact

Equations (2.4) and Table 2.4 describe the impacts that diamagnetic, paramagnetic, and ferromagnetic materials have on the magnetic fields in nearby tissues. Figure 3.19c illustrates an extreme *susceptibility artifact* from dental fillings containing magnetizable metals.

CHAPTER 4

Magnetization of a Voxel

The concept of transitions among proton spin states is the basic notion upon which the quasi-QM picture of NMR is built. It leads directly into the construct of the *nuclear magnetization,* $\boldsymbol{m}(x,t)$, arising from all the protons in the *voxel* at position x. And this thought, in turn, underlies two critically important processes: the behavior of $\boldsymbol{m}(x,t)$ in various sorts of static and changing magnetic fields, and the particle-interaction mechanisms that allow a disturbed system of protons to relax. Both of these ideas are fundamental to clinical MRI.

4.1 The Magnetization, $m(x, t)$, of the Voxel at Location x

The very weak collective magnetic field that a group of protons together produce per voxel (or per mm^3 or per gram—there has to be prior agreement on which) of material is called their *voxel nuclear magnetization*, and denoted $\boldsymbol{m}$. The magnetization has both magnitude and direction, and is thus a vector. When we have cause to note that this is the net magnetization in the voxel at 1D position x, we shall write it $\boldsymbol{m}(x)$. (It doesn't matter whether we say x or $\boldsymbol{x}$ here since, in 1D, they represent the same thing.) When its magnitude or its orientation or both are varying over time, t, as well, it becomes $\boldsymbol{m}(x, t)$. The local magnetization in each of a number of voxels will be the central player in much of the story of MRI.

The data one can obtain directly from the body at any instant, however, is not the set of separate voxel magnetizations, $\{\boldsymbol{m}(x,t)\}$, one for each x-position, but rather their vector sum. The *net magnetization* coming from all of them combined is

$$\boldsymbol{M}(t) \equiv \sum_x \boldsymbol{m}(x,t). \tag{4.1}$$

The information provided by MRI comes from, and only from, monitoring the behavior of $\boldsymbol{M}(t)$ over time under different, carefully controlled circumstances. In the remainder of this chapter, however, we shall consider what is occurring in only a single voxel over time, $\boldsymbol{m}(t)$. So for now we can do away with the x and the $\boldsymbol{M}$ altogether and focus on $\boldsymbol{m}(t)$. Later we shall explore ways to undertake the difficult and critical task of separating out each $\boldsymbol{m}(x,t)$ from $\boldsymbol{M}(t)$.

In re-introducing $\boldsymbol{m}(t)$ here, we first adopt the quasi-QM model of chapter 2. Suppose there are N protons in a single voxel (Figure 4.1). At any time, the number $N_-(t)$ of them will happen to be in the lower-energy state at time t, on average, and $N_+(t)$ in the higher. And, of course, the total number of protons in the voxel is

$$N \equiv [N_-(t) + N_+(t)], \tag{4.2a}$$

which is time-independent.

But at any instant, there may be a slight *excess* in either the lower or the upper state of magnitude $[N_-(t) - N_+(t)]$. Since $[N_-(t) - N_+(t)]$ will recur frequently in our considerations, we shall name this measure of the difference something shorter:

$$n(t) \equiv [N_-(t) - N_+(t)]. \tag{4.2b}$$

Each proton contributes an amount $+\boldsymbol{\mu}_z$ or $-\boldsymbol{\mu}_z$ to the composite field, where $\boldsymbol{\mu}_z$ is the projection of $\boldsymbol{\mu}$ onto the *z-axis*. The *z*-component of the total voxel magnetization may, therefore, be defined as

$$\boldsymbol{m}_z(t) \equiv [N_-(t) - N_+(t)] \times \boldsymbol{\mu}_z = n(t)\boldsymbol{\mu}_z. \tag{4.3a}$$

In the virtually impossible situation in which all the spins are aligned fully along or against $\boldsymbol{B}_0$, then $n(t) = \pm N$, and $\boldsymbol{m}_z(t) = \pm N \times \boldsymbol{\mu}_z$. But after a real system has had time to settle down and come to rest in a state of *thermal equilibrium*, why should there exist a slight excess of spins in the lower energy state? Also, just how great is that difference? In other words, how large is the magnetization in a voxel of tissue, or just plain water, when sitting in a strong magnetic field, $\boldsymbol{B}_0$, but with nothing else is going on? This voxel magnetization at thermal equilibrium is honored with its own special designation, with the subscript 0 added on:

$$m_0 \equiv m_z(t = \infty). \tag{4.3b}$$

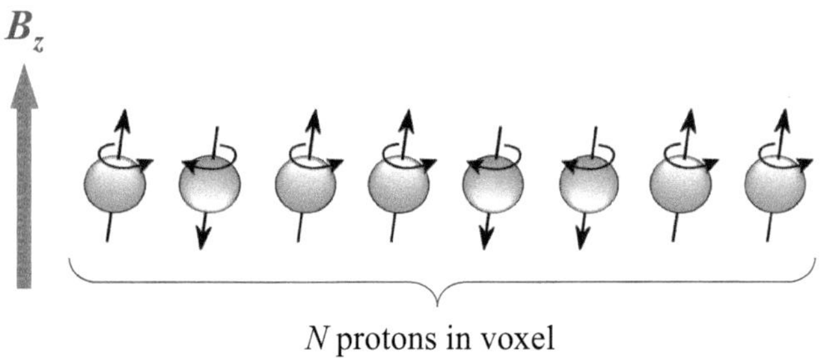

Figure 4.1 N proton spins in a voxel aligned either along or against the externally applied magnetic field $\mathbf{B_0}$. At thermal equilibrium, there will be a slight excess in the lower-energy state, resulting in an equilibrium magnetization $\boldsymbol{m}_0$.

4.2 Thermal Equilibrium: The Battle between Energy and Entropy

Let's begin the discussion of $\boldsymbol{m}(t)$ and $\boldsymbol{m}_0$ with an analogy (Figure 4.2). Suppose a box contains small, white aluminum balls with slippery surfaces, and the same number of frictionless black balls of the same size, but made of iron, much denser. Shaking the box vigorously, with the input of much random mechanical energy, results in great disarray. The system is constantly roiling, with balls continually banging about and interchanging places and seeking a condition of *maximum entropy*, one of greatest disorder. When, on the other hand, the box is tapped very gently for a while, the heavier black balls all settle slowly into the bottom rows, a state of *minimal* gravitational potential *energy*. In between, when the disruptive forces nearly overwhelm gravity, but not completely, the result of this battle between entropy and energy is that there is a small excess of black balls lower down. (Would the average numbers in the various rows be the same with identical amounts of shaking if you carried out this exercise on the Moon, where gravity is only a sixth as strong?)

Here's another analogy. Suppose that a lightweight tray covered with small compasses, together representing the spins in a single voxel, resides within a shielded room where no external magnetic field can reach, and the compasses are too far apart to influence one another much. The tray is agitated gently by a loud-speaker that pumps in acoustic noise energy. The needles jiggle and jump about, pointing randomly every which direction, but the net magnetic field that they themselves produce collectively, their magnetization, $\boldsymbol{m}(t)$, averages to zero over time (Figure 4.3a).

Now apply a modest but constant external field, $\boldsymbol{B}_0$, and the needles will tend to align along it somewhat and, over time, $\boldsymbol{m}(t)$ moves toward its equilibrium value, $\boldsymbol{m}_0$. Individual needles may flop back and forth, but on average they will create a weak but finite net magnetization (Figure 4.3b), the pale green arrow. Pro-

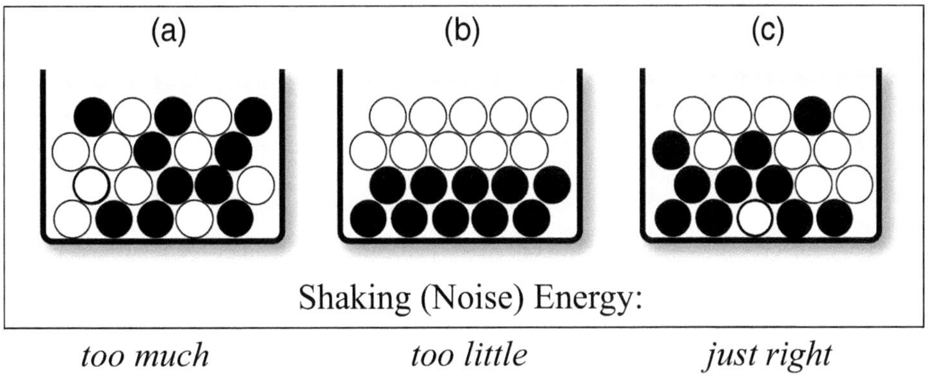

Figure 4.2 An analogy for the struggle between energy and entropy: black, dense frictionless balls in a gravitational field competing for spaces with the same number of less dense white ones of the same size. (a) When there is much energy input through shaking, then that process dominates, and the system becomes fully disordered—into a state of *high entropy*—with the balls continually switching places with one another. Roughly equal numbers of both kinds of balls populate all four energy levels. (b) With very gentle agitation, however, it can end up in a state of *low energy*. (c) Finally, in the Goldilocks zone, some but not too much shaking settles the system into an intermediary Boltzmann configuration, the nature of which depends both on the level of agitation (for NMR, read *temperature*) and on the strength of the gravitational (*magnetic*) field.

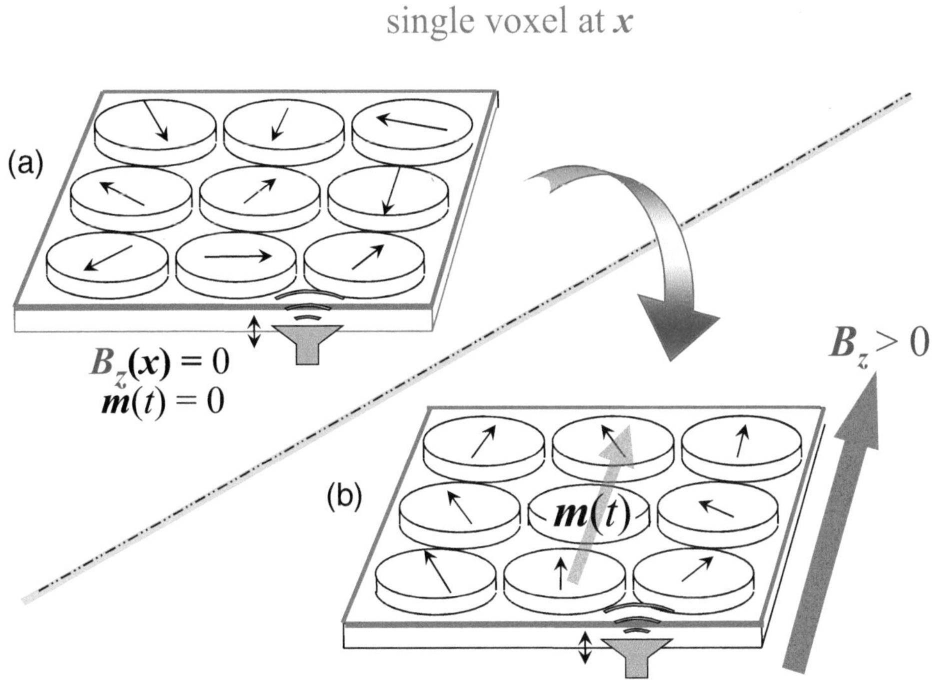

Figure 4.3 Another analogy: compass needles within a single square 'voxel' on a tray agitated by a loud-speaker. (a) With no external $\boldsymbol{B_0}$ field present, they point all over. (b) When an external field is added, they continue to jostle about, but they do tend to align along it, and they themselves together generate a weak *voxel magnetization*, $\boldsymbol{m}(t)$, the pale green arrow.

tons don't act precisely this way, but their behavior is rather similar.

The same kind of struggle between energy and entropy occurs for the protons in a voxel of water or soft tissue. If the externally applied static field is initially off (i.e., $B_z = 0$ at $t = 0$), then the proton spins are aligned randomly in all directions (Figure 4.4a). The magnetic moments of the individual protons, each of magnitude μ, point every which way, and if you add together all their individual (very small) magnetic fields, the sum is very close to zero: $\boldsymbol{m}(t = 0) = \sim 0$.

Suppose you now place the voxel between the poles of a weak, vertically aligned magnet and switch on the field at $t = 0$ (Figure 4.4b). The protons, subject to the sometimes counterintuitive dictates of quantum mechanics, immediately snap into an alignment in which the z-components of their angular momentum vectors lie either *along* or *against* the external field—but with no other orientations possible! During that initial moment of mass confusion at time $t = 0+$, just after $t = 0$, it happens that almost exactly equal numbers of them end up in the lower- and higher-energy states. The process is stochastic and governed by random molecular motions, so we have no idea what any individual spin will do. This is again a configuration of great disorder and high entropy—like when the black and white balls are all mixed up. Although this situation is totally different from what was happening before the magnet was switched on, the magnetization of a voxel will initially again be zero: $N_+ \sim N_-$, and $\boldsymbol{m}(t = 0+) = \sim 0$. With ordinary compasses, all the needles would rapidly swing into the same, lowest-energy

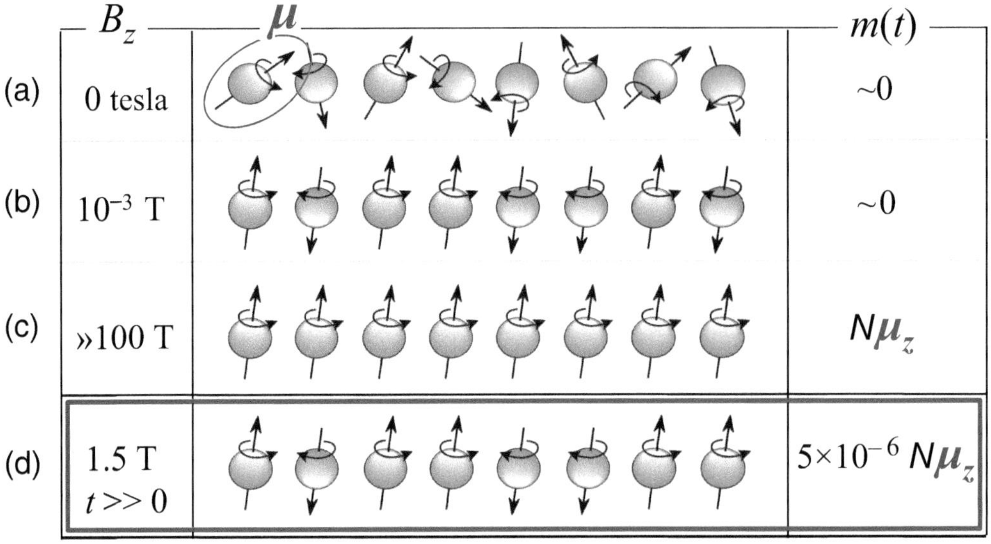

Figure 4.4 The proton spins in a voxel of water. The magnetization **m**(*t*) is the weak magnetic field that is created by all the protons themselves within it. (a) With no magnetic field present, the spins align every which way, and the net magnetization is zero. (b) Immediately after a major disruption—such as abruptly turning on a relatively weak external field—there will now be equal numbers of protons in the up and down states, for reasons that can be explained with quantum mechanics. The voxel magnetization is zero here again briefly just after $t = 0$, where $n \equiv (N_- - N_+) = 0$, but now for a very different reason. (c) With an unrealistically high magnetic field or low temperature, on the other hand, nearly all spins will align in the orientation of lower energy, and **m** will be about $N \times \mu_z$. (d) But at body temperature and clinical field strengths, the effects of thermal jostling are still important, and the outcome of this balancing act is a slight excess of protons in the lower energy state, so that $\mathbf{m} = n \times \mu_z$. In practice, with a typical 1.5 T MRI magnet, there is typically a five parts per million (5×10^{-6}) excess of protons in the lower energy state.

EXERCISE 4.1 Does it make sense that m(t = 0) = ~0 in both Figures 4.4a and 4.4b, given that the physical situations are so dissimilar?

orientation. But protons behave differently from compass needles. Bottom line, again $\mathbf{m}(0) = 0$. Once again, quantum mechanics doesn't place much stock in common sense.

Now imagine ramping the magnetic field up to an unrealistically high strength, such as 100 T, and reducing the temperature to near absolute zero, −273 °C. A condition of nearly complete order arises, in which nearly all the spins drop into their lower-energy orientation. The system descends as a whole into a state of lowest possible energy (Figure 4.4c). In laboratory experiments carried out on samples of ice at temperatures near absolute zero and in a strong $\mathbf{B}_0$, the spins all line up like bowling pins. With all *N* protons aligned parallel to one another in the voxel, and each contributing its own magnetic field μ_z to the effort, the net magnetization will be of maximum possible magnitude, $\mathbf{m}_z = [N \times \mu_z]$.

Finally, at body temperature and in a standard-strength MRI field, the case we're really interested in, the population of protons will settle eventually into a communal configuration in which *almost* but *not exactly* (and that's critically important!) the same numbers of protons point up and down (Figure 4.4d). There is, in fact, a slight excess of them in the lower-energy state, as expected from Boltzmann statistics. Thermal jostling may induce individual spins to intermittently flip up and down, to some extent, but once *thermal equilibrium* is achieved, as many go one way as the other. The overall dynamic balance is quite stable, so the equilibrium magnetization does *not* have to be written as a function of time, and in this case it is commonly designated $\mathbf{m}_0$. Perhaps a better symbol would be $\mathbf{m}_\infty$, which suggests a very long time after any disturbance has occurred and the system has settled back down. But for MRI, by convention, $\mathbf{m}_0$ it is!

EXERCISE 4.2 Back of the envelope calculation: what fraction of the spins end up in the lower energy level at thermal equilibrium?

EXERCISE 4.3 Which one(s) of the four parts of Figure 4.4 has(have) the lowest entropy?

4.3 The Magnitude of $m(t)$ at Thermal Equilibrium, $m_0 \equiv m(\infty)$, and Its Impact on Image Quality

In a voxel of water or tissue at thermal equilibrium at temperature T and in a magnetic field suitable for imaging, slightly more than half the protons will be in the lower-energy spin state, with a small but definite excess of protons pointing upward. This business is described nicely by the well-known *Boltzmann distribution* of statistical mechanics which, incorporating Equation (2.9c), here assumes the form

$$\begin{aligned} N_+ / N_- &= e^{-\Delta E/k_B T} = e^{-2\mu_z B_0/k_B T} \\ &\sim [1-(2\mu_z B_0 / k_B T)], \end{aligned} \quad (4.4a)$$

where k_B is Boltzmann's constant, and T is the temperature of the heat reservoir in thermal contact with the spin system, i.e., body temperature. N_+ is the number of protons in the voxel that inhabit a *higher energy state.* The last step comes from a Taylor-series expansion of the exponential. Multiply both sides by N_-, subtract N_+, and approximate (very closely) N_- as ½ N, and the difference between populations at equilibrium drops out:

$$n_0 = N\mu_z B_0 / k_B T. \quad (4.4b)$$

n_0 is the n of Equation (4.2b) for the special case of thermal equilibrium.

EXERCISE 4.4 Demonstrate that at body temperature, the expansion of $e^{-2\mu_z B_0/k_B T}$ *is legitimate, i.e., that* $(2\mu_z B_o/k_B T)$ *is much smaller than 1.*

EXERCISE 4.5 $n_0/N \sim 10^{-6}$*. How about* $n(t)/N$*?*

This gives rise to a spontaneous *equilibrium magnetization* of magnitude

$$m_0 = n_0\mu_z = N\mu_z^2 B_0 / k_B T. \quad (4.4c)$$

This is the case for a spin system sitting quietly, entirely on its own, in a steady external magnetic field.

Three important points

First, at equilibrium at body temperature, there is a very small excess of proton spins pointing north. In a 1.5 T field, the magnitude of this equilibrium surplus is $n/N \sim 5 \times 10^{-6}$, or about 5 parts per million (ppm). That may not seem like much, until we recall that there are something like 40 billion *billion* (4×10^{19}) protons in a cubic millimeter of water. In a single 1 mm^3 voxel, there will be an *excess* of something like 200 trillion (2×10^{14}) of them in the lower-energy state. Again, this margin may sound slender, but it does give rise to the detectable net magnetization that is ultimately responsible for any sort of MRI signal coming from it. The magnitude of the voxel magnetization under conditions of thermal equilibrium is, quite simply, the ultimate determinant of the maximum possible strength of the NMR signal that gives rise to *any* kind of MR image.

EXERCISE 4.6 What is the proton excess in a 3 T magnet, in a $2 \times 2 \times 2$ *mm*^{3} *voxel of water?*

EXERCISE 4.7 Should image quality depend on B_0*?*

Secondly, the equilibrium magnetization for any particular voxel, $\boldsymbol{m}_0(x)$, is proportional to $N(x)$, the total number of protons in it. So it should be the case that

$$\boldsymbol{m}(x,t) \propto \mathrm{PD}(x), \quad (4.5a)$$

even away from equilibrium. Likewise for the strength of the signal, $s(x,t)$, coming from the voxel

$$s(x,t) \propto \mathrm{PD}(x), \quad (4.5b)$$

EXERCISE 4.8 $\boldsymbol{m}(x,t)$ *is a vector and a function of time in Figure 4.5a, but PD(x) is neither. Is this a problem?*

The third point is that at body temperature, according to Equations (4.4), m_0 is almost exactly linear in B_0. (The effects of any gradient fields are too small to worry about here.) As we shall see, there is a fundamental trade-off for MRI among voxel size (and resolution), data acquisition time, and SNR (and apparent contrast). Any of these can be made better, but only at the detriment of one or both of the other two. One obvious way to improve this balance is to increase the main magnetic field strength—as long as you have the money to do so.

4.4 Effect of Field Strength, $|B_0|$, on Image Quality

Machines that operate at 1.5 T are still the norm, largely because of their lower purchase cost, but sales of 3 T devices are taking place briskly. It is estimated that as of 2014, moreover, there were already several hundred whole-body clinical research devices in the world that operate at 7 T [Balchandani et al. 2014]. Switching to a system with double the field strength—from 1.5 T to 3 T, say—will increase the excess number of spins in the lower energy level by a factor of 2, hence $\boldsymbol{m}_0$ as well. You might suspect that the strength of the MR signal coming from a tissue voxel in the patient, $s(x,t)$, and picked up by the receive coils is proportional to B_0. For reasons pursued later, the signal actually rises more closely with the square of the field, as B_0^2. But the SNR, which is often as important as signal strength in imaging, increases almost linearly with B_0 instead, at least for the fields at which MRI is usually carried out.

Jumping from a 1.5 T to a 3 T system will essentially double the excess number of spins in the lower energy level (from Equations (4.4)) and $\boldsymbol{m}_0$, improving the signal-to-noise. Figure 4.5 demonstrates the improvement in resolution and general clarity that accompanies switching from a 1.5 T to a 7.0 T system (along with the addition of gadolinium contrast agent, white areas). Indeed, microvasculature that is not even suggested at 1.5 T can appear clearly in 7 T images. The change will also raise values of tissue T1 (but not so much for T2) and present more of a challenge with respect to certain artifacts.

If Maxwell's equations consider wave propagation within media, they indicate that when electromagnetic energy enters a body, the wavelength becomes shorter, electric currents are generated in the partially conducting tissues, and reflection and refraction may occur at the interfaces. The interaction of the wave's electric field $\boldsymbol{E}$ of Equations (2.1) with matter is known as the *dielectric effect*. For MRI at 1.5 T, the wavelength is of the order of 1.2 meters in air, and shorter in dielectric materials, where it is comparable to the dimensions of the regions being imaged. This may lead to the *dielectric* or *RF standing-wave MRI artifact* arising at 3 T and above. Wave interference may give rise to central darkening (from constructive interference) and lateral brighter bands (constructive), separated by a quarter wavelength, in the case of Figure 4.6 [Collins et al. 2005]. The problem is not as conspicuous at 1.5 T, where the RF wavelength is twice as great, unless the

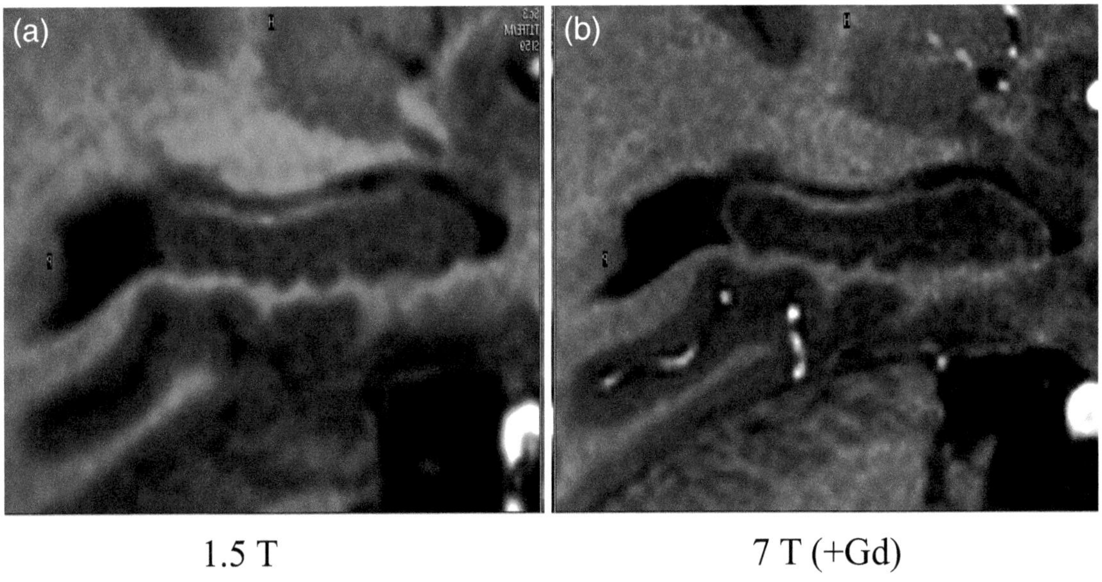

Figure 4.5 The effects of higher field strength. $\boldsymbol{m}_0$ is proportional to $\boldsymbol{B}_0$, and MR signal strength also increases with field strength. The difference in image quality obtained at (a) 1.5 T and (b) 7 T is striking, with better SNR and resolution in the latter (which also follows administration of intravenous contrast agent, responsible for the small hyper-intense regions). [Courtesy of Adam Anderson and John C. Gore, Vanderbilt University.]

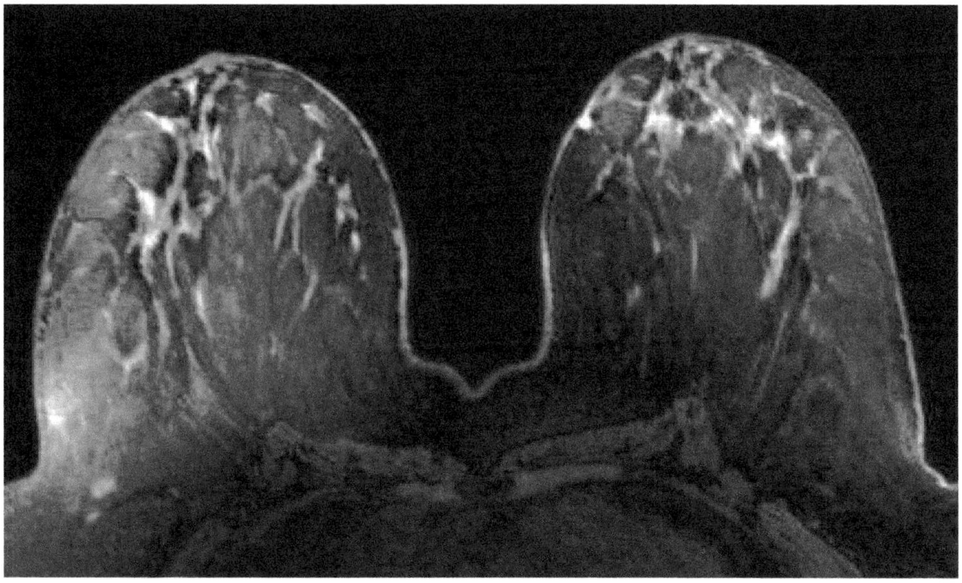

Figure 4.6 The *dielectric* or *RF standing wave artifact* is a problem encountered occasionally at 3 T and higher fields. Here it is the shadowing on the medial sides of the breasts. [Reproduced from Zhuo J. and R. P. Gullapalli, "AAPM/RSNA physics tutorial for residents: MR artifacts, safety, and quality control." *Radiographics* 2006(26): 275–97, with permission of the Radiological Society of North America (RSNA)].

dielectric content of the patient dramatically changes, such as at fluid accumulation from ascites in the peritoneal cavity. The shorter wavelengths at 7 T induce an even larger variation of brightness and darkness over a smaller distance.

There are other biological interactions and effects that can occur above 3 T, including photophosphenes (flashes in the eye that the patient may experience), altered cardiac electrophysiology (pulse sequences that gate with the heart may require new methodologies), and mechanical hazards from stronger static field gradients near the ends of the magnet bore. The development work is ongoing, however, and suggests that these and other obstacles may well be surmountable.

A research area where the experimental complications are much less severe, and the biophysics possibilities unlimited, is in small animal studies. These are currently ongoing at fields typically from 7 T to near 20 T, with accompanying resolution of down to 10 microns. There are numerous reports of the imaging of amyloid plaques in transgenic mice that display Alzheimer's disease, for example, and of the movement of individual cells in real time. This is stimulating the development of new contrast agents specific to particular cell-types and individual proteins involved in gene expression, and the behavior of intra-cellular structures. And that's just the beginning.

4.5 Spin Population Dynamics

Figure 4.4d presents the simplest of interesting spin systems, a population of protons in a constant and uniform field of several tesla. This set of spins is, in theory, in permanent contact with a thermal reservoir that itself remains at constant temperature. The bottom-line message of the figure is that when nothing else is going on, the proton system will also be at thermal equilibrium, both within itself and with the external reservoir.

Various phenomena, however, can work to disrupt that equilibrium. An important one is *stimulated radiative transitions* brought about by exposure to Larmor frequency electromagnetic waves, that can drive the system away from thermal equilibrium. A second is *non-radiative transitions*, brought about by other effects such as interactions with the sea of phonons (quanta of vibrational energy) that can move a system back toward thermal equilibrium; this involves the direct transfer of energy between the spin system and the reservoir, in particular through T1 spin-relaxation. (It will soon become apparent that T2 processes, on the other hand, involve *no* transfer of energy!) There exists a third mechanism, *spontaneous transitions*, that is important for the electrons of excited atoms; but the rate at which it occurs is proportional to the cube of the frequency, and it is virtually non-existent for nuclei in NMR.

We consider separately stimulated radiative transitions brought about by Larmor-frequency photons and non-radiative transitions from other effects. Either way, the rate at which the population of the lower-energy spin level changes depends, of course, on the numbers of protons entering and leaving it. The former is proportional both to the number of protons currently in the upper level, N_+, and also to the *probability* per unit time that a proton there will drop down into the lower level, which will be written W_{+}. At the same time, N_- decreases at the rate $N_- W_{-+}$, as lower-state protons are excited upward. While in some situations W_{+-} and W_{-+} may be equal, in general they depend separately on the physical process ongoing, which may or may not involve RF photons. They will depend on the rate at which ν_{Larmor} RF or other power is being applied; on the amounts of other forms of energy that are being pumped into or dissipated from the system; and on other factors. On balance,

$$dN_- / dt \;=\; N_+ W_{+-} - N_- W_{-+} \qquad (4.6a)$$

where, again, $N_- \sim \frac{1}{2}(N + n)$. A comparable equation applies to the upper level. Subtracting one from the other gives

$$dn(t) / dt \;=\; 2[N_- W_{-+} - N_+ W_{+-}]. \qquad (4.6b)$$

This can be rearranged to give the useful general bookkeeping expression

$$dn(t) / dt \;=\; N(W_{-+} - W_{+-}) - n(t)(W_{-+} + W_{+-}). \qquad (4.6c)$$

What is happening physically—i.e., pumping RF power into the spin system, loss of energy through transfer either way between the system and the reservoir, whatever—is detailed in W_{-+} and W_{+-}. And the above general dynamic equations apply to either stimulated radiative transitions or to non-radiative transitions.

EXERCISE 4.9 Explain the meanings of the above three equations (4.6) in words.

EXERCISE 4.10 If no RF power is being applied from a transmitter, what might contribute to W_{-+}?

Equations (4.6) have several applications, and here we shall mention two of them. Let's turn first to stimulated radiative transitions.

4.6 Stimulated *Radiative* Transitions and Spin System Saturation

The present paragraph will make sense only if you have spent some time playing with quantum mechanics; if you haven't, just jump ahead to the next one. QM tells us that a time-dependent perturbation, such as the effect of the magnetic field of an RF photon on a proton dipole magnetic moment, can be expressed as a potential energy contribution, *V(t)*, to the system's Hamiltonian. Then there is a calculable probability that *V(t)* will cause transitions between the proton states $|\uparrow\rangle$ and $|\downarrow\rangle$, which are more commonly written $|+\rangle$ and $|-\rangle$. That transition rate is proportional to the matrix element $|\langle +|V|-\rangle|^2$. Of concern here is the special situation of *radiative* transitions between the proton spin states that are being stimulated by an electromagnetic field in which $hf \;=\; (E_+ - E_-)$. In that case it happens that $|\langle +|V|-\rangle|^2 \;=\; |\langle -|V|+\rangle|^2$, which states that the probability rates of upward and downward transitions are equal. It may not be intuitively obvious why, but the probability of a photon with energy $h\nu_{Larmor}$ inducing a proton to transition from the higher spin state to the lower, $W_{+\to-}$, is exactly the same as the converse. We can, therefore, call them both just W_{RF}:

$$\textit{RF-stimulated:} \qquad W_{RF} \;\equiv\; W_{-+} \;=\; W_{+-}. \qquad (4.7a)$$

It follows from Equation (4.6a) that the rates of change of the two spin-populations, during a time that they are being stimulated by enough externally applied Larmor-frequency RF power, are

$$\begin{aligned} dN_+ / dt \;&=\; [W_{RF} N - (t)] - [W_{RF} N_+(t)] \\ &=\; -n(t) W_{RF}, \end{aligned}$$

and similarly for dN_- / dt. Subtracting these two gives

$$dn / dt \;=\; -2n(t) W_{RF} \qquad (4.7b)$$

which could have come, of course, directly from Equation (4.6c). The solution to this is

$$n(t) \;=\; n(0) e^{-2W_{RF} t}, \qquad (4.7c)$$

regardless of $n(0)$. It indicates that although more spins may have originally been in the lower level, the two populations approach one another in numbers over time as RF energy is being continuously added to and absorbed by the spin system. Eventually $N_-(t)$ would come to equal $N_+(t)$, a situation of *spin saturation*, after which further absorption of power ceases. Saturation clearly leaves the system far from the condition of thermal equilibrium.

EXERCISE 4.11 Can this happen unless as much power is being lost in spin relaxation or any other manner as is being supplied to the system?

In reality, of course, you cannot just continue to pump power into a system. At some point it must start getting rid of that energy, eventually at a rate equal to that of its input. And that's where spin relaxation comes in.

4.7 Stimulated *Non-Radiative* Spin Transitions and T1 Relaxation

A spin system that is in contact with a thermal bath and left to its own devices will tend to move toward, or remain in, dynamic thermal equilibrium with the reservoir. Any tiny thermal fluctuations of the system away from that condition will be compensated for through a *non-radiative* (*not* involving EM radiation) flow of minute amounts of heat energy to or from the reservoir.

This result entices us to consider the situation for *non*-radiative stimulation by oscillating magnetic fields at a water proton caused by, say, the erratic and random RF-frequency fields from its partner proton, as resulting from the molecule's rotations. Some of these may happen to occur at ν_{Larmor}, and they could bring about *spin-lattice relaxation* at a rate at or near 1/T1.

Reconsider Equation (4.6) for a proton population in contact with its thermal reservoir but without any RF energy transfers. At thermal equilibrium, $dN_-/dt = 0$, so $N_+W_{+-} = N_-W_{-+}$ and

$$W_{+-} / W_{-+} = N_- / N_+ \quad \textit{at thermal equilibrium.} \tag{4.8a}$$

By Boltzmann theory and Equations (4.4), $N_- \neq N_+$, so it must also be that

$$W_{+-} \neq W_{-+}. \tag{4.8b}$$

Then Equation (4.6c) can be re-written as

$$dn/dt = [n_0 - n(t)]/T1, \tag{4.9a}$$

where n_0, the population difference at equilibrium, is

$$n_0 = N \times [(W_{-+} - W_{+-})/(W_{-+} + W_{+-})].$$

With a bit of foresight, we choose to define

$$\boxed{1/T1 \equiv (W_{-+} + W_{+-}),} \tag{4.9b}$$

and the T1 rate is just the sum of the up and down rates.

When Equation (4.7b) is modified to include both RF input and relaxation, it becomes

$$dn/dt = -2W_{RF}n(t) - [n(t) - n_0]/T1, \tag{4.10a}$$

with solution

$$n(t) = n_0 / (1 + 2W_{RF} \times T1), \tag{4.10b}$$

$$dE/dt = \Delta E\, n_0 / (1 + 2W_{RF} \times T1). \tag{4.10c}$$

EXERCISE 4.12 Is this definition of 1/T1 compatible with earlier ones?

EXERCISE 4.13 Verify Equations (4.10a–c). Explore and discuss these important results.

EXERCISE 4.14 Demonstrate Equation (4.10c) and discuss. What does a low-power input with small W_{RF} imply?

This all gets rather involved when one tries to evaluate W_{-+} and W_{+-} (and T1) from first principles. Bloembergen et al. [1947; 1961] developed a comprehensive theory of relaxation, which includes issues like the quantum mechanical treatment of systems affected by both periodic and temporally random perturbations $V(t)$, autocorrelation functions such as the statistical average $[\overline{V(t+\tau)V(t)}]$ and the *power spectrum* derived from it, its exponential decay at a characteristic *correlation time*, and other matters. We shall leave these to more advanced tomes, but can assure the reader that the only aspect essential for clinical work is the final product, T1 itself. We shall return to this topic in chapter 9.

C H A P T E R 5

Mathematical Machinations

This chapter provides a concise review of a few mathematical tools that many students of medical physics or engineering bring with them to graduate school, but others do not. College physics major programs differ considerably. Take what you need here, and pass by the rest.

We begin by going over some elementary properties of vectors and complex numbers. In reconstructing an MR signal into an image, it is convenient to manipulate the data in an abstract but simple space known as ***k**-space.* We shall populate this construct, also known as *reciprocal-space*, at first in 1D with $\boldsymbol{k}$-vectors lying along the k_x-axis, each associated with a spatial frequency. In chapter 7, the same $\boldsymbol{k}_x$-vector will assume a second, surprisingly different but equally important meaning, as a surrogate for time, t.

Much of physics deals with temporal and spatial waves in vector spaces, whether they be acoustic, electromagnetic, optical, quantum mechanical, or others. One of the most powerful implements for dealing with them is Fourier analysis, which allows us to decompose a labyrinthine periodic waveform down into a unique set of orthogonal components.

The MR data acquisition process generally involves sampling an analog (continuous) signal voltage at multiple discrete times and translating the measured values into digital form with an analog-to-digital converter (ADC).

Meanwhile, it is critical to maintain as high a signal-to-noise ratio (SNR) as possible, and noise reduction commonly employs statistical methods. In addition, when the computational conditions are not fully copacetic, cyber demons can give rise to MR artifacts, as in Figure 3.19; here are illustrated several more, arising from under-sampling of the MRI signal and from patient motion.

5.1 Two Simple but Essential Functions

Exponential functions assume critical roles in describing numerous physical, biological, and other processes in medical imaging, such as

- radionuclide decay;
- the dampening of a jolted element in an ultrasound transducer or the discharge of an electronic capacitor through a resistor (RC circuit);
- the decay of a pulse of fluorescent light after stimulation;
- the dilution of a tracer in a chemical or biological compartment;
- the attenuation of high- and lower-energy photons passing through absorbing or scattering materials;
- the growth of a well-fed population of bacteria;

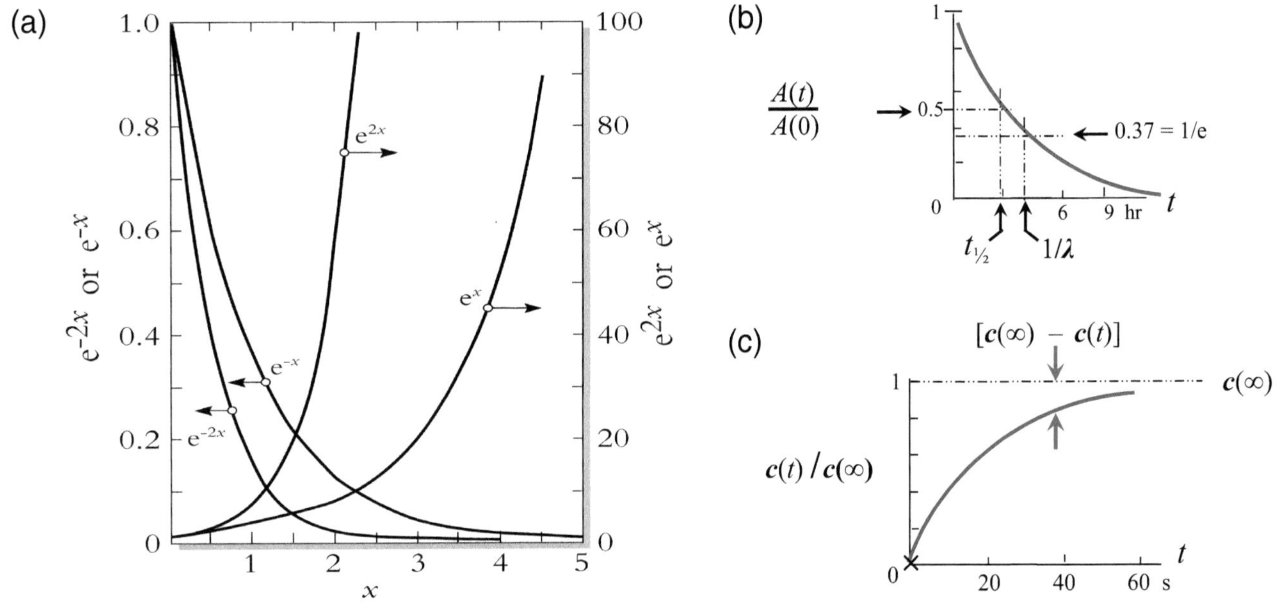

Figure 5.1 Exponential functions of x or t. (a) Examples with either a positive or negative exponent. (b) Decay in the activity of a radionuclide sample. After the characteristic time $1/\lambda$, activity has declined to $e^{-\lambda(1/\lambda)} = 37\%$ of its initial value. (c) Growth in the concentration of sugar, $c(t)$, in hot tea as a spoonful of it dissolves. The relaxation toward thermal equilibrium of a cohort of proton spins in a magnetic field, following a disturbance, has the same general shape!

- the probability of cell survival following various levels of alpha-particle irradiation; and
- the relaxation of a cohort of proton spins in a magnetic field.

While everyone has had occasion to manipulate $e^{\alpha t}$ and $e^{-\beta t}$ (Figure 5.1a), it is sometimes important to also keep in mind how and when such exponentials may arise. That will prove to be the case as we delve more deeply into NMR and MRI theory, and much of what is said here will be applied then.

Exponential decay and growth

The decay of a radionuclide comes as close as just about anything to being perfectly exponential. (A lone proton, by the way, is considered to be perfectly stable by everyone except string-theorists.) A sample initially containing $n(0)$ unstable nuclei has, the time t later, only $n(t)$ of them remaining, the rest having metamorphosed into stable or unstable progeny. As you would expect, the rate at which they transform, $dn(t)/dt$, is proportional to the number that remain unaltered, still there and available to transmute:

$$A(t) \equiv dn(t)/dt = -\lambda n(t), \qquad (5.1a)$$

where $A(t)$ is the sample's *activity* and λ is its *decay* or *transformation rate*. This first order ordinary linear differential equation has the well-known exactly *exponential* solution

$$n(t)/n(0) = e^{-\lambda t}, \qquad (5.1b)$$

where $n(t)/n(0)$ is known as the *survival fraction* at time t, among other names (Figure 5.1b).

There is a function that is *inverse* to the exponential of Equation (5.1b), the *natural logarithm* to *base* e, $\log_e$ (or ln(), in shorthand), with the defining and extremely useful property that

$$\ln(x) = y \quad \text{if and only if} \quad e^y = x. \qquad (5.1c)$$

From this, $\ln(e^{\alpha}) = \alpha$. 'Semi-log' graph paper was designed in the days before computers, and on it exponential data plot out as a straight line. A number of functions look almost exponential on linear paper, but straight is unequivocally straight on semi-log! The logarithm to bases other than 'e,' such as the *common logarithm* to base 10 ($\ln_{10}$), have found applications as in the definitions of Optical Density (OD) and the acoustic decibel (dB), but are seen much less frequently in physics.

To end up precisely with an exponential, Equation (5.1b), the decay rate

$$\lambda \equiv \left|-dn(t)/dt \,/\, n(t)\right|, \qquad (5.1d)$$

must be absolutely *constant*, the same tomorrow as it is today. ($|-q| = q$ means *absolute value* of whatever is between the pair of vertical bars. And, of course, λ does *not* refer to wavelength here.) This implies that the *probability* of a transformation occurring per second is *time independent*; it is not at all influenced by any happenings in its past, and also totally unaffected by the experiences of other nuclei in the sample.

EXERCISE 5.1 What is the relationship between λ and a sample's half-life, $t_{1/2}$, where $n(t_{1/2})/n(0) = 1/2$? What is $A(t)/A(0)$ at the characteristic time $t = 1/\lambda$? See Figure 5.1b.

EXERCISE 5.2 How does one construct semi-log graph paper (on paper or a monitor screen)?

EXERCISE 5.3 Re-interpret Equations (5.1) probabilistically.

The decay constant λ is radionuclide-specific, and its value depends only on the specific isotopic species to which it happens to belong, nothing else. Its magnitude—indeed, the very fact that it is absolutely, unequivocally a constant—is determined entirely by the details of the physics of the strong nuclear, weak nuclear, and electromagnetic forces affecting the protons and neutrons within its nucleus. The nucleus is in a state of incessant agitation, with its protons and neutrons roiling about in response to totally *random* fluctuations in these interactions. And their motions, in turn, cause further fluctuations. But no aspect of this physical situation changes systematically over time, including the decay rate:

$$\lambda \neq \lambda(t). \qquad (5.1e)$$

Here λ is not at all influenced by how long a radionucleus has already been with us, nor even by how many times it has *almost* had its moment of glory. In other

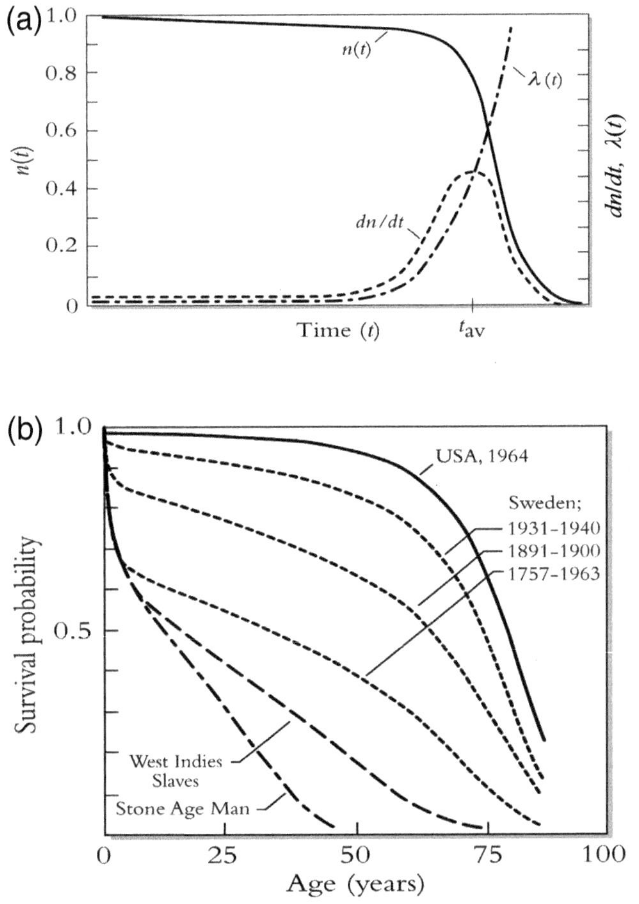

Figure 5.2 Survival as a function of age for several populations.

words, individual radionuclei have no separate, personal histories. From the moment it springs into being until it meets its fate; a nucleus keeps no individual record of what has gone on since its own creation. Each one passes through its existence in isolation—every radionucleus is an island—with no awareness of the others in the sample or of their decays. It is because of the deep-seated randomness with which decays occur, and because of the complete lack of interaction or communication among the individual nuclei, that λ is, to the extent determinable, constant.

Not everything in nature will, or even should, have fixed half-lives. Imagine, for example, a bunch of identical batteries in identical clocks. For a good while they all work fine, and then they fail at more or less the same time (Figure 5.2a), quite unlike items with a constant half-life. Likewise, patterns of mortality in groups of humans in different eras do not display half-lives (Figure 5.2b).

The take-home message is that an exponential function of time, distance, etc., will pop up whenever the *rate of change of something* is proportional to *however much of the something is left*, Equation (5.1a).

And, oh yes! The longitudinal relaxation along an external magnetic field of a disturbed voxel of spins occurs nearly exponentially with the roughly constant time parameter T1, as in the $[1 - e^{-t/T1(x)}]$ term in Equation (1.5a) and Figure 5.1c. Likewise, the dephasing as $e^{-t/T2}$ of a packet of protons precessing in the ***x-y*** plane does, as well. These actions are actually a good deal more subtle than the decaying ring of a bell, but they are similar enough to it to display exponential behavior.

EXERCISE 5.4 A spoonful of granulated sugar is dumped into a cup of hot tea (Figure 5.1c). Show that the concentration increases as $(1 - e^{-t/T})$. What is the meaning of the time constant T?

EXERCISE 5.5 The population of microscopic organisms in a nutrient-filled petri dish starts off growing as $n(t)/n(0) = e^{+qt}$ with a nearly constant parameter q. Why? And why, on the other hand, might that growth pattern change later?

EXERCISE 5.6 Is it a priori obvious that λ for a group of identical radionuclei is not a function of time? Or does that have to be found empirically?

Sinusoidal oscillation

For a weight of mass, *m*, oscillating back and forth on a horizontal, frictionless table, the spring opposes any small displacement, *x*, from its equilibrium position, $x = 0$, with the Hooke's Law force (Figure 5.3a). As with any other system that is elastically deformed not too much, the restorative force is linear in the strain. The equation of motion $F = ma$ assumes the specific form

$$m\,d^2x/dt^2 = -Kx(t), \tag{5.2a}$$

where the constant K is characteristic of the spring's stiffness. A solution to this elementary linear second-order differential equation, and of other Hookean systems, is well known to be, for sufficiently small oscillations,

$$x(t) = A\sin[2\pi\nu t + \varphi],$$

or, equivalently, (5.2b)

$$x(t) = A\sin[\omega t + \varphi].$$

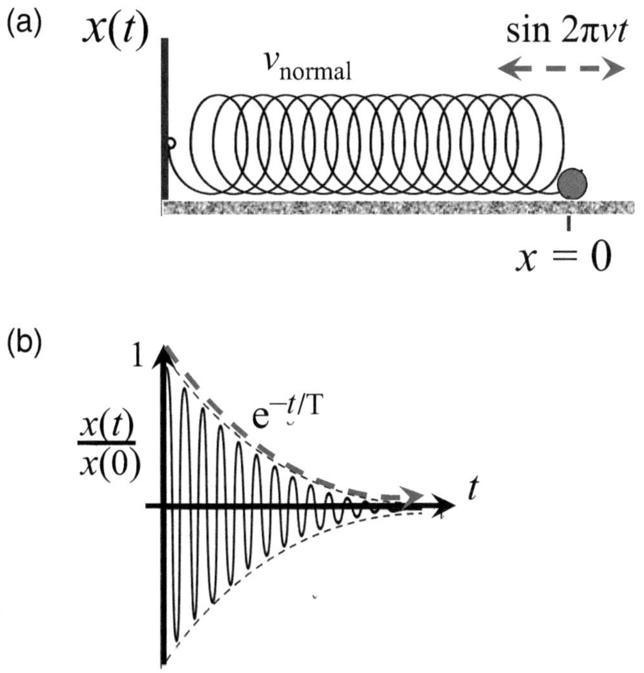

Figure 5.3 Motion of a mass on a frictionless table at the end of an elastic spring. (a) The normal mode of motion for the system is sinusoidal and about the equilibrium position, $x = 0$. (b) With friction present, the amplitude of the oscillation declines exponentially with characteristic time, T, that depends on the rate of dissipation of energy.

The frequency is ν, where $\omega = 2\pi\nu$ is the *angular frequency* and φ is the *phase*. The amplitude and phase of the wave-form for a particular situation are set by the initial (boundary) conditions.

Once the system is underway, the mass will oscillate back and forth about its equilibrium point over time, its *normal mode* of motion, until the movement is eventually damped out by non-conservative (frictional) forces. Its *normal mode* or *resonant frequency* is

$$\nu_{res} = (K/m)^{½}/2\pi. \quad (5.2c)$$

More generally, the *period* for temporal oscillations, the duration of one cycle, is

$$\boxed{\mathrm{T} \equiv 1/\nu,} \quad (5.3)$$

in *seconds per cycle*, where the period is the time for the phase to move forward by one cycle, i.e., a change by 2π. If the system is not frictionless, then the amplitude of the oscillations will decay over time in a manner that will be at least roughly exponential in time (Figure 5.3b).

Similarly, a function that happens to be periodic over *space* may be expressed as

$$y(x) = A\sin[2\pi k\, x + \varphi]. \quad (5.4a)$$

Here the time is fixed, and the wave is viewed as a function of position along the, say, $\boldsymbol{x}$-axis. The *wavenumber k*, named after the German physicist Heinrich Kayser, can be defined as the rate of change of *phase* with distance, *in radians per meter*. Equivalently, $2\pi k$ is the *rate of change of phase in cycles per meter*. The wavelength, λ, is just the peak-to-peak separation of two adjacent cycles—or, again, the distance over which the phase of the wave advances by 2π. (It muddles matters that some authors incorporate a factor of 2π into the definition $k \equiv 2\pi/\lambda$, where λ is the *wavelength*, a spatial analog to the temporal period.)

A wave propagating along the x-axis and periodic in both time and space, may be written

$$\boxed{y(x,t) = A\sin[2\pi k\, x + 2\pi\nu\, t + \varphi].} \quad (5.4b)$$

*EXERCISE 5.7 (Challenging) A string of length **l**, linear density ρ, and tension T is set to vibrating. What is its fundamental frequency? Compare with Equation (5.2c).*

EXERCISE 5.8 A champagne bubble undergoes small radial oscillations. On what factors might its resonant frequency depend? [Spratt et al. 2018]

While it is true that $\lambda \times \nu = c$ for EM radiation in free space, for other wave situations there is *not* necessarily any *a priori* special relationship between k and ν of this sort. As you would expect, Feynman provides an interesting discussion of this slightly confusing business [vol I, 29-2, 1963].

We close out this section with two identities that will prove useful on a number of occasions: the first of these, known as a sum-to-product identity, concerns the superpositioning of two sine functions, with angles α and β:

$$\boxed{\sin\alpha + \sin\beta = 2[\sin ½(\alpha+\beta)] \times [\cos ½(\alpha-\beta)].} \quad (5.5a)$$

Similar constructs apply for ($\sin\alpha - \sin\beta$) and ($\cos\alpha \pm \cos\beta$). A simple example of this appears in Figure 5.4, where two sine waves of the same amplitude but

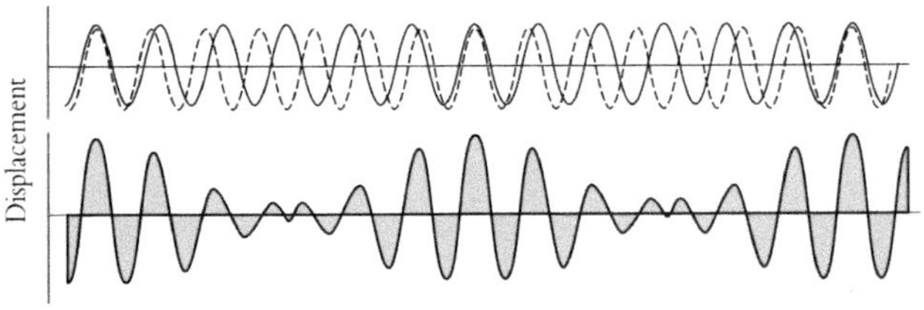

Figure 5.4 Interference between two waves that differ slightly in frequency.

different frequencies are combined, resulting in a 'beat' pattern.

Equation (5.5a) can also explain the dissimilarity between the interference patterns of [sin x + ⅓ sin 3x] and [sin x + ⅓ sin (3x + $\pi/2$)]. The two higher-frequency terms differ by a phase factor of 90°, and both have an amplitude that is one third of sin x (Figure 5.5). The two summations are clearly distinguishable, both visually and by way of Fourier analysis. The take-home message is that when waves combine, *phase matters*—the phases of two sine waves are just as important in their summation as their frequencies and amplitudes. This will become fully evident when we turn to 2D MR imaging in chapter 12. The same sort of thing appears in 2D in Figure 5.6, where a pair of wires tap a water surface at the same frequency and in phase.

Other trigonometric identities that will come in handy are

$$(\sin\alpha)\times(\sin\beta) \;=\; \tfrac{1}{2}[\cos(\alpha-\beta)]-[\cos(\alpha+\beta)] \tag{5.5b}$$

and related product-to-sum formulae.

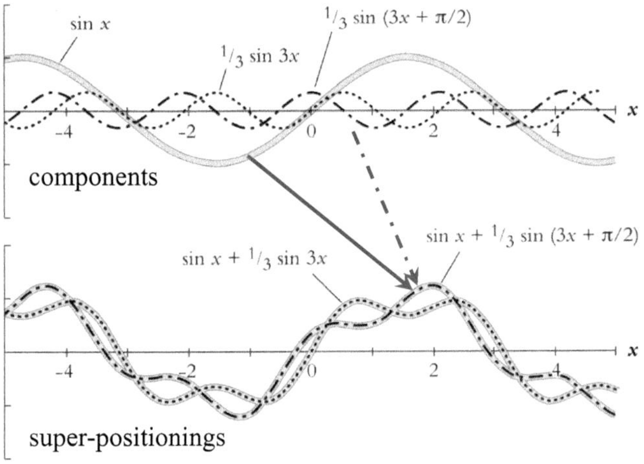

Figure 5.5 Interference of sin x with either [⅓ sin 3x] or [⅓ sin (3x + $\pi/2$)], which differ in phase by $\pi/2$. *Phase matters!*

5.2 Scalar (Dot) Product of Vectors and Functions

The three-dimensional *real space* in which a patient's anatomy, and clinical MR images of it, reside is a *vector space* that can be spanned by a triad of orthonormal Cartesian *basis* or *unit* vectors $\boldsymbol{x}_1$, $\boldsymbol{x}_2$, and $\boldsymbol{x}_3$ (or, equivalently, **i**, **j**, and **k**) aligned along x-, y-, and z-axes, respectively. The basis set

$$\{x_q,\ q = 1,2,3\} \tag{5.6a}$$

is *orthonormal* with regard to the standard real-space *scalar* or *dot product*, '·', in that

$$\textit{normalized} \qquad \boldsymbol{x}_q \cdot \boldsymbol{x}_q = 1$$

$$\textit{orthogonal} \qquad \boldsymbol{x}_q \cdot \boldsymbol{x}_{q'} = 0\,.$$

The *orthonormality* of this basis set can be summarized succinctly with the aid of the *Kronicker delta* function, δ_{ij}, as

$$\boldsymbol{x}_q \cdot \boldsymbol{x}_{q'} \;=\; \delta_{qq'} \tag{5.6b}$$

where

$$\delta \mathbf{q}\mathbf{q}' \;\equiv\; \begin{cases} 1, & q' = q \\ 0, & q' \neq q \end{cases} \tag{5.6c}$$

Spectral analysis of real vectors

Real 3D space has the marvelous attribute that any position, velocity, magnetic field, or other real vector $\boldsymbol{w}$ in it can be *decomposed* as the linear sum:

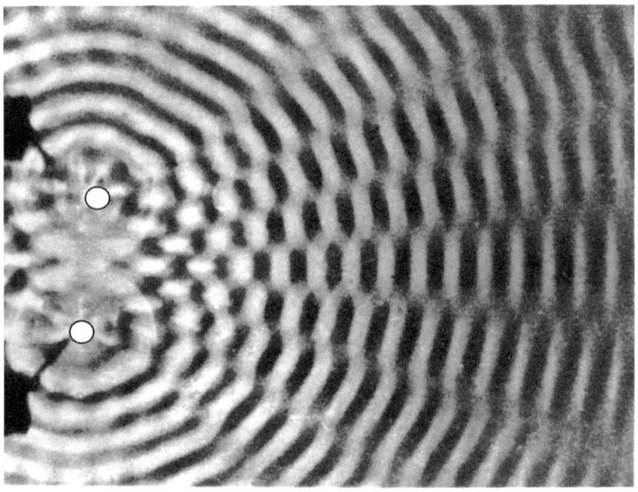

Figure 5.6 Simplest interference pattern in 2D.

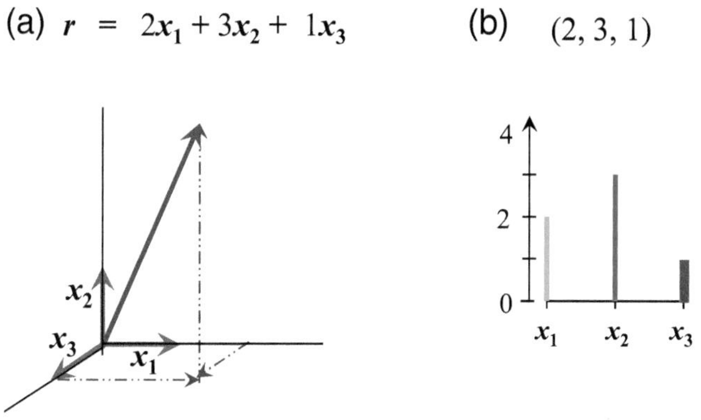

Figure 5.7 A vector and its spectral decomposition. (a) Breaking the vector $\boldsymbol{r} = 2\boldsymbol{x}_1 + 3\boldsymbol{x}_2 + 1\boldsymbol{x}_3$ down into its constituent components along the **x**-, **y**-, and **z**-axes. (b) Spectral decomposition of the vector, relative to the $\{\boldsymbol{x}_1, \boldsymbol{x}_2, \boldsymbol{x}_3\}$ basis system.

$$\boldsymbol{w} = w_1\boldsymbol{x}_1 + w_2\boldsymbol{x}_2 + w_3\boldsymbol{x}_3 = \sum_{q} w_q \boldsymbol{x}_q, \quad (5.7a)$$

relative to the particular basis set of Equation (5.6a). By way of

$$w_q = \boldsymbol{w}\cdot\boldsymbol{x}_q, \text{etc.,} \quad (5.7b)$$

the vector $\boldsymbol{w}$ can be expressed also in terms of its components alone,

$$(w_1, w_2, w_3). \quad (5.7c)$$

The particular vector $\boldsymbol{r} = 2\boldsymbol{x}_1 + 3\boldsymbol{x}_2 + 1\boldsymbol{x}_3$ appears in Figure 5.7, along with its *spectral decomposition*, or set of components (2, 3, 1) taken relative to this basis system. While trivial here, the notion of spectral decomposition turns out to be crucially important when we begin reconstructing MR images out of incoming RF signals.

EXERCISE 5.9 What does the spectral composition of **w** *become if the basis system of Figure 5.7 is replaced with one that is rotated 90° about the z-axis to the left? To the right?*

The scalar product of two real vectors $\boldsymbol{v}$ and $\boldsymbol{w}$ assumes the form

$$\boldsymbol{v}\cdot\boldsymbol{w} \equiv v_1w_1 + v_2w_2 + v_3w_3. \quad (5.8)$$

Into the complex plane

MR spectral decomposition is carried out by way of Fourier analysis, which commonly makes use of complex numbers and functions. For the *complex number z*,

$$z = x + \mathrm{i}y, \quad (5.9a)$$

where "i" is defined through $\mathrm{i}^2 = -1$ or, $\mathrm{i} \equiv \sqrt{-1}$. The real numbers x and y lie along the two Cartesian coordinates of z in the complex plane, and are known as the *real* and *imaginary parts* of z, respectively: $\mathrm{Re}(z) = x$ and $\mathrm{Im}(z) = y$ (Figure 5.8). Nothing here is really 'imaginary,' of course, and it would be just as valid and perhaps less confusing to name x and iy the 'Pink' and 'Floyd' parts of z. The complex plane is essentially a 2D vector space, but it has additional properties regarding functions that are *holomorphic* or *analytic*, which is to say *differentiable*, within it. But that is not of interest here.

Complex numbers or vectors can be expressed in polar coordinates as $x = |z| \cos\theta$ and $y = |z| \sin\theta$, where $z = |z|(\cos\theta + \mathrm{i}\sin\theta)$ (Figure 5.8). Here $|z| = (x^2 + y^2)^{1/2}$ is known as the *modulus* of z; and $\theta = \tan^{-1}(y/x)$ is its angle or *phase*. It is left as an exercise to confirm the renowned *Euler's identity*,

$$z = |z|(\cos\theta + \mathrm{i}\sin\theta) = |z|e^{\mathrm{i}\theta}. \quad (5.9b)$$

The *complex conjugate* of $z = x + \mathrm{i}y$, written $\bar{z}$ (and sometimes z^*), is defined as

$$\bar{z} \equiv x - iy, \quad (5.9c)$$

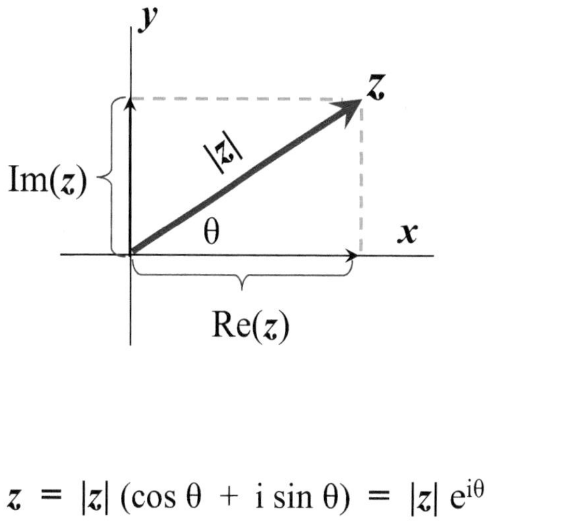

Figure 5.8 Definitions of $|z|$, θ, $\mathrm{Re}(z)$, and $\mathrm{Im}(z)$, and $e^{\mathrm{i}\theta}$, for the complex number z, and the Euler identity.

EXERCISE 5.10 Demonstrate the validity of the useful Euler's formula with Taylor-series expansions of the relevant functions.

The scalar product of two *complex* numbers $\boldsymbol{v}$ and $\boldsymbol{w}$ is a generalization of Equation (5.8):

$$\boldsymbol{v}\cdot\boldsymbol{w} = \sum_{q} v_q \bar{w}_q \,. \tag{5.9d}$$

One can extend the above argument regarding real-valued vector spaces to the decomposition of *complex-valued functions* in more abstract *function spaces*, and many of the basic ideas carry over. Given some basis set $\{f_n(x), n = 0, 1, 2,\ldots\}$ defined over a suitable interval, a general inner or scalar product between any pair of them is commonly taken to be

$$f_m(x)\cdot f_n(x) \equiv \int f_m(x)\overline{f_n(x)}\, u(x)\, dx. \tag{5.10a}$$

The form of the real and positive *density* or *weight function*, $u(x)$, is dependent on the set of basis vectors selected; for the vector spaces considered herein, $u(x) = 1$. Basis functions in such function spaces are said to be *orthonormal* if they are orthogonal and of unit length:

$$f_m(x)\cdot f_n(x) = \int f_m(x) f_n(x)^* dx = \delta_{mn}. \tag{5.10b}$$

The Taylor series expansion, $f(x) = \sum a_n x^n$, provides a familiar example of an infinite, countable, ortho-normalizable basis set, $\{x_n\}$.

Orthonormal sinusoidal basis vectors

Nearly everything oscillates, and among the most widely encountered 1D basis sets of physics is

$$[1, \cos x, \sin x, \cos 2x, \sin 2x, \cos 3x, \ldots], \tag{5.11a}$$

defined over an interval like $[-\pi, \pi]$. A space spanned by such trigonometric basis vectors is known as a *Fourier space*.

The sine and cosine functions provide a solution space for the familiar vibrational *wave equation*, $\partial^2 Q/\partial t^2 = c^2 \partial^2 Q/\partial x^2$, that follows from combining Newton's equation of motion with Hooke's Force Law, Equation (5.2a). The speed of wave propagation is denoted c. This wave equation is a simple version of the general *Sturm-Liouville* partial differential equation that appears in various guises throughout physics. Other forms of Sturm-Liouville have as solutions the *Bessel*, *Laguerre*, *Legendre,* and *Hermite* polynomials. When properly prepared by the Gram-Schmidt ortho-normalization process, the set of solutions to any of these can form an orthonormal basis set for spanning the associated function space.

It is straightforward to show, with the aid of Equation (5.5b), that the basis set of Equation (5.11a) is orthogonal and can be normalized as

$$\int_{-\pi}^{\pi} \left[\left(1/\sqrt{\pi}\right)\sin mx\right]\left[\left(1/\sqrt{\pi}\right)\sin nx\right] dx = \delta_{mn}. \tag{5.11b}$$

EXERCISE 5.11 Prove Equation (5.11b).

We will later be interested in the case in which the generic complex function $z(t)$ rotates in the complex plane over time, t, with angular velocity $2\pi\nu$; equivalently, $\theta = 2\pi\nu t$. Then

$$z(t) = |z|(\cos 2\pi\, \nu t - i \sin 2\pi\, \nu t) = |z| e^{-2\pi i \nu t} \tag{5.12}$$

describes the vector $z(t) = \{x(t), y(t)\}$ rotating in the complex x-y plane around its origin like, as we shall see in the next chapter, the transverse magnetization vector, $\boldsymbol{m}_{xy}(t)$.

5.3 Fourier Series Expansion of a Function Periodic in Time or Space

In the early 1820s, Jean-Baptiste-Joseph Fourier demonstrated that just about any function of time or space, no matter how irregular in shape, can be untangled and decomposed into a combination of sine functions of the appropriate frequencies, phases, and amplitudes (or, equivalently, of sine and cosine functions of the correct frequencies and amplitudes). Fourier and those who followed have devised an elegant formalism for carrying this out mathematically, and now it is an essential tool in the reconstruction algorithms of MR, CT, and other images from raw data, among countless other applications. Although we shall not here cover the details of how to actually carry out its calculations, some of its principal results will appear here frequently. A discussion of how actually to carry out the Fourier Transform (FT) may be found in texts on advanced calculus or mathematical methods of physics [Titchmarsh 1962; Dennery et al. 1995; James 1995; Buck 2003]. The *Fast Fourier Transform* (FFT) is an offshoot of the FT that is widely employed in computer-based calculations [Duhamel et al. 1990; Rockmore et al. 2000; Pres et al. 2007].

The decomposition of a function or curve turns out to be especially easy, no matter how complicated it may seem, if it is *periodic*. That is, the entire function is exactly repetitive with temporal *fundamental frequency* ν_1 (cycles/second, Hz), or spatial frequency k_1 (cycles/meter). Imagine a function periodic in time with a repetition frequency ν_1, for example. It can be represented as a simple *Fourier sum* of sine or cosine terms or both, whose frequencies are *harmonics*, or integer multiples, of the *fundamental*, ν_1. The harmonics assume the values $\nu_p = p \times \nu_1$ for integer values of p: $p = 1, 2, 3, 4,\ldots$yield $\nu_1, 2\nu_1, 3\nu_1, 4\nu_1,\ldots$. For a spatial function, the harmonic frequencies are $k_p = p \times k_1$. For a *non-periodic* function, on the other hand, ν or k assumes a continuum of values, and the sum evolves into a *Fourier integral*, to be considered shortly.

For simplicity, let us consider here periodic functions of space that are *odd*, like sine functions, and for which

$$f_{odd}(-x) = -f_{odd}(x). \qquad (5.13a)$$

The corresponding *even*, cosine-like function obey

$$f_{even}(-x) = +f_{even}(x). \qquad (5.13b)$$

Imagine the odd square-wave, $s_{\text{sq-odd}}(x)$, of Figure 5.9a, which repeats itself along the x-axis in a 1D real-space with the spatial frequency of k_1 copies per millimeter. This simple example will involve only odd functions, and only sine terms will partake in the Fourier sums. (Why?) The square-wave might correspond to a pattern of equally wide, equally spaced dark and light stripes, the apparent brightness of which repeats with a wavelength of $1/k_1$.

The simplest Fourier approximation (granted, not a very good one) to the odd square-wave $s_1(x)$ of unity amplitude is

$$s_1(x) = (4/\pi)\sin(2\pi k_1 x), \qquad (5.14a)$$

shown as the waveform in Figure 5.9b. (The Fourier formalism provides a prescription for finding this contribution to the sum, and all others.) Because of the factor of $4/\pi$, the sum function sticks up a bit above where the flat top of the square-wave should be. We commonly view frequency in terms of oscillations per unit of time, but it is just as valid to think of a snapshot of a sine wave, here of wavelength $1/k_1$ mm extending through space along the $\boldsymbol{x}$-axis. By the way, the corresponding term in a temporal series would read

$$s_1(t) = (4/\pi)\sin(2\pi\nu_1 t),$$

where ν_1 is the fundamental frequency and $1/\nu_1$ is the period.

EXERCISE 5.12 Demonstrate that sin $(2\pi k_1 x)$ is a periodic function. What is its period? How about its frequency, in cycles/millimeter, or radians/m?

The *Fourier* or *frequency spectrum*, also called the *Fourier decomposition*, or *Fourier expansion*, for this series approximation of our square-wave appears on

Fourier analysis in your body

Your auditory system, incidentally, is very effective at carrying out one kind of Fourier analysis without any electronic help. You can pick out the individual notes in a piano chord, for example, even though they arrive as a single mixed sound that results from periodic compression and rarefaction of air caused by several strings. But the ear and brain together can decompose it into its fundamental frequencies plus harmonics (by virtue of the ear's cochlear design), corresponding to the multi-nodal combinations of vibrations of all the strings.

The eye and brain, by contrast, cannot do the same; they perceive a superpositioning of blue and yellow light as being green. But a glass prism breaks sunlight into its optical constituents, providing a spectral decomposition.

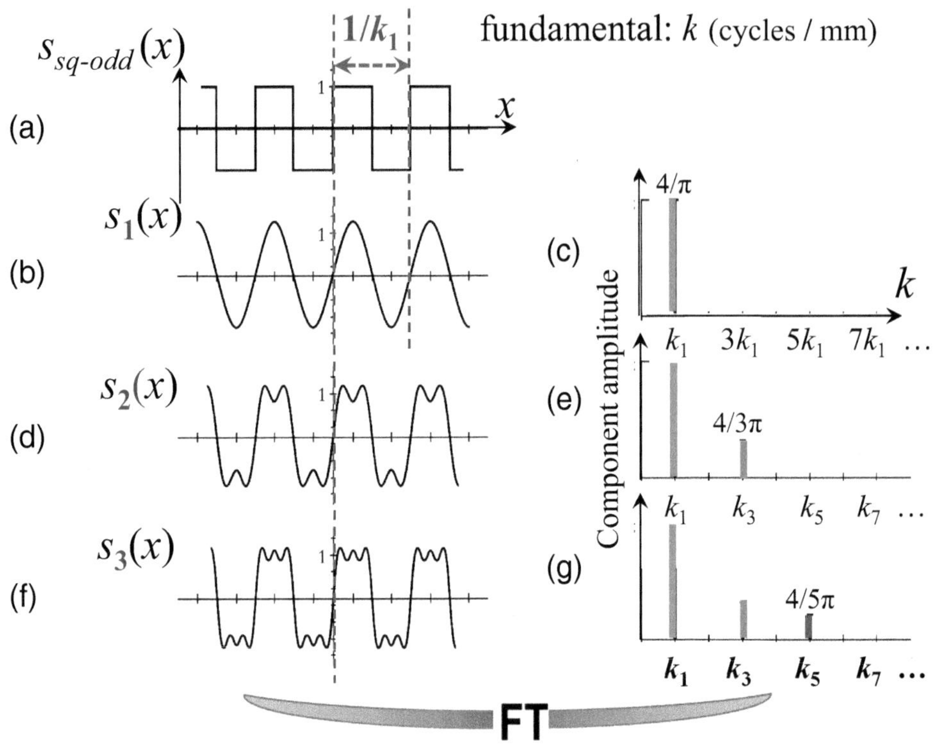

Figure 5.9 Complicated periodic patterns can result from the superpositioning of two or more monochromatic spatial waves. Any odd (for simplicity here) periodic pattern like the odd square wave, $s_{sq\text{-}odd}(x)$, can be built up through interference among sine waves at *integer multiples* of the *fundamental* frequency, known as its *harmonics*, with the right amplitudes. Conversely, Fourier decomposition can break a messy odd periodic waveform down into its constituent monochromatic sine components, which act like an orthogonal basis set. (a) A repeating spatial square waveform, $s_{sq\text{-}odd}(x)$, with the *fundamental* spatial frequency k_1. (b) A sine wave consisting only of the fundamental, and (c) its Fourier spectrum, where the $\boldsymbol{k_x}$-axis is marked off in integer multiples of the fundamental. (d) The fundamental plus a 'third harmonic' contribution of the appropriate amplitude, and (e) the corresponding spectrum, now with a new but related measure of position along the k_x-axis: $k_p \equiv p \times k_1$, with integer p. (f) The fundamental plus the third and fifth harmonics, and its (g) spectrum, with frequency now expressed as a **k**-vector in a newly contrived 1D ***k**-space*. As more harmonics are added in, the sum more closely resembles the original square wave.

the right-hand side in Figure 5.9. We consider relating spatial frequency contribution to position along a $\boldsymbol{k_x}$-axis by labeling it with marks designating 1 cycle/mm, 2 c/mm, 3 c/mm, and so on. It is simpler in Fourier analysis, however, to indicate it as integer multiples of the fundamental spatial frequency, k_1, as is done in Figure 5.9c. So far, the spectrum reveals only a sine wave function of frequency k_1 and amplitude $4/\pi$, but it will improve as we include more terms.

The somewhat better approximation,

$$s_3(x) = (4/\pi)[\sin(2\pi k_1 x) + \tfrac{1}{3}\sin(6\pi k_1 x)],$$

contains an additional *third harmonic* contribution (Figure 5.9d). The new sine term is chosen to have a spatial frequency of $3k_1$ and an amplitude that is one-third that of the first sine term. Its crests and troughs fill in the gaps and diminish the peaks of sin $(2\pi k_1 x)$, resulting in something a little closer to a square-wave. The Fourier spectrum, up to this point, appears in Figure 5.9e.

EXERCISE 5.13 What does [sin $6\pi k_1 x$] signify?

Including another harmonic (Figures 5.9f and 5.9g) is better yet:

$$s_5(x) = (4/\pi)[\sin(2\pi k_1 x) + \tfrac{1}{3}\sin(6\pi k_1 x) + \tfrac{1}{5}\sin(10\pi k_1 x)].$$

As more and more terms are added, the resulting wave interference patterns (which is what they really are) will come increasingly to resemble the original square wave. And the graph of the *infinite* Fourier series expansion

$$s_{\text{sq-odd}}(x) = (4/\pi)[\sin(2\pi k_1 x) + \tfrac{1}{3}\sin(6\pi k_1 x) + \tfrac{1}{5}\sin(10\pi k_1 x) + \ldots], \quad (5.14b)$$

is indistinguishable from it. A one-dimensional spatial picture such as $s_{\text{sq-odd}}(x)$ varies along the x-axis, and we could describe it as residing in a 1D image space.

The frequencies of the sine terms that contribute to the periodic function are distinctly separate from one another. The function's Fourier spectrum consists of distinct spikes at discrete values of the frequency, and is said to be *discrete*, as opposed to *continuous* like Planck's black-body spectrum.

EXERCISE 5.14 An infinite series in which the terms can be put into a 1-to-1 correspondence with the set of all integers is said to be countably infinite. Does Equation (5.14b) represent such a series?

It is inviting to redefine the units of the k-axis as explicit multiples of the fundamental (Figure 5.9c). It proves to be conceptually useful to introduce the labels k_1, $k_3 \equiv 3\times k_1$, $k_5 \equiv 5\times k_1$, $k_7 \equiv 7\times k_1$, etc., along a newly introduced k_p*-axis*, where

$$k_p \equiv p\times k_1. \quad (5.15a)$$

With this newer notation, Equations (5.14) can be re-expressed as

$$s_{\text{sq-odd}}(x) = (4/\pi)[\sin(2\pi k_1 x) + \tfrac{1}{3}\sin(2\pi k_3 x) + \tfrac{1}{5}\sin(2\pi k_5 x) + \ldots], \quad (5.15b)$$

More generally, for *any* odd 1D periodic function,

$$s_{\text{odd}}(x) = \alpha_1 \sin(2\pi k_1 x) + \alpha_2 \sin(2\pi k_2 x) + \alpha_3 \sin(2\pi k_3 x) \ldots = \sum_{p=1}^{\infty} a_p \sin 2\pi k_p x, \quad (5.15c)$$

where the *amplitudes* α_p are determined by the shape of the repeated odd pattern in real space. The separation of discrete, neighboring points along it is $1\times k_1$ in general, although for the specific case of the odd square wave, the spacing happens to be $2\times k_1$.

Note that [cos $(2\pi\, k_p x)$ is an *even* function of x, and *any* periodic function of x, whether even or odd or neither, can be expanded as

$$s(x) = \alpha_0 + \sum\left[\alpha_p \sin 2\pi k_p x + \beta_p \cos 2\pi k_p x\right]. \quad (5.15d)$$

This can also appear as

$$s(x) = s_0 + \sum s_p \sin(2\pi k_p x + \varphi_p), \quad (5.15e)$$

where s_p is the amplitude of the pth harmonic and φ_p its relative phase. For the time being, for simplicity, however, we shall continue to work only with odd periodic functions, like $s_{\text{sq-odd}}$, where $\varphi_p = 0$.

EXERCISE 5.15 Show that Equations (5.15d) and (5.15e) are equivalent.

EXERCISE 5.16 Find an expression for an even square-wave function of time. Why does the expansion of an odd function not contain a constant?

We can even consider the $\boldsymbol{k_x}$-axis as comprising a 1D vector space, with $\boldsymbol{k_1}, \boldsymbol{k_3}, \boldsymbol{k_5}, \boldsymbol{k_7}$, etc., as vectors in it, as in Figure 5.9g. This, in fact, leads to the approach we shall be taking when we move on to 2D MRI.

To summarize: a number can be rendered equally well in either decimal or binary form. Likewise, the spatial patterns in Figure 5.9 contain exactly the same information as do the Fourier sums of Equations (5.14) and their corresponding spectra (or equivalently, their 1D **k**-space spectral representations). One can jump back and forth readily between the spatial pattern in real-space and its representation in **k**-space *via* Fourier transforms, and we shall be doing that in image reconstruction.

$\{e^{-2\pi i\, k_p x}\}$ basis vectors; the Dirac delta function, $\delta(x-x')$

It is possible to adopt, instead of sines and cosines, the complex exponentials as basis vectors:

$$\left\{e^{-2\pi i\, k_p x},\ p = 0,1,2\ldots\right\}. \quad (5.16a)$$

Equation (5.15c) then assumes the more compact and manipulable form

$$s(x) = \sum_{p=1} \xi_p\, e^{2\pi i (p/\lambda)x} = \sum_{p=1} \xi_p\, e^{-2\pi i\, k_p x}, \quad (5.16b)$$

summed over the integer index p. The second equality of this recognizes λ as the spatial periodicity of the square waveform and incorporates Equation (5.15a). It will soon be apparent with application of the Fourier transform that the amplitude of each term is

$$\xi_p = k_p \int s(x) e^{2\pi i k_p x} dx. \tag{5.16c}$$

The scalar product of Equation (5.10b) can demonstrate orthonormality (here we replace k_p with any two choices of p, such as k and k'):

$$\int_{-\infty}^{\infty} \left[e^{-2\pi i k x}\right]\left[e^{-2\pi i k' x}\right]^* dx = \int \left[e^{-2\pi i k x}\right]\left[e^{+2\pi i k' x}\right] dx \tag{5.17a}$$

$$= \int \left[e^{-2\pi i (k-k') x}\right] dx$$

$$= \delta(x - x').$$

Similarly,

$$\int \left[e^{-2\pi i k x}\right]\left[e^{+2\pi i k x'}\right] dk = \delta(x - x'). \tag{5.17b}$$

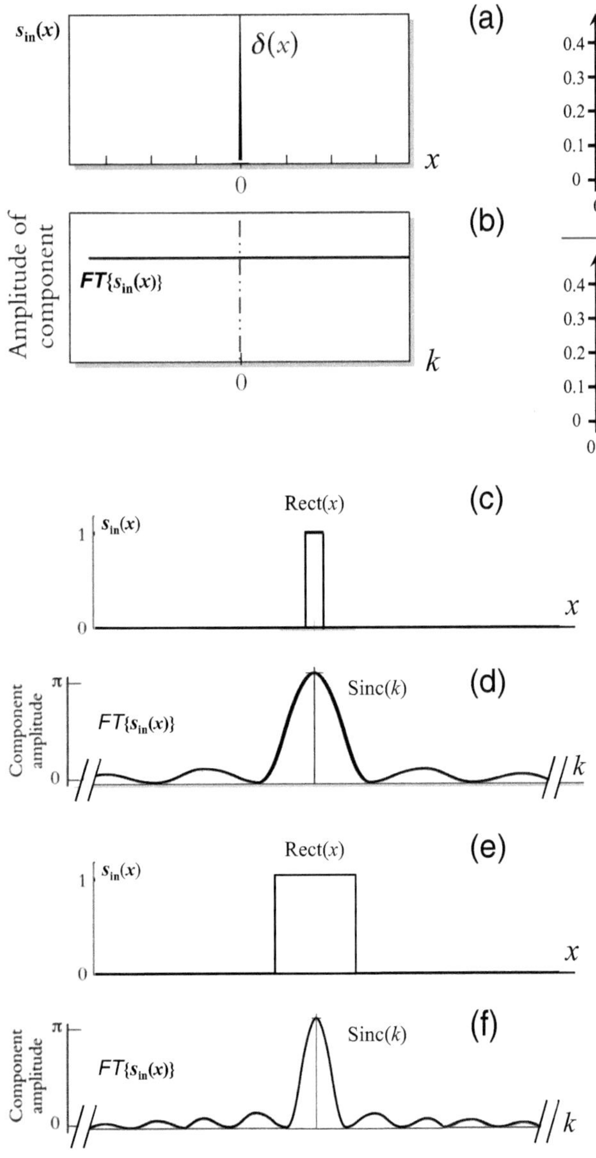

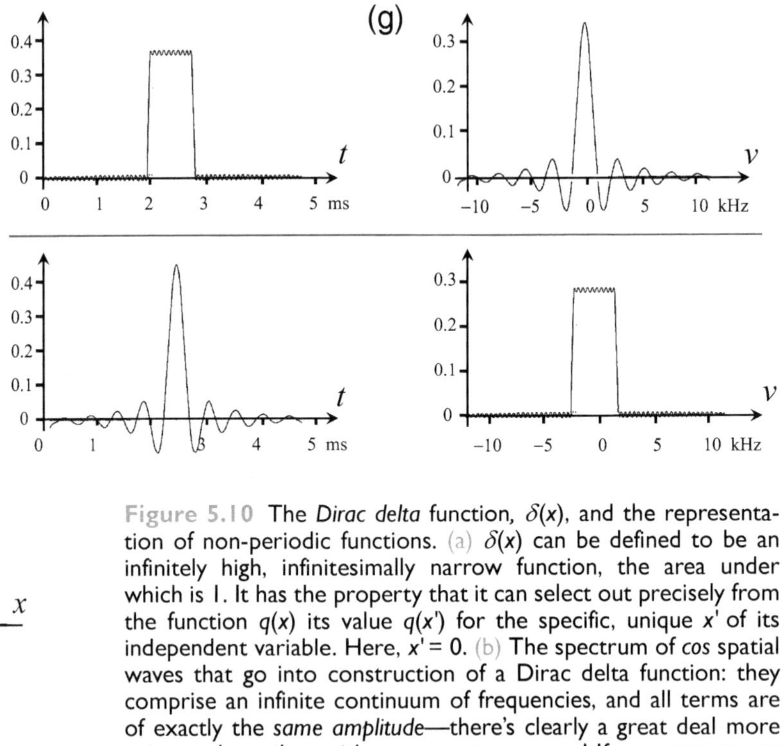

Figure 5.10 The *Dirac delta* function, δ(*x*), and the representation of non-periodic functions. (a) δ(*x*) can be defined to be an infinitely high, infinitesimally narrow function, the area under which is 1. It has the property that it can select out precisely from the function *q*(*x*) its value *q*(*x'*) for the specific, unique *x'* of its independent variable. Here, *x'* = 0. (b) The spectrum of *cos* spatial waves that go into construction of a Dirac delta function: they comprise an infinite continuum of frequencies, and all terms are of exactly the *same amplitude*—there's clearly a great deal more going on here than with an even square wave! If you want to see how a system responds to stimuli over a wide range of frequencies, just strike it with a delta function impulse in time or space, and examine (perform a Fourier analysis on) the output signal. (c) through (f) demonstrate that as an entity in real-space (such as, here, the Rect(*x*) function) grows broader, its representation in ***k***-space grows narrower—with δ(*x*) being the extreme example. The spectrum for a non-periodic curve such as Rect(*x*), has component frequencies so close together that their tips appear to meld together into a smooth, continuous spectrum. One way to deal with Rect(*x*) mathematically is to view it as part of a periodic square wave in which the squares continually grow farther apart; as an object becomes narrower, its Fourier transform becomes correspondingly wider in ***k***-space. (g) The Rect(*t*) and Sinc(*v*) functions are Fourier transforms of one another.

The 1D *Dirac delta functions*—$\delta(x)$ and $\delta(k)$ of Equations (5.17a) and (5.17b) and Figure 5.10a—may also be defined as having the shape of an infinitely high and infinitesimally narrow spike that encloses an area of 1 under the curve. (Numerical approximations of it for computer-based calculations are not so extreme.) *All* spatial frequencies contribute *equally* to its make-up, as seen in its Fourier spectrum (Figure 5.10b). In other words, $\delta(x)$ is as narrow a function as one can find, and the *bandwidth*, BW, of its spectrum is infinitely wide, with all component contributions of the same amplitude.

Because $\delta(x)$ has the remarkable and invaluable property that for a well-behaved function $w(x)$,

$$\int_{-\infty}^{\infty} w(x)\,\delta(x - x')\,dx = w(x'), \qquad (5.17c)$$

it is invaluable in many physics calculations. Just don't ask a professional mathematician about it!

Series truncation

All good things must come to an end, of course, and that includes Fourier sums. As evident from Figure 5.9, the more and higher the frequencies of the harmonics included in creating an image, the closer it comes to the original. The harmonics will be of progressively shorter wavelengths, and you may be inclined to incorporate increasingly finer detail up until no further improvement in the image is noticeable to the eye, or to the artificial intelligence program analyzing the image. But the Fourier sum, even with large numbers of harmonics, still will not *exactly* reproduce the original wave in Equations (5.14).

It is usually not possible, of course, to manipulate individually more than a small number of the terms in an infinite series. And in MRI, we can only acquire a finite amount of data in a finite amount of time, and we are able to employ only a relatively small number of harmonics to mimic $s(x)$. So one must *truncate* the series of Equations (5.14) and (5.15), cutting off the summation at some frequency, thereby converting it into an approximate, *finite* Fourier series. In the clinic, unfortunately, such approximation errors may manifest in an image if the truncation occurs too early.

One such *artifact* can be present to some degree, whether dramatic or hardly noticeable, in MR images. The *truncation* or *Gibbs' ringing* artifact, is a by-product of trying to approximate an infinite sum with too short a finite Fourier sum. This truncation artifact, somewhat like Figure 5.9f, leads to wiggles in what otherwise should be a smooth function. The irregularities appear predominantly near sharp edges where the function is undergoing abrupt jumps in amplitude, such as the leading and trailing edges of a square pulse. This is where the Fourier series approximation can be most in error, because higher frequencies are needed to represent the fine detail of the change accurately. Similarly, truncation artifacts may appear in the image of a white disc as sets of bright and dark lines parallel to the boundary (Figure 5.11a) or within vertebral bodies (Figure 5.11b). With the disk, adoption of a larger sam-

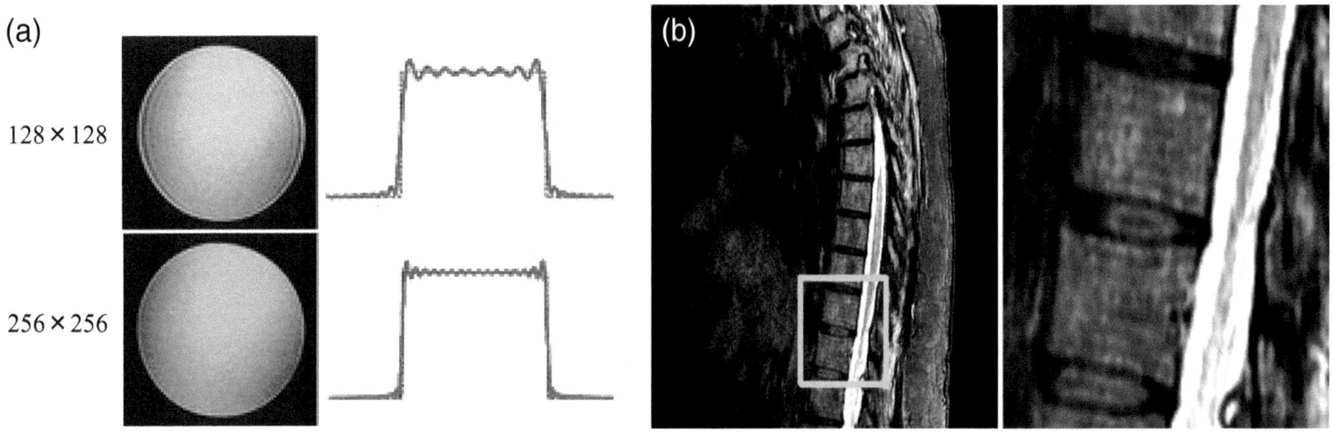

Figure 5.11 *Truncation artifacts* arising from limitations on how sharp an edge the MRI reconstruction algorithm, which includes Fourier techniques, is capable of reproducing. (a) Truncated Fourier approximation of a uniformly bright white disk. The smaller sampling or display matrix implies earlier truncation and less resolution where that is critical. (b) Within vertebral bodies, early truncation leads to a series of nearly-horizontal dark, thin lines. With increased resolution, the artifact would decrease at the cost of a longer scan time. [Courtesy of Jason Stafford, MD Anderson.]

pling or display matrix corresponds to later truncation in, say, Equations (5.14), hence a more accurate representation and a smoother image.

If this artifact is present in an MR image, even to a small extent, because of the approximate nature of a finite series representation, what can be done to prevent it from hindering clinical diagnoses? As apparent from Figure 5.9, the more harmonics that are included, the less the error. So we can increase the number of s_p and φ_p values sampled in Equations (5.15), which also improves resolution. Unfortunately, that usually requires more scanning time, so a balance must be determined regarding the ultimate resolution desired and the magnitude of the artifact that is tolerable. We shall return to this important issue later.

Another approach involves the process of *low-pass filtering* the data to de-emphasize the ringing. One particular variant of this approach is called *windowing*, which filters out the high-frequency contributions to the s_p and φ_p values before performing the Fourier sum. While windowing can decrease the truncation artifact, it may also lead to some loss of detail.

So in general, the more Fourier components to be included in the representation of an incoming MRI analog voltage signal, the higher the upper end of the frequency spectrum that the signal receiver must be able to handle. A larger number of Fourier components translates into more memory needed to store the data and more computation power and time to reconstruct images—and that ultimately means more expense to create a sufficiently fast scanner. In capping a series, the truncation frequency must be high enough to prevent the introduction of unacceptable distortions or irregularities. Fortunately, higher-frequency contributions to a Fourier sum almost always become small, and one can terminate the series after the terms no longer have a discernible impact on the sum, retaining only the lower-frequency terms. Pixel size places a limit on resolution for any digital representation, and so also does the spatial frequency at which the truncation is made. But how high do you have to go?

An immediate ceiling is imposed on an imaging system by its inherent powers of resolution. Resolution is often expressed rather loosely as the distance between points that can just barely be distinguished as being separate, e.g., 0.5 mm (Figure 3.8). Alternatively, it can be depicted as a spatial frequency, which is the inverse of a distance, such as 2 line pairs per millimeter (2 lp/mm), a measure found in digital and film radiography. These are not exactly the same thing—indeed, they refer to somewhat different aspects of detail and sharpness. More about that later.

The Modulation Transfer Function method of Equation (3.13) is an excellent quantitative measure of a system's ability to preserve contrast and spatial resolution as it processes images. Common sense and the MTF method suggest a rule of thumb: if the weakest link in an imaging chain can process or display information with a spatial (or temporal) resolution of about R (Figure 3.9), then the rest of the system needs to perform well only up to a spatial frequency of a few times $1/R$. Any improvement in the other components is wasted effort.

It is also sometimes not obvious what frequency should play the role of the 'fundamental' in a spatial Fourier analysis, but generally the longest wavelength considered is taken to be the width of the original or repeated pattern. The Field of View is normally chosen to be large enough to cover the entire required image, or a single complete repetition of it if it is periodic (Figure 3.3). In other words, there is no need for the Fourier representation of an image to contain wavelengths longer than the entire FOV—so it is usually a safe bet to set the *longest* possible wavelength, $\lambda_{max}\,(\equiv 1/k)_{max}$, equal to it. If there were any waves contributing to the sum that are of a wavelength longer than this, then the whole train of square waves (or whatever) would float on a slowly undulating background spatial wave, which means that the chosen FOV is not wide enough.

We have been considering the creation of square waves or other odd functions out of cleverly selected sums of sine waves. As coming chapters will show, it is actually the opposite process—the Fourier decomposition of the MR signals received from a patient—that is the real concern in MRI.

5.4 Fourier Integrals and Transforms

By an extension of the concept of the Fourier series, it is possible to represent an odd curve that is *not* periodic with an *integral* of sine terms, rather than a sum.

Imagine that the rectangles of Figure 5.10 become broader, the required harmonics become far more numerous and more closely spaced, and eventually the upper tips of the spectrum appear to blend into a seemingly smooth and continuous spectral curve, as in Figure 5.10. In the limit, the Fourier representation of a single rectangular block in space contains a continuum of contributions at *all* frequencies, not just of a fundamental and its discrete harmonics: the Fourier sum has morphed into a *Fourier integral* of the form

$$w(x) \;=\; \int_{-\infty}^{\infty} a(k)\sin(2\pi k\,x)\,dk. \qquad (5.18)$$

The *Fourier transform* is a mathematical device that allows the decomposition of any signal that is a continuous, non-periodic function of time or space, and to determine its spectrum. It plays numerous invaluable roles throughout physics and engineering, providing general linkages between time and frequency, or between space and spatial frequency.

Mathematicians have shown that it is legit to decompose a complex non-periodic function $w(x)$ into its Fourier components as

$$W(k) \;=\; \mathbf{FT}[w(x)] \;\equiv\; \int_{-\infty}^{\infty} w(x)\,e^{-2\pi i(k\,x)}dx. \qquad (5.19a)$$

The complex function $W(k)$, which provides the kth spectral component of $w(x)$, is said to be its *Fourier transform.* And just as we broke $w(x)$ into its Fourier components $W(k)$, likewise it is possible to reverse the process and construct the original $w(x)$ out of its spectral components $W(k)$ by way of the *Inverse Fourier Transform*, FT^{-1}, with a '+' sign in the exponent:

$$w(x) \;=\; \mathbf{FT}^{-1}[W(k)] \;\equiv\; \int_{-\infty}^{\infty} W(k)\,e^{+2\pi i(k\,x)}dk. \qquad (5.19b)$$

Just as $W(k)$ is the spectrum of $w(x)$ in **k**-space, likewise $w(x)$ can be viewed as the spectrum, in image space, of $W(k)$. That is, $w(x)$ and $W(k)$ are a Fourier transform pair. They are merely two quite different-seeming but fundamentally equivalent representations of the same bundle of information. A very useful, and self-evident, theorem states the FT of an inverse FT of a function gives back the original function

$$\mathbf{FT}^{-1}\{\mathbf{FT}[w(x)]\} \;=\; w(x), \qquad (5.19c)$$

and, conversely, Figure 5.10g. That is, the Fourier transform is invertible, with no loss of information, apart from that caused by any numerical truncation. Such a notion proves central to image reconstruction calculations.

For a function, $w(\boldsymbol{r})$, in two- or three-dimensional space, the FT assumes the form

$$W(\boldsymbol{k}) \;=\; \mathbf{FT}[w(\boldsymbol{r})] \;=\; \int w(\boldsymbol{r})\,e^{-2\pi i(\boldsymbol{k}\cdot\boldsymbol{r})}d\boldsymbol{r}, \qquad (5.19d)$$

with 2D or 3D vectors $\boldsymbol{k}$ and $\boldsymbol{r}$.

The single rectangle is important enough to have its own moniker, the Rect(x) function (Figures 5.10c and 5.10e). Its Fourier representation, or spectrum, is called Sinc(k) (Figures 5.10d and 5.10f),

$$\mathrm{Sinc}(k) \;\equiv\; (\sin k)\,/\,k, \qquad (5.20a)$$

Rect(x) and Sinc(k) (Figure 5.10g) carry exactly the same information; indeed, each is the Fourier transform (FT) of the other, and it would be just as meaningful to view Rect(k) as the spectrum of Sinc(x)

$$\mathrm{Rect}(v) \;=\; \mathbf{FT}[\mathrm{Sinc}(t)], \qquad (5.20b)$$

*EXERCISE 5.17 Demonstrate that in the time-frequency domain, Rect(v) = **FT**[Sinc(t)] and vice versa.*

EXERCISE 5.18 Is there a connection between $\int q(x)\,\delta(x-x')\,dx$ *and a Fourier integral?*

The progression in Figure 5.10 illustrates an important point. Here $\delta(x)$ is the limiting case of a commonly seen phenomenon: as an entity or function grows narrower in time or space, the bandwidth of temporal or spatial frequencies required to represent it accurately increases.

In certain situations, other transforms (e.g., wavelet, Discrete Cosine, Laplace, Z-transform, etc.) can also represent acquired data, but with fewer coefficients/parameters. This is discussed in more advanced treatments.

If it seems that a lot of mathematical machinery is required to undertake a Fourier analysis, just bear in mind that the ear does so in listening to Bach, and without having to resort to any calculus.

FT [PSF(x)] = MTF(v)

Back to the MTF of Equation (3.9) and Figure 3.14 for a moment. If an imaging system had absolutely no

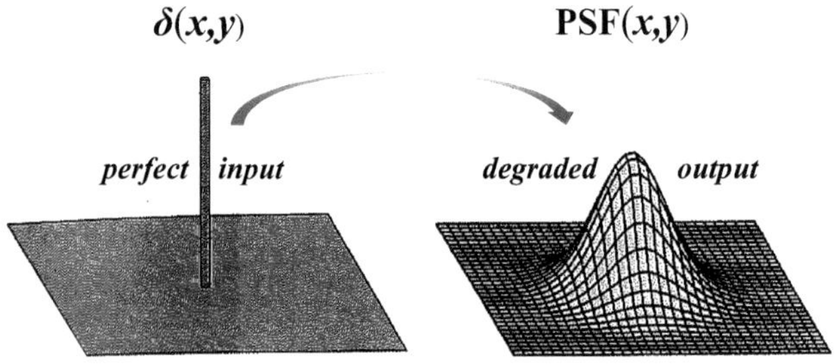

Figure 5.12 When a 2D Dirac delta function acts as input to a linear system, what emerges as output is the system's 2D PSF. The contributions to $\delta(x)$ at all frequencies are of the same amplitude, and each one passes through the system independent of the others, by the definition of *linearity*. One important application of the PSF is indicated by Equation (5.21).

degrading effects on a very slender input signal, say, the output would also be a line delta function, as with the LSF marked 'perfect' in Figure 3.14. Its spectrum would show the amplitudes of the contributions at all frequencies still being unity, just as with the input.

But any real imaging device will diminish the quality of a signal, and the result is a broadening and reshaping of the LSF(x). The Fourier spectrum of the output signal no longer displays components all of the same unit amplitude; the amplitude falls off, rather, at higher frequency. Indeed, the curve is none other than the MTF(ν).

The Fourier components of all frequencies pass through an ideal linear system independently, by definition of linearity. But the effects of the system on waves of the various frequencies do differ from one another, and this can introduce distortions in the signal being processed. In the case of a delta-function input, where all the input components are of the same amplitude, the outputs interfere constructively at some points and destructively at others, which leads to the creation of the output *point spread function*, PSF (Figure 5.12). That's fine, since we can distinguish the output at each frequency separately—which is exactly what a Fourier decomposition can provide! The Fourier spectrum of the LSF(x) reveals the extent to which input signal amplitude is lost at any frequency—which is to say, the value of $M_{out}(\nu)$. Their ratio is, in short, the MTF(ν), and

$$\mathbf{FT}[\text{PSF}(x)] = \text{MTF}(\nu). \tag{5.21}$$

While we haven't formally proven Equation (5.21) rigorously, hopefully it is intuitively plausible.

EXERCISE 5.19 Compare the MTF of the LSF with that of the PSF.

Patient motion artifacts and the Fourier Shift Theorem

Two common forms of *physiological motion*, breathing and heartbeat, may smear out or otherwise degrade an image. Because they are both at least somewhat periodic, they can also give rise to distinct *ghost artifacts,* i.e., erroneous multiple copies of a moving piece of anatomy that appear in an image (Figure 5.13).

Dealing with motion effects in a manner that preserves the needed contrast, resolution, and SNR is one of the challenges that has spawned a host of ingenious innovations in MRI. Of the ways to deal with the consequences of respiration, the simplest is just the breath-hold, but this may either alter the cardiac cycle or make its effects on the image all the more apparent. Others that have been developed include respiratory and cardiac gating, altering of ***k***-space acquisition trajectories (discussed later), and the application of ultra-fast pulse sequences, which may ameliorate the problem.

What if, in the middle of collecting data, a patient on the imaging table twitched to the left? Now, all of the future Fourier coefficients will differ from what we would have calculated before! If we try to build an image by combining coefficients drawn from before and after the movement, we're going to get a smeared-out something. What's to be done? The "fix" is in two parts. Both before and after the problem, it is necessary

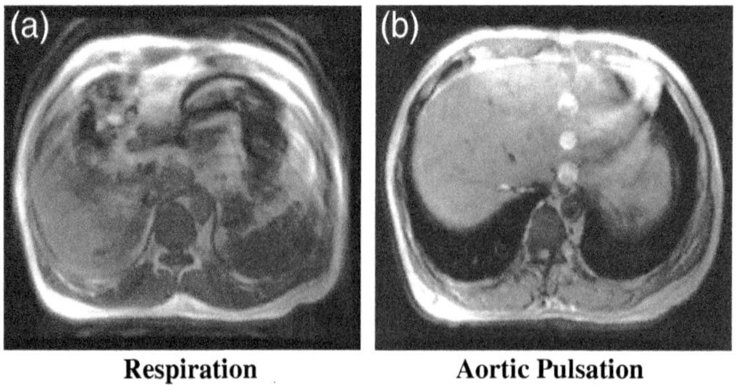

Figure 5.13 Artifacts from respiratory and cardiac motion. (a) Breathing and the cardiac cycle cause the blurring and ghosting in a thoracic image. Breath-holding eliminates most of the effects of chest movement, but (b) that makes more visible cardiac pulsation effects, such as the multiple copies of the major vessels.

to re-measure data in a few redundant places in ***k***-space—basically, we want to repeat measurements near the center of ***k***-space before and after the motion. The second involves use of a special property of Fourier transforms, described by the *Shift Theorem.* It describes a way to modify the data if, say, the patient shifts a little during a scan.

Here is a sketch of how the theorem works. In moving the function $q(x)$ in real space with a small displacement of Δx, from x to $x' = x+\Delta x$, say, the impact on the Fourier transform of $q(x') = q(x+\Delta x)$ is

$$\mathbf{FT}\left[q(x+\Delta x)\right] = \int q(x+\Delta x)\mathrm{e}^{-2\pi i k x}dx$$
$$= \int q(x+\Delta x)\mathrm{e}^{-2\pi i k (x+\Delta x)} \times \mathrm{e}^{+2\pi i k \,\Delta x)}dx.$$

Since $dx = d(x+\Delta x) = dx'$,

$$\mathbf{FT}\left[q(x+\Delta x)\right] = \mathrm{e}^{2\pi i k\,\Delta x}\int q(x')\mathrm{e}^{-2\pi i (kx')}dx' \quad (5.22)$$
$$= Q(k)^{*}\,\mathrm{e}^{2\pi i k\,\Delta x}.$$

That is, after the wave, $q(x)$, is shifted in the spatial domain, the Fourier transform looks the same as the un-displaced transform multiplied by a *phase factor* of $\phi = [2\pi\, k\, \Delta x]$. The phase factor will be different for all locations in k-space, but it always depends on the patient's shift, Δx. With a special MRI pulse sequence, some motion artifacts can be resolved by rescaling the Fourier transform using these phase factors, along with other Fourier transform properties, before performing the inverse transform and creating an image. The amount of rescaling is determined by examining how the redundantly measured data in k-space differs before and after the shift. If the data are the same, then there was no motion; otherwise, you can determine Δx from a data comparison and use it to rescale via a phase factor everywhere else. This motion "fix" only works for global shifts of the patient, where the whole anatomy remains rigid and moves together. If only part of the body moves (e.g., cardiac motion), the change in ***k***-space from one time to another is more complicated, and the shift theorem is not adequate for dealing with the motion artifact.

5.5 Vectors in a 1D *k*-Space

The various terms in Equations (5.15) can be expressed in a manner better suited for MRI, albeit more abstractly, in terms of a particular vector-space known as ***k****-space*. This formulism directly represents waves in the 'reciprocal' spatial-frequency space in which data are collected during MRI. ***k***-space in MR imaging is like a matrix for organizing the addresses where sampled signal data are to be stored. Most commonly, the boxes are arranged along one (or in 2D scanning, multiple) horizontal rows, each with its positions labeled with values of k_x. In 2D scanning, every parallel row is itself indexed with a unique value of k_y.

In the simple case of 1D imaging, the matrix consists of one row of same-sized, equally spaced address boxes tagged sequentially with integer values of k_x. These addresses might run from $-k_{max}$ to $+k_{max}$, say. After the voltage from the pickup coil enters the receiver, it is sampled and digitized, and the sequentially obtained voxel values placed in adjacent k_x 'boxes.'

Imagine the 1D ***k***-space associated with Figures 5.9c, 5.9e, or 5.9g as comprising the narrow white line along the k_x-axis in Figure 5.14. It might be designed to run along the axis from $-k_{max}$ on its left, through $k = 0$ at the center, and on to terminate at k_{max}. As in Figure 5.9g, the length of any (horizontal) wave *vector*, $\mathbf{k}_p$, indicates the spatial frequency of the wave under consideration. The datum value that resides at that point in *k-space*, e.g., α_p, is associated with its (vertical) amplitude (Figure 5.14 again). For other curves that may require odd and even functions as Fourier components to represent the curve accurately, k-space stores the amplitudes for both components, α_p and β_p. So each place in ***k***-space stores a complex number.

Two example points along the $\mathbf{k}_x$-axis of Figure 5.14, at $\mathbf{k}_p$ and $k_{p'}$, correspond to specific sine waveforms in 1D *real*-space of spatial frequencies k_p and $k_{p'}$, and they describe sine-waves with amplitudes of, say, α_p and $\alpha_{p'}$ (reflected also in their distances above or below the white line). It is apparent, moreover, that adding the real-space waves associated with points indicated in k_x-space is essentially the same as adding the ($\alpha_p \sin 2\pi\, k_p x$) and ($\alpha_{p'} \sin 2\pi\, k_{p'} x$) terms together in Equation (5.14a) and Figure 5.9d.

Combining the data residing at two different locations in 1D ***k***-space (pointed to by two 1D $\mathbf{k}_x$ 'vectors'), such as k_p and $k_{p'}$, does *not* mean that we are calculating their vector sum of length $|k_p + k_{p'}|$. That would be incorrect, and ($k_p + k_{p'}$), in fact, is meaningless. Rather, it indicates the *superpositioning* of their two real-space waveforms, as in Figures 5.5a and 5.9.

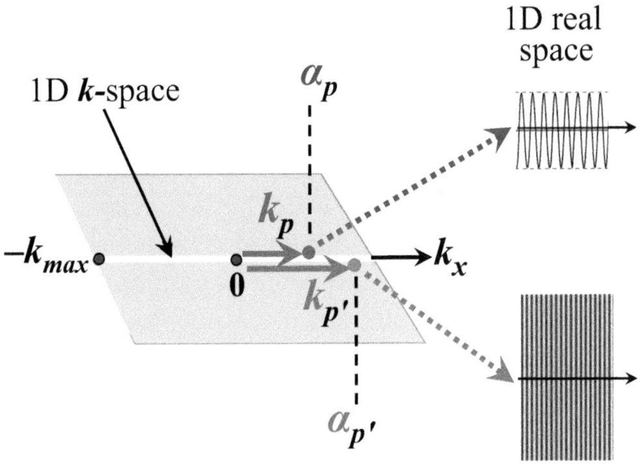

Figure 5.14 The white line illustrates a 1D vector k_x-space, and the blue region is non-existent. The line runs along the k_x-axis from $-k_{max}$ on its left to k_{max}, passing through $k = 0$ at the center. Two specific vectors are shown as k_p and $k_{p'}$, along with their associated sinusoidal planar wave fronts in real space as viewed from above, aligned along the real x-axis. The coefficients for these wave components, α_p and $\alpha_{p'}$, record their amplitudes, which are reflected here also by their distances above or below the k_x-axis. This image is actually just a re-casting of Figure 5.9g in a slightly different form.

Suppose that $S(x)$ is an original MRI signal from a patient (or even just a square wave) sampled, digitized, and expressed as a sequence of data points in a k_x-space spanned by the basis vectors $\{\ldots, -k_3, -k_2, -k_1, 0, k_1, k_2, k_3, \ldots k_{max}\}$, for example. Once such a basis set is agreed upon, then *all* pertinent information about a sampled signal is stored at various locations in k_x-space.

The rules will be different, however, and a little more complicated, when we move into a 2D k-space with both k_x- and k_y-axes in chapter 12.

5.6 Digital Representations of Symbols and Images

Modern mathematical operations, including Fourier transforms, and nearly all data processing, storage, and transmission involve the manipulation of digital information. Not only are digital computers essential for these jobs, but also the *image acquisition*, *processing*, *storage*, and *display* processes can be treated largely as if they are unlinked—unlike, say, the taking of an x-ray or photograph film. With a digital system, each of these tasks can be optimized separately. Image receptors tend to be linear, moreover, with the output signal proportional to the magnitude of the stimulus, and of broad latitude (the range over which it operates in a linear fashion). Digital systems have imaging processing capabilities to improve contrast, sharpness, and noise characteristics, such as windowing, panning, zooming, and distance measurement, and live or rapid display, repeat, or freeze. And they can feed directly into a *picture archiving and communications system* (PACS) with its immediate, loss-less storage (archiving, retrieval); electronic communications (local, teleradiology); databases for statistical analysis; artificial intelligence (AI); and more.

A digital computer handles numbers and other symbols as strings of bits. A bit (for *b*inary dig*it*) is the smallest quantity of information that a digital computer, or most anything else, can store, process, or transmit. A bit has exactly two states, which can be called "on" or "off," "yes" or "no," "1" or "0," or the like. (This is not the case for quantum computers, but it will be a while before they are capable of assuming a role in MRI.) A co-joined pair of bits can represent four states (distinguished as 00, 01, 10, and 11), and three bits can stand for 8 different things (labeled 000, 001, 010, …, 110, 111). Indeed, a collection of N bits can be used to represent 2^N different numbers, letters of the alphabet, symbols, states, etc.

Bits are commonly assembled and managed in groups of eight, called *bytes*. There are 256 distinct entities that one byte could represent, which is more than enough to cover all the 52 standard capital and lower case letters, the 10 digits, and much else. One byte would allow you to count from 1 to 256, or 0 to 255, or even –128 to 127, depending on the convention you choose. A two-byte *word* could get you up to 65,536, and so on.

EXERCISE 5.20 A protein molecule normally consists of a long, linear sequence of amino acids, uniquely determined by a corresponding sequence of codons in a strand of DNA. A codon is a short string of nucleic acid bases that codes for a single amino acid. There are 20 different commonly occurring natural amino acids out of which proteins are built, and four different bases for making codons (thymine, adenine, guanine, and cytosine). What is the smallest possible size of (number of bases in) a codon, given that codons all contain the same number of bases?

Likewise, it is possible to represent an image digitally. Creating a digital representation of (digitizing) an ordinary film radiograph, for example, is essentially the converse of painting by numbers. The computer partitions the film into an imaginary N_x by N_y matrix (like ordinary graph paper), such as 1024×1024. This grid comprises about a million tiny, square film picture elements, or *pixels*. Data entry might be accomplished by scanning the film with a laser beam point-by-point in a raster pattern, with the light transmissivity being monitored with a light detector. The system keeps track of and tabulates every *pixel address* and *pixel value* in this fashion (Figure 5.15).

Suppose a neurosurgeon in San Diego needs a patient's pre-digital-era film radiograph that was acquired and archived in a New York City hospital a while ago, and she needs it fast. It took considerable time and effort to locate the film, since it was never loaded onto the computer-based PACS. It turned up buried in a stack of films for a study in Dr. Jones' office, and he's off on a month-long lecture and golf tour of Burkina Faso.

When eventually retrieved, the radiograph was entered into film digitizer. It was placed flat on a horizontal glass plate within the light-proof box of a computer-controlled scanning-laser film densitometer, the two green dashed boxes in Figure 5.16a. The laser points down at the radiograph from above, and a photodetector looks upward at the beam from below it. An electro-optical device attached to the laser can change the direction in which the beam is pointing, shifting its position by the tiny amount, Δx or Δy in the x- or the y-direction, with the application of voltages. Because it is the computer itself that provides those voltages, it always knows exactly where on the film the beam is striking—and where the photodetector is watching. The address and the measurement of the transmitted laser light is recorded for each pixel (Figure 5.16b).

It may be helpful, when considering MRI signal sampling and digitization later, to go over this in a little more detail here. To begin, the computer directs the laser beam to the start of the narrow (Δy-wide), horizontal band of film centered at $y = y_1$ (Figure 5.17a). As it sweeps slowly in the x-direction along the line, the photodetector generates a continuous analog signal voltage, $V(t)$, proportional to the amount of laser light transmitted through the film there (Figure 5.17b). The signal voltage from the photodetector is sampled at the ith pixel, at time t_i and for a *dwell time* interval of duration Δt_{dwell}. This lets $V(t)$ briefly charge up a capacitor, and the total charge on it for the pixel is then read electronically by an analog-to-digital (A/D or ADC) converter (Figure 5.17c). The digital result, which is

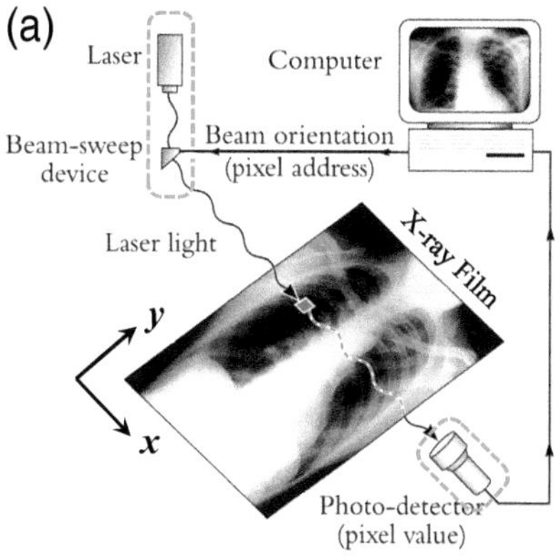

(b)

Pixel address x	Pixel address y	Pixel value
-	-	-
-	-	-
-	-	-
172	49	1.05
173	49	1.06
174	49	1.03
175	49	1.06
176	49	1.83
177	49	1.80
178	49	1.83
-	-	-
-	-	-
-	-	-

Figure 5.16 Film digitization. (a) The system's computer directs a laser beam to scan the film from above, one line at a time, as a light-sensitive detector on the other side monitors the film's optical transmissivity (two dashed green boxes). The detector can either move with the laser beam or take in the whole process with a film-sized view (mechanically simpler but more noise). The continuously varying (analog) signal voltage generated for a single scan line is detected, sampled, digitized and recorded, and the process is repeated for the next lines. (b) Film digitization data, tabulated as pixel addresses and values.

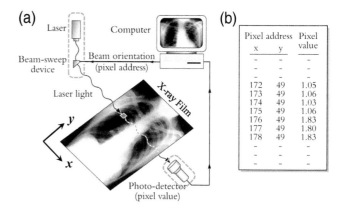

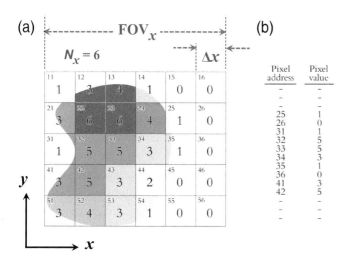

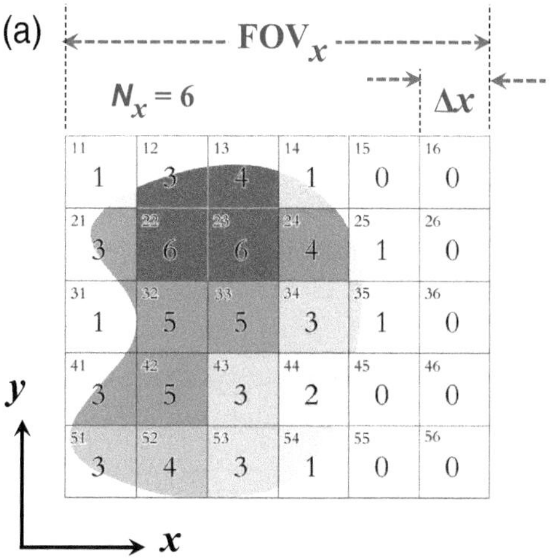

(b)

Pixel address	Pixel value
-	-
-	-
-	-
25	1
26	0
31	1
32	5
33	5
34	3
35	1
36	0
41	3
42	5
-	-
-	-
-	-

Figure 5.15 Representing an image in digital form. (a) This image is being partitioned by an imaginary grid into 30 square pixels, each Δx on a side. Every one of the squares is assigned a unique *pixel address*, and the depth of its shade of gray is expressed as the *pixel value*. (b) The image can then be represented as a listing of pixel addresses and the corresponding pixel values. If the number of pixels lying along the x-axis is N_x, then the Field of View in that direction is of width $FOV_x = N_x \times \Delta x$, Equation (3.2).

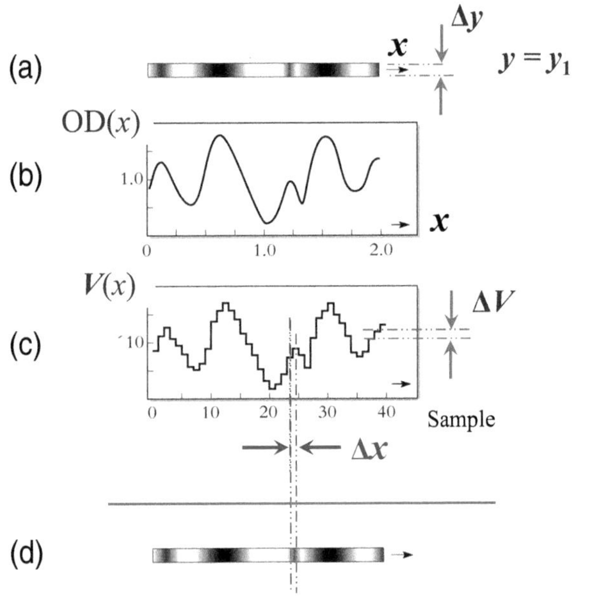

Figure 5.17 *Digitizing* an analog signal, such as that contained within a sheet of x-ray film. The computer partitions the whole film into an imaginary matrix of hundreds of thousands of film pixels, assigning a film pixel address to each. (a) Starting at the top left-hand corner, the optical densitometer scans a thin horizontal row of film, and (b) translates the optical transmissivity into a continuous, analog voltage. (c) This voltage is sampled twice for every film pixel, and transformed by an analog-to-digital converter (ADC) into a millivolt digit (or with some other appropriate degree of precision). Then on to the next row. The relevant information is stored in a digital listing of all pixel addresses and associated values, and entered into a PACS. (d) From there it can be transmitted to a distant clinic and retransformed by a digital-to-analog converter back into analog voltages to drive a display, etc.

proportional to $V(t_i)$, is the *pixel value*, and it is accurate to within some ±½ ΔV. (A glass thermometer and you, together, nicely illustrate the process of digitization; the length of the mercury column varies continuously with the temperature but, in sampling it, you read it off and record it as a discrete number, to the nearest degree. But is ±½° good enough accuracy for your needs?)

EXERCISE 5.21 Suppose that for some clinical purpose, an aperture error, Δt or ΔV or Δx, exceeds what is needed. What thing(s?) can be done to improve the situation?

The computer stores the *pixel address* and value in digital form, and then the laser beam proceeds to the next pixel location. The photodetector samples the transparency there, too, and generates the corresponding pixel value. The laser and detector are stepped from pixel to pixel across the film, and they move once downward at the end of each ***x***-row in a well-defined raster pattern. The two-dimensional matrix of numbers produced in this fashion is called a *bitmap*. The final digital representation in the computer's memory can be displayed with the same two-dimensional matrix pattern of pixel ***x-y***-addresses as in the film.

Once the digitized representation of the radiograph arrived at a computer in San Diego, the image was made to reappear on a display by reversing the digitization process (Figure 5.17d). The computer pulls up all the transmitted pixel addresses and pixel values for the image and places them in their proper positions in the matrix, pixel by pixel. For any pixel, its value controls the display brightness on the flat screen monitor there, and perhaps false color. This happens for all the pixels extraordinarily quickly, of course, but the individual steps are straightforward. The resulting array of pixel intensities produces two- or three-dimensional images that reflect the characteristics of the tissues within the region being examined.

But is it good enough to do the job?

Pixel size and grayscale

Pixels must be smaller than the details that have to be resolved; minuscule features can be incorporated into an image only if the pixels are sufficiently little. The smaller the pixels and the greater their number (and the greater the number of gray levels employed), the more faithfully can a representation capture the original image—but also, the greater the cost. In addition, The Shannon-Nyquist Theorem, discussed below, drives the required pixel size even lower. A related but separate issue is the number of pixels required to represent the image. It increases with both the dimensions of the field of view (FOV) and with the desired spatial resolution, or pixel size, in accord with Equation (3.2a).

The number of bits needed to address any particular pixel depends on the total number of pixels. If an image is electronically partitioned according to a 1024×1024 grid, for example, the intensity is specified at a million sites. The *x*- and *y*-coordinates can each be expressed as a 10-bit binary number, so that every site can be accessed by a single 20-bit address, or a 3-bite word that includes some wastage.

Fewer pixels (or gray levels) per image means faster image reconstruction and processing, and lower storage and communication costs. Thus an important objective in designing or purchasing imaging equipment is to create a system with more than enough ability to accomplish the requisite clinical tasks, but not much beyond that, so as to keep cost and time down.

Figure 5.18 illustrate what occurs if pixel size and grayscale are not handled properly. Figure 5.18a is a typical, state-of-the-art clinical MRI image produced with a 512×512 pixel matrix and 256 shades of gray. In Figure 5.18b, the same image is reconstructed on a 64×64 grid, in which the pixel dimensions are a factor of 8 larger and the resolution far worse. This is exaggerated for emphasis, but such pixilation might be noticeable even at 256×256.

Even with adequate resolution, there must also be enough distinct levels of gray available (Figure 5.18c). The number of different grayscale values, which determines the fineness of possible shading, is called the *image depth*. One might choose to have a pixel value of 0 correspond to pitch black on the display, and some suitably large value (such as 255) to white, but other choices of grayscale depth and centering may be preferable. The gray level step-size generally should be less than what can be distinguished by the eye (about 64 levels), even though a typical *liquid crystal display* (LCD) has a much finer 8-bit depth. If every picture element is designed to display 264 levels of gray (from an 8-bit deep display), the image is fully described in a listing of a quarter-million bytes, one for each address. It is possible to adjust the average brightness of the voxels, and the range of intensities that they span, by the process of *windowing*, in which the center pixel value and the slope of the linear relationship between the level of gray and the relative value of T1, for example, are shifted. This is done badly in Figure 5.18d.

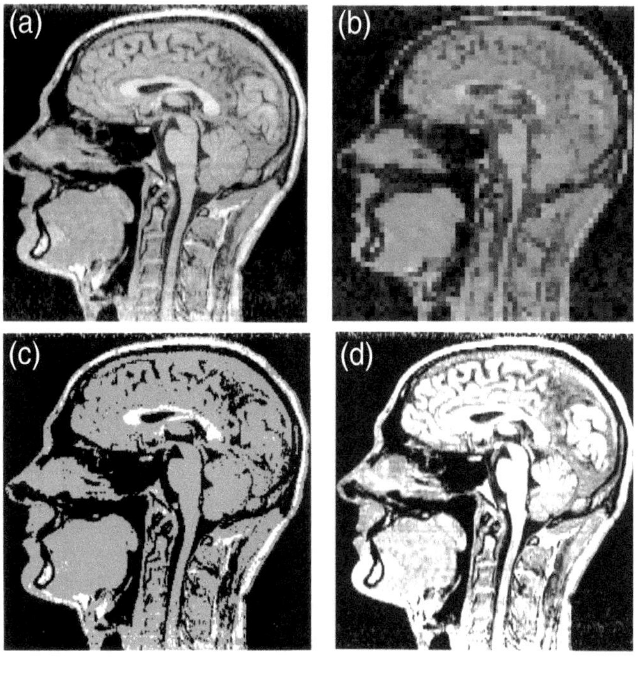

Figure 5.18 The importance of many, small pixels, and of enough shades of gray or color. (a) A standard sagittal MRI slice of the head with a 512×512 pixel matrix and 256 shades of gray. (b) A 64×64 matrix, again with 256 shades of gray. The poorer resolution, and the pixilation of the image, are readily noticeable. (c) With only four shades of gray, even a high-resolution slice is worthless. (d) The viewer can adjust the *windowing*, the slope and the center of the linear relationship between the level of gray and the relative value of T1, etc. This *grayscale* has 256 shades, but the windowing is far from optimal. [Courtesy of W. S. Kiger, III, Massachusetts Institute of Technology.]

For greater contrast resolution or better dynamic range, for use with an artificial intelligence (AI) program, say, more than one byte (e.g., 10 bits) would be used as the pixel value at any address. Little is gained, however, by making the separation of gray levels correspond to a difference smaller than the random fluctuations in intensity that occur naturally. Thus, the noise level imposes a natural lower bound on the useful brightness level step-size. At the other extreme, too small a number of grayscale levels can give rise to artifacts, such as the false appearance of sharp edges.

EXERCISE 5.22 How small must the pixels be to produce a clinically good enough digital representation of the San Diego radiograph?

Data compression

It should be apparent, when all is said and done, that it can become extremely expensive and time consuming to store, manipulate, or transmit huge volumes of data.

To encode a single 1-byte-deep 2000×2000-pixel bitmap image of, say, a digital radiograph, four million pairs of bytes, representing the addresses and shades of gray of all the pixels, are needed. If the addresses are maintained in some standardized order and don't have to be listed explicitly, this can be cut in half by listing only the grayscale values, and not the addresses.

Efficient *image compression* techniques can reduce the number a good deal more, decreasing considerably the computer memory and time needed to store or transmit an image. The reverse process, of restoration of an image to its perfect or near-perfect original form, is known as decompression.

A *qite* simple way to *compres* written English somewhat, for example, would be to replace every *ss* with a single *s* and to eliminate every *u* after a *q*. (You wouldn't even have to make special allowances for Iraq.) Image data compression techniques, going by such esoteric names as discrete cosine transform and fractal compression, are more complex, but in a few moments of computer time they can reduce the number of bits required to represent an image by an order of magnitude or more.

Suppose we are using a 10-bit grayscale (2^{10} = 512), with possible pixel values running between 0 and 1023, to digitize an image. The computer discovers that pixels at the six consecutive addresses 5283 through 5288 happen to have the same gray level, namely, 5 (Table 5.1a). One approach is to store this information as it is. Alternatively, we could cleverly define a special command that lets the computer drop all but the first and last addresses for a series of identical gray-level values in between: we could remove the particular value 1023 from the grayscale itself, say, and reserve it as a *flag* to indicate that all the addresses from the flagged one to the next one listed contain the same pixel value (Table 5.1b).

This is an example of perfect or *noiseless* or *lossless* encoding, in which there is no loss of information or introduction of noise. It may be possible to obtain more (or cheaper or faster) compression with slightly noisy (or *lossy*) methods, in which there is some loss of information and degradation of image quality; such methods can be adopted only if images that are decompressed and re-displayed remain clinically adequate.

Errors can, of course, slip into any assembly of large numbers of zeros and ones, especially when there is noise around, and various *error-detecting* algorithms have been devised to check for them. The simplest is through the use of a *parity bit* in binary numbers: the information content of a byte comprises the first seven ones and zeros. For an *even parity* check, add the seven together, and if the sum is odd, place a 1 in the eighth position. The sum of all eight will now be even, indicating that the byte is probably correct. More complex codes can do more, locating and even fixing erroneous bits (most of the time correctly), even if they do take up space in a byte. But they add a layer of confidence in large bank transfers, decisions on responding to warnings of a nuclear attack, etc., when you'd rather not make a little mistake.

Table 5.1 Data compression. (a) Uncompressed. (b) With this form of compression, the particular pixel value "1023" is reserved as a '***FLAG***.' It indicates that the addresses between the current and the next one that is listed all contain the same pixel value (in this case 005.)

Address	Value
5281	620
5282	623
5283	005
5284	005
5285	005
5286	005
5287	005
5288	005
5289	614
5290	612
....	

Address	Value
5281	620
5282	623
5283	(*FLAG*) 1023
5288	005
5289	614
5290	612
....	

5.7 Noise, Statistics, and Probability Distribution Functions

The signal induced within a pickup coil and sent to the RF receiver vacillates continuously over time. It comprises the combination, [$S(t) + N(t)$], of pure MRI spin information, $S(t)$, with *stochastic noise*, $N(t)$. Some people refer to the combination of the two as 'signal,' but for now we shall reserve that term for $S(t)$ alone.

As suggested in the discussion of image quality in chapter 3, the patient's body generates *noise*, such as that accompanying inductive, dielectric, and resistive losses. In addition, small and rapidly varying numbers of electrons flow through the MR receiver electronics, independent of any signal present, spontaneously generating weak random noise voltages. If the magnetic shielding around the electronics is inadequate, any extraneous varying magnetic fields from outside will create electric noise that may be picked up by the system's RF receiver as noise. Even impromptu electrical jolts in the instrument itself, and imperfections in its computer reconstruction algorithms, may add to the problem. To extract as much valuable information as possible from what reaches the receiver, it is necessary to analyze the noise quantitatively and eliminate as much of it as possible. That operation is usually carried out by means of filtering mechanisms that draw upon applicable statistics.

Here we shall review a few rudimentary ideas about statistics, with a focus first on the most commonly encountered form, *Normal* or *Gaussian statistics*. After touching on Poisson noise, we then extend this discussion to the specialized *Rician* statistics that describe the noise in reconstructed MR images more comprehensively than Gaussian statistics, especially when the detected MR signal is relatively low.

Elementary statistical notions

Imagine a pathway of many flagstone rectangles of the same size. Just after a brief drizzle, a first-year radiology resident is invited to determine how many drops hit each square. Rather than count them all, he *samples* N of them, and finds that n_i events are counted in the ith square (Figure 5.19a).

The results are compiled as a *histogram* (Figure 5.19b), and the *sample mean* and *sample standard deviation* are easily calculated from the data with a calculator and a little tedium. These two particular statistics provide important general descriptions valid for all kinds of sampling processes involving random, statistically independent events.

Suppose N of the squares are examined in the sample, and n_i drops strike the ith one. The *sample mean* or

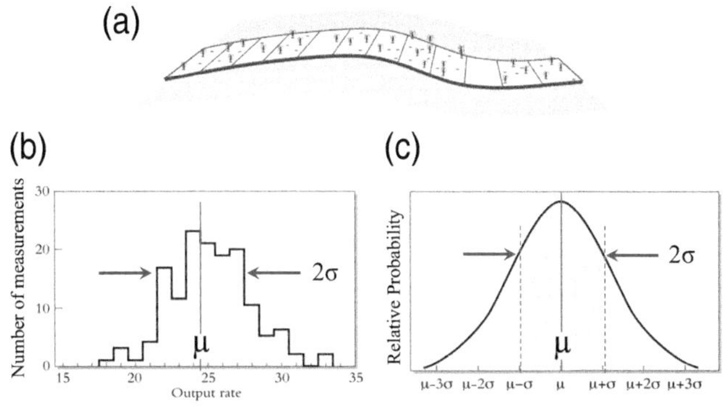

Figure 5.19 Statistical distribution of water droplets landing on paving stones during a light sprinkle. (a) The locations of random events occurring in the squares of a real (or imaginary) matrix. This might reveal, alternatively, the relative numbers of x-ray photons striking separate pixels in a 'uniformly exposed' portion of a digital image receptor. (b) A histogram summarizes the relative numbers of events counted in separate pixels. The mean value, μ, and the standard deviation, σ, were obtained from Equations (5.23). (c) When μ is not too small, a smooth, symmetric quasi-Gaussian probability distribution can sometimes be fitted approximately to the data.

average, commonly represented as μ, is the customary measure of the *central tendency* within a sample:

$$\mu \equiv 1/N \sum_{i}^{N} n_i . \qquad (5.23a)$$

Most likely, this provides a good estimate for the entire population as well.

Even if the patio appears to be fairly uniformly exposed to the sprinkle, the number of raindrop hits will vary from stone to stone, as is apparent from the histogram. The *sample standard deviation*, σ, quantifies the *scattering* of the data about the mean:

$$\sigma \equiv \sqrt{1/N \sum (n_i - \mu)^2} . \qquad (5.23b)$$

Another related and frequently seen parameter, incidentally, is the *sample variance*, which is just the standard deviation squared.

All of this is fine in a descriptive sense, if what we care about is the *mean* number of raindrops per stone in the whole population. If we are instead concerned with finding both the mean and the amount of variation from stone to stone, it becomes necessary to make a big assumption: that a sample of a finite number of observations of raindrops hitting stones is enough to indicate what to expect when examining the entire population of an infinite number of stones. Here we generally wave our hands in the air and say that our *sample mean* and *sample standard deviation* adequately represent the corresponding *population mean* and the *population standard deviation*. It turns out that the *sample standard deviation* (commonly written s) has the same form as in Equation (5.20b), but with the N replaced with an $(N-1)$.

While the standard distribution increases with the square root of the mean, the relative width of the distribution, or the *relative variation*, actually decreases as the mean grows larger:

$$\text{Relative variation} \equiv \sigma / \mu . \qquad (5.23c)$$

The larger the number of counts that occur in the average square, the smaller will be the *relative* magnitudes of the fluctuations in the individual readings about that average, and the more likely it is that the number of counts in any particular square will lie relatively close to the mean. In MRI, this usually means a smooth-appearing image.

EXERCISE 5.23 What can one say about the relative variation of a distribution centered on the origin?

The prospect of a future raindrop hitting in any particular square is, of course, completely independent of what has happened previously there, or anywhere else. The numbers of squares hit with n = 0, 1, 2, 3, 10, or any other integer number of raindrops may be described by a statistical distribution that is fitted to the data histogram. In particular, when μ is not too small, and with a given value of σ, the probability or relative number, $P_{\mu,\sigma}(n)$, of seeing exactly n of them in any particular square can often be approximated or predicted from the *Gaussian* or *Normal* distribution (Figure 5.19c), pretty much your generic bell-shaped curve:

Gaussian/ Normal
$$P_{\mu,\sigma}(n) \equiv (2\sigma^2\pi)^{-1/2}\, e^{-\frac{1}{2}[(n-\mu)/\sigma]^2}. \quad (5.24)$$

The mean number of events, μ, can be any real number (though for our raindrops, $\mu \geq 0$), and σ can be any non-negative real number, but n must be zero or a positive integer. You cannot have a square with –5 or $\sqrt{7}$ hits! Three examples of the Normal distribution appear in Figure 5.20.

A remarkable number of physical phenomena display Gaussian distributions. An explanation of this is provided by the *central limit theorem* of statistics, which says that the *summed behavior* of a large number of independent random processes (such as various sources of stochastic noise) tends to affect a measurement in a balanced way, even if each individual source may be biased toward lower or higher readings. The combined effect of multiple sources of electronic noise, for example, will spread out measurements symmetrically around the average values of the voltages that are detected by the RF receiver, with larger deviations from the expected value being rarer. More specifically, noise fluctuations in the raw signal in the MRI system, including from both patient and electronics, will follow a Normal distribution.

One of the most useful properties of any process describable by a Gaussian curve is that about two thirds of all events will have results lying within one standard deviation of the mean, between $(\mu-\sigma)$ and $(\mu+\sigma)$. Likewise, 95% of the events will lie between $(\mu-2\sigma)$ and $(\mu+2\sigma)$, with counts falling outside of that range only 5% of the time, and so on. Table 5.2 records this important and widely applicable observation.

The voltage signal entering a receiver consists of legitimate information into which is admixed random noise. We shall associate *S*(*t*) with the 'true' input signal, which at any moment centers on $\mu(t)$, and the random noise, *N*(*t*), is likewise parameterized by the standard deviation, $\sigma(t)$ (Figure 5.19 again). The signal component entering the receiver becomes less apparent as the magnitude of *N*(*t*) increases, and an important measure of this noise-degradation is the *Signal-to-Noise Ratio*, or *SNR*, introduced in Equation (3.12), and the inverse of the relative variation:

$$\text{SNR} \equiv S(t)\,/\,N(t) = \mu\,/\,\sigma. \quad (5.25)$$

EXERCISE 5.24 The field of statistics seems to be natural for dealing with separate, countable items or events. How can statistics be applied to a continuous signal S(t) or noise, N(t)?

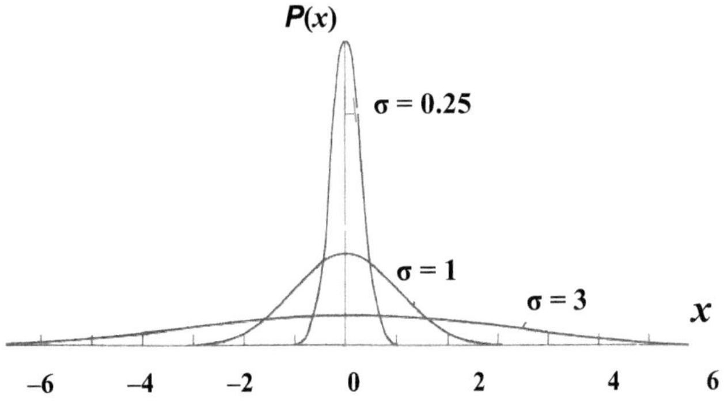

Figure 5.20 Three examples of the Gaussian/Normal distribution, all centered on μ = 0, but with σ = 0.25, 1, and 3, respectively.

Table 5.2 If a curve happens to be of a Normal shape, or even close enough to one, then about ⅔ of the measurements will lie between $(\mu-\sigma)$ and $(\mu+\sigma)$, and about ⅓ outside that range; 95% between $(\mu-2\sigma)$ and $(\mu+2\sigma)$; etc.

Value of *q*	$P_{\mu,\sigma}(q)$
$(\mu-\sigma) < q < (\mu+\sigma)$	0.683
$(\mu-2\sigma) < q < (\mu+2\sigma)$	0.950
$(\mu-3\sigma) < q < (\mu+3\sigma)$	0.997
$(\mu-4\sigma) < q < (\mu+4\sigma)$	0.9999
$(\mu-5\sigma) < q < (\mu+5\sigma)$	0.999999

Notes on Poisson statistics

A process in which discrete and countable identical events occur independently, randomly, and at a constant rate over time, distance, area, or other continuous parameter, is known as a *Poisson* process. Such is the case for a few raindrops falling on patio squares during a light sprinkle, electrons passing through a critical transistor per microseconds, direct or scatter x-ray photons striking the pixels in the matrix of a photodetector image receptor, decays in a sample of long-life radionuclei each minute, and the proton densities and T1-values of voxels in a region of homogeneous tissue. Poisson processes and their statistical distributions are, like the Normal, ubiquitous.

A form of stochastic noise familiar in medical imaging is *quantum mottle*, which may sometimes be clearly visible in x-ray and radionuclide studies. This is not so much a case of there being too many interfering high-energy photons (as with Compton scatter), but rather of an insufficient number of quanta of information being detected, as in Figure 3.16c. This problem can almost always be corrected with an increase in x-ray exposure (mA-s) or dose of radiopharmaceutical (Bq or Ci), but presumably at the cost of greater radiation risk.

For these imaging modalities and countless other situations, the relative numbers of random events that will occur in equal intervals of time, length, area, etc., are describable in terms of *Poisson statistics*. Go back to the raindrops on the patio squares. With the rain falling lightly, briefly, and evenly, the individual events are random, and the future prospect of a hit in any particular square is completely independent of what has happened previously, there or in any other squares. Such a situation is a *Poisson process*, and the actual numbers of squares hit with $n = 0, 1, 2, 3, 10$, or any other integer numbers of raindrops, may be described by the *Poisson statistical distribution*. If the average number of drops per square is μ, then quite remarkably, the probability or relative number, $P_\mu(n)$, of seeing exactly n of them in any particular square, pixel, etc., can be predicted from

$$P_\mu(n) = \mu^n e^{-\mu}/n!\ ,\ n = 0,1,2,\ldots,\qquad n! \equiv n\times(n-1)\times(n-2)\times\ldots\times 1\ . \qquad (5.26a)$$

$n!$ is called *n factorial*, and a good approximation to it is provided by *Stirling's formula*. Poisson statistics are applicable to any Poisson process occurring in time or space, as long as μ is relatively small. It's rather like the way we find exponentials whenever the rate of change of some quantity, $q(p)/dp$, is proportional to how much of it is still around, $q(p)$, at the time, distance, dose, whatever, parameterized by p, as in Equation (5.1a).

Three characteristics of the Poisson distribution may be particularly surprising, noteworthy, and useful. One is that $P_\mu(n)$ is fully characterized with the single parameter, the mean, μ. It is rather astounding that a broad range of seemingly disparate physical events and processes in time or space can all be described with the same general functional form—not only that, but each with only a single, unique physically determined rate constant.

The second, and probably just as unexpected, is that the standard deviation, σ, in a Poisson distribution always happens to be equal to the square root of the mean. If μ raindrops fall per square, on average, then the standard deviation, σ, will be

$$\sigma = \sqrt{\mu} \qquad \text{(Poisson process)}. \qquad (5.26b)$$

In some imaging modalities, the statistic being viewed is the signal strength $S(t)$, and the random variations in $S(t)$, i.e., the noise, $N(t)$, are nearly equal to $\sqrt{S}$:

$$\boxed{\sigma \approx \sqrt{S}} \qquad \text{(many kinds of noise)}. \qquad (5.26c)$$

EXERCISE 5.25 A sidewalk consists of concrete squares, all 1.5 m on a side. After a brief and light rain, our radiology resident finds that a total of about 2200 drops landed on 22 squares. Roughly, how likely is it that a typical square will be hit with between 80 and 120 raindrops? More than 130?

EXERCISE 5.26 Show that, for the same sidewalk, the number of drops in 2/3 of the stones differs from the average by less than 1σ. Why?

The third is that for a sufficiently large mean μ, the Poisson distribution comes to closely resemble the Gaussian (Figure 5.21). With a mean above 100 or so, the two curves become virtually identical. For $\mu > 10$, a small correction to the Poisson formula is needed to make the curves overlap.

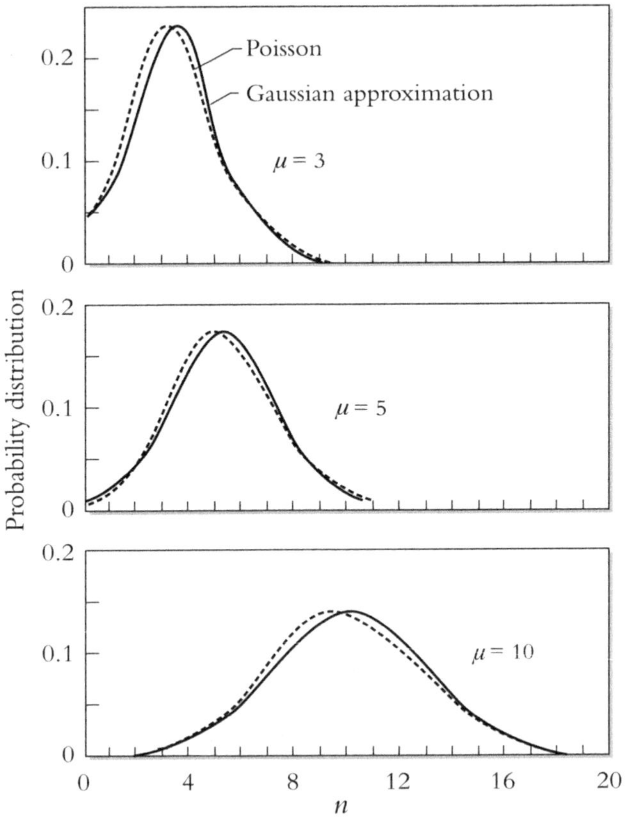

Figure 5.21 As the mean in a Poisson grows larger, its curve comes to resemble the Gaussian/Normal distribution.

Quantifying the stochastic noise level as the average standard deviation in the signal, this yields an SNR for systems undergoing Poisson, or nearly Poisson, random processes as

$$\text{SNR} = S/\sqrt{S} = S^{1/2}. \tag{5.27a}$$

Even for non-Poisson noise, something like this is often the case. So the signal-to-noise ratio happens to improve, at least approximately, with the *square root* of the signal strength. The cleaner image in Figure 3.16, for example, was made up of a hundred times as many photon interactions with the pixels of the image receptor as the other, and its SNR is √100 = 10 times greater.

√NEX×SNR

Equation (5.25a) hints at another way to improve the quality of an MRI study, by taking essentially the same study multiple times, all under the same conditions, and then averaging the raw results. The scanner technologist selects the *number of excitations* to be greater than 1: NEX image data sets are obtained one right after the other and all with exactly the same imaging parameter settings. These independent images should be identical were it not for the noise. The NEX sets of data obtained are averaged, voxel by voxel, and reconstructed to create the final product. The information content for a voxel doesn't change from one image to the next, but the average noise goes down by a factor of √NEX.

How much better is the quality of the averaged image than of the individual ones? When NEX = 2, which is regularly adopted in the clinic as a compromise between image quality and speed, there are twice as many information-bearing quanta that add together coherently, and the two signals from a voxel sum *linearly*, $S = S_1 + S_2$. The randomly distributed noise quanta, on the other hand, do not simply add—to a large extent they cancel out one another by chance, rather, as often as they sum together. The temporally uncorrelated noise, therefore, adds in *quadrature*, and $\sigma_{\text{average}} = (\sigma_1^2 + \sigma_2^2)^{1/2}$. The result is that the SNR improves by a factor of $(\text{NEX})^{1/2}$:

$$\begin{aligned}\text{SNR}_{\text{average}} &= (S_1+S_2)/(\sigma_1^2+\sigma_2^2)^{1/2} \approx (2S_1)/(2\sigma_1^2)^{1/2}\\ &= 2^{1/2}(S_1/\sigma_1) = 2^{1/2}\text{SNR}_{\text{single}}\end{aligned}$$

or

$$\text{SNR}_{\text{NEX}} = \text{NEX}^{1/2}\,\text{SNR}_{\text{single}}. \tag{5.27b}$$

The SNR of a pair of images averaged is thus a factor of √2 higher than for one alone, and the improvement in appearance may be immediately noticeable. One price to be paid, however, is a doubling of the imaging time and, for x-ray imaging, another is twice the radiation dose.

Rician Distribution

As you can doubtless imagine, the incorporation of noise minimization techniques into signal processing methodologies is subtle and involves advanced statistical and other mathematical approaches. Here we have only touched on the tip of the iceberg, and we recommend that the interested reader turn to Barrett's masterful and deep text [Barrett et al. 2004].

To reduce noise, there are benefits to averaging in ***k***-space rather than with real-space images at a later stage of reconstruction, and this is where Rician statistics come in. The properties of the noise within raw data (i.e., in ***k***-space, which is filled with complex numbers) differ from those in images on a workstation display. A raw signal generates a voltage in the receiver that can take both positive and negative values, and the noise in ***k***-space data follows a near-Gaussian distribution. A Fourier transform, moreover, preserves the noise distribution. On the other hand, an MR image in real-space—an image that shows the magnitude of received signal over a region—displays pixel brightness that can assume only non-negative values. The transform from complex ***k***-space image data to real-space information is nonlinear, and the noise distribution changes to a *Rician* distribution. For a high-signal *region of interest* (*ROI*), the noise distribution is virtually the same as a Gaussian. In a low-signal ROI in a reconstructed MR image, however, the noise distribution deviates strongly from the Gaussian and obeys the Rician distribution formula,

$$P_{q,\sigma}(x) \;=\; (x/\sigma^2)\,e^{-(x^2+q^2)/2\sigma^2}\, I_0(xq/\sigma^2). \quad (5.28)$$

where x is the measured intensity and q is a parameter that represents the pixel intensity in the absence of noise. As q increases (and pixel intensity with it) the shape of the distribution morphs toward the Gaussian. $I_0(xq/\sigma^2)$ is the modified Bessel function of the first kind with order zero. This all becomes rather knotty, and probably a good course of action for the interested reader would be to turn to the original source on MRI noise [Gudbjartsson et al. 1995].

Advances in MRI are adding more layers of processing/reconstruction on top of what we have already described. For example, multiple images from different coils can be combined in various ways to optimize SNR and reduce scan speed. Signal and noise properties of the coils themselves are likely to be nonuniform over the tissue being imaged. And some pulse sequences utilize phase instead of magnitude information (e.g., phase-contrast MRA) or nonlinear post-processing (e.g., susceptibility-weighted imaging) that further twists Rician noise into something rather unpleasant; a comprehensive treatment of MRI noise properties for the plethora of reconstruction techniques is beyond the scope of this book. In any case, the concept of signal-to-noise ratio in MRI is evolving, and one must pay attention to how and where it is measured, as well as how the image was generated.

5.8 Sampling an Analog Signal

Some would argue that the original *Sgt. Pepper* on vinyl (26 May 1967...you had to be there) has a warmer tone than the CD released a decade later. While the debate remains unsettled, the *Nyquist-Shannon Sampling Theorem*, developed in the 1940s by those two gentlemen, does shed some light on it. This well-known pillar of signal processing places a lower limit on how much sampling of an analog signal of finite bandwidth is required to capture it fully. To put it another way, it reveals how little sampling you can get away with in digitizing the signal in such a manner that the original continuous-time waveform can later be recovered fully from the digital representation [Smith 1997].

Shannon-Nyquist sampling theorem

Imagine a temporal analog signal that contains components up to a maximum frequency of $\nu_{\text{sig-max}}$ Hz, in the sense of Equations (5.7). To fully capture this signal digitally, according to the Sampling theorem, it is necessary to sample it at a *sampling rate*, ν_{samp}, that must be at least <u>*twice*</u> $\nu_{\text{sig-max}}$:

$$\nu_{\text{samp}} \;\geq\; 2 \times \nu_{\text{sig-max}}\,. \quad (5.29)$$

In other words, if the highest frequency (and shortest-wavelength) contributor to an analog waveform is of frequency $\nu_{\text{sig-max}}$ Hz, then a sampling rate of twice that would be needed to avoid either a loss of information or the introduction of misinformation.

Healthy young people are capable of hearing audio tones from about 20 Hz up to something like 20 kHz. That is the highest frequency at which the eardrum can vibrate with sufficient amplitude to trigger the auditory nerves. So if an analog-to-digital converter (ADC) is to prepare a CD well enough to retain all of the crispness of "Fixing a Hole" so that it can then be reproduced exactly back to analog form, then it has to sample it at a bit more than 40,000 times per second. Indeed, CDs are designed to play back sounds sampled at 44.1 kHz.

In general, the first step in converting an analog signal to digital is to sample the magnitude of the original periodically, every T second, where T and t_{dwell} are standard notations for the *sampling period*, *effective sampling time*, or *dwell time*. The input voltage is sampled at, and only at, the times $t = n\mathrm{T}$ for a sequence of integer values of n, and it becomes $V(n\mathrm{T})$. The *sampling frequency* or *sampling rate*, ν_{samp}, is the number of Samples obtained per second (S/s), and is just the inverse of the dwell time for each sample,

$$\nu_{samp} = 1/\mathrm{T}, \qquad (5.30)$$

reminiscent of Equation (5.3).

The ADC is an electronic counterpart to the *Dirac comb* function, which comprises a succession of Dirac delta functions, $\delta(t)$, all spaced T apart:

$$Ш(t) \equiv \sum_{n=-\infty}^{n=\infty} \delta(t = n\mathrm{T}). \qquad (5.31a)$$

$Ш(t)$ is also known as the *shah* or *sha* function, because its common symbol resembles the letter sha (Ш) of the Cyrillic alphabet.

The sha works through simple multiplication with the continuous signal voltage, $S(t)$, being sampled:

$$S(n\mathrm{T}) = Ш(t)S(t) \equiv \delta(n\mathrm{T})S(t). \qquad (5.31b)$$

This sampling of the input transforms $V(t)$ into a train of discrete impulses indicating the value of the function at every spike of the comb, the red dots in Figure 5.22a. The bit depth of the system, i.e., the number of bits employed in representing the values of $S(n\mathrm{T})$, and a depth of one byte allows 256 values for the sampled voltages. If that is not adequate, one can turn to 2- or 3-byte words, 16 or 24 bits long.

Because it is periodic, the sha operator can also be represented as a Fourier series,

$$Ш(t) \equiv (1/\mathrm{T})\sum_n e^{2\pi i n(t/\mathrm{T})}. \qquad (5.31c)$$

This offers a related but somewhat different route to reconstruction, one that involves use of the *convolution* operation on a signal rather than just straight multiplication by it.

There is some random noise inherent in any analog signal, and some more is introduced by the sampling and digitization processes themselves. The 'noise' or inaccuracy introduced during digitization is rarely a problem, though, when the measurements are taken with sufficient frequency and bit depth.

The *bandwidth* of an apparatus commonly refers to the frequency range over which it can operate properly. Meanwhile, the Nyquist-Shannon theorem, Equation (5.29), demands that if the MRI signal contains Fourier components up to $\nu_{sig\text{-}max}$, then the sampling frequency must be at least twice that to preserve the signal's fidelity during digitization.

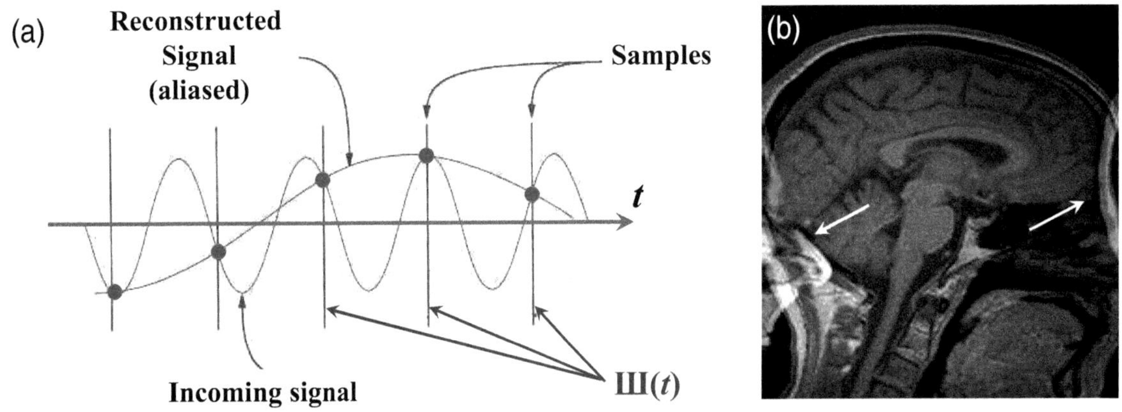

Figure 5.22 Sampling, the Shannon/Nyquist theorem, and aliasing. The system must sample the incoming temporal signal from the MRI pickup coils and receiver enough times and sufficiently quickly. If the highest-frequency sine component of importance inherent within the signal oscillates at ν_{max}, then the signal must be sampled (red dots) at a rate that is at least as fast as $2 \times \nu_{max}$. (a) The incoming short-wavelength signal curve is sampled at a rate that is only about ¾ times that of its frequency. This is too infrequent, with misleading and wrong results, and would introduce *aliasing* into the original image of interest. A longer-wavelength curve reconstructed from the sampled data is a sine wave, and it appears to fit the sampled data nicely. Which is the correct curve-fit? (b) An example of *aliasing* or *wrap-around artifact*, in which the nose to the left is wrapped around from the front of the patient. A clear case of sticking one's nose where it doesn't belong. [Reproduced from Zhuo J., Gullapalli R.P., "AAPM/RSNA physics tutorial for residents: MR artifacts, safety, and quality control." *Radiographics* 2006(26):275–97, with permission of the Radiological Society of North America (RSNA)].

Aliasing artifact

Acquiring too many samples is wasteful of time, which is a prized commodity for MRI, and possibly also of money spent on equipment that is faster than needed. But if not enough samples are obtained, in violation of the theorem's requirement, there will be insufficient information to re-create a high-quality image. How serious a problem can that be?

Suppose we need to deal with an MR signal that seems to be sinusoidal in time, as with the higher-frequency curve of Figure 5.22a. We sample the curve at some temporal frequency, shown as the red dots in the figure. We then give the digitized data (the timing of the red dots) to our research-oriented radiology resident, telling her that we think the incoming signal (which unfortunately is no longer available) is probably a sine curve, and that her job is to try to reconstruct it by fitting a suitable sine curve to the samples.

She begins by selecting and examining a large number of sine curves of different amplitudes, wavelengths, and phases. Over several sleepless nights, she locates the longer-wavelength one in the figure. Success! But being both conscientious and intent on making a good impression, she keeps on going, and eventually finds a second, shorter-wavelength curve. Uh oh! Two curves both seem to fit the data, and there is no way of telling, from the data alone, which is the correct one, if either. If the sampling had occurred at a faster rate, however, at a frequency at least more than twice that of the original, no confusion would have arisen, and it would have been possible to reproduce the correct sine curve unambiguously from the data points.

EXERCISE 5.27 Is that true? What if the original signal, from which the data were obtained, had been of a much higher frequency?

To make this complicated tale quite simple: failure to sample a spatial signal sufficiently rapidly can lead to an *aliasing* or *wrap-around artifact* (Figure 5.22b), and one can end up with the patient's nose out of joint.

5.9 Resolution vs. Detectability

An entire 1D image can be constructed out of a Fourier sum of sine and cosine waves, and the sampling theorem applies here as well. Suppose that the highest spatial frequency component in an analog image is k_{max}, and $1/k_{max} = \lambda_{min}$ is the associated shortest wavelength. How does this relate to the ideas of pixel size and resolution?

Resolution

A *pixel*, square and Δx on a side, is normally taken to be the region of the smallest entity that can be seen on a computer monitor or other display if the image is properly zoomed. With a dot-matrix printer, for example, a pixel might be what covers the environs of a single dot, which is the smallest feature discernible in an image. We might have suspected a few moments ago, before hearing of Nyquist-Shannon, that $\lambda_{min} = 1/k_{max}$ would be of the order of the dimensions of a pixel, and in particular,

$$\lambda_{min} = 1/k_{max} = \Delta x\,, \qquad \text{\underline{\textit{WRONG!}}}$$

This is almost, but *not* quite, right. For an image to be digitized properly, the Nyquist-Shannon theorem requires, rather, that sampling must occur at a rate high enough so that at least *two samples* are collected for any feature in it. A Fourier representation of a pattern must include sine waves short enough to reproduce a feature the size of a pixel, which means that there must be at least *two samples per pixel.*

There's another way to look at this. If two adjacent pixels differ in intensity—one being bright, the other dark—then the wavelength of a sine wave that can reproduce this situation must have a peak in the center of one pixel and a valley in that of the other. The separation of the *centers* of the two is thus *half* a wavelength, on the one hand, and one pixel wide, on the other, which leads to

$$1/k_{max} = \lambda_{min} \leq 2 \times \Delta x\,. \qquad (5.33)$$

(Shorter wavelengths could be included as well, but they should have no bearing on the image.)

If, on the third hand, we ask what is the smallest wavelength *required* for a faithful representation of an entity, then the answer is that the above condition must be met. The shortest wavelength relates to the largest value of $\boldsymbol{k}$, that is, k_{max}. This leaves Equation (5.33), where the factor of two makes all the difference! By analogy, the finest resolution a digital x-ray apparatus can capture corresponds to the width of a *pair* of lines (black and white) in a line-bar test pattern.

EXERCISE 5.28 Show that if the smallest object to be resolved is narrower than half the wavelength of a sine wave, then an equivalent statement of the Shannon sampling theorem is that the wavelength of the smallest sinusoidal wave needed to represent the data should be less than this half-wavelength.

EXERCISE 5.29 If one full wavelength (i.e., one sine cycle) can represent one radiographic line pair, how many line pairs per mm would k = 2 cycles/cm be? What voxel size would this be when the film is digitized?

EXERCISE 5.30 What is the relationship between pixel size and resolution?

So what's a good way to relate resolution to sampling rate and pixel size? Equation (3.6) and Figure 3.9a suggest that resolution refers to the separation between two points that can just barely be distinguished from one another which, of course, also involves both the viewing conditions and a subjective assessment. Figure 3.9b implies that the resolution of a digital image relates to the size (hence, inversely to the number) of pixels needed to produce an image that does not appear to be pixelated. Again, it depends on the eye of the beholder. Something more quantitative could be the MTF of Equations (3.8) and (5.18), where Δx might be taken as the *full width at half maximum* (FWHM) of the associated LSF. And then chapter 7 will demonstrate that the limit of image resolution can be defined in terms of the maximum spatial frequency, k_{max}, required in $\boldsymbol{k}$-space,

$$\Delta x = 1/(2k_{max}), \qquad (5.34a)$$

which, in turn, depends on the size of the feature being viewed. This also follows from Equation (5.33) above if you divide both sides by 2 and ignore the "greater-than" condition. Chapter 7 will also show that the resolution for an image is *inversely proportional* to the *size* of the region of $\boldsymbol{k}$*-space* needed to represent it; in 1D, $\boldsymbol{k}$-space extends from $-k_{max}$ to $+k_{max}$, or over $2k_{max}$, which defines the width of the $\boldsymbol{k}$-space field of view ($FOV_{\boldsymbol{k}}$)

$$\Delta x = 1/\mathrm{FOV}_k. \qquad (5.34b)$$

All of this suggests that while there may be several overlapping ways to approximately define the resolution of an imaging system, or of an image from it, no single one of them serves all purposes.

Detectability requires enough size, sufficiently high contrast, and low noise

The bottom-line issue is that images are intended to help the physician to detect a pathology and to arrive at a correct diagnosis. Fortunately, a wealth of experience acquired over the years provides us with a set of operational ground rules for producing good images, and some of these have already been discussed. In addition, there exist empirical methods that can help in attempts to improve on those practical rules. The best known of these is a transplant from electrical engineering, the *receiver operating characteristics* (*ROC*) curve method introduced in chapter 3.

On a related but different note, what about the ability to detect the presence of a particular smallish anatomic feature in a region, whether or not one can make out its shape? The *detectability*, or *perceptibility* of a feature in an image depends on several factors. It must be of sufficient size and contrast relative to the local background, and it must differ from it at least enough so as not to be lost in the random variations of stochastic noise. But how much is that?

Figure 5.23 illustrates the case of a digital radiograph of a 1-mm^3 nodule embedded within soft tissue. The tissue voxels are 1-mm^3 each, and the face of the radiation detector is partitioned into 1-mm^2 pixels. In a 1-second exposure, 10,000 photons make it through the soft tissue voxels, on average, but the nodule attenuates the x-ray beam by an extra 2%; that is, 200 fewer photons pass through that one voxel than for soft tissue value everywhere else.

In the physically impossible, hypothetical case of Figure 5.23a, there are no random Poisson fluctuations in the number of x-ray photons per pixel. Exactly 10,000 RF photons pass through each soft tissue voxel (except the one with the nodule) and trigger the corresponding x-ray detector pixel. The voxel containing the nodule passes through a signal intensity that is lower than background by 2% and, in the absence of noise, the lesion is in plain sight.

In the more realistic Figure 5.23b, 10,000 photons/mm^2 per voxel make it through to the detector, on average, but the nodule still transmits 2% fewer of them. There is noise present now and, assuming that it is at least roughly Poisson in nature, then the standard deviation σ is the square root of the mean, or $\sqrt{10{,}000}$

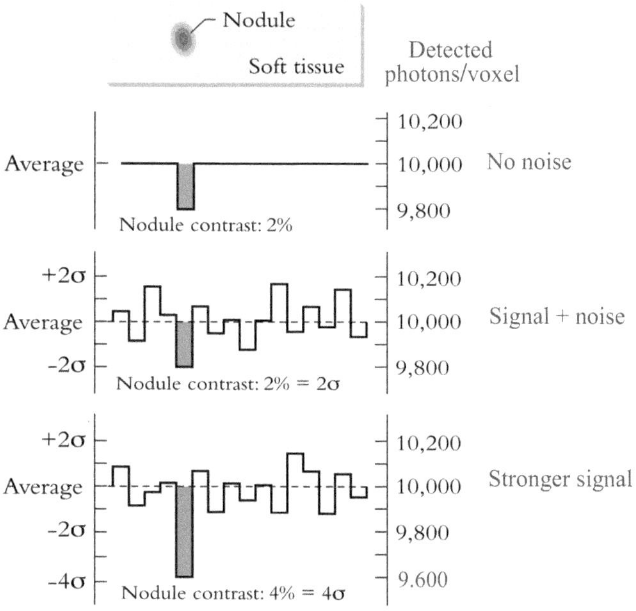

Figure 5.23 A simple example of the statistical detectability of the signal from a small nodule against that of background noise. (a) Hypothetical, completely unrealistic case of no background noise. (b) Signal from nodule here is comparable to the noise level. (c) The drop in amplitude of signal from the nodule is twice as great as the noise level.

= 100, Equation (5.26b). The vertical axis on the right records the number of photons per square millimeter per second that eventually contribute to the x-ray image, and on the left this is expressed in multiples of the standard deviation, σ. With a standard deviation of 100 per mm^2 here, one could say that the nodule transmits less than its surroundings by an amount equivalent to 2σ. But according to Table 5.2, about 5% of all 1-mm^2 areas differ from noise by this amount or more, which means that the nodule can easily be missed.

But as we continue our nodule hunt, we find another with greater contrast, one that transmits 4% less radiation than background, rather than just 2%. It falls to 4σ below the noise level and is easily noted (Figure 5.15c). Alternatively, we can increase the visibility of both by increasing the intensity of the beam by a factor of 4.

EXERCISE 5.31 Explain this last statement.

It would be most helpful if we could combine contrast, resolution, noise, and the capabilities of the eye together in a neat little package, labeled something like *image conspicuity*, to serve as an overall figure of clinical utility of images or imaging systems. The fact that each of these first three fundamental image attributes alone can be analyzed quantitatively suggests that such an integration might be possible. But despite much research activity in the area, and in the closely related field of image analysis by computers, this tantalizing synthesis is yet to be fully achieved.

It complicates matters that different characteristics of images are of primary importance in different diagnostic situations. When looking for hairline fractures, radiographic contrast between bone and soft tissue may already be far more than what is necessary, and only the sharpness of edges is of real interest. In an MRI or ultrasound search for a tumor, on the other hand, the contrast may be of paramount significance. The diagnostic utility of the image is limited by the amount of noise produced by the patient, by the computer data-reconstruction algorithm, etc.

In the late 1940s, Albert Rose produced several papers on the problem of signal detection in noise. He introduced the DQE methodology, Equations (3.13), and soon thereafter proposed what is now known as the *Rose criterion* [Burgess 1999; Barret et al. 2004]. Rose used the *contrast-detail phantoms*, like those of Figure 3.17, to study the ways in which the interplay of contrast, noise, and resolution affects the ability to discern objects. The phantom indicates the levels of contrast required to detect items of different sizes and inherent contrasts, as in the low- and high-noise cases shown. For noise-limited, low-contrast images, the minimum contrast necessary for detection, C_{min}, is related to the area of the object, A, and the signal-to-noise ratio roughly through the semi-empirical relationship

$$(C_{min})(A^{1/2})(\text{SNR}) = \text{constant}. \qquad (5.35)$$

As would be expected, less contrast is required for larger objects and when the noise level is low.

It is frequently posited that a simpler version of this also works: in non-extreme situations, an object is generally detectable if the SNR > 5, that is, if its brightness exceeds the background noise standard deviation by at least a factor of 5.

CHAPTER 6

'Classical' Approach to Proton NMR in a Single Voxel

NMR has been described so far by way of a highly simplified two-state *quasi-quantum* model, with which we derived the Larmor equation. We also described a method of slowly sweeping the frequency of continuous wave (CW) RF power while watching for energy absorption peaks to image the spatial distribution of proton density in a one-dimensional phantom. This is not the way that MRI images are acquired in the clinic, however, so now we turn to an approach that is more realistic.

As indicated in Table 2.6, there is a completely different but complementary view of proton behavior. The *classical* model is based on the observation that a voxel's net magnetization, $\boldsymbol{m}(t)$, produced by the sizable cohort of the protons in it, behaves in a uniform and constant magnetic field much like an ordinary gyroscope moving about in the Earth's gravitational field. The classical picture proves helpful in discussing some of the more subtle aspects of MRI, including T2 relaxation and the various pulse sequences employed in medical procedures.

In practice, MRI involves the switching on and off of carefully designed and timed gradient magnetic fields, along with the application of multiple brief, precisely sculpted *pulses* of radiofrequency energy, each of which is centered at the Larmor frequency. Again, this is radically different from the swept-frequency CW approach employed up until now. Instead, various pulse methods underlie the Free Induction Decay (FID), Spin-Echo (SE), Inversion-Recovery (IR), Gradient-Echo (GE), and other sequences employed clinically. And all of this can be discussed "classically" without any mention of proton quantum states or spin flips.

This chapter is concerned with what happens within a single voxel. We shall name the local external field $\boldsymbol{B}_0$ throughout, rather than $\boldsymbol{B}_z(x)$, and assume a homogeneous magnetic field within the voxel. It is an easy matter to generalize to two dimensions and 3D later.

6.1 Normal Modes of Oscillation

As discussed at the start of chapter 5, entities that experience certain kinds of forces may undergo oscillations that are periodic, and perhaps even nearly sinusoidal. A tuning fork, a pendulum, an electronic *tuned circuit*, and a weight attached to a spring on a table are familiar examples (Figure 6.1a). For these, the magnitude of the oscillations, once begun, will decrease nearly exponentially over time because of frictional or friction-like loss of energy, each system with its own *characteristic time*, T. This time is defined, somewhat arbitrarily, as how long it takes the oscillations to fall to $e^{-1} = 0.37$ of their initial amplitude.

Likewise, a child on a swing is pulled downward by gravity and upward, at a rapidly changing angle, by the ropes. At the peak of its arc, the system has gravitational potential energy; at the bottom, all of that is converted to kinetic energy—which is dissipated by friction if not replenished. This simple pendulum displays a natural periodic motion—its *normal mode* of oscillation—that occurs at its own *natural resonant frequency*.

A conical pendulum in which a weight travels in a circle (Figure 6.2a) undergoes a more complex motion than does an ordinary swing, but it, too, exhibits a normal mode of oscillation. In this case, the resonant fre-

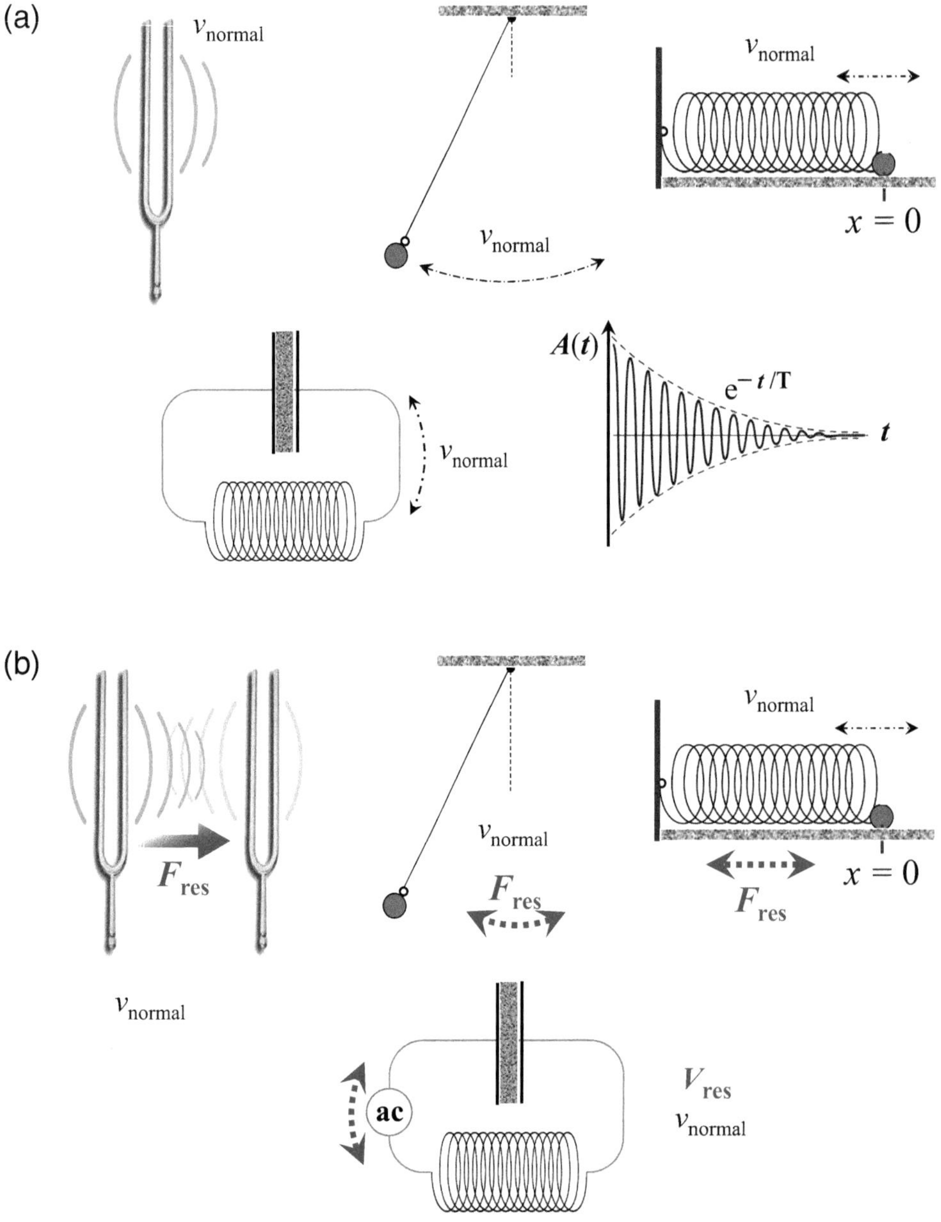

Figure 6.1 Normal-mode oscillations and resonance. Every one of these systems has a unique, sinusoidal *natural periodic mode* of oscillation, at a *natural resonant frequency*, ν_{res}, determined by the values of the physical parameters affecting it. (a) When such a system is disturbed, it will oscillate at its *normal mode*, or *natural resonant frequency*. The amplitude of the cycles decreases nearly exponentially because of frictional or friction-like forces, with a characteristic time, T. (b) On the other hand, when a weak force, $\mathbf{F}_{res}(t)$, is being applied at ν_{res} and in phase with the motion, the amplitude of its oscillations continues to grow for a while, resulting in large and exaggerated motions that indicate the resonance condition. If more energy is being pumped into the system than it can disperse, it begins more complex, non-sinusoidal behavior.

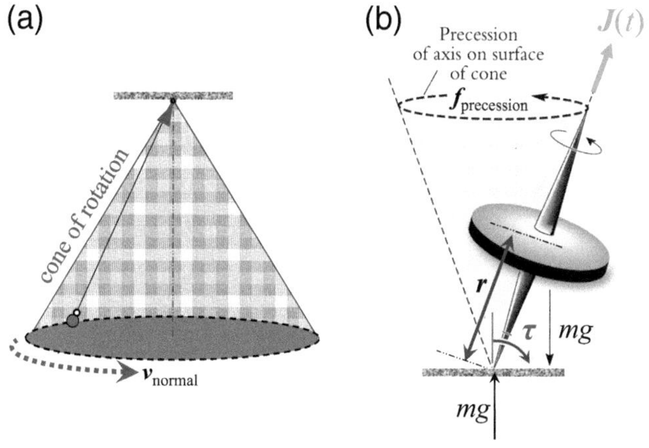

Figure 6.2 Other normal modes of oscillation. (a) Conical pendulum, in which the string holding the swinging weight carves out a hypothetical upside-down vertical cone in one of its normal modes of motion. What might be another? (b) A gyroscope, in which a gravitational torque, $\boldsymbol{\tau}$, is applied perpendicular to the axis of the angular momentum vector, $\boldsymbol{J}$. The normal-mode behavior is *precession* of $\boldsymbol{J}(t)$, and the spin axis, about the vertical. Its natural resonant frequency, or rate of *precession*, is determined by its mass and shape, its instantaneous rate of rotation about its own axis, and the strength of the force of gravity.

quency is determined by the length of the string and by the strength of the gravitational field.

What this is all leading up to is the gyroscope.

If a space station far above the Earth is spinning freely about an axis, with no torques or friction acting on it, neither the rate of rotation nor the orientation of the axis will change perceptibly over time, in accord with the law of the conservation of angular momentum.

But a gyroscope on Earth, supported at the bottom of its axis of rotation, is not in a state of unchanging motion. Two distinct forces act on it (apart from any friction): gravity pulls straight down at its center, in effect, and the pedestal supporting its lower end pushes upward (Figure 6.2b). This pair of forces exerts a *torque*, $\boldsymbol{\tau}$, or twisting force, on it that is perpendicular to the axis of rotation, and which attempts to topple the gyroscope over. But its spin, and conservation of the angular momentum of its wheel, $\boldsymbol{J}$, resist its falling. Its *equation of motion* is

$$d\boldsymbol{J}/dt = \tau, \qquad (6.1a)$$

where expressing $\boldsymbol{\tau}$ in terms of the forces acting on the gyroscope. Solving the equation of motion is left as an exercise.

EXERCISE 6.1 Find the equation of motion for a frictionless gyroscope, and find a solution.

The result is a new, and perhaps unanticipated, kind of motion: the gyroscope undergoes *precession*, its normal mode of motion, a swinging of its axis of rotation along the surface of a hypothetical vertical cone. The rate at which it is spinning about its own axis at any moment does not change, in the absence of friction, nor does its rate of precession. It may not be obvious why this should happen—but it is easily explained by working through the Newtonian equation of motion. As you might suspect, the stronger the gravitational field, the faster the rate of precession—a gyroscope on the Moon would take six times as long to make a complete circuit as when on Earth. We'll find the same kind of thing happening with the magnetization coming from a cluster of protons precessing in a magnetic field, only now the angular momentum vector of a proton is written $\boldsymbol{I}$, from Equations (2.5b) and (2.8), and now

$$d\boldsymbol{J}/dt = \tau, \qquad (6.1b)$$

The torque is provided by the magnetic field acting on it, $\boldsymbol{B}_z$.

6.2 Resonance at the Normal Mode Frequency

Many kinds of mechanical, electrical, acoustic, optical, and other systems display resonances at their natural resonant frequencies. Even a weak external driving force, $\boldsymbol{F}(\nu_{drive})$, can cause the eventual buildup of large-amplitude oscillations, if applied in phase and at a natural resonant frequency (Figure 6.1b).

$$\boxed{\nu_{drive} = \nu_{res}\,.} \qquad (6.2)$$

That is, energy is transferred to it very efficiently through application of a periodic, resonance-frequency force in phase with the motion. Only if the system loses energy as fast as it is supplied will the motion stabilize.

A girl on a swing offers a happy example. If you provide many little shoves in very rapid succession, the swing will barely move. Likewise, if you lean into her slowly, the swing will just move along with your hand. But if applied at the resonant frequency of the daugh-

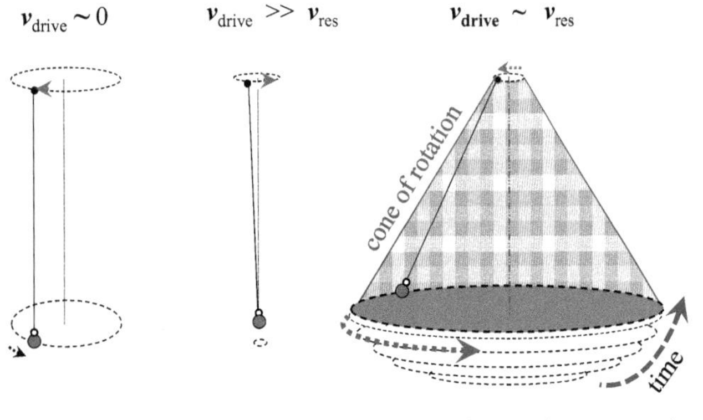

Figure 6.3 Resonance. Only when the driving force is at the resonant frequency ν_{res} and in phase with the motion of the mass will power be transferred to a circular pendulum. When that does begin to happen, however, the orbit (and the cone) will open up, and the plane of the orbit will rise. Eventually, the bob is whizzing around in a horizontal plane and ever-increasing angular velocity—a condition beyond resonance referred to as *saturation*.

ter-plus-swing system and in phase with its natural oscillations, even modest pushes will cause the eventual buildup of large-amplitude oscillations. Resonance, and squeals of delight!

Continuous wave (CW) NMR was carried out in earlier chapters in essentially this manner. A transmitter provided a steady supply of RF of constant amplitude, but with slowly increasing frequency. Absorption of RF power goes through a peak only very near and at the resonance frequency (Figure 2.7).

Let's return to a circular pendulum (Figure 6.3). If one moves the upper end of the string slowly in a circle at very low frequency, then the bob will simply follow along, tracing out the same circle. If the circular driving force is of high frequency, then the bob will have little time to follow, and it will remain nearly immobile. But if you move the string around at the system's normal mode frequency, then the weight will move in ever higher, larger-diameter circles. The string travels on the surface of an inverted cone, and the cone opens up slowly over time up until energy dissipation equals its input. You did this a thousand times as a child.

Resonance of a swing (or an electronic tuned circuit, or anything else) can be seen in a very different, but closely related, way. You can give the kid one big push, and she and the swing will oscillate back and forth at the same resonant frequency as before. Same as with clanging a bell. The pulsed NMR technique, which forms the basis for virtually all clinical MRI work, involves the brief application of a pulse of resonance-frequency RF power, and watching what happens after.

Precession and nutation of a gyroscope

Now we are ready to move on to resonance of a precessing gyroscope, which is analogous to the case of NMR. Resonant transfer of energy to it, and its response, can be demonstrated by pulling very lightly on a thread attached to it (Figure 6.4a). A particular kind of pull is required, though: the tension in the thread should be held constant, but the direction from which the pull comes must move in a circle, in phase with and tangent to its line of motion of precession at any instant. There are now *two* torques acting on the system: a strong one due to gravity, and the other caused by the thread—and these together cause the gyroscope to undergo a more complicated sort of motion. In addition to the axis of rotation traveling on the surface of a vertical cone, precession, the system now *also* undergoes much slower *nutation*: the conical surface *itself* opens up very slowly, relative to the rate of precession, like the glorious welcoming of a morning glory at dawn (hey, this is supposed to be a serious work of literature, too). The wheel's axis of rotation now spirals gently down toward the horizontal (Figure 6.4b). If the direction of the force exerted by the thread changes more slowly or rapidly than the natural precession rate, it will have little effect on the gyroscope's

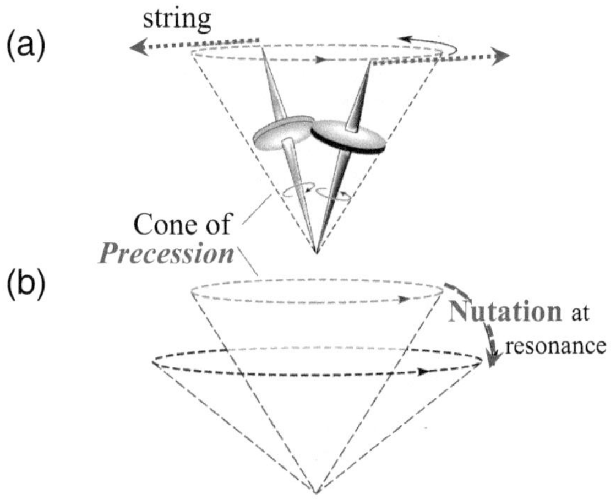

Figure 6.4 It is possible, via a somewhat nonintuitive way of applying torque to a gyroscope (a) to provide it with resonant-frequency power, resulting in an also nonintuitive response: *nutation*, (b) in which the cone of precession itself slowly opens up.

long-term behavior. But when applied at the resonance (like Larmor) frequency—voilà! Newton's laws explain it all.

For a wide range of oscillatory phenomena in all branches of physics, incidentally, the variation in amplitude, $A(\nu)$, with frequency of applied power, is much the same. As the system approaches and passes through the resonant frequency, it can often be described by an approximately symmetric *universal resonance function*, also known as the Lorentz or Cauchy distribution (Figure 6.5):

$$A(\nu) \sim (\Gamma/2\pi) / \left[(\nu - \nu_{res})^2 + (\Gamma/2)^2 \right]. \quad (6.3)$$

With this formalism, Γ is the *linewidth* of the resonance, commonly taken to be the full width at half maximum (FWHM), which is determined by the magnitude of the relaxation, friction, or other damping effects. A weakly damped periodic system has a small Γ—that is, the resonance curve has a narrow linewidth—and one says that its "Q" ($\equiv 1/\Gamma$) is high. A high-Q church bell typically has a powerful, narrow dominant frequency at resonance, a nearly monochromatic, sharp sound of a distinct pitch, and of long duration. Analogously, the widths of NMR lines for a tissue are determined largely by the rates 1/T1 and 1/T2 at which relaxation processes take place, as will be seen.

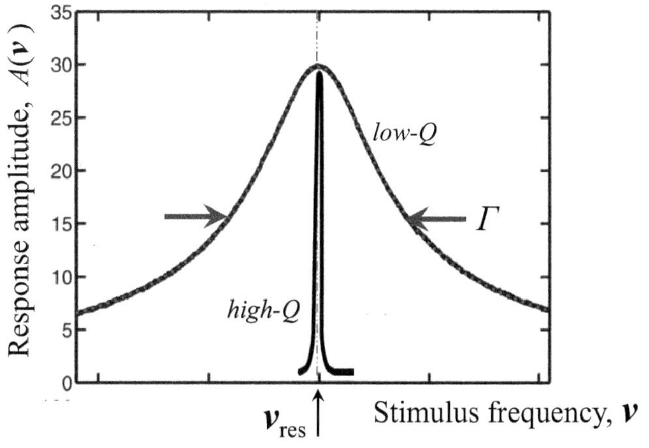

Figure 6.5 Lorentz/Cauchy distribution, known also as the 'universal resonance function,' describes the amplitude of a system's response as the frequency of a sinusoidal driving force of frequency ν passes through resonance, Equation (6.3). A 'high-Q' system has a narrow resonance. The linewidth of the spectrum of the ringing of a bell, Γ, increases with the amount of damping, or relaxation, present, and the system's 'Q' is said to decrease.

EXERCISE 6.2 What happens to $A(\nu)$ when $\nu = \nu_{res}$, and why?

6.3 In a Voxel, Classical Precession of $m(t)$ about B_0 at ν_{Larmor}; The Bloch Equations

As noted already, perhaps a few too many times, a nucleus acts somewhat like a spinning charged body, and so it possesses both angular momentum, $\boldsymbol{I}$, and a magnetic moment, μ. The greater the amount of spin of a nucleus, the greater the magnitude of its angular momentum, I, and the stronger will be the field that it itself produces, Equations (2.5).

QM allows us to describe the *average* behavior (or in the quantum vernacular, the *expectation value*) of a *large population* of spins in a 1-mm^3 voxel, a reasonable fraction of the 10^{20} or so protons in it. Further, the quantum formalism says that we can even do so classically and completely, according to old-fashioned Newtonian physics, as long as we consider only expectation values for *clusters* of nuclei, and not individuals. Because of this, virtually all the practical physics of NMR and MRI are enshrined in a system of *Newtonian* expressions known as the *Bloch equations*. Although they are fully and ultimately derived from quantum theory, these end up as completely *classical* descriptions of the phenomenological kinetics of nuclear magnetic resonance. In particular, similar sets of equations capture the detailed motions both of a gyroscope in a gravitational field and of a cohort of protons in a magnetic field.

A spinning body experiencing an external torque is governed by its equation of motion, $d\boldsymbol{J}/dt = \boldsymbol{\tau}$, Equation (6.1). For a spinning charged body, it becomes $d(\boldsymbol{\mu}/\gamma)/dt = \boldsymbol{\tau}$ by Equation (2.5b). When summed and averaged over a cluster of protons in a voxel, it yields an expression for the dynamics of the expectation value of the nuclear magnetization vector, $d(\boldsymbol{m}/\gamma)/dt = \boldsymbol{\tau}$, where $\boldsymbol{m}$ was defined at the beginning of chapter 4 and in Equation (4.3a) as the expectation value for a sufficiently large assembly of protons. The *Lorentz* torque on a magnetic moment $\boldsymbol{m}(t)$ in the field $\boldsymbol{B}_0$ is the vector cross product $\boldsymbol{m} \times \boldsymbol{B}_0$. In the absence of relaxation or other processes, the equation of motion for the ensemble becomes the bare-bones, three-dimensional *Bloch equation* (Table 6.1):

Table 6.1 The *Bloch equations*, when $\mathbf{B_0}$ is the only magnetic field present (i.e., no gradient or ν_{Larmor}-RF energy is being applied), and no relaxation processes are in effect.

$$
\begin{aligned}
d\boldsymbol{m}/dt &= \gamma\,(\boldsymbol{m}\times\boldsymbol{B}_0)\\
dm_x/dt &= \gamma\, B_0\, m_y\\
dm_y/dt &= -\gamma\, B_0\, m_x\\
dm_z/dt &= 0
\end{aligned}
$$

$$
\begin{aligned}
m_x(t) &= m_{xy}(0)\cos 2\pi\nu t\\
m_y(t) &= -\,m_{xy}(0)\sin 2\pi\nu t\\
m_z(t) &= m_z(0)
\end{aligned}
$$

$$d\boldsymbol{m}(t)/dt = \gamma\boldsymbol{m}(t)\times\boldsymbol{B}_0. \tag{6.4a}$$

The Cartesian components of this are

$$
\begin{aligned}
dm_x/dt &= \gamma B_0 m_y\\
dm_y/dt &= -\gamma B_0 m_x\\
dm_z/dt &= 0,
\end{aligned}
\tag{6.4b}
$$

where, as before, $\boldsymbol{B}_0$ is aligned along the z-axis. The first two of these three simultaneous first-order differential equations are coupled, but the third may be addressed separately. The solution to them, apart from initial phase angles, is

$$
\begin{aligned}
m_x(t) &= m_{xy}(0)\cos 2\pi\,\nu_{Larmor}\,t\\
m_y(t) &= -\,m_{xy}(0)\sin 2\pi\,\nu_{Larmor}\,t\\
m_z(t) &= m_z(0),
\end{aligned}
\tag{6.5}
$$

where

$$\nu_{Larmor} = \gamma B_0/2\pi = 42.58\,B_0.$$

The interpretation of Equations (6.5) is routine (Figure 6.6). The first two say that the projection of $\boldsymbol{m}(t)$ onto the horizontal $\boldsymbol{x}$-$\boldsymbol{y}$ plane, namely the *transverse magnetization* vector $\boldsymbol{m}_{xy}(t)$, rotates at the Larmor frequency. The third reveals that $\boldsymbol{m}_z(t)$, the *longitudinal magnetization* vector, does not vary over time. So as $\boldsymbol{m}(t)$ precesses, just like the gyroscope in the gravitational field, it traces out the surface of a cone aligned along the external magnetic field. The angle of the cone—that is, the angle that $\boldsymbol{m}(t)$ makes with the vertical—can assume any value at all. If there are no relaxation mechanisms at work, the height and width of the cone remain constant.

EXERCISE 6.3 Modify the Bloch equations so as to account for T1 relaxation.

One aspect of this figure may have left you wondering. While it sounds fine for the magnetization of a bunch of protons to be precessing in a cone, the earlier spin-up/spin-down picture suggests, to the contrary, that the 10^{-6} fraction of excess spins should be aligned pointing fully upward along the +z-axis. How, then, could it be that any component of $\boldsymbol{m}(t)$ could be rotating in the $\boldsymbol{x}$-$\boldsymbol{y}$ plane? How do we explain $\boldsymbol{m}(t)$ skating around on the cone surface ν_{Larmor} times per second? Here we need to undertake a giant leap of faith, and maybe a dash of quantum slight-of-hand, and even then it might require some smoke and mirrors. If you're feeling somewhat perplexed about all this, just bear in mind Richard Feynman's mollifying dictum: "I think I can safely say

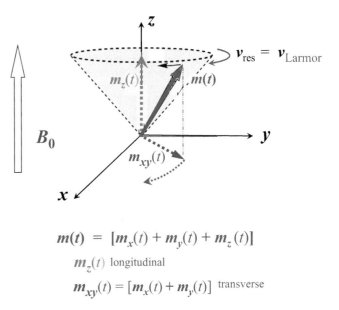

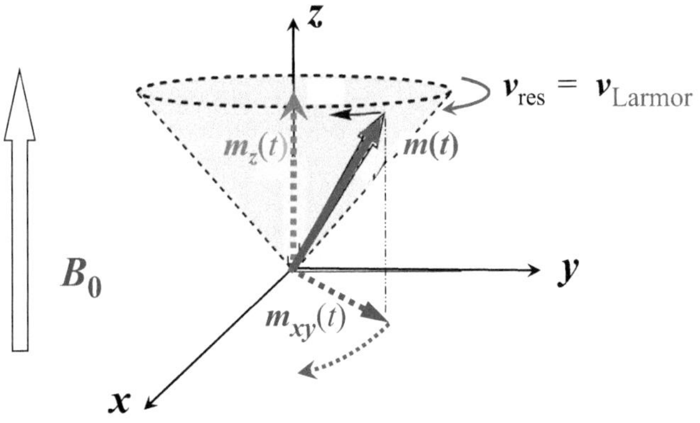

$$\boldsymbol{m}(t) = [\boldsymbol{m}_x(t) + \boldsymbol{m}_y(t) + \boldsymbol{m}_z(t)]$$

$\boldsymbol{m}_z(t)$ longitudinal

$\boldsymbol{m}_{xy}(t) = [\boldsymbol{m}_x(t) + \boldsymbol{m}_y(t)]$ transverse

Figure 6.6 The normal mode behavior of the *net magnetization* of a cohort of protons in a voxel, $\mathbf{m}(t)$, in an external magnetic field, $\mathbf{B_0}$. Its precession about the field is very much like that of a top or ordinary gyroscope in a gravitational field. As described by the Bloch equations, the precessional (Larmor) frequency is proportional to the strength of the field, $\nu_{Larmor} = 42.58\,B_0$, which is exactly the same Larmor equation that was found with the simple quasi-quantum approach! (What if that were not so?) With this precession about the $\mathbf{z}$-axis, it is possible to consider the $\mathbf{z}$- (longitudinal) and the $\mathbf{x}$-$\mathbf{y}$ (transverse) components of $\mathbf{m}(t)$, namely $\mathbf{m_z}(t)$ and $\mathbf{m_{xy}}(t)$, separate from one another, and independent.

that nobody understands quantum mechanics." [Feynman 1965].

6.4 Classical View of NMR: Nutation of $m(t)$ about B_1 at $(B_1/B_0) \times \nu_{Larmor}$

Life is pretty straightforward when there is only a static, main magnetic field present, $\boldsymbol{B}_0$. But the situation becomes more interesting when the protons in a voxel are exposed to an additional, relatively weak magnetic RF field, $\boldsymbol{B}_1(t)$, designed specifically to circle in the horizontal plane at the Larmor frequency and with the correct phase. It must always, moreover, point toward (*not along*!) $\boldsymbol{B}_0$ and the z-axis. The RF will have the same effect on $\boldsymbol{m}(t)$ as does the thread on the gyroscope (Figure 6.4a). During the brief application of resonant energy, the general Bloch equation becomes

$$d\boldsymbol{m}/dt = \gamma\, \boldsymbol{m}(t) \times \left[B_0 + B_1(t)\right], \quad (6.7a)$$

where

$$|B_1(t)| = \mathrm{Re}\left[B_1 e^{-2\pi i \nu_{Larmor} t}\right]. \quad (6.7b)$$

$|\boldsymbol{B}_1(t)|$ refers to the real part of the complex function $[B_1 e^{-2\pi \nu_{Larmor} t}]$, where the magnetic component of the RF field is applied perpendicular to $\boldsymbol{B}_0$.

Typically, $\boldsymbol{B}_1(t)$ produced by a simple coil is *linearly* polarized and aligned along the $\boldsymbol{x}$-axis (or $\boldsymbol{y}$-axis) (Figures 6.7a and 6.7b). But it can be viewed or decomposed as the superposition of two *circularly* polarized magnetic fields rotating in opposite directions (Figure 6.7c). Only half of the overall power is turning in synchrony with the proton precession, and thereby capable of causing nutation (or spin transitions, in the quasi-QM picture). The other half is turning in the 'wrong' direction and has negligible effect on anything, apart from the deposition of heat in the tissue, a topic that will be discussed in the chapter on MR safety. Quadrature transmission, coming up in two chapters, and other techniques can reduce the wastage of RF and unnecessary tissue heating.

Solving Equations (6.7) is a bit more challenging when $\boldsymbol{B}_1 \neq 0$, but the meaning of the results is clear. If the RF energy being generated is much above or below ν_{Larmor}, then $\boldsymbol{m}(t)$ would simply continue to precess in its normal-mode of motion. But RF that is resonant with the spin system will cause the net magnetization to undergo *nutation*, in addition to precession, and *the cone of precession opens up*, reminiscent of Figures 6.3

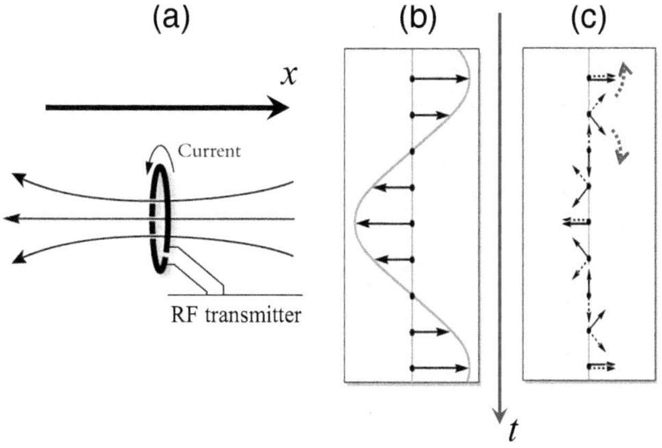

Figure 6.7 RF fields produced by a simple coil (a) and (b) are linearly polarized along the **x**-axis, seen here from above. (c) This linearly polarized field can be resolved into constituent parts as the superposition of two circularly polarized fields turning in opposite directions. Only the one rotating *with* the proton spins affects them, and causes them to undergo NMR or MRI.

and 6.4b. Here Figure 6.8 provides two similar depictions of the single 'classical' picture of the NMR phenomenon: near thermal equilibrium, and with $\boldsymbol{m}_0 = \boldsymbol{m}_z$ starting out lying along $\boldsymbol{B}_0$, say, the hypothetical cone itself on which $\boldsymbol{m}(t)$ is precessing opens up slowly, while $\boldsymbol{B}_1$ is on, and at some point, it briefly coincides with the horizontal plane. After that, it begins folding up again, but now below the $\boldsymbol{x}$-$\boldsymbol{y}$ plane, where it resembles Figure 6.3. The nutation continues until the magnetization points due south, and then it starts heading back upward again, eventually returning to where it started. Then the whole cycle repeats itself. But this process takes place *only while the RF power is still being applied*, and that happens only fleetingly in MRI, during an RF pulse that is milliseconds in duration. In fact, nearly always in clinical MRI, RF power application and the nutation occur fleetingly—typically about the time it takes to drive $\boldsymbol{m}(t)$ from along the z-axis down into the $\boldsymbol{x}$-$\boldsymbol{y}$-plane—and sometimes much less.

The idea of $\boldsymbol{m}(t)$ spiraling down from the vertical into the horizontal plane (or perhaps through some other angle) is an essential linchpin of the theory of NMR. So let's go through it again a bit differently, and consider the centrally important 90° RF pulse.

6.5 Grand Entrance of the 90° Pulse

Suppose a voxel of protons starts off at thermal equilibrium in a strong, polarizing magnetic field. The slight

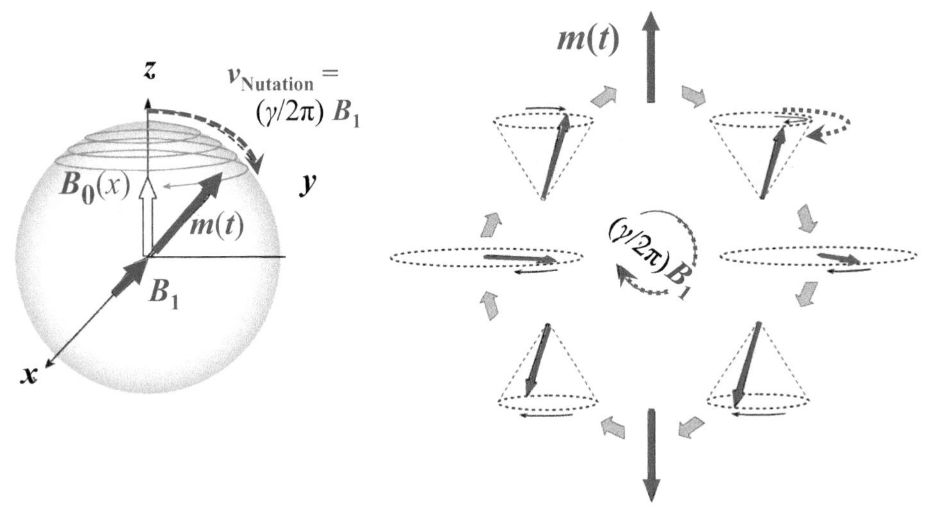

Figure 6.8 Two complementary views of $\mathbf{m}(t)$ *precessing* at the Larmor frequency in a static $\mathbf{B}_0$ field, during the occasional instances that a Larmor-frequency field, $\mathbf{B}_1(t)$, is being applied. During those brief moments, $\mathbf{m}(t)$ undergoes *nutation* as well, with the cone of precession opening up and refolding at the rate $\nu_{\text{nutation}} = 42.58\, B_1$. This is just the Larmor equation, again—but now the relevant magnetic field is $\mathbf{B}_1(t)$, which is thousands of times weaker than $\mathbf{B}_0$, perhaps only 25 microtesla (μT) or so.

excess of spins aligned along $\boldsymbol{B}_0$ leads to a net *longitudinal magnetization* only along the *z*-axis, $\boldsymbol{m}_z(t)$, initially of magnitude m_0. The spin vector components in the transverse plane sum to zero, on average, creating no transverse magnetization, and $\boldsymbol{m}_{xy}(t) = 0$ at this point.

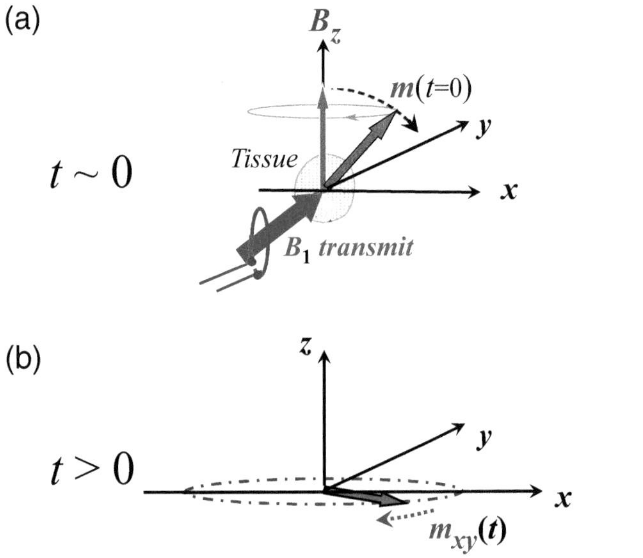

Figure 6.9 NMR of protons in a voxel in a uniform 1.5-T field. Suppose the spins start out in a state of thermal equilibrium, with the magnetization $\mathbf{m}_z(t = 0_-) = \mathbf{m}_0$, just before $t = 0$. (a) At $t = 0$, a 90° pulse of RF energy at the Larmor frequency for water or tissue protons is pumped through the *transmit coil*, thereby creating $\mathbf{B}_1(t)$ (*red arrow*) which is orthogonal to $\mathbf{B}_0$. The pulse causes the magnetization, blue arrow, to nutate rapidly from the vertical down into the horizontal **x-y** plane. (b) At time $t = 0_+$, soon *after* completion of the 90° pulse, this voxel magnetization begins precessing in the **x-y** plane and is renamed $\mathbf{m}_{xy}(t)$ (blue arrow again), with $|\mathbf{m}_{xy}(t = 0_+)| = |\mathbf{m}_z(t = 0_-)|$. In the absence of relaxation or other disruptive processes, it will continue circling there indefinitely, at constant magnitude.

The plan is to apply a 90° pulse at $t = 0$ and keep track of what happens. Just *before* the 90° pulse, at $t = 0_-$, the only component of net coherent magnetization in existence points fully in the *z*-direction, and is of magnitude $m(t) = \boldsymbol{m}_z(0_-) = \boldsymbol{m}_0$.

At $t = 0$, the transmitter pumps a carefully sculpted '90° pulse' of Larmor-frequency power into a *transmit coil* (Figure 6.9a, red arrow) arranged so that the RF magnetic field is normal to the *z*-axis. $\boldsymbol{B}_1(t)$ is designed to be of exactly the right combination of intensity and duration to rapidly swing the voxel's magnetization, $\boldsymbol{m}_z(t)$, (upper blue arrow) from its vertical orientation through 90° down into the horizontal plane, where it becomes $\boldsymbol{m}_{xy}(0_+)$. The RF field is then switched off immediately. The magnetization, renamed $\boldsymbol{m}_{xy}(t)$, begins to precess freely in the horizontal plane (Figure 6.9b, lower blue arrow) as the *transverse magnetization*.

To reiterate this important point: just before the saturation pulse, the net coherent magnetization $\boldsymbol{m}(t)$ was aligned vertically and was of magnitude $|m_z(0_-)|$, and we called it $\boldsymbol{m}_z(t)$. Immediately after, it has been transmuted into $\boldsymbol{m}_{xy}(0_+)$. It is precessing at the Larmor frequency in the transverse *x-y* plane, and is still of exactly the same magnitude (Figure 6.9b),

$$\left|\boldsymbol{m}_{xy}(0_+)\right| = \left|\boldsymbol{m}_z(0_-)\right|, \tag{6.8}$$

The radial *direction* in which $\boldsymbol{m}_{xy}(t)$ points rotates repeatedly through 360° at ν_{Larmor}, but its *magnitude* hardly varies at all (still assuming very little relaxation). The physical metamorphosis described by this equation is a simple but critically important one, and variations on it will reappear on several occasions.

EXERCISE 6.4 What does Equation (6.8) say is going on physically between the times 0_+ and 0_-?

EXERCISE 6.5 Given some value for B_1, find $t_{90°}$. With a typical $B_1 = 25\ \mu T$ and any B_0, how long does it take for m(t) to spiral through 90° from the vertical down into the x-y plane?

One creates a 90° pulse by activating an RF field of magnitude B_1 for a period of time Δt such that

$$\xi B_1 \times \Delta t \ = \ 90°, \tag{6.9}$$

where ξ is an appropriate conversion factor. Thus both pulse strength and duration can be adjusted to find a suitable combination. This assumes that the envelope of the RF pulse is rectangular in form.

EXERCISE 6.6 In reality, the shape of its envelope is almost always more complex. How might one explain and express that? Now generalize Equation (6.9).

6.6 Laboratory Frame of Reference vs. One Rotating at ν_{Larmor}

An outside observer seated in the *Laboratory* or *Fixed Frame of Reference* would see a magnetization vector precessing at the Larmor frequency about the main field B_0. When a weak resonant RF field is turned on, the magnetization would also appear to be nutating, as in Figure 6.8—a rather complicated picture.

Suppose you're at Coney Island, watching the children bob up and down on the merry-go-round horses. The motion is too complex for you to determine clearly which boys are throwing things at whom, so you jump on board, too. The children are still oscillating vertically, but now the circular motion disappears, in effect, and everyone is at a fixed position on the carousel. The action is much easier to untangle.

Likewise, the behavior of ***m***(*t*) over time is most readily followed by someone in a frame of reference that rotates at the precessional frequency with it. Consider first the situation in which $B_1 = 0$. From the vantage of an observer going around in the *rotating frame of reference*, ***m***(*t*) seems to do nothing at all when it experiences only $\boldsymbol{B}_0$. It just sits there, unchanging over time (Figure 6.10a). It's as if the act of transforming to the rotating frame switches off the main magnetic field, B_0, just as jumping onto the carousel seems to make the rotational motion go away. The main field is really still present, of course, but when observing its influence on the magnetization from the vantage point of the rotating frame, we can temporarily pretend that it is not. The mathematics of this transformation from fixed to rotating frame is not particularly illuminating, so we'll not bother with it here.

EXERCISE 6.7 How much faster is ν_{Larmor} than $\nu_{nutation}$?

Now turn on $B_1(t)$ (Figure 6.10b). Seen from the Laboratory Frame, $\boldsymbol{B}_1(t)$ is circling horizontally at the Larmor frequency, the red arrow in the *x′-y′* plane. Viewed from the Rotating Frame, on the other hand,

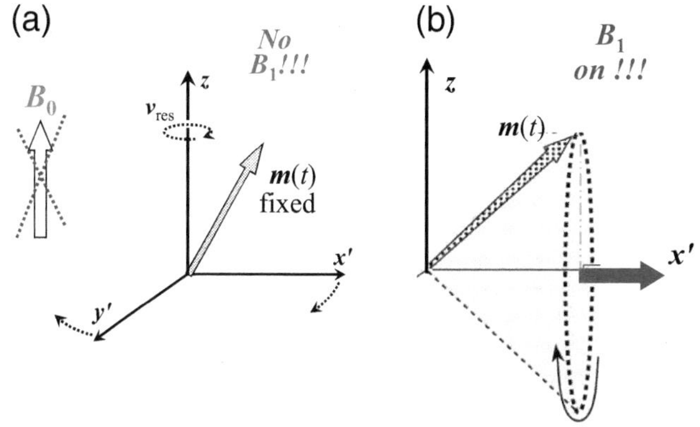

Figure 6.10 Shifting ***m***(*t*) from the Laboratory Frame of Reference to the Rotating one, assuming no (or extremely slow) relaxation. The laboratory frame **x-**, **y-**, and **z-**coordinate system transforms into the **x′-**, **y′-**, **z-**system which rotates about the **z-** = **z′**-axis at ν_{Larmor}. (a) When you are on the carousel, the ponies go up and down, but they don't move forward. Similarly, when viewed from a frame rotating at the Larmor frequency, $\mathbf{B_0}$ magically appears to vanish. Thereafter, when $\mathbf{B_1}$ is off, which is the case nearly all the time, then ***m***(*t*) seems static in the rotating frame, not precessing or doing anything else. (b) But during those brief moments that the resonant RF is turned on, a new *fixed* (not oscillating) field, $\mathbf{B_1}$, aligned along the rotating **x′**-axis, materializes, and ***m***(*t*) nutates ('precesses') about *it*, at the rate 42.58 × B_1.

$\boldsymbol{B}_1(t)$ *seems* to be a fixed, constant field pointing along the (rotating) x'-axis—in fact, it's the only field present in that frame (still pretending that $\boldsymbol{B}_0$ has vanished). Then $\boldsymbol{m}(t)$ 'precesses' around *it* (rather than around B_0, which has 'vanished') with a *nutational* frequency of

$$\nu_{\text{nutation}} = (\gamma/2\pi)B_1 = (B_1/B_0)\nu_{\text{Larmor}}. \quad (6.10)$$

This is just a new and somewhat different application of the Larmor equation. Just as the *precession* rate is proportional to the strength of the main field, B_0, likewise the frequency of *nutation* also is linear in the RF field strength, B_1.

B_1 is typically something like 25 μT, five or so orders of magnitude weaker than B_0. So while $\boldsymbol{m}(t)$ may *precess* at a rate of 63.87 MHz in a 1.5 T field, the magnetization will *nutate* at a frequency of only 42.58 [MHz/T] × 25 × 10^{-6} [T] ~ 1 kHz, give or take. As will soon become evident, it is the variations over time of this nutation phenomenon that carry the patient-specific intelligence needed for the creation of an MR image. That is, although the MRI signal emerges from the patient's body and is detected by the device's RF receiver at around 64 kHz in frequency, the portion of this that conveys the information material to image reconstruction is contained within a band only a few kilohertz wide!

Suppose that the magnetization starts off aligned parallel to the external field, but a 90° pulse is applied. The duration of the time that B_1 is to be applied, $t_{90°}$, may be determined from Equation (6.8) and from the condition

$$t_{90°} \times \nu_{\text{nutation}} = 0.25\,\text{cycle}. \quad (6.11)$$

The stronger B_1 is, and the higher ν_{nutation}, the shorter the time required. A pulse that nutates the magnetization from the vertical down to the horizontal plane is called a 90°, or π/2, pulse.

EXERCISE 6.8 *In a 1.5-tesla field and with B_1 ~ 25 μT, what duration should a rectangular 90° pulse have? Roughly how many precessions of* $\boldsymbol{m}(t)$ *occur during this time?*

So much for causing NMR to occur, but how is it detected? How can one tell that NMR is actually taking place? The net magnetization might be nutating like crazy, but how would anyone know? The NMR process can be made to disclose itself through Faraday induction in a 'receive' coil, and utilizing technologies developed for AM radio. Read on!

CHAPTER 7

Free Induction Decay Imaging of a 1D Patient (without the Decay)

Most of the standard sequences of RF and gradient pulses employed for clinical imaging are variations on the Spin-Echo (SE) or Gradient-Recalled Echo (GRE or GE) techniques, to which we shall turn in chapter 11. But before that, here we shall examine the simple spin excitation and recovery *free induction decay* (FID) sequence, also known as *saturation-recovery.*

We shall carry out FID first in a single voxel, and it simplifies matters a little to drop the position variable, x, for now. We shall make the major and *un*realistic assumption (to make matters easier at first) that relaxation effects in tissue water are slow enough to be ignorable. Later we perform *proton density*-MRI on a multi-voxel, 1D phantom, and employ Fourier analysis to untangle the frequency spectrum of the MRI signal. This leads directly to an image in 1D real-space. A variation on the technique, going by way of $\boldsymbol{k}$-space, is a bit more subtle and abstract, but shows how to extend the approach into 2D and 3D imaging.

This chapter is, in a sense, the lynchpin of the book, and we suggest that you become comfortable with it. Much that follows just extends and generalizes what is here.

7.1 Saturation-Recovery in a Single Voxel, Assuming Extremely Slow Spin Relaxation

Figure 6.9 demonstrated the ability of a 90° pulse to swing the net magnetization for a single voxel, aligned along the *z*-axis, $\boldsymbol{m}_z(t)$, down into the ***x-y*** plane (Figure 7.1a). Instantaneously morphed at $t = 0$ into transverse magnetization, $\boldsymbol{m}_{xy}(0_+)$, it begins precessing at ν_{Larmor} (Figure 7.1b). The invocation of quantum mechanics is required to provide an explanation of what goes on during that transition, but its impact can be summarized as the essential Equation (6.8), reproduced here as

$$|\boldsymbol{m}_{xy}(0_+)| = |\boldsymbol{m}_z(0_-).| \tag{7.1}$$

EXERCISE 7.1 Will the signal in Figure 7.1c be strictly sinusoidal, and so monochromatic?

This suggests a method of detecting a resonance whenever it might occur. Figure 7.1b includes a *receiver antenna*, also known as a *detection* or *pickup coil*, aligned normal to the ***y***-axis. By the Faraday law, Equation (2.1b), as $\boldsymbol{m}_{xy}(t)$ precesses in or near the ***x-y*** plane, its magnetic field cuts periodically through the coil, inducing a weak ν_{Larmor}-voltage in it that feeds the system's RF receiver (the curved green arrow in Figure 7.1b and in Figures 7.1c and 2.1c). Fourier analysis of $s(t)$, or simple inspection in this case of an almost monochromatic signal, reveals a near-sinusoidal voltage at 63.87 MHz (Figure 7.1d).

The magnetic field, $\boldsymbol{B}_{xy}(t)$, that arises from the rotating transverse magnetization will Faraday-induce a signal, $s(t)$, in a receiver coil. One might suppose that by Boltzmann's Equation (4.4c), $s(t)$ should be directly proportional to $\boldsymbol{B}_{xy}(t)$, hence to $\boldsymbol{m}_{xy}(t)$, and ultimately to $\boldsymbol{B}_0$, but that would account for only part of the story.

A voxel signal $s(t)$ is commonly expressed in terms of the *rate of change* of the *magnetic flux* through the coil (Figure 2.1b again). The *flux* through a loop of area A is defined as the surface integral of the field

$$\Phi(A,t) \equiv \int_{\boldsymbol{A}} \boldsymbol{B}_{x,y}(t)\cdot\hat{\boldsymbol{n}}\,dA \tag{7.2a}$$

where "·" indicates the scalar (dot) product of a pair of vectors, $\hat{\boldsymbol{n}}$ is the unit vector normal to the surface, and the integral is taken over the area $\boldsymbol{A}$ of it. As the transverse magnetization, $\boldsymbol{m}_{xy}(t)$, precesses, the magnitude of the signal voltage Faraday-induced in a pickup coil is

$$s(t) = d\Phi(t)/dt,$$

or

$$s(t) \propto \partial \boldsymbol{m}_{xy}(t)/\partial t, \tag{7.2b}$$

which increases with the rate of precession, ν_{Larmor}, and that, in turn, is proportional to the strength of the main field, $\boldsymbol{B}_0$. From this it follows that

$$s(t) \propto B_0^{\,2} \tag{7.2c}$$

In practice, the strength of the RF signal produced by $\boldsymbol{m}_{xy}(t)$ is a nasty function of the transmitter and receiver antenna designs, including their shapes, sizes,

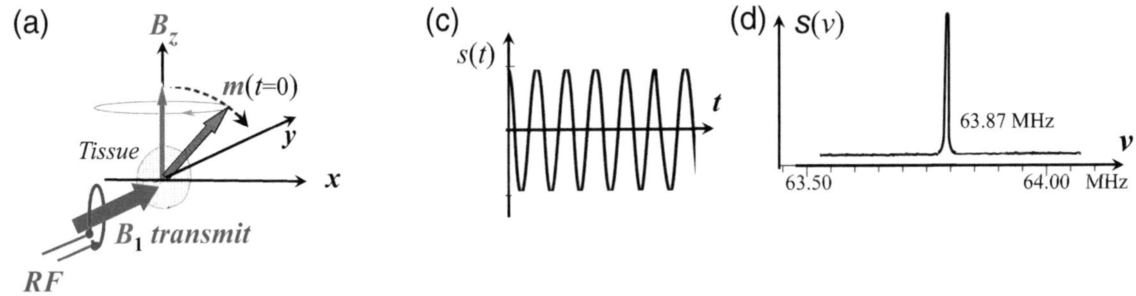

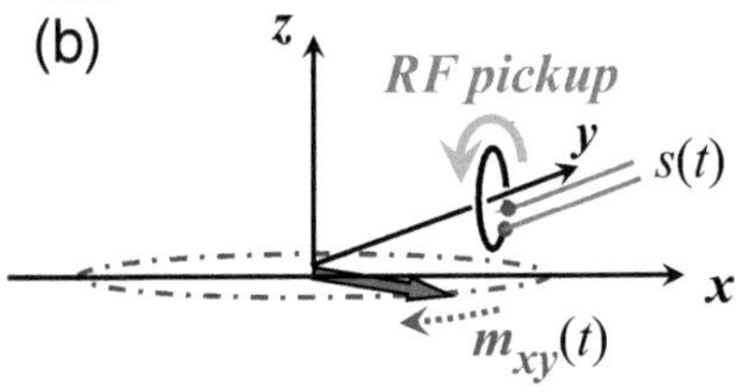

Figure 7.1 Creation and detection of an NMR or MRI signal from the materials in a voxel. (a) Following a 90° pulse, (b) a *changing* magnetic field generates an electric field that can drive a current around a nearby wire loop (*curved green arrow*), as described by Faraday's fundamental law of magnetic induction. (c) This can create a near-sinusoidal voltage signal, $s(t)$, varying at ν_{Larmor} in the MRI *receiver.* (d) The signal reveals a peak at 63.87 MHz, which drops off rapidly in intensity on either side of resonance.

and positions relative to the patient, and other considerations, as well as of B_0. More complete analyses of signal strength, and of noise from the electronics and from the patient herself, have been undertaken by RF communications engineers; they tend to become much more complicated than Equation (7.2c) alone would suggest. Suffice it to say here that the *signal-to-noise ratio* ends up lower than the factor of B_0^2 implies; in fact, it is closer to B_0. Still, a principal motivation for carrying out MRI at higher fields is that it leads to a general improvement in that important imaging parameter.

Again, we assume here that the ongoing proton spin relaxation processes are very slow, so that the detected FID signal will remain nearly constant in amplitude, not decaying appreciably. If the voxel's magnetization happened to be pointing nearly up or down, with almost no component in the ***x-y*** plane, then the receiver coil would be oblivious to it, and $s(t) = 0$. Indeed,

The ***only signal*** you ***ever*** see
in MR comes from
$\boldsymbol{m}_{xy}(t)$,
the *component of* $\boldsymbol{m}(t)$
precessing in the ***x-y*** *plane*.

It has just been argued that the amplitude of the voltage signal that is Faraday-induced in the pickup coil is related to the magnitude of the magnetic flux, $\Phi(t)$, passing through the coil. The value of the flux, at any instant, is proportional to the *x-y* component of the precessing magnetization. From Equation (6.5), the solution to the simplest Bloch equation (no RF, no relaxation, no nothing) can be expressed, in complex-number form, as

$$\boldsymbol{m}_{xy\ \text{fixed}}(t) = \boldsymbol{m}_{xy}(0)\mathrm{e}^{-2\pi i \nu_{\text{fixed}} t}, \tag{7.3a}$$

in the notation of Equation (5.6) and from the perspective of the fixed frame of reference. $\boldsymbol{m}_{xy\ \text{fixed}}(t)$ can be separated as

$$\begin{aligned} m_{x\ \text{fixed}}(t) &= \mathrm{Re}\left[\boldsymbol{m}_{xy}(0)\mathrm{e}^{-2\pi i \nu_{\text{fixed}} t}\right], \\ m_{y\ \text{fixed}}(t) &= \mathrm{Im}\left[\boldsymbol{m}_{xy}(0)\mathrm{e}^{-2\pi i \nu_{\text{fixed}} t}\right], \end{aligned} \tag{7.3b}$$

where Re and Im refer to the real and imaginary parts of what follows. These can be expressed also as

$$\begin{aligned} m_{x\ \text{fixed}}(t) &= |\boldsymbol{m}_{xy}(0)| \sin(2\pi \nu_{Larmor} t), \\ m_{y\ \text{fixed}}(t) &= |\boldsymbol{m}_{xy}(0)| \sin(2\pi \nu_{Larmor} t + \pi/2). \end{aligned} \tag{7.3c}$$

The extra $\pi/2$ means that the two signals differ in phase by 90°, and one of them will be detected a quarter of a cycle after the other. Keep that point in mind when we turn to quadrature signal detection in the next section.

The signal $s(t)$ is directly proportional to the transverse magnetization, and it appears only when $\boldsymbol{m}_{xy\ \text{fixed}}(t)$ happens to be briefly aligned nearly along the ***y***-axis, and normal to the pickup coil,

$$s(t) \propto \boldsymbol{m}_{xy\ \text{fixed}}(t)\cdot\hat{\boldsymbol{y}} = \boldsymbol{m}_{xy}(0)\mathrm{e}^{-2\pi i \nu_{\text{fixed}} t}\cdot\hat{\boldsymbol{y}}. \tag{7.4a}$$

We could carry on with the '$\cdot\,\hat{\boldsymbol{y}}$' vector notation, but there is no harm in dropping it. So let's accept the shorthand expression that

$$s(t) \propto \boldsymbol{m}_{xy\ \text{fixed}}(t) = \boldsymbol{m}_{xy}(0)\mathrm{e}^{-2\pi i \nu_{\text{fixed}} t}. \tag{7.4b}$$

$s(t)$ is nearly monochromatic for a single voxel (Figure 7.1c) with a spectrum, $S(\nu)$, that consists of a single narrow peak (Figure 7.1d). The same result can be obtained also, and is done so in practice, by subjecting the signal to Fourier analysis.

We shall soon proceed on to the multi-voxel linear 1D phantom in the presence of an ***x***-gradient field. The sum of the signals from all of them together gives rise to a far more complicated signal, $S(t)$, composed of a near-continuum of resonance frequencies, each of which is a function of the *x*-position from which it comes, and Fourier analysis is definitely needed to untangle them.

A final point before leaving the single voxel at *x*. It has been assumed here that any relaxation or other disruptive processes at work are very weak, so that $\boldsymbol{m}_{xy\ \text{fixed}}(t)$ precesses in the transverse plane with a magnitude that decays only very slowly. For any mechanical or electronic system whose oscillations are decaying exponentially in amplitude, as in Figure 6.1, the standard convention is to define, somewhat arbitrarily, a *characteristic* damping *time* T, defined as the interval over which the ringing (or swinging, or...) falls to 1/e = 0.37 of its initial value.

The same applies to the decay of a spin signal. Earlier we mentioned T1 relaxation, albeit from the spin-flip perspective, and we will address it and T2 in a pair of chapters coming up. Both T1 and T2 relaxation pro-

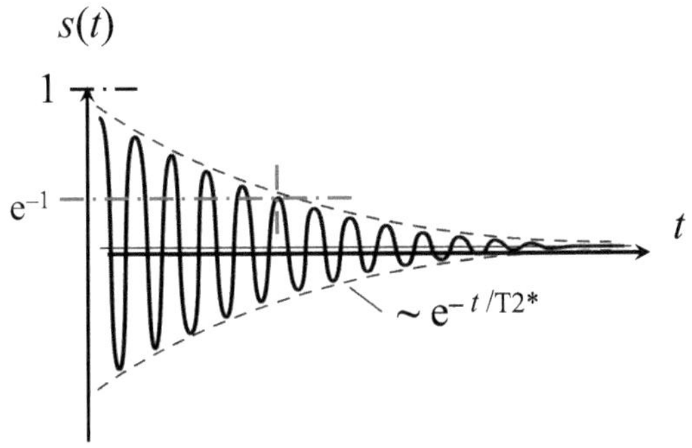

Figure 7.2 An FID signal in which there actually are relaxation processes at work. This shows a signal decay that appears to be driven by T2* relaxation. The relaxation time T2*, called 'tee-two-star,' is rather arbitrarily defined as the time it takes this signal to fall to 1/e = 0.37 of its initial value, as will be discussed in chapter 10.

cesses cause the signal amplitude to fall off nearly exponentially over *characteristic times* T1 and T2, respectively, which are typically only tens or hundreds of milliseconds long. This causes $s(t)$ to decrease at some overall *decay rate* of 1/T2* that is dependent on the biophysics of the tissue and on the imaging pulse sequence (Figure 7.2). We'll explain what T2* represents in chapter 10.

In beginning the analysis of pulse sequences, however, it simplifies matters considerably to ignore relaxation processes. And we shall continue to do so here.

7.2 Improving the SNR by a Factor of √2 with Quadrature Detection

Up until now, the MR receive antenna has consisted of a single coil, positioned on the ***x-y*** plane being scanned, with its face pointing normal to the ***y***-axis (Figure 7.3a). While single-coil imaging was common in the past, Equation (7.3c) suggests turning to *quadrature detection*, in which *two independent* coil elements face at right angles to one another (Figure 7.3b). This can lead to an improvement of the SNR by a factor of

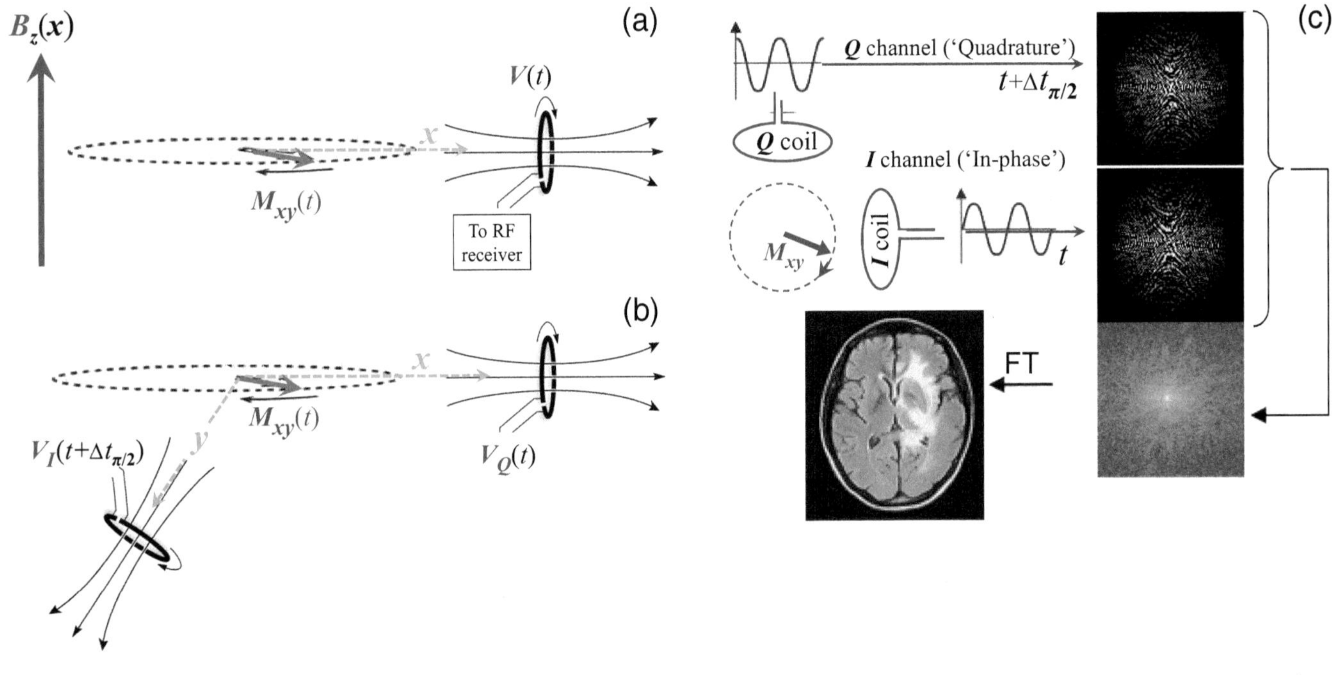

Figure 7.3 Quadrature detection simultaneously picks up the signals from two coils that are aligned perpendicular to one another. (a) Signal detected by a single antenna coil. (b) With two coils aligned perpendicular to each other, the information-bearing signal components of the two will differ only slightly from one another because of the 90° phase shift resulting from the time, $\Delta t_{\pi/2}$, it takes $\mathbf{M}_{xy}(t)$ to move on from one coil to the other. Otherwise the information content of the two signals will be essentially the same, and strongly correlated, and intentionally introducing a 90° shift between them electronically, and then adding them in phase, would yield an information signal of about twice the strength as from either alone. Most of the *noise* from either of the two, on the other hand, is of high frequency and will have plenty of time to change completely during the $\Delta t_{\pi/2}$, and consequently this is *un*correlated. (c) Recombining the two signals after the 90° phase shift thus doubles the received signal strength over what either alone would have yielded, but increases the noise only by a factor of √2. Most importantly, this increases the SNR by 2/√2 = √2. To the right are the two **k**-space images before and after the two signals are merged. A Fourier Transform of the combined, complete **k**-space image reveals a standard MR image.

√2—which is enough to make the difference between an image that is diagnostic and another that is worthless.

The net magnetization, $\boldsymbol{M}_{xy}(t)$, precessing in the $\boldsymbol{xy}$-plane, sweeps through one of the two coils one quarter-cycle (and a fraction of a nanosecond) ahead of the other. The signals are 90° out of phase, but their information content changes hardly at all over the quarter of a cycle of precession. The two channels are referred to as the "real" or ***I*** channel ('In-phase') and the "imaginary" or ***Q*** channel ('Quadrature'). The words 'real' and 'imaginary' are completely meaningless here, of course, and just a carry-over from the theory of complex numbers, where one could express $\boldsymbol{M}_{xy}(t)$ as a 'phasor,' $[\boldsymbol{M}_x(t) + i\boldsymbol{M}_y(t)]$. Because the signal from each coil element is separate but virtually identical (apart from the 90° phase shift and uncorrelated noise), delaying one coil's signal by 90° leaves the voltages from the two almost fully in phase.

Simply combining the two now will double the amplitude of the information-bearing component of the signal. The high-frequency stochastic *noise* experienced by each, conversely, fluctuates randomly. The noise from the two will be completely *un*correlated, and will destructively interfere with one another. Unfortunately this cancellation is not complete, and when the two signals are added together, the remaining noise component is still √2 greater than that from either separate channel alone. Combining the ***Q*** and ***I*** signals now results in a SNR increase by a factor of 2/√2 (Figure 7.3c),

$$\textit{Quadrature detection SNR improvement factor} = 2/\sqrt{2} = \sqrt{2}. \tag{7.5}$$

Alternatively, demodulating the signals in the ***Q*** and ***I*** channels separately yields 'real' and 'imaginary' information that, after a Fourier transform, produce so-called *real* and *imaginary* images (Figure 7.4a,b). These can be immediately combined in such a way as to lead to *magnitude* and *phase* images (Figure 7.4c,d). Almost always it is only the magnitude image that is needed, revealing anatomical detail modulated by image contrast.

Quadrature coils and detection can be implemented in several fashions, and what has just been described is perhaps the simplest of them. When moving on from a single, two-channel coil pair to *parallel imaging* with multiple antennae, as in chapter 13 on Fast Imaging, things become significantly more complex.

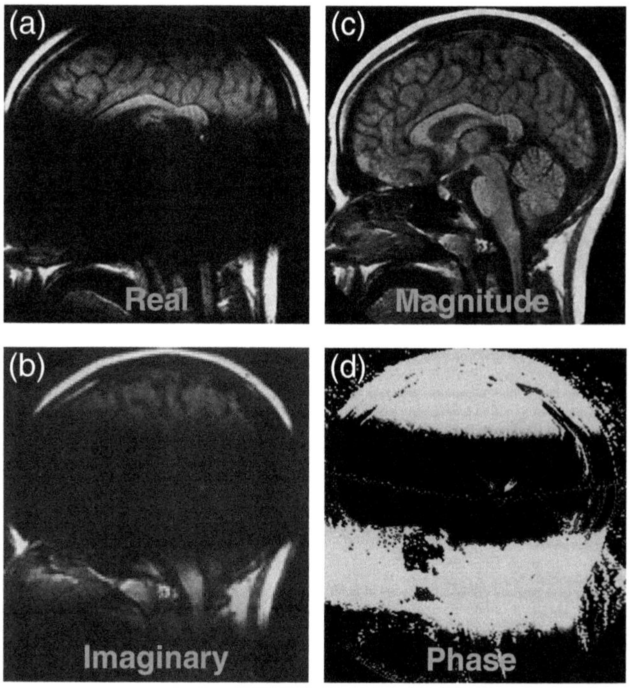

Figure 7.4 Quadrature detection and then separately processing what is in the real and imaginary channels. Sampling and Fourier transforming what is in either the (a) real or (b) imaginary channel alone makes possible only partial reconstruction of an image. (c) One method of combining the ***I*** and ***Q*** data results in a standard *magnitude image*, the kind needed for nearly all clinical studies, while (d) *phase images* have limited value.

7.3 Net Nuclear Magnetization from *All* the Voxels in the 1D Phantom Together: $M(t) \equiv \sum_x m(x, t)$

Let's turn from one voxel to our 1D phantom lying along the $\boldsymbol{x}$-axis, and add an $\boldsymbol{x}$-gradient field of 0.2 mT/m, as in Figures 3.5 and 7.5. The phantom consists of a row of identical small rectangular voxels containing various amounts of water or tissue—that is, it has a non-uniform PD(x). The voxels will be centered at discrete positions along the $\boldsymbol{x}$-axis, at $x_n = n \times \Delta x$, with the integer n ranging from $-½N_x$ to $+½N_x$, where N_x might be 256, say. If a row of N_x voxels has a specified Field of View (FOV_x) along the $\boldsymbol{x}$-axis of 25 cm, then the width of each one of them is $\Delta x = \text{FOV}_x/N_x$, Equation (3.2a) and Figure 3.5.

The configuration of the antennae is new here, and the planes of the transmit and receive coils are parallel

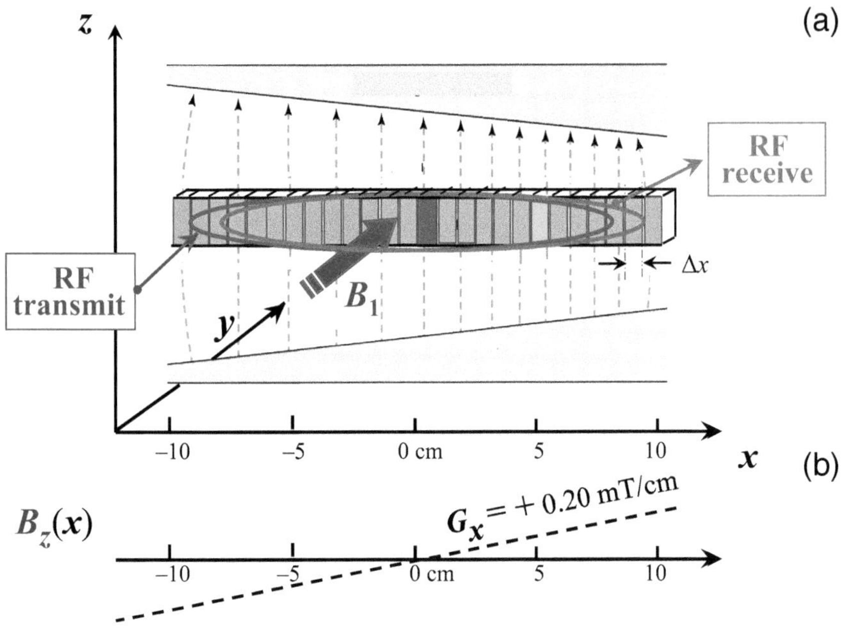

Figure 7.5 1D phantom, (a) in which two voxels containing water sit in a 1.5 T main field, (b) during application of a gradient field, G_x = (1.502 T − 1.500 T) / 0.1 m = +20 millitesla/meter (mT/m) = 0.20 mT/cm, as in Figure 3.5. The near-vertical dashed field lines come closer together toward the right, indicating that the field is stronger there, and that the ***x***-gradient is positive. The transmit and receive antenna coils have a new configuration, different from that of Figure 2.6.

to the phantom. This differs significantly from the continuous-wave cases of Figures 3.4 and 3.5. With the earlier, pseudo-quantum spin-flip picture, the transmit and receive coils sat at opposite ends of a row of water chambers, and a power meter indicated the *absorption* of RF energy at resonance by the water protons between them. Here, by contrast, the RF transmit and receive coils largely cover the region of interest, but from the sides of the phantom, not its ends. And the detection coil now picks up RF signals *emanating from* it. The transmitter antenna is designed to produce the same intensity of $\boldsymbol{B}_1(t)$ in all the voxels being examined, and the receiver antenna is, ideally, equally sensitive to all of the voxels, unaffected by their positions and RF frequencies. Both demands are far more easily stated than achieved in practice.

At thermal equilibrium, every voxel's magnetization starts out aligned along or against $\boldsymbol{B}_0$ and the ***z***-axis. At $t = 0$, the transmitter antenna produces a 90° saturation pulse. The water proton magnetizations from all the voxels together combine to create a composite transverse field

$$\boldsymbol{M}_{xy\ \mathrm{fixed}}(t) \equiv \sum \boldsymbol{m}_{xy\ \mathrm{fixed}}(x_n, t). \tag{7.6}$$

$\boldsymbol{m}_{xy}(x_n,t)$ represents the *mean value* of the magnetization averaged over the volume of the *single voxel* centered at x_n; it precesses in the ***xy***-plane and generates a nearly monochromatic voltage in the receive coil. $\boldsymbol{M}_{xy}(t)$ is a much more complicated entity that refers to the *net magnetization* from all voxels together, measured at the receiver coil. Because of the externally applied ***x***-gradient, the voxel magnetizations all now precess at different Larmor frequencies, and $\boldsymbol{M}_{xy}(t)$ becomes quite messy.

7.4 Selecting the z-Slice (or in 1D, the Row of Voxels)

We want to examine the single row of voxels located along the x-axis. They obviously must lie within the thin plane normal to the z-axis at some $z = z'$, so an early objective is to pick out that plane. To image a transverse slice of body tissues of very small but finite thickness, Δz, at the level $z = z'$, it is necessary to create a 90° pulse that affects all the protons in the slice, and only them. This slice selection entails briefly turning *on* the ***z***-gradient, $\boldsymbol{G}_z \equiv d\boldsymbol{B}_z/dz$, while applying a *narrow*-band 90° pulse centered at $\nu_{\mathrm{Larmor}}(z')$, where the Larmor frequency at $z = z'$ is the function

$$\nu_{Larmor}(z') = \nu_{Larmor}(B_0 + G_z z') \tag{7.7}$$

of $B_0 + G_z z'$. (Unlike what we have been doing up to now, here we cause the ***x***-gradient to remain off.) The band of frequencies of the pulse, known as its *transmit bandwidth*, comprises a continuum of RF frequencies of equal power over the narrow span from $\nu(z' - ½\Delta z)$ to $\nu(z' + ½\Delta z)$. This range, usually on the order of a few kilohertz wide, is centered at the resonance frequency for the slice of interest. For example, it would be around 63.87 MHz for 1.5 T for a slice at isocenter.

Similar arguments pertain for coronal, sagittal, or other image planes, but to minimize confusion, we shall stick with transverse slices, normal to the z-axis.

EXERCISE 7.2 What is the transmission bandwidth *needed to excite all the 1-mm³ voxels in the row if only the $\boldsymbol{G_x}$ = 0.2 mT/cm $\underline{\boldsymbol{x}}$-gradient is applied?*

7.5 Signal from the 1D Phantom

Immediately after excitation with a 90° pulse, $\boldsymbol{M_{xy}}(t)$ from all the voxels in the row phantom (which, for the moment, means the two containing water) precesses in the x-y plane. It induces, in the receiver pickup coil, the multi-voxel signal, $V(t)$, the amplitude of which is proportional to $\boldsymbol{M_{xy}}(t)$:

$$\begin{aligned} V(t) \propto \boldsymbol{M_{xy}}_{\text{ fixed}}(t) &= \sum_x \boldsymbol{m_{xy}}_{\text{ fixed}}(x,t) \\ &= \sum_x \boldsymbol{m_{xy}}_{\text{ fixed}}(x,0)\, e^{-2\pi i\, \nu_{fixed}(x)t}, \end{aligned} \tag{7.8}$$

where $2\pi\boldsymbol{\nu}$ is the angular frequency of the precession at x. Spin precession in the particular voxel at x may be expressed as $\boldsymbol{m_{xy}}_{\text{ fixed}}(x,t) = \boldsymbol{m_{xy}}_{\text{ fixed}}(x,0)\,e^{-2\pi i\,\nu_{\text{fixed}}(x)\,t}$ when viewed from the *fixed* (laboratory) frame of reference. The value of $\boldsymbol{\nu}_{\text{fixed}}(x)$ is determined by $\boldsymbol{B_0}$, which is constant and uniform, and by the $\boldsymbol{x}$-gradient field, which affects all the voxels in the row by different amounts, depending on the voxel's value of x. It takes a brief time for the $\boldsymbol{x}$-gradient to turn on and off, but for now we shall assume that it happens instantly, so that any gradient is a binary function of time.

$\boldsymbol{G_x}(t)$ is activated while the signal voltage $V(t)$ is being read out by the receiver, but it is *off* at all other times. In the fixed frame and while the $\boldsymbol{x}$-gradient is *on*, the Larmor equation for the voxel at the location x is

$$2\pi\nu_{fixed}(x) = \gamma[B_0 + G_x(t)]. \tag{7.9a}$$

This looks like Equation (7.7), but it is saying something altogether different. Equation (7.9a) has *nothing* to do with selecting the z-plane, but rather it returns us to what is happening along the $\boldsymbol{x}$-axis. The protons in the voxel at position x precess at the local Larmor frequency, and from the perspective of the fixed frame,

$$m_{xy\text{ fixed}}(x,t) = m_{xy}(x,0)\left[e^{-i(\gamma B_0)t}\right]\left[e^{-i(\gamma G_x \times x)t}\right]. \tag{7.9b}$$

EXERCISE 7.3 How do the stories of Equations (7.7) and (7.9) differ?

It simplifies matters if we now continue the tale from the perspective of the *rotating* frame of reference. $\boldsymbol{B_0}$ vanishes, in effect, as in Figure 6.10, and Equation (7.9b) becomes

$$m_{\boldsymbol{xy}\text{ rotate}}(x,t) = m_{\boldsymbol{xy}}(x,0)\left[e^{-i(\gamma G_x\, x)t}\right], \tag{7.10a}$$

in which

$$2\pi\,\nu_{\text{rotate}}(x) = \gamma G_x x. \tag{7.10b}$$

$\nu_{\text{rotate}}(\boldsymbol{x})$ is called the *offset*, or distance in frequency-space away from the central resonance frequency, $\gamma B_0 / 2\pi$. The resonances for the voxels at all values of x are evenly spread out by G_x along the $\boldsymbol{x}$-axis. And for any one of them, the frequency of rotation of the frame is proportional to its position $\boldsymbol{x}$. Equation (7.2a) is modified to

$$\begin{aligned} m_{\boldsymbol{xy}\text{ rotate}}(x,t) &= m_{\boldsymbol{xy}}(x,0)\left[e^{-i(\gamma G_x\, x)t}\right] \\ &= \boldsymbol{m_{xy}}(x,0)\left[e^{-2\pi i\,\nu_{\text{rotate}}(x)t}\right]. \end{aligned} \tag{7.10c}$$

The magnetization of this one voxel, precessing in the $\boldsymbol{x}$-$\boldsymbol{y}$ plane, intersects the antenna only along the $\boldsymbol{y}$-axis (Figure 7.1b) where it produces the contribution $\boldsymbol{m_{xy}}(x,0) \times e^{-2\pi i\,\nu_{\text{rotate}}(x)t}$ to the overall signal.

Summing over the N_x discrete voxel locations in our 1D $\boldsymbol{x}$-space yields the net magnetization, $\boldsymbol{M}(t)$, and the net signal

$$V(t) \propto \boldsymbol{M}(t) = \left[\sum \boldsymbol{m_{xy}}(x,0)\, e^{-2\pi i\,\nu_{\text{rotate}}(x)t}\right]. \tag{7.11a}$$

$\boldsymbol{B_{xy}}(\boldsymbol{x})$ from each voxel is distributed throughout space, but only the part of the signal that affects the change in total flux in the coil actually generates the MRI signal. So the location and orientation of the receiver coil are critical.

Finally, since $\boldsymbol{m_{xy}}(x,0)$ is proportional to the proton density at x, PD(x), this can be rephrased as the very important

$$\boxed{V(t) \propto \sum \text{PD}(x)\, e^{-2\pi i\,\nu_{\text{rotate}}(x)t}.} \tag{7.11b}$$

as detected by the receiver coil. The ultimate intention is, after picking up $S(t)$ while the $\boldsymbol{x}$-gradient is on, to find the magnitudes of the PD(x) for all x. Since we have $\nu(x)$ from the Larmor equation, and also knowledge of the strength of the gradient field, $\boldsymbol{B}_z(x) = \boldsymbol{B}_0 + \boldsymbol{G}_x\, x$, you can solve the above equation for the set of unknowns {PD(x)}. This amounts to constructing the inverse to Equation (7.11b), in effect, and that will require the services of a Fourier Transform (actually, an *inverse* Fourier Transform).

We shall now carry out this program in two related but somewhat different ways. The first has the virtue of being simpler and easier to describe. But only the second, the ***k***-space approach presented in section 7.8, still works for 2D and 3D patients. The first of these two mimics the discussion in chapter 3 of 1D spin-up/spin-down states, but here it utilizes RF pulses, rather than continuously sweeping the frequency of the RF.

7.6 FID Imaging of the 1D Phantom

Chapter 3 created a PD MR image of a 1D N-compartment phantom by applying an $\boldsymbol{x}$-gradient (Figure 3.5). Back then, the central frequency of a narrow band of continuous-wave (CW, *not* pulsed) RF was slowly swept upward, and the frequencies and amplitudes of any resonances were recorded one at a time. The frequencies of the peaks indicated the local magnetic fields, which in turn are rigidly linked, through the $\boldsymbol{x}$-gradient, to x-positions along the phantom. In addition, the brightness of any part of the MR image was made to be proportional to the proton density in the corresponding voxel. That was everything needed to generate the MR spectrum, hence the 1D PD MR image. But it was immensely slow and difficult to execute with precision in a real sample.

Here we will repeat the exercise, but this time with FID and in a manner that is much closer to what actually happens in clinical MRI (Figure 7.6a). Again, there is water in only two of the voxels of the phantom, at x = 0 and x = 5 cm. The uniform main field is again from a 1.5 T magnet, and the strength of the $\boldsymbol{x}$-gradient is $\boldsymbol{G}_x$ = 0.20 mT/cm, as in Figure 3.5b.

We start off with a single, brief ***narrow***-band 90° RF pulse, turned on while a $\underline{z}$-gradient is being activated for a short while (Figures 7.6b and 7.6c). The magnitude of $\boldsymbol{G}_z$ and the precise value of the ν_{Larmor} employed will together determine the z-position of the $\boldsymbol{x}$-$\boldsymbol{y}$ plane that will be examined, namely the one containing the 1D phantom. All the protons in the entire phantom are driven down into the $\boldsymbol{x}$-$\boldsymbol{y}$ plane, and the z-gradient is then removed. The time dependence of the RF power generated by the transmitter, $P_{RF}(t)$ [we could also call it $P_1(t)$] with its several 'lobes,' surely does look peculiar. But what we really want is to transmit a band with the same *amplitude of RF power* at *every frequency* over a specified narrow range of them —that is to say, energy with the 'rectangular' function of *frequency*, Rect(ν) of Figure 5.10c,e. So it is necessary to cleverly design the shape of the power output over *time*, $P_{RF}(t)$, to achieve that end. What is needed happens to be the so-called 'Sinc' function of Figure 5.10d,f, the Fourier transform of Rect(ν), which indeed has lobes.

Then, the $\underline{\boldsymbol{x}}$-gradient is turned on *during* (and *only* during) *signal readout* (to the right in Figure 7.6c). As with the examples of Figures 3.5b and 7.3b, the imposed $\boldsymbol{x}$-gradient is $\boldsymbol{G}_x$ = (1.502 T – 1.500 T) / 0.1 m = 0.20 mT/cm. The water-containing voxel situated at the phantom's center, where $B_z(x = 0)$ = 1.5 T, is the only place where the gradient is not felt. Resonance occurs at the corresponding Larmor frequency, $\nu(x = 0)$ = 63.87 MHz. The second voxel, only half filled with protons, is located at x = +0.05 m, where locally $B_z(0.05) = B_0 + G_x \times 0.05$ = 1.501 T, so that $\nu(0.05)$ = 63.91 MHz.

Spins in the various voxels precess *independently*, each at the frequency determined by the local field strength (Figures 7.6d and 7.6e). The magnetization in each voxel will precess at its own specific Larmor frequency, as determined by its position, x, and the associated local magnetic field strength, ($\boldsymbol{B}_0 + \boldsymbol{G}_x\, x$) (Figure 7.6d). (Also shown are the hypothetical precessions for the other voxels, as if there were water in them, as seen from above in the *fixed* frame of reference.) The $\boldsymbol{x}$-gradient allows us to differentiate the two water compartments by the local resonant frequency.

EXERCISE 7.4 Confirm that $\nu(0.05)$ = 63.91 MHz.

For this situation, how wide a range of RF frequencies, i.e., what *receiver bandwidth*, BW_{rec}, must the system be able to handle? To hear signals from all the voxels for our 20 cm phantom with the 0.20 mT/cm x-gradient present, the receiver must be sensitive to RF energy from about $\nu(-10\text{ cm})$ = 63.79 MHz up to

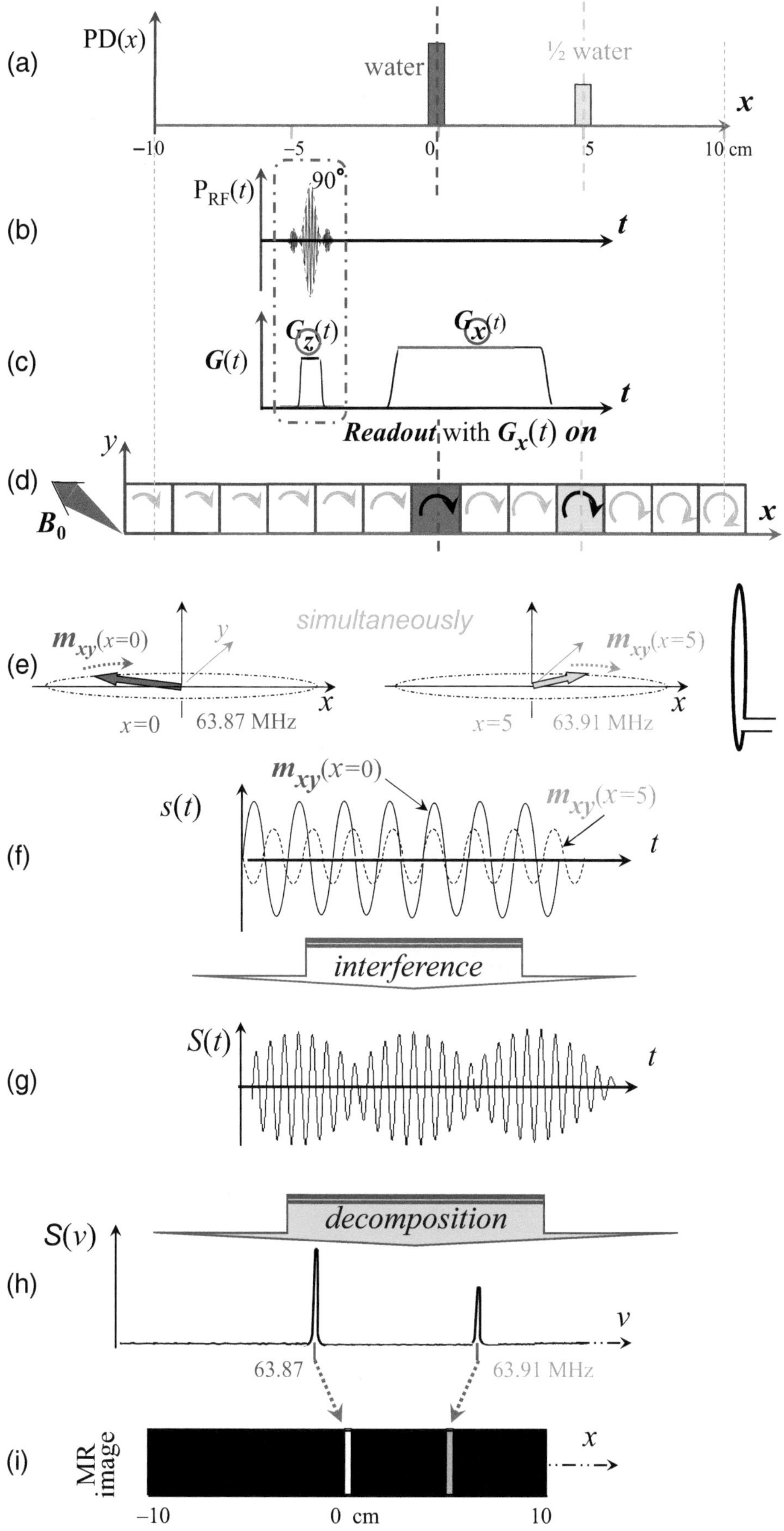

Figure 7.6 FID PD MRI in a 1D phantom in an **x**-gradient field, like that of Figure 3.5, with frequency encoding for voxel **x**-position. (a) PD(**x**) in the phantom of our example. The voxel at **x** = 0 is full of water, and that at **x** = 5 cm is half filled. (b) and (c) Application of a narrow-band 90° pulse, with the ***z*-*gradient*** on briefly for slice selection, but *not* the **x**-gradient. After the RF excitation, the **x**-gradient is then activated during readout (only), to spread out the ν_{Larmor} values for the voxels during readout, and it is turned off at other times. The water protons in each voxel will precess in the **x-y** plane at their own unique Larmor frequency during readout, as determined by the local value of $|\mathbf{B}_z(\mathbf{x})|$. (d) The hypothetical precessions that the other voxels would have if there were water in them, as seen from above in the *fixed* frame of reference. (e) The magnetizations of the protons at x = 0 and x = 5 cm will precess concurrently but independently, with nearly monochromatic frequencies of 63.87 and 63.91 MHz, respectively. These are about 40 kHz apart, which tells us that all the voxels are separated in frequency by 40 kHz per 5 cm, or 1 kHz/mm! (f) The magnetizations in the two voxels separately induce nearly sinusoidal voltages at their resonant frequencies in the RF pickup coil, the amplitude of one of which is twice that of the other. (g) But the coil senses only the overall, net changing magnetic field, not that from each voxel separately. So it ends up sending to the receiver a single, compound *beat* (at about 40 kHz) *interference* voltage pattern that arises from the superpositioning of the two single-voxel signals. (h) Fortunately, the digital machinery of the Fourier transform (**FT**) can untangle the mess, working to reverse the interference process, and to *decompose* the beat signal into its two (here) constituent parts. (i) That, in turn, makes possible the construction of an MRI rendering of the phantom's water distribution.

63.95 MHz at +10 cm, a span of about 16 kHz. This has obvious implications for the design of the RF receiver. Be forewarned that the expression 'receiver bandwidth' has another connotation as well, to be discussed later, but the context will allow you to distinguish which meaning is which.

Each of the two magnetization vectors shown here independently induces its own ν_{Larmor} voltage in the pickup coil (Figure 7.6f). But what the coil actually experiences is the overall *net* changing magnetic field there, not the distinct, separate fields from the individual voxels. What the antenna sends to the RF receiver is the overall signal generated as the sum, or superpositioning, of all the (here, only two) magnetizations acting together, Equation (5.5a), re-written here as

$$\sin \nu_1 + \sin \nu_2 = 2[\sin(\nu_1 + \nu_2)/2] \times [\cos(\nu_1 - \nu_2)/2]. \quad (7.12)$$

This describes the interference of two separate sine waves oscillating at frequencies ν_1 and ν_2: the composite signal displays a *carrier*-like signal at the higher frequency $\nu_{carrier} = (\nu_1 + \nu_2)/2$, which is *modulated* at the much lower *beat frequency* $\nu_{beat} = (\nu_1 - \nu_2)/2$, yielding the 'beat' pattern, *S*(*t*) (Figure 7.6g).

The remaining task is to somehow decompose the MR signal, *S*(*t*), of Figure 7.6g so as to determine the relative numbers of protons precessing at each frequency—i.e., to acquire the frequency spectrum, S(ν) (Figure 7.6h). Most likely you noticed that the notation S of S(ν) is presented in Arial font, somewhat different from, *S* of *S*(*t*). But even if you didn't note this, you could tell that the MR signal, *S*(*t*), and its spectrum, S(ν), are dissimilar because of their distinct and disparate independent variables, ν versus *t*.

Again, because of the ***x***-gradient, every frequency is associated with a precise local magnetic field by way of the Larmor equation, hence to a unique position along the phantom. With our 2-voxel example and the beat pattern of Figure 7.6g, it may be possible to guesstimate the frequencies and amplitudes of the two contributing waves simply by inspection (Figure 7.6h). The positions and amplitudes of the peaks within the frequency spectrum thus provide enough information for the production of a direct anatomic map of proton density (Figure 7.6i) yielding a complete PD(***x***) MRI image!

Figures 7.6g and 7.7a both display *beat* patterns characteristic of the interference between two sinusoidal signals that differ in frequency by a small amount,

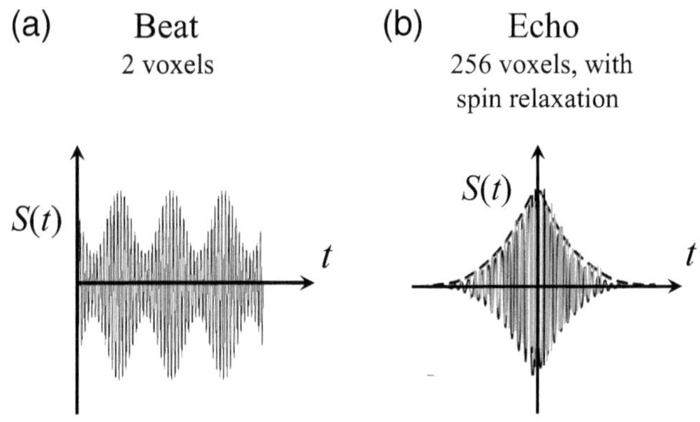

Figure 7.7 Beat vs. echo MRI signals. (a) Simple beat signal from the interference of two monochromatic signals of nearly the same frequency. (b) More realistically, the signal comes from hundreds of voxels, rather than two, and spin-relaxation processes are at play.

Equation (7.12). Such idealizations, however, do not appear in MRI studies in practice. Indeed, the standard clinical pulse sequences do not give rise to simple beat signals, but rather to *echoes* or echo-like signals (Figure 7.7b). Interference occurs among the signals from hundreds or thousands of voxels that contain various amounts of water, each resonating at its own frequency and phase. This combines with spin relaxation effects to produce MR signal that are far more complicated. Indeed, with a partitioning of the 1D patient into 256 or 512 voxels in a row, the MR signal is commonly like the *echo* of Figure 7.7b, such as that produced by a real *spin-echo* or *gradient-echo* clinical pulse sequence. This is usually the case, and here heroic Fourier measures are called for to untangle things.

Clinical MR signals are thus inherently very convoluted and nasty-looking, but computer-based FT and other powerful mathematical techniques are still able, amazingly, to untangle them. Indeed, that kind of *decomposition* is exactly what Fourier analysis excels at. Following sampling and digitization of the MR signal, the digital *Fast Fourier Transform* (FFT) undertakes the converse of interference: it quickly separates the MR signal into its constituent parts, with a peak for every voxel and its frequency, in effect. Problem solved!

The overall magnitude of echo signals from conventional spin-echo and gradient-echo MR studies is crudely symmetric in time, and the outline, or envelope, of each half is nearly exponential. This is like drawing Figure 7.2 twice, the second time flipped back

Table 7.1 Measured T2 values for some biological materials, to be discussed in chapter 10. For now, they provide rough upper bounds on echo half-widths.

Tissue	T2 (ms)
H20	2000
Adipose	70
Blood, venous	250
Brain	
White matter	80
Gray matter	100
CSF	160
Muscle	40
Liver	50
Kidney	60
Lung	80

over across the vertical at $t = 0$, and stitching the two parts together. We will need to sample and digitize a good part of the echo to extract and analyze the inherent MR information content, so we need a rough estimate of the time period over which the echo is presented.

The duration of your generic echo depends on several factors. The spin-echo signal from a single voxel of water falls off nearly exponentially with a time constant of about T2 (Table 7.1). For a region of tissue comprising many voxels, however, signals coming from various locations will undergo interference as well, and that can speed up echo decay. These and other effects indicate that the measured and listed T2 values are rough upper bounds on echo half-widths.

7.7 1D Image Reconstruction by Way of Frequency-Space

The bottom-line message of Figure 7.6 is that the signal $S(t)$ picked up in the receiver coil contains considerable information about the MRI processes ongoing along the length of the 1D phantom. So a simple way to arrive at PD(x) would be just to Fourier Transform the incoming MRI signal, $S(t)$, from 1D time-space to 1D frequency-space, yielding the frequency spectrum, S(ν), (again, note the Arial font for the S), and then to exploit the linkage between the frequency, ν, and the voxel position, x, along the phantom. This procedure works nicely for a 1D patient.

Figure 7.8 diagrams this in block fashion. We begin again with Equation (7.11b), which contains the sum $\sum \mathrm{PD}(x)e^{-2\pi i\, \nu(x)\, t}$, but we shall alter its form somewhat. As we have already seen and used, the frequency $\nu(x)$ is linearly proportional to x. The converse is also true, so we can convert Equation (7.11b) into a corresponding sum over the discrete values of the voxel RF *frequencies*, ν, as the index, *rather than over x*:

$$S(t) = \sum_{\nu} \mathsf{S}[\nu(x)]\, e^{-2\pi i\, \nu(x) t}. \qquad (7.13a)$$

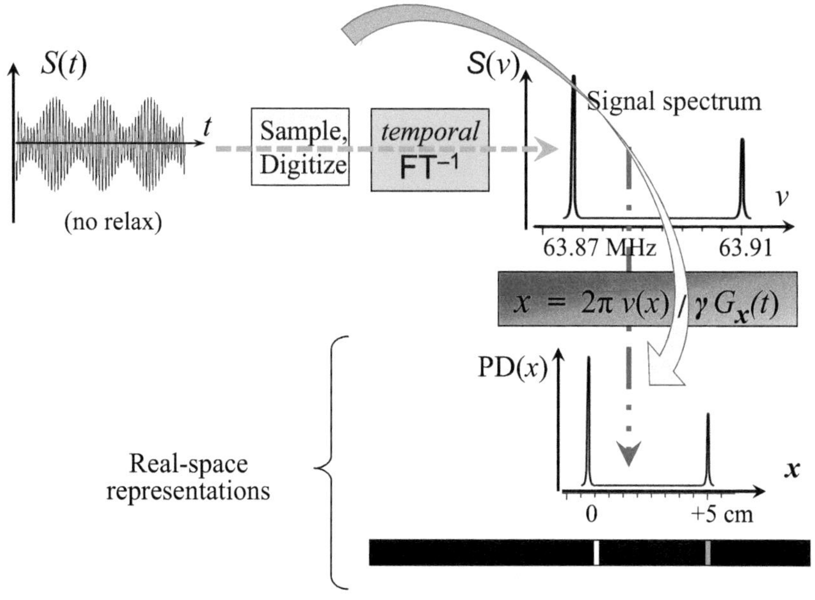

Figure 7.8 A simple and direct procedure for reconstructing a *spatial* image, PD(x), out of *temporal* MR raw signal data, $S(t)$. It involves converting a digital representation of $S(t)$ into its Fourier decomposition, i.e., its spectral representation, S(ν), with a digital Fast Fourier Transform (FFT), followed by the change of variables, $x = \nu / (\gamma/2\pi)G_x$. This has the benefit of largely paralleling what went on in chapter 3, but it also has the slightly problematic downside of working only for a 1D patient. Both x and ν are scalars, and they cannot be extended into two and three dimensions. But the general approach does point toward a similar methodology that can do so. In addressing 2D and 3D phantoms we shall, instead, have to operate with vectors in **k**-space, discussed soon.

The new term $S[\nu(x)]$ is just a revised version of PD(x) that records proton density as a function of local Larmor frequency, instead of directly in terms of x. And this is allowed here because of the existence of the 1-to-1 correspondence between x and ν.

If we were to reduce the voxels' thicknesses to the infinitesimal, and increase their numbers so as to maintain the same FOV, then Equation (7.13a) can be re-expressed as an integral rather than a sum. The signal from all the voxels together then appears as

$$S(t) \;=\; \int_{\nu} S(\nu)\, e^{-2\pi i\, \nu(x) t}\, d\nu. \qquad (7.13b)$$

Equation (7.13b) *happens*, quite astoundingly, to look a lot like a *Fourier transform*! And we didn't even go out of our way to make things come out like that! This serendipitous turn of events is nevertheless extremely helpful, since an *inverse* Fourier transform lets us untangle this messy complex of sine waves. Here it will allow us, in fact, to extract $S(\nu)$ from $S(t)$, and ultimately to find PD(x). We simply have to carry out the inverse Fourier operation.

Here is the crux of the matter: the inverse Fourier Transform (**FT** $^{-1}$) of a regular FT of a function just gives back the original function, Equation (5.13a). So take an *inverse* temporal Fourier Transform of Equation (7.13b) and, employing Equation (5.17b), find

$$\begin{aligned} \mathrm{FT}^{-1}[S(t)] &= \int S(t)\, e^{+2\pi i\, \nu' t} dt \;=\; \iint \Big[S(\nu)\, e^{-2\pi i\, \nu\, t} d\nu \Big] e^{+2\pi i\, \nu' t} dt \\ &= \int S(\nu) \Big[\int (e^{-2\pi i\, \nu\, t})(e^{+2\pi i\, \nu' t})\, dt \Big] d\nu \qquad (7.14) \\ &= \int S(\nu)\, \delta(\nu = \nu')\, d\nu \;=\; S(\nu), \end{aligned}$$

shown along the top row in Figure 7.8. This most important result,

$$\boxed{S(\nu) \;=\; \mathrm{FT}^{-1}[S(t)],} \qquad (7.15a)$$

provides a channel that leads directly from the data of the sampled incoming signal, $S(t)$, to PD(x), which we wished to find all along. The only other thing needed is the straightforward identity/change of variables, essential in its own right,

$$\boxed{x \;=\; \nu\,/\,(\gamma/2\pi) G_x,} \qquad (7.15b)$$

from a rearrangement of Equation (7.10b). This provides an effortless linkage between frequency space and image-space, shown on the right-hand side of Figure 7.8. It brings PD(x) out, and makes it possible to create and display the MR image. Voilà!

It may be helpful to summarize all this succinctly. The method of image reconstruction outlined in Figures 7.6 and 7.8 has the virtue of being relatively linear and transparent. It begins as a narrow-band 90-degree pulse (with the z-gradient on) nutates $\boldsymbol{m}(t)$ for all the voxels in the row (or, for 2D, in a plane) from along the z-axis down into the $\boldsymbol{x}$-$\boldsymbol{y}$ plane. After the $\boldsymbol{x}$-gradient is turned on for readout, the spins will be precessing there at different rates. The separate signals from all the voxels combine to create the composite signal, $S(t)$, in the pickup coil. The task of determining PD(x) consists essentially of inverting this interference process, performing the spectral decomposition of $S(t)$ with a time-to-frequency inverse FT, Equations (5.12). The upper half of Figure 7.8 reveals the system's spectrum, the number of protons resonating at each frequency. Following the arrow sweeping downward on the right-hand side of Figure 7.8, $S(\nu)$ is then transformed into PD(x) by means of the change of variables, Equations (7.15). This formalism closely follows the one introduced back in chapter 3, but it employs the pulse technique rather than the swept-frequency continuous-wave (CW).

Lest we be taken aback by all the formalism, let us ask ourselves, "But what does this mean, intuitively? What is actually going on physically during these mathematical machinations?"

By turning on a gradient, we are tuning the precessional frequency of the transverse magnetization in each voxel to a value that depends entirely, and linearly, on the voxel position along the $\boldsymbol{x}$-axis. The receiver picks up all the voxel signals, each of a different frequency, from the complete line of voxels combined, resulting in a complex amplitude-modulated signal centered at the middle of the narrow band of Larmor frequencies (which was spread apart by the x-gradient during readout). The modulation, or wave envelope in Equation (7.12), varies *temporally* in a particular manner that depends on the *spatial* distribution of the signal sources (i.e., the transverse magnetizations). There is no tuned-circuit like that of an AM radio to separate out the contributions from all the voxels but, rather, the inverse Fourier transform does the job instead, and allows mathematical extraction of the spatial information from the temporal signal.

7.8 1D Image Reconstruction by Way of *k*-Space

The above reconstruction formalism is built around manipulation of the temporal frequency, $\nu(x)$, a scalar entity. While this has been fine for a 1D patient, where voxel position may be fully defined without the introduction of vectors, it cannot be applied directly in a patient who happens to be two- or more-dimensional. It does, however, suggest an alternative route to proceed, by way of so-called **k**-space. This construct may appear rather abstract at first, but much of it is actually very much like what we have just done. In one dimension, it appears to involve nothing more than just the introduction of a new variable, the wavevector $\mathbf{k}$, to replace time, t. While for a real patient it offers much more, here we shall stay with the 1D phantom, and continue to think of $\mathbf{k}$ as essentially an *alter ego* for the time variable.

As before, the MR signal incoming from the receiver antenna, $S(t)$, can be decomposed by a temporal inverse FT, and the same information content expressed as the temporal frequency spectrum $\mathsf{S}(\nu)$ (Figure 7.9a). Figures 7.5 and 7.9b indicate that in 1D, $\mathsf{S}(\nu)$ and PD(x) are *isomorphic*—of the same general shape, albeit of different independent variables, units, and dimensions. After all, they differ only because of the change of variables, Equation (7.15b), and the associated re-labeling of axes.

This suggests that we explore what happens if we continue moving farther clockwise around the loop.

The lower half of the diagram indicates a second FT, in which the real-space function PD(x) is decomposed into (i.e., its spectrum expressed as a superpositioning of) *spatial waves*, each of some spatial frequency k and wavelength $1/k$ (Figure 7.9c). That is, we are Fourier Transforming PD(x), in real x-space, into the function $\mathcal{S}(k)$ defined in abstract ***k**-space*,

$$\mathcal{S}(k) = \int \mathrm{PD}(x)\, e^{-2\pi i\, k\, x} dx. \qquad (7.16a)$$

For the time being, this is just a prescription for expressing PD(x) as a combination, $\mathcal{S}(k)$, of spatial waves. To emphasize the connection between k and x in our 1-D example, we'll give k the subscript 'x' as a reminder.

So far, we have carried out a straightforward inverse FT followed by a change of variables, and then a second FT (Table 7.2). This means that, as the figure shows, the shape of $\mathcal{S}(k_x)$ must be isomorphic to that of the $S(t)$ directly above it (Figure 7.9d). It differs from it, in essence, only in terms of the independent variable; so it is therefore possible, with the right change of variables, to transmute from **k**-space back to the MR signal $S(t)$ in t-space, where we began! The key take-home concept is that the four functions, $S(t)$, $\mathsf{S}(\nu)$, PD(x), and $\mathcal{S}(k)$, carry basically the *same information*, but convey it in four dissimilar ways.

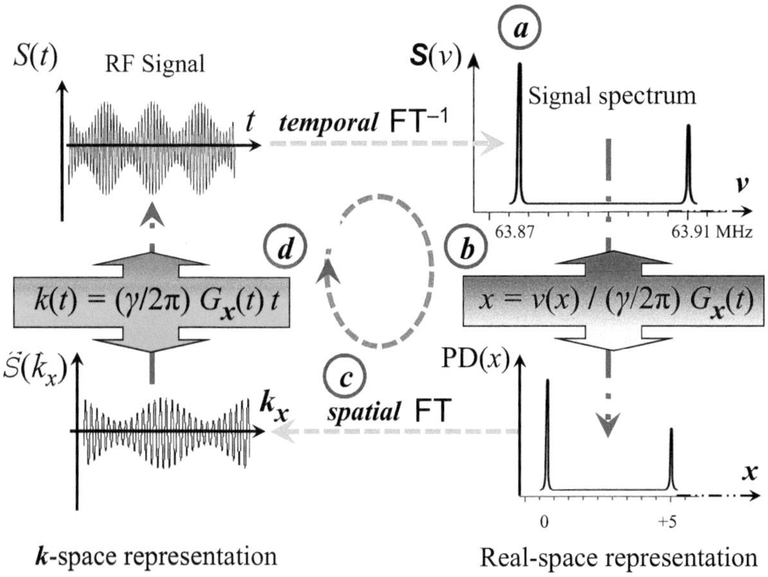

Figure 7.9 With the two-voxel phantom, it is possible to continue traveling clockwise around a complete loop, (a) from $S(t)$ to $\mathsf{S}(\nu)$ by way of an *inverse temporal* FT. (b) The change of variables, $x = \nu(x) / (\gamma/2\pi)G_x(t)$ leads on to PD(x). (c) Then a *spatial* FT converts PD(x) into $\mathcal{S}(k)$ in **k-space.** (d) Finally, back to $S(t)$ with a second change of variables; this one introduces new meaning for the parameter **k**, namely $k(t) \equiv (\gamma/2\pi)\, G_x(t)\, t$. The four functions—$S(t)$, $\mathsf{S}(\nu)$, PD(x), and $\mathcal{S}(k)$—all carry essentially the *same information*, just expressed in different forms.

Table 7.2 $S(t)$, $S(\nu)$, PD(x), and $\mathcal{S}(k)$ in Figure 7.9 all convey substantially the same MR information, albeit in separate ways.

90° Excitation ($G_z(t)$ *on!*) : ***z-slice***
Read/sample echo ($G_x(t)$ *on!*) : $\boldsymbol{S(t)}$
Signal spectrum : $\boldsymbol{S(\nu)}$
Set $k_x(t) = (\gamma/2\pi)\, G_x(t)\, t$: $\boldsymbol{\mathcal{S}(k_x(t))}$
Spatial FT$^{-1}\{\mathcal{S}(k_x(t))\}$: **PD(x)**

The factor of $e^{-i(\gamma G_x x)}$, Equation (7.10c), endows k_x with a second pivotal interpretation. Not only is $\boldsymbol{k}$ a *spatial wave frequency,* as discussed in chapter 5, but also it can be viewed as the newly defined, physically meaningful, vector parameter,

$$k_x(t) \equiv (\gamma/2\pi)G_x(t)t. \tag{7.16b}$$

EXERCISE 7.5 Show that the right-hand side of Equation (7.16b) has units of cycles/mm.

EXERCISE 7.6 $e^{-i(\gamma G_x x)t}$*, above, switches the order of x and t from that in Equation (7.10c). It does not matter mathematically, but does it conceptually?*

The subscript $\boldsymbol{x}$ on k_x ties it to the direction of the gradient variation, along the $\boldsymbol{x}$-axis. The fact that k_x is *linear* in t is a paramount idea: while the $\boldsymbol{x}$-gradient remains on, $k_x(t)$ is evolving. After $\boldsymbol{G_x}(t)$ is turned off, $k_x(t)$ is still defined, but it stops changing.

Now comes the really cool part. To arrive at PD(x) straight from the incoming RF signal, we can reverse the flow of actions of Figure 7.9. Beginning with $S(t)$, we go around the loop *counter*-clockwise, by way of k_x-space (Figure 7.10). The *mathematical* definition of $\boldsymbol{k}$—or more broadly, of $\boldsymbol{k}(t)$, Equation (7.16b)—leads to the crucial physical insight that the signal $S(t)$ can be re-expressed not only through its *temporal* frequency components, including harmonics, $S(\nu)$, but alternatively as a superpositioning of *spatial waves,* $\mathcal{S}(k_x)$. This is, in fact, what is needed to construct the 1D image, PD(x).

A key tenet of MRI pulse sequences is that once spin relaxation is introduced into the system, acquiring an optimal image with good contrast and signal-to-noise and with minimal artifacts depends to some extent on the art of selecting the order in which data-points in $\boldsymbol{k}$-space are to be obtained. So far, data acquisition has been linear, step-by-step along the $\boldsymbol{k_x}$-axis, but other kinds of paths through $\boldsymbol{k}$-space are possible, and perhaps even preferable. Careful sequencing of gradient pulses after an RF excitation can be used to navigate to specific points in $\boldsymbol{k}$-space, record the signal, and then move to another location.

All of this information-processing is done digitally, of course, starting with the electronic *sampling* of the MR signal voltage *over time* (Figure 7.11a). Then, following the change of independent variable from t to

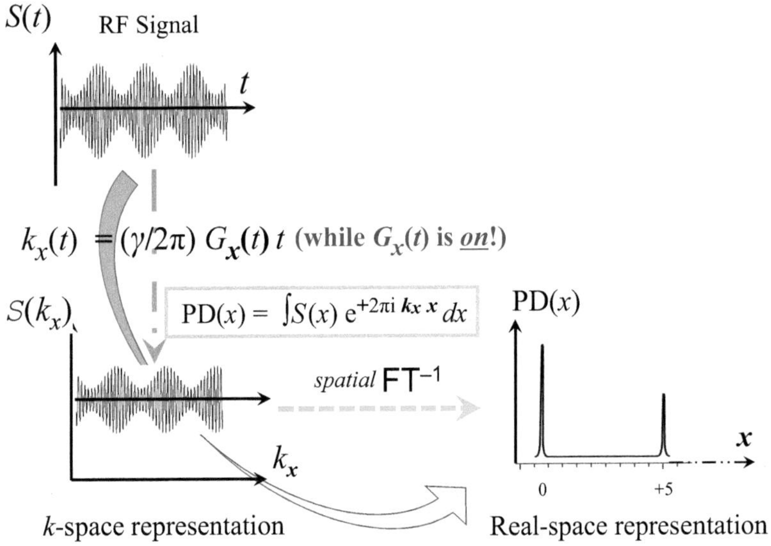

Figure 7.10 Another route from $S(t)$ to PD(x), by way of $\boldsymbol{k}$-space and $\mathcal{S}(k_x)$. Application of the definition of $k_x(t)$, Equation (7.16b), causes $S(t)$, the MR signal coming into the scanner over time, to metamorphose into the isomorphic $S[k_x(t)]$ in k-space. Upon a digital spatial inverse Fourier transform, this emerges as the real-space MR image, PD(x). Because $k_x(t)$ and $\boldsymbol{x}$ can act as vectors (rather than as scalars like time and frequency), this approach can be made to carry over to a 2D or 3D analysis.

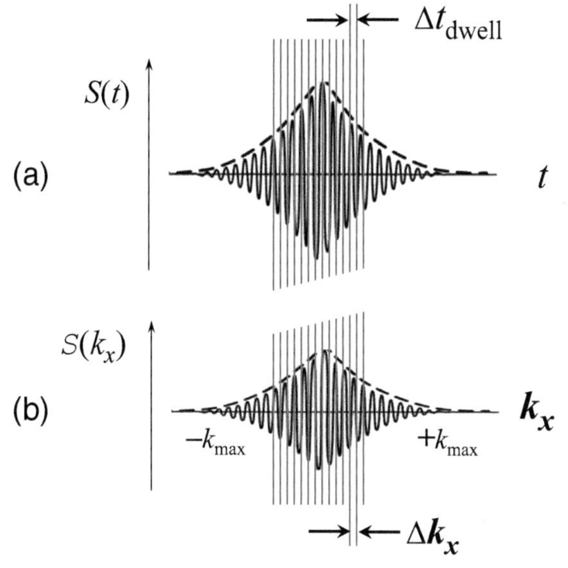

Figure 7.11 Sampling the signal. (a) It takes a finite period of time, Δt_{dwell}, for an analog-to-digital converter (ADC) to take a sample of $S(t)$ and present it as a digital number. (b) Associated with each value of time is a unique value of $\boldsymbol{k}_x$.

$k_x(t)$, each sample is associated with a spatial wave of frequency $k_x(t)$ cycles/mm (Figure 7.11b). As samples of $S(t)$ are made at many consecutive, closely spaced times, the system takes corresponding little steps, Δk_x apart, along the k_x-axis. That is, the system marches along the k_x-axis linearly with time, as long as $\boldsymbol{G}_x$ is being activated, and associates the present measured value of $S(t)$ with that of a unique value for $k_x(t)$. And for every one of these k_x values, we measure and record the corresponding signal strength, $\mathcal{S}(k)$. Again, a new font for $\mathcal{S}$, with the new argument, k_x.

Hundreds of samples of $S(k(t))$ are obtained along a horizontal line in 1D $\boldsymbol{k}_x$-space, as k_x moves typically from $-k_{x\text{-max}}$ through 0 and on to $+k_{x\text{-max}}$. In this fashion, the signal information is said to *fill* or *populate* a *line in k-space*. For the simplest case, of 1D imaging, only one such line is created. In 2D, one or a few hundred such lines are filled, one at a time, for an image — so all that effort spent on examining the 1D phantom was not wasted!

Again, morphing this way from an MRI signal, $S(t)$, into an image in real-space, PD(x), involves two steps. The first is a change of variables, replacing t and $S(t)$ with $k(t)$ and $\mathcal{S}(k_x)$. $S(t)$ consists of raw patient data that typically is digitized immediately, so $\mathcal{S}(k_x)$ is physically meaningful. Then a digital inverse Fourier Transform from k_x-space to real-space, along the bottom of Figure 7.10, yields the proton density, PD(x), along the x-axis,

$$\text{PD}(x) \;\propto\; \text{FT}^{-1}[\mathcal{S}(k)] \;=\; \int \mathcal{S}(k)\, e^{+2\pi i\, k_x\, x} dk, \qquad (7.16c)$$

which can trace its provenance directly back to $S(t)$! This is similar to but far more flexible and powerful than Equations (7.11), since $\boldsymbol{k}$ can become a 2D or 3D vector while t cannot.

Another view of all this is that gradients can be used to encode frequency and spatial information into an image by manipulating the *phase* of the magnetization vector as a function of space and time, as suggested by Figure 7.6. The magnetization can be re-expressed in terms of 'phase' by replacing $m_{xy}(x,t)$ with

$$m(x,k_x) \;\propto\; \text{PD}(x) e^{-2\pi i\, k_x(t) x}, \qquad (7.17a)$$

where the independent variable is now $k_x(t)$, rather than t—but where $k_x(t)$ is made to be linear in time. The exponent can be thought of as a *phase angle*, $\psi(x,t)$, of the magnetization vector at the voxel at x:

$$e^{-2\pi i k_x(t)\, x} \;\equiv\; e^{-i\psi(x,t)}, \qquad (7.17b)$$

where

$$\psi(x,t) \;=\; 2\pi\, k_x(t)\, x. \qquad (7.17c)$$

This indicates that $k_x(t)$ may be seen as the change in phase, or amount of rotation that the magnetization vector has undergone by time t per unit distance along the $\boldsymbol{x}$-gradient. Then discussion of what is going on along the k_x-axis can be couched in terms of the phase ψ_x rather than the k_x-value. This phase reflects the variation of the direction in which the magnetization points as we move along the x-axis, after the $\boldsymbol{x}$-gradient has been turned on for a particular time. This parallel perspective will be seen as helpful as we consider details of pulse sequences and related matters. We return to it in the final section of this chapter.

7.9 Connections between Image Space and *k*-Space

Over time, voxel magnetizations that precess at different frequencies when $\boldsymbol{G}_x$ is active will come into and out of phase with one another. If we took a snapshot, the voxel magnetizations would appear to be pointing in various directions (Figure 7.6d,e). For a 1D patient,

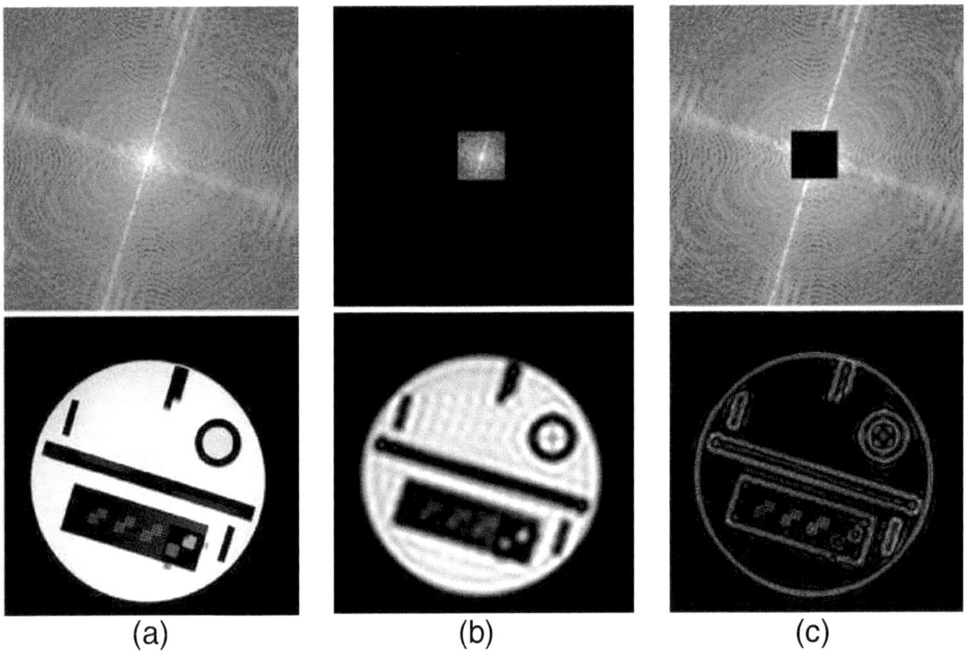

Figure 7.12 Effects of contributions to an MR image from (a) all of **k**-space, (b) long-wavelength **k**-vectors near their origin, and (c) high-spatial-frequency vectors toward the peripheries. [Zhou 2010].

the k_x-vector at any particular moment describes how this pointing varies spatially along the x-axis. The value of a k_x-vector at a single time during readout, moreover, depends on the history of the application of gradients of distinct strengths for specific durations of time. Both the individual values of k_x across the imaging region and the extent of k_x-space (which is $2 \times k_{max}$ wide) depend on choices of the gradient strength. Knowing how k_x relates to spatial position in real-space helps us understand how gradients affect FOV and resolution of an image.

The concept of **k**-space has been introduced as a way to reconstruct MR scans in two and three-dimensions, and also to help conceptualize image formation. A well-known triad of images nicely demonstrates (in 2D) the contributions from various regions of **k**-space to an MR. For a high-contrast, high-resolution, real-space image, *all* of **k**-space must be incorporated in the Fourier transform (Figure 7.12a). If we only use the **k**-vectors near the origin, we will get a high-contrast but low-resolution MR image (Figure 7.12b). Consideration only of the peripheral parts of **k**-space away from the origin, where the short spatial wavelength contributions may be found, provides fine detail but almost no overall contrast (Figure 7.12c).

Resolution in real space is inversely proportional to the width of the FOV_k in k-space

Visually, an image is a collection of pixels that form a picture. But a digital image is equivalent to an array of numbers. A conventional 2D image on a computer display uses integers or rational numbers to control brightness, which may vary across the image because of some physical processes and mechanisms taking place within the patient's body that control contrast. That matrix of numbers has certain dimensions—the number of pixels in width and height. If our image is a "representation" of something, like a camera photograph or an MRI image, that array of numbers also spans a physical space—the FOV.

To create an image of a certain resolution and size, we need to carve out a region in **k**-space to fill with raw data. At this point, **k**-space may still seem rather confusing and non-intuitive (after all, what does **k**-space actually represent?). However, since we can transform back and forth between spatial-based *image*-space and frequency-based **k**-space, let's just accept that they are both 'spaces.' And since we think we know how to describe a 'space' in a spatial sense, we start exploring the properties of **k**-space by saying they are *analogous* to image space. For example, just as a physical image has an extent called the 'FOV' and an intrinsic resolution defined by the pixel size (both given in units of length), our $\boldsymbol{k}$-space also has a measure defining the largest values that $\boldsymbol{k}$ will assume during imaging. This defines the extent of the FOV_k, and a unit of resolution in $\boldsymbol{k}$-space (the spacing between samples, or Δk), where FOV_k and $\Delta \boldsymbol{k}$ are in units of spatial frequency (Figure 7.11b). We control these by turning our gradients on and off at particular times and by adjusting our sampling time. So a big question is this: how do the characteristics of our physical image rely on the characteristics of $\boldsymbol{k}$-space, and vice versa?

One datum in $\boldsymbol{k}$-space contains information about a component in an image at a single, unique spatial frequency. In an extreme case, we can create an array in $\boldsymbol{k}$-space in which every position contains a value of zero except for one, and then perform an inverse Fourier transform. The resultant image would have *non-zero* values throughout the whole FOV. So a single datum in $\boldsymbol{k}$-space has the power to affect all the pixels in image space.

It may come as a surprise that the *resolution* for an image is *inversely proportional* to the *size* of the region of $\boldsymbol{k}$*-space needed to represent it.* In 1D, a region of $\boldsymbol{k}$-space might extend from $-k_{max}$ to k_{max}, a span of $2 \times k_{max}$, which defines its $\boldsymbol{k}$-space field of view (FOV_k). Back in Equations (5.15) we demonstrated that the extreme limit of image resolution (Δx) is defined in terms of the maximum possible spatial frequency, k_{max}, in $\boldsymbol{k}$-space:

$$\Delta x = (1/2k_{max}). \tag{7.18a}$$

Rearrangement leads to

$$\Delta x = 1/(FOV_k). \tag{7.18b}$$

In other words, image resolution is determined by the size of the image representation in $\boldsymbol{k}$-space! Shrinking the region of $\boldsymbol{k}$-space associated with a real-space image leads to increased pixel size and decreased resolution. From the relationship between k_x, G_x, and t, one can reduce the region in $\boldsymbol{k}$-space by applying gradients that are weaker or shorter in duration.

What property of $\boldsymbol{k}$-space determines the FOV_x of our real-space image? As discussed in Section 5.2, the Fourier representation of an image need not contain any wavelengths longer than the entire imaging dimension, FOV. That is, the longest wavelength, λ_{max}, which corresponds to the minimum $\boldsymbol{k}$ ($=\Delta k$), is equal to the FOV, and

$$\lambda_{max}(=1/\Delta k) = FOV_x. \tag{7.18c}$$

Our image FOV is determined by the separation of points (or 'resolution') of the array in $\boldsymbol{k}$-space used to create the image. Connecting this back to gradients again: to decrease our FOV, the point separations in $\boldsymbol{k}$-space must be larger, which would generally require increasing the gradient or leaving it on for a longer period of time.

7.10 Phase Waves

As suggested by Equations (7.17), there is an another way to view the 1D frequency-based imaging example we have just worked through. $S(t)$ reflects the *spatial variation* in *temporal* frequencies created when the x-gradient is on. The magnetizations precess at different rates, either clockwise for voxels at $x > 0$ or counter-clockwise for $x < 0$, when viewed from above in the rotating frame—in which $\boldsymbol{B}_0 = 0$, in effect. And, of course, the precession rate for a voxel depends on its x-position (Figure 7.13a).

Imagine a row of small balls lying 1 mm apart along the x-axis (Figure 7.13b). One face/hemisphere of every ball is colored white, and the other is blue. Each one represents a cohort of protons in its voxel just after the 90° pulse. Each now lies with its magnetization vector, $\boldsymbol{m}(x,t)$ lying in the horizontal plane and pointing like an arrow from the middle of the ball outward through the center of its white face. (The arrows are not shown in the figure.) Every such magnetization vector rotates around its vertical z-axis at the *local* Larmor frequency, $\nu(x)$.

A snapshot of the orientations of the magnetization arrows, at any given moment, will reveal a particular spatial configuration of the directions they are individually facing. This is the basis for the concept of the spatial 'wave,' covering multiple voxels, that keeps note of the *phase*, $\varphi(x,t)$, of the magnetization for each voxel/ball. The magnetizations from the individual voxels, which point in different directions, add together both constructively and destructively in production of their net magnetization, $\boldsymbol{M}(t)$. And this evolves over time while the gradient is left on, as does the captured MRI signal.

Beginning at time $t = 0\_$ nanoseconds, just before $G_x(t)$ is turned on, consider the top row in Figure 7.13b. All voxel/ball magnetizations are in phase with one another, and with the white hemisphere directly facing us.

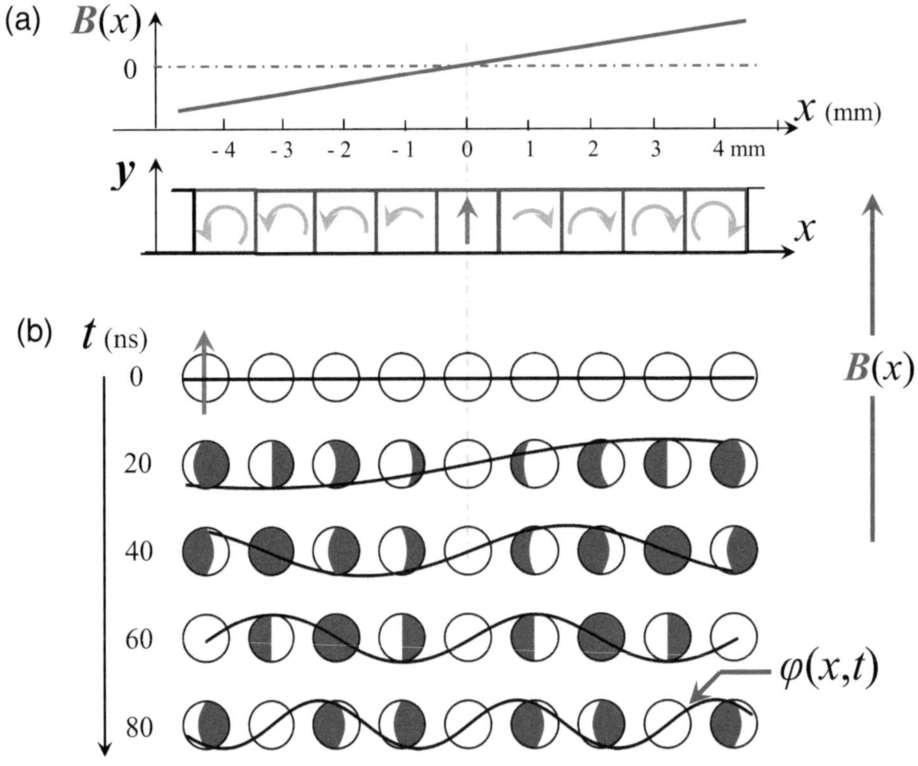

Figure 7.13 The spatial-frequency makeup of an MR signal changes continuously over time. As the signal is read out, it contains, quite remarkably, *increasing* amounts of high-*spatial*-frequency components. (a) While the **x**-gradient is on, each voxel magnetization precesses (viewed here from above, in the rotating frame of reference) at a rate proportional to x, its distance from the center, $x = 0$. (b) Picture a precessing $\mathbf{m}_{xy}(x,t)$ as a ball rotating on a vertical axis, when viewed from the side. Half its face is white, the other blue. The top row of balls displays the phases of their magnetization, $\varphi(x,t)$, exactly at $t = 0$. The farther any ball is from $x = 0$, the more rapidly it precesses, and the faster it accumulates phase. The differences in acquired phase between adjacent voxels also increases with time, t. Plotting $\varphi(x,t)$ against x yields a sine wave, but its value is meaningful only at integer values of x. Its spatial wavelength, $1/k_x(t)$, at any t is the distance between two of its peaks. This wavelength shortens over time, and the corresponding k-value increases.

If at $t = 20$ ns, and in the rotating frame, we take a snapshot photo and thereby briefly freeze the rotational motion. By then the ball at $x = +1$ mm will have rotated by an amount proportional to the local field, and incremented its phase by, for this example, 30°. Its neighbor to the right, which sits in a local gradient field twice as great, will have acquired a total phase of 60° relative to that at $t = 0$, and the one beyond that will now be at 90°, and so on. To the left of $x = 0$, the action is in the opposite direction. The phase of the one at $x = 0$ remains unchanged at 0°.

Repeating this at $t = 40$ ns, each ball will have spun twice as far as at $t = 20$ ns, and correspondingly for the phases at 60 ns, 80 ns, and at later times. At each of these times, the Larmor frequency increases with the distance, $|x|$, from the central voxel. The rate of rotation remains the same within each voxel over time, but the *phase differences* among voxels within a row grow. This is made more apparent by plotting the *phase* as a function of position for each instant of time, t, as a sine wave. The *spatial wavelength*, $1/k$, is the distance along the x-axis that it takes for the phase to increment by 360°, one full cycle. The spatial frequency, $k(t)$, is seen here to increase linearly with time.

EXERCISE 7.7 Is the wavelength of each of these sine waves $1/k(t)$?

The qth sample of the MR signal, $S(t_q)$, is read out at the time $t_q = q \times \Delta t$ after $G_x(t)$ is turned on at $t = 0$, which is when the sampling began. It is associated with a wave whose spatial frequency, k_q, is proportional to t_q.

As a result, the information content of an MR signal evolves smoothly while the $\boldsymbol{G}_x(t)$ is on. In this particularly hyper-simplified example, all magnetizations begin in phase at $t = 0$. With this assumption, the early signal consists only of long spatial wavelength components, while the tailing end of each comprises high-k contributions. QED.

As we will see in a clinically realistic case, we never start with magnetizations in phase at $t = 0$. So, the actual link between readout time, t_q , and long or short spatial wavelengths is something more complicated than this (and a little less intuitive).

CHAPTER 8

MRI Instrument

The structure of a modern superconducting-magnet MRI system is sketched in Figure 8.1 and Table 8.1, and it is described elsewhere in greater detail [e.g., Liang et al. 2000; Elster et al. 2001; Chen et al. 2009; Webb 2012; Brown et al. 2014]. It consists of three major pieces of highly specialized hardware, plus computers.

(a) Main magnet

The patient is carefully positioned within the stable *main magnetic field*, which polarizes the tissue protons along and against the *z*-axis. This field is normally generated either by a powerful horizontal, "closed" cylindrical superconducting magnet, like the one shown, or much less frequently, by a vertically aligned "open" electromagnet or permanent magnet. A *superconducting magnet* must have a bore large enough to accommodate the patient comfortably, and yet be capable of providing a steady 1.5 T or 3 T field. The fields from *permanent* and *electromagnets* are considerably weaker; indeed, these units are losing market share due to their lower signal-to-noise, poorer magnetic field homogeneity, and lower image quality.

(b) Gradient magnets

Three independent electromagnet *gradient* coils residing within the bore of the main magnet generate the rapidly switchable ***x*-**, *y*-, and *z*-gradient magnetic fields (G_x, G_y, G_z) for spatial encoding of the spin-containing voxels. These are energized intermittently by their respective *gradient drivers* which are, in effect, highly specialized audio amplifiers that supply waveforms to the coils under the control of the *pulse programmer*, which receives its instruction set from the console computer.

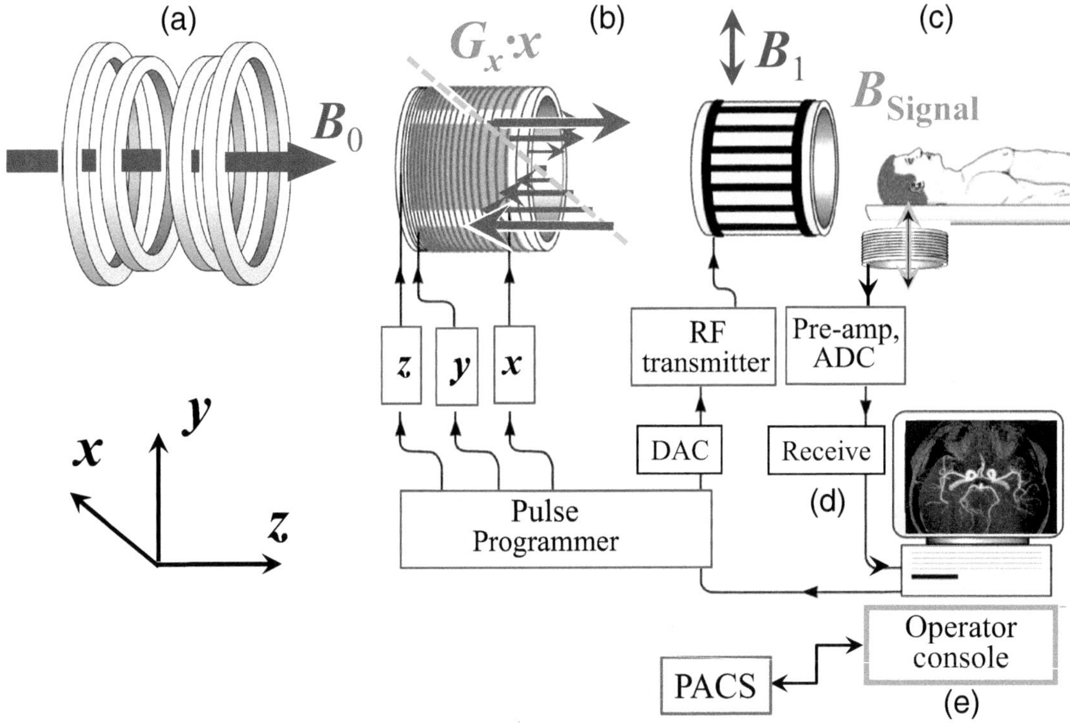

Figure 8.1 In this typical 1.5 T MRI device, the patient is carefully positioned within (a) the highly uniform and static main field, $\mathbf{B}_0$, that points longitudinally along its bore and the horizontal **z**-axis. Produced by several superconducting Helmholtz coil pairs, two of which are shown here, these are capable of creating an intense, very uniform field over a large central volume. Often, additional coils are included to improve homogeneity or to reduce the extent of the fringe field. The main magnet system is encased in a liquid-helium Dewar flask, which maintains the tightly wound superconducting wire near absolute zero. (b) Three room-temperature electroagnet coils fit within the bore of the main magnet and establish independent, orthogonal, nonuniform magnetic field gradients, $\mathbf{G}_x$, $\mathbf{G}_y$, $\mathbf{G}_z$. For each, the field itself points (ideally) only in the **z**-direction, parallel to $\mathbf{B}_0$, but its strength varies linearly according to **x**-, **y**, or **z**-position, respectively. (c) An RF transmit coil, placed within the gradient coils and close to the patient, generates a *near-field*, Larmor-frequency magnetic field, $\mathbf{B}_1$, aligned perpendicular to $\mathbf{B}_0$. The RF signals emerging from the body immediately thereafter convey the relevant MRI information. (d) These analog RF signals are detected during the data acquisition/READOUT period by RF receive coils, and sampled and digitized by an analog-to-digital converter (ADC). They go through image reconstruction, and they are then sent to the console computer for further computer processing and display. (e) Images are then ready for entry into a Picture Archiving and Communication System (PACS) for reading by local physicians, transmission to more distant monitors, archiving, etc.

Table 8.1 The major components of a typical MRI system deal, respectively, with the static, uniform main magnetic field; the three orthogonal gradient fields that are rapidly and independently switched on and off; the RF system that generates and immediately thereafter detects and samples brief pulses of RF near ν_{Larmor}; and the computer systems that control all operations, carry out reconstruction calculations, and interface with the PACS.

Main Magnet, B_0	*RF System, B_1*
1.5T, 3T supercon magnet	transmitter, receiver
inhomogeneity correction	transmit/receive RF coils
passive shim plates,	head, body, knee
active shim coils	*or* receive-only RF coils
self-shielding windings	surface, vol., phased-array,
	quadrature detection
Gradient System, G	
gradient field coils	*Computer Control*
gradient amps	operator console
	pulse programmer
	reconstruction

(c) RF pulse transmission and signal reception

In a way, the MR RF equipment resembles a highly specialized amplitude modulation (AM) radio transmitter-receiver pair operating near 64 MHz (1.5 T) or 128 MHz (3 T), along with the associated signal data processing system. RF pulses, ideally of the same amplitude in every voxel, are applied to perturb the polarization of the proton magnetization, and the resulting disturbances give rise to detectable RF signals emerging immediately from the patient.

In the *transmitter*, the sculpting and sequencing of the brief bursts of RF are also managed by the pulse programmer. They are of *narrow bandwidth* (*BW*) and centered at the local Larmor frequency and, for a whole-body scanner, are amplified to the kilowatt level for delivery to the transmitting coil. The resulting outgoing electromagnetic wave contains a ν_{Larmor} magnetic field part (in the micro-tesla range), of which the near-field component penetrates into the patient. Transmission is usually performed using transmit coils that fit within the gradient coils and can accommodate the patient's body.

Following transmission, a *quadrature* receive coil or an array of individual coils (generally with their elements placed close to the body for greater sensitivity) pick up a weak resonance echo from the excited volume of tissue. This signal, nano- to micro-volts in strength, is induced by the precessing magnetization in the excited tissue voxels. The signal typically passes through a low-noise pre-amplifier, after which the chain of receiver electronics can take various forms. In all new MRI devices produced today, though, the RF is further amplified and is directly sampled and digitized at a rate of tens of MHz by the *analog-to-digital converter* (ADC). After that come extraction of the MR information from the 64 MHz or 128 MHz background, more digital processing, and image reconstruction.

(d) Console computer, pulse programmer, and reconstruction computer

Several different computers are required to drive an MRI scanner. The *console computer* is the primary interface directing the MR data acquisition process. Under the management of the MRI operator, it also sends protocol instructions to the *pulse programmer* (sequencer) to execute a specific sequence of RF and gradient pulses. It also displays and post-processes images, managing them in the Digital Imaging and Communications in Medicine (DICOM) format for entry into a *Picture Archiving and Communications System* (PACS). From there they can be displayed, archived, transported across town or around the world, fed into an Artificial Intelligence (AI) network, and so on.

The amount of data collected during a scan is computationally intensive and requires substantial memory, so a separate computer known as the *reconstruction engine* is typically responsible for capturing and processing the acquired data for the actual image reconstruction.

Several of the more important properties and characteristics of a typical 1.5 T system are summarized in Table 8.2. The present chapter describes in more depth certain aspects of the design and construction of the typical modern MRI instrument just sketched above.

Table 8.2 Typical parameters and characteristics of a 1.5 T superconducting-magnet system

$\boldsymbol{B_0}$	1.5 T
homogeneity	<0.02–0.1 ppm <20 cm
shielding	passive and active
cryogen	1000–2000 liters liquid He
boil-off	0–2 liters He/year
weight	3–10 tons
Gradient dB/dx	20–80 mT/m max
rise time	0.1–0.5 ms
slew rate	100–200 mT/m/ms
gradient on-time	milliseconds (ms)
RF transmission	10–30 kW
B_1	10–50 μT
90° pulse on-time	~1 ms
typical SAR	0.01–4 W/ kg
shielding	Faraday cage
Spatial resolution	<1 mm in-plane
2D slice thickness	1–5 mm
max FOV	45–50 cm
acq, recon matrices	256×256 typical

But be cautioned that nearly everything said here will have, in practice, variations and exceptions. As materials and engineering solutions address critical limitations of extant design, systems will evolve to increase sensitivity, specificity, and stability, as well as to reduce size, weight, and cost.

EXERCISE 8.1 Compare the 90° pulse time here with that from Exercise 6.5.

8.1 Main Magnet

The *main magnet* produces a spin-polarizing magnetic field, $\boldsymbol{B_0}$, that is strong, constant over time, and highly uniform (to within a few parts per million) throughout a volume large enough to accommodate a good portion of a body. MRI main magnets are of three different types: superconducting, electro-magnets, and permanent magnets, in order of decreasing achievable field strength. Only superconducting magnets are capable of achieving homogeneous, stable, large-volume fields above 0.5 T or so.

MRI superconducting magnets

Superconductivity is a property of some materials that is characterized by two remarkable phenomena. When cooled below a material-specific critical *temperature*, the electrical resistance in it drops abruptly to zero. Not only that, but also all magnetic flux is expelled from it (*Meissner* effect). It is the first of these that is of interest to MRI.

Nineteenth-century physicists knew that the *electrical conductivity* of metals decreases at lower temperatures, and they speculated as to what would happen near absolute zero—in particular, would it continue downward, or level off, or perhaps even stop altogether as the electrons 'froze in place.' In 1908, the Dutchman Kamerlingh Onnes managed to liquefy helium at 4.2 ° kelvin (–268.93 °C). Onnes must have been astounded shortly thereafter, in 1911, to discover that the conductivity of mercury metal behaved the opposite way. It seemed to become *infinite* at helium temperature: its resistivity simply vanished! Onnes called this bizarre phenomenon *superconductivity*. At the time, he also noted the appearance at a point 2° K lower of *superfluid* helium, although this second phenomenon had far less of an impact on him. In 1957, John Bardeen, Leon Cooper, and John Schrieffer explained low-temperature superconductivity as a superfluid of Cooper pairs of electrons interacting via phonons. Analogous to water molecules and ocean waves, the individual electron pairs travel only a tiny amount, but their collective behavior propagates nearly at the speed of light.

Meanwhile, intense research has led to materials with critical points far above 4.2 °K, wires of which can create magnetic fields of strengths up to 10 T or so, and remain superconducting in such fields. Current magnets developed in fusion-energy work, comprised of tapes of rare-earth barium copper oxide (ReBCO), can achieve fields greater than 20 T at liquid nitrogen temperature (77 °K). But materials that become superconducting here generally still do so only well below room temperature, and they have been brittle and difficult to produce in the form of wire. At present, an alloy of equal parts niobium and titanium (NbTi) remains widely used in MRI magnets. The wire conducts hundreds of amperes of current with virtually no resistive losses below –263 ºC or so, only 10 degrees above absolute zero. Because of imperfections in the wire and solder joints between wires, however, a residual resistance on the order of 10^{-9} ohms produces minuscule thermal losses of heat energy and electric current.

The great majority of modern MRI devices employ a *closed-bore* cylindrical superconducting magnet. Three quarters of them operate at 1.5 tesla. Those at 3 T provide the potential for better signal-to-noise ratio, finer spatial and temporal resolution, and enhanced contrast across a range of applications. The costs to be paid for these benefits are greater specific absorption rate (SAR, local tissue heating), more acoustic noise, and a significantly higher purchase price. These trends are being extended by experimental laboratory and clinical imaging at 7 T.

In any case, a wire of NbTi as thick as a hair embedded in a copper matrix and typically 50 km-long is precisely wound into a set of coils. Somewhat similar to Helmholtz pairs, there are typically three or four pairs of them of different diameters and thicknesses (Figure 8.1a). These are designed and wrapped so as to maximize field homogeneity and to minimize the fringe field at either end of the bore, and they can produce fields much more uniform than that of a simple solenoid or single Helmholtz pair. They are supported on strong and rigid fiberglass or non-ferrous stainless-steel frames capable of counteracting the magnetic forces between and within coil sections.

Once the electrical current is initiated in the superconducting wires, there is no need for power input to keep it and its magnetic field going virtually indefinitely, nor is any significant amount of heat given off. But to maintain the superconducting condition, the entire coil set must be immersed in liquid helium within a *Dewar flask* or *cryostat* (Figure 8.2). In previous generations of MR scanners, the Dewar of helium was surrounded by another Dewar of liquid nitrogen as an important source of thermal insulation. The nitrogen slowed helium evaporation, although both needed periodic replenishing (typically once a year for helium, and more frequently for nitrogen). Cost and scarcity of helium have driven modern instruments to reinvent the helium Dewar as a more efficient sealed cryogenic refrigeration system—including a compressor and a helium re-condenser referred to as a *cold head*. All of this is surrounded by a vacuum and radiative reflecting material for optimal thermal isolation. The newest machines advertise a 'zero-boil off rate,' so that a Dewar may require a refill of He only one or two times during its clinical lifetime (and maybe even never, depending on the life of the magnet). For the future, medical equipment companies are working on a novel magnet design with a sealed cryostat of much smaller volume, with literally no need to refill. The precision manufacture of the magnet/cryostat assembly and wire accounts for a large part of the high initial price of an

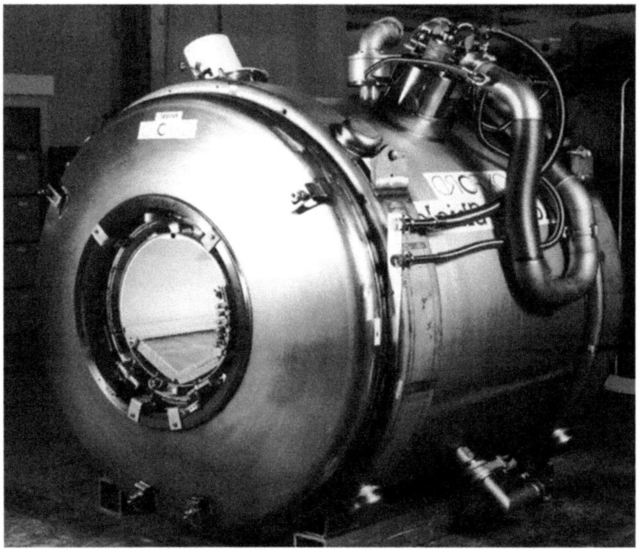

Figure 8.2 A 1.5 T main MRI magnet with the covers off. The superconducting wire coil is cooled in a bath of liquid helium (typically on the order of 2000 L) within a stainless steel Dewar/thermos vessel, which maintains it at an extremely low temperature, near absolute zero. [Courtesy of Oxford Magnet Technology Limited, Oxford, UK.]

MRI, which tends to scale with magnetic field strength and bore diameter.

EXERCISE 8.2 Devise a persistent switch circuit to inject current from a power supply into the wire of a superconducting magnet to get the field going. Hint: it involves a small heater.

An interesting engineering problem arises here: how to inject the high current into a superconducting magnet's coil? After all, one end of the wire is ultimately attached to the other end, as a closed circuit, so that current can flow freely around the windings. But how to get that state of affairs going? The solution involves a short section of the current pathway consisting of a superconducting (*persistent*) switch paired with a heater, where one can access connection points on both sides of this switch from outside the magnet. A source of high current is connected across those points, and the heater is turned on, which eliminates the superconductivity of the switch and throws open a current loop from the power supply through the magnet. The current source is slowly "ramped" up to several hundred amps, after which the heater is turned off, which renders the switch (and the entire circuit) superconducting, with current now running unhampered around the magnet.

Two critical issues in the design of any MRI magnet are field temporal stability and spatial homogeneity. The first of these is no problem for a healthy, modern superconducting magnet, where the field strength, averaged over the central imaging region, should not change more than 0.1 part-per-million (ppm) per hour.

Homogeneity is another matter. If the field strength, B_0, is plotted at a number of points within the bore, then one definition of the homogeneity is expressed as

$$\text{homogeneity} \equiv (B_{max} - B_{min}) / B_{central}, \tag{8.1}$$

where B_{max} is the largest reading, and so on. A *homogeneity* of 1.5 ppm may be achievable over a spherical imaging region 50 cm in diameter, but an acceptable value for this parameter depends on magnet design, field strength, and applications. Inadequate homogeneity of the static field can cause diminished signal-to-noise ratio and spatial resolution, leading to image distortions and artifacts. Field homogeneity is initially tuned off-site at the manufacturing plant and later on-site by carefully positioning *passive shim plates* of steel in the bore. The geometry of the field is often designed to achieve good homogeneity within an ellipse-shaped region in the bore, since supine patients tend to be elliptical in cross section. The balance of slight currents in various additional *active shim coils* surrounding the superconducting magnet produce slight field corrections to the static field that improve upon the passive shimming alone. Active coils can be either outside the liquid helium Dewar (*resistive shims*) or within it (*superconducting shims*). The latter coils, made of NbTi wire like the main magnet, are difficult to adjust and are seen with decreasing frequency.

The patient's body shape and inherent tissue properties cause its magnetic susceptibility to be nonuniform, and the shimming process will require additional patient-specific adjustments. One technique is to apply and adjust a small DC current to each of the gradient coils during *every* automatic pre-scanning procedure. For complicated anatomical geometries, such as the breast or orbital regions, some scanners have special coils that vary more strongly with distance than linearly. These coils are known as *higher-order shim* coils, and they can help achieve especially high homogeneities in localized regions of tissue for applications such as for MR spectroscopy.

Fortunately for shim designers, it happens that the spatial dependence of any field (magnetic, electrical, gravitational, etc.) can be decomposed into a particular combination of orthonormal polynomials, the *spherical harmonics*, as with Equations (5.10).

In the case of electromagnetism, these are also solutions to Maxwell's equations. Akin to the elements of a Fourier expansion, they have specific geometric configurations: the first term accounts for a spatially uniform field component, the next three portray fields that vary linearly in space, and subsequent members describe components of quadratic, cubic, etc. geometric structure. It is possible, moreover, to construct electromagnets that can generate approximations of such fields. So after measuring the dependence on position of an MRI magnet's own real field, $\boldsymbol{B}_0(x,y,z)$, it is possible to adjust the currents in these higher-order shim magnets to counteract irregularities present, thereby achieving greater homogeneity.

Smaller and far less costly magnets, superconducting and otherwise, have been designed to examine the extremities, rather than the whole body. These can be fully adequate for studying bone fractures, neoplasia, and infections, and also arthritis, neuronal irregularities, impact injuries, etc. They are being employed routinely by veterinarians, moreover, to scrutinize all kinds of problems in pets (as are full-size scanners for valuable large investments such as race horses and breeding stock).

Fringe fields

One of the problems with siting an MRI is that a superconducting main magnet may have a field that extends over a large area outside the instrument itself. This could require a very large imaging suite, but that could be cost-prohibitive in a hospital. Alternatively, one can pull in this *fringe field* to a smaller volume with either passive or active shielding. The passive variety involves applying a thick layer of steel to one or more of the walls, ceiling, or floor. Active shielding, consisting of additional superconducting coils surrounding the main magnet, is available for modern machines. By winding the coils in the direction opposite to that of the main magnet, the shielding coils counteract the field outside the scanner, greatly diminishing its field footprint.

While a uniform magnetic field exerts no force on a magnetizable object, an *inhomogeneous* gradient field (like that at the end of a cylindrical magnet) does (Figure 1.11). Calculations of gradient forces tend to be knotty, so we shall note here in passing that for the simple case of an object with a permanent magnetic dipole moment, $\boldsymbol{m}$,

$$|\boldsymbol{F}| = \nabla(\boldsymbol{m} \cdot \boldsymbol{B}). \tag{8.2}$$

When the magnitude of $\boldsymbol{m}$ itself is affected appreciably by the strength of $\boldsymbol{B}$, as with a paramagnetic material or a ferromagnet, things begin to get complicated. The *fringe magnetic gradients* outside the main magnet can exert strong translational forces on objects, such as flying magnetizable scalpels, wheelchairs, or oxygen bottles, which can be extremely hazardous (Figure 1.11). $\boldsymbol{B}_0$ at the center of the bore is relatively homogeneous in space, on the other hand, and instead of a strong pull, it creates only a *torque*, ($\boldsymbol{m} \times \boldsymbol{B}$), on a ferromagnetic object, such as the wrong kind of aneurism clip.

Active shielding reduces the magnetic footprint of the magnet, and it also compresses and increases the spatial field gradient strength around it. That means that patients, staff, or visitors who might have ferrous objects with or within them may feel no force when first entering the room, only to have the magnetic object yanked away with little warning nearer the scanner. Entry to the MRI suite requires careful screening, along with direct supervision (or training in the risks, if appropriate).

It is generally considered safe to approach the 0.5 mT *isofield surface* enveloping the magnet (Figure 8.3); the field strength is taken as a surrogate for the gradient. Coming closer may cause problems for people with magnetically sensitive medical implants (e.g., pacemakers). The '5 Gauss' line, which is the Food and Drug Administration regulatory limit, should be, and commonly is, contained between or within the walls of the scan room. A spatial footprint of the 0.5 mT (5 gauss) line can be mapped on the floor, in any case, and regions both inside and outside the scan room need appropriate warning signage.

Recent trends in superconducting magnet design have been toward bore-holes that are higher in field strength, shorter in length, and wider in diameter, while maintaining current standards of homogeneity. These are primarily for patient comfort and to reduce claustrophobia, along with greater flexibility in patient positioning, but they make it physically more demanding

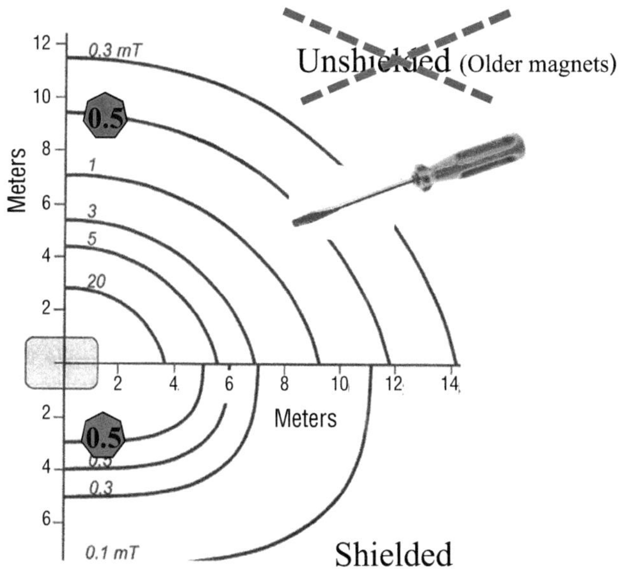

Figure 8.3 One quadrant of a contour map of the fringe fields surrounding a 1.5 T magnet, either without (upper) or with (below) passive plus active shielding, which is virtually always in place. Shielding greatly cuts down the area enclosed by the 0.5 mT surface, which delimits the zone taken to be safe for most people, and to which access is normally controlled. Entry to regions above 0.5 mT should be denied to *all* but those known to be safe in a strong field.

and expensive to achieve homogeneity. Still, it has been possible to construct 1.5 T magnets with field configurations compact enough for them to be fitted onto trucks for transport to distant imaging sites, while the superconducting magnetic field remains on.

Permanent magnets and electromagnets

Some permanent magnets and resistive electromagnets still exist in operation (Figure 3.1b). They usually produce vertically aligned, lower-strength fields. The air gap between the two poles is comforting to some patients with fears of confinement, and it also allows physicians to carry out certain procedures relatively easily. Their use has greatly diminished, though, because of their somewhat lower SNR and general image quality, the rise of wide- and short-bore cylindrical superconducting systems, and the excessive weight (permanent) or operational costs for electricity (electromagnet).

Most non-superconducting MRI systems employ open *permanent magnets* in which the fields are frozen into up to a hundred or so tons of large C-shaped (horseshoe-shaped) blocks of ferromagnetic alloy. They consume no electric power or cryogens and are largely maintenance-free from this standpoint, but they are capable of producing only a few tenths of a tesla. Despite a preference among many radiologists for higher fields—largely because of better SNR, resolution, and scanning times—the more modest acquisition costs of these magnets have made them highly cost-effective in certain clinical settings that required a tighter budget.

A few percent of clinical machines use open, room-temperature *electromagnets*. The poles of one of these are commonly held by supports above and beside the patient, which leaves a relatively spacious and non-confining patient area, a configuration that can look much like that of a permanent magnet. Each of the two coils might consist of copper or aluminum tubes wound on an aluminum frame, which can together produce a 0.6 T main field. The conductor may consume 50 kW of electrical power in resistive heating and require the removal of heat at that rate, such as by pumping cooling water at 50 L/min through it.

8.2 Gradient Coils

The linear *x*-, *y*-, *z*-*gradient magnetic fields* are produced by three independent *gradient coils* situated within the bore. The gradient fields are designed to always point in the *z*-direction, parallel to the main field, but the *strength* of each can vary linearly with distance in any direction from the center of the main magnet, Equation (3.1) and Figure 3.2. These are intermittently energized by the *gradient drivers*, highly specialized high-power audio amplifiers. They are much weaker than the main field, with *gradients* on the order of 50 mT/meter—so that over a 0.4 m wide FOV, the field strength can vary by 0.02 T. Coil design is still an area of active research [Poole et al. 2007]. Like the coils of a superconducting main magnet, the gradient coils and their amplifiers are also quite costly.

It is difficult to maintain linearity over a large field of view, and the design of coil geometries can become quite sophisticated (Figure 8.4). A nonlinearity in the gradient field can produce an artifact that, fortunately, is easily recognized and mostly corrected with *gradient distortion correction* software (Figure 3.13). Other problems are associated with gradient magnet cooling, which is neither instantaneous nor perfectly efficient. The coils take time to reach a steady-state temperature which, in turn, affects their resistance, hence current. (This is true of any non-superconducting coil with rea-

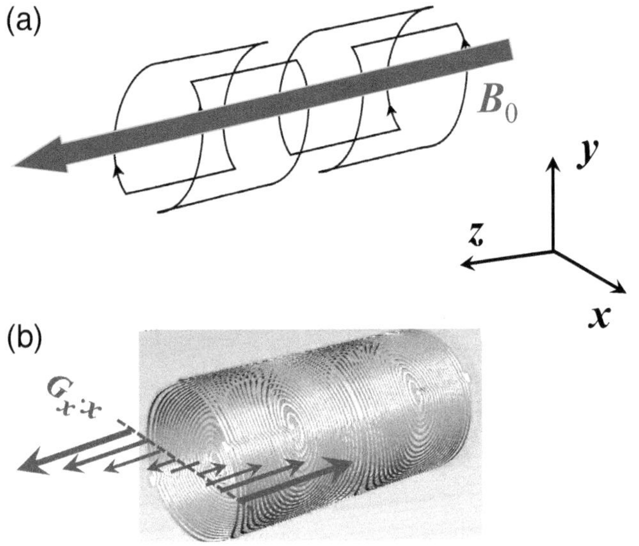

Figure 8.4 Gradient fields. (a) Three independent non-superconducting gradient coils fit just inside the bore of a cylindrical superconducting magnet. For the G_x coil shown here, the magnetic field everywhere points along the main field, B_0, but its strength increases linearly with position in the **x**-direction, as $G_x \cdot x$. The general flow of current in an **x**-gradient coil is simpler in concept than (b) in practice; seen here is the design for one of multiple layers of windings. Important parameters describing the capabilities of a gradient coil are the *gradient* or slope of the field, dB_z/dx, which ranges typically from 20 to 80 mT/meter. The *rise time* is about 0.25 ms, and the *slew rate*, 50 to 200 mT/meter/ms. Simple discussions of MRI commonly make the approximation that gradient switching occurs instantaneously, replacing $\int G(t)\,dt$ with a binary pair, $G = \{0,1\}$.

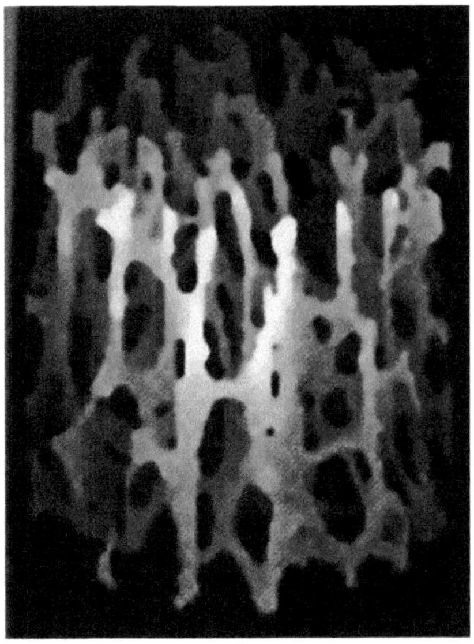

Figure 8.5 A 3D view of a cylinder of bone 5 mm across and high. Magnification MRI requires the use of gradient fields much higher than normal. [Courtesy of Scott Huang, Felix Wehrli, John Williams, and the journal *Medical Physics*.]

sonable current passing through it, such as the RF transmit antenna.) Any other metal in the scanner may be warmed as well, such as the passive shims. Such effects make the field from the gradient coils drift in strength, along with any process dependent on them. Excitation of the tissue and spatial localization during image formation will be affected by the drift. This can lead to signal loss or artifacts, such as apparent motion between repeated image acquisitions.

Gradient fields have to be rapidly switchable and highly reproducible. A *rise time* (to approach the maximum gradient value) of ¼ ms (250 μs) is normal, and a gradient may be pulsed 'on' for only milliseconds at a time. The two measures of gradient performance, magnitude and rise time, are combined to give a *slew rate*, which can approach 200 mT/m/ms (or T/m/s). Gradient field strength and rise time capabilities have a significant effect on instrument flexibility and performance. In general, a low maximum magnitude can limit the minimum slice thickness or in-plane field of view of a 2D scan, as we'll investigate later. Fast slew rate and maximized imaging speed allow short echo times, fluid flow compensation and encoding, and good performance of echo-planar imaging applications. It takes substantial current to generate gradient fields, up to ~500 amperes, and the system must, therefore, be constantly cooled, usually via a chilled water supply. The coolant may have to extract 25 kW of heat for the system to operate at 100% duty cycle, albeit briefly, and the problem only gets worse with the trend toward larger bore size.

Magnification imaging is used in clinical and anatomical research, incidentally, and it requires especially strong gradients (Figure 8.5) [Harnsberger 2007].

Another gradient-based imaging effect occurs because the rapid switching of a magnetic field on and off generates reactive *eddy currents* in nearby metal components. Most scanners require metal bracing to keep the magnet from imploding because of the magnetic forces, and to keep the vacuum intact within the helium Dewar. Electrons in this conductive material will attempt to resist any rapid changes in the magnetic field via their own motion, which generates a current, hence a magnetic field, that opposes the change. Chapter 7 showed that to traverse ***k***-space, it may be necessary to turn on a gradient and leave it on for a while; but if eddy currents shift the gradient field slightly, it will seem that data are being collected at one location in ***k***-space, while actually it is happening at another. This error leads to several distinct artifacts that lower the image quality, such as a *Nyquist ghost* of the original image, shifted by a factor of ½ FOV.

A standard approach to overcoming eddy current effects is to shield each gradient coil with a second that is wired in reverse, and activate to the amount needed. This approach doesn't work perfectly, however, and sometimes more complex *eddy current compensation* has to be called upon.

A factor that contributes to the discomfort of some patients is the loud clicking, buzzing, and rattling that results from periodic interactions between the fields of the gradient and main magnets. While this can be muffled considerably by piping Ravel's *Bolero* in through earphones, researchers continue to seek other solutions.

8.3 The MRI RF System Is Like an AM Radio Transmitter/Receiver Pair

In a sense, the RF system of an MRI is somewhat like a commercial *amplitude modulation* radio system. Transmission and reception of an AM signal involves producing and detecting RF electromagnetic radiation that falls within a narrow band of frequencies. Let's set MRI aside for the moment and focus first on how ordinary AM radio works.

An AM radio *transmitter* generates a monochromatic, constant-amplitude *carrier* RF signal which, for commercial broadcasting in the United States, must lie between 540 and 1610 kHz (Figure 8.6a). (FM stations operate between 88 MHz and 108 MHz.) Let's suppose that the carrier is chosen to be 1 MHz in our current AM example. The audio information of interest, of much lower frequencies (typically 50 Hz to 10 kHz), is embedded onto the carrier by *modulating* its *amplitude*. Suppose that the information to be conveyed is to be a pure 4,000-Hz tone—that is, the amplitude of the carrier is made to vary at that frequency. With a *modulation index* of 100%, the signal height ranges from 0 to twice the carrier amplitude. The highly amplified oscillatory voltage from the transmitter ends up sloshing electron current back and forth in the antenna at and near the carrier frequency. Because the electrons are rapidly accelerating and decelerating charged particles, they radiate EM radiation in accord with the Maxwell

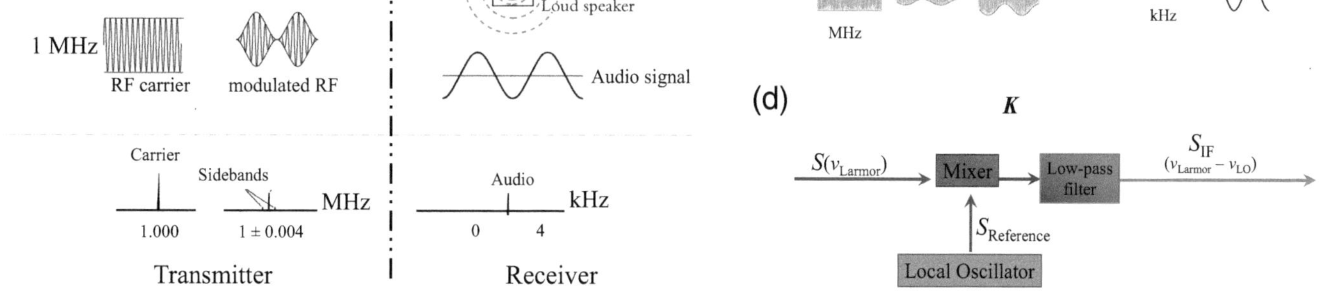

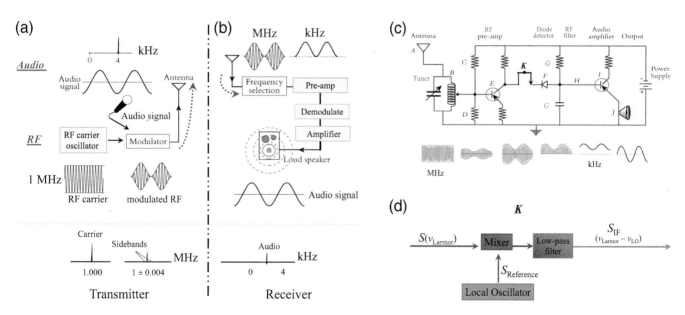

Figure 8.6 Operation of the radiofrequency electronics of an MRI scanner is like the transmission and reception of an amplitude modulated (AM) radio signal. (a) In an AM transmitter, an audio signal to be transmitted (in this case, a pure 4-kHz tone) modulates an RF carrier (1 MHz, here), and the resulting high-voltage radiofrequency wave drives electrons in the antenna back and forth. (b) A receiver reverses the process. From the confusion of multitudinous EM waves being sent in from the antenna, the *tuned circuit* selects out the one of interest and rejects all others. The resulting RF signal $S_{RF}(t)$ carries the audio information. The demodulator can be as simple as a diode detector that recovers the audio component signal, which a smoothing filter and second-stage amplifier end up driving through the speaker. (c) A rudimentary AM radio receiver, the components of which are described in the text. This very simple design does not shift the incoming signal down to an intermediate frequency, as the vast majority of real receivers do. (d) Creation of an Intermediate Frequency (IF) version of the signal. This circuit goes to point *K* in (c).

theory. A spectral (Fourier) analysis of the resultant signal would reveal not only a peak at the carrier frequency, but also a pair of others—4 kHz above and below it respectively, its *sidebands* at 0.996 MHz and 1.004 MHz. The reason this comes about is suggested by Equations (5.5a) and (7.12). All the audio information of interest happens to end up entirely in the sidebands; although the carrier comes along only to drive the bus, it does remain as part of the signal. The electronics must be able to process all frequencies between and including the sidebands, a range of 8 kHz centered at 1 MHz.

A concept that arises frequently in discussions of signal processing is *bandwidth*. While it is applicable also to spatial waves, *BW* usually refers to temporal frequencies. The bandwidth of a radio, for example, is the range of frequencies that it can handle without introducing distortions. Throughout most of the world, commercial broadcast transmission signals are spaced 10 kHz apart.

EXERCISE 8.3 What is the bandwidth of a standard AM radio receiver? What is the range of frequencies that the input of the radio must be able to accommodate?

AM *reception* reverses the process (Figure 8.6c). A *tank-* or *tuned-circuit*, or *narrow-band-pass filter*, ***B***, comprised of a capacitor and an inductor (coil), has a normal-mode resonant frequency in the RF range. Either its capacitance (usually) or its inductance can be adjusted so that the resonant frequency equals that of the original broadcast carrier, selecting out a suitably slim band around it, and rejecting all the rest of the RF coming in from the antenna. After passing along a simple wire (for now), ***K***, the receiver then *demodulates* (*rectifies* and *smooths*) the resulting audio-modulated signal. The result should be a replica of the original audio signal, in our example the pure 4 kHz tone again, which is amplified and sent to the loudspeaker.

Super-heterodyne AM radio

Some younger readers have had the misfortune of growing up in the era of integrated circuits and digital hardware, and were thus sadly deprived of the opportunity to acquire an affection for analog audio and RF electronics. But for those of you who are nonetheless curious, let's explain a bit more.

Figure 8.6c diagrams a hobbyist's build-it-yourself AM radio receiver. All the nearby radio stations separately set the virtually free electrons in the antenna wire to sloshing back-and-forth at and near its own unique carrier frequency, ***A***, and the instantaneous input voltage is the superposition of all these incoming signals. They do not interfere with one another, in effect, and the one corresponding to the particular carrier frequency of interest, ν_{RF}, can be singled out with a tank circuit, ***B***, which resonates at the unique frequency determined by the precise setting of the capacitance and the coil inductance (Figure 6.1). Somewhat miraculously, just as you can distinguish any note of a piano from the others, the radio's tuned circuit can allow the modulated carrier of your favorite classical station to pass through undeterred, while rejecting all the others. This selected lone signal is fed into a low-noise RF pre-amplifier, the transistor ***E***; the settings of resistors ***C*** and ***D*** determine the average *bias voltage* on the *base* electrode of the transistor for best operation. After the signal passes through the wire ***K*** (again, for now), it is extracted by rectifying and demodulating the RF with a silicon diode, ***F***. Smoothing it with a capacitor-resistor filter, ***G***, recovers the original audio signal, ***H***, that modulated the carrier in the first place. A second transistor, ***I***, amplifies this audio signal and—lo and behold!—the Trout Quintet splashes joyously out of the speaker, ***J***.

Nearly all modern AM and FM radio receivers, however, benefit from the inclusion of an *intermediate frequency* (IF) stage, based on a trick known as *heterodyning*. The desired RF input frequency, ν_{RF}, is selected out by the tank circuit (*B*, Figure 8.6c). At point ***K***, after the pre-amp stage but before passing on to the rectifier, audio amplifiers, and smoothing filters, the newly isolated, now very narrow-band RF signal is *mixed* or *heterodyned* with the output of an adjustable-frequency *local oscillator* (Figure 8.6d). The monochromatic *reference* sine wave, $s_{LO}(t)$, from the LO is tuned such that the frequency difference from the carrier is *fixed*—455 kHz for AM radio. The mixing of $S_{RF}(t)$ and $s_{LO}(t)$ does not affect the information content of the incident RF but, by Equation (5.5a), it does result in a new, much lower *beat* or *intermediate frequency* (IF) waveform centered at

$$\nu_{IF} \equiv \nu_{RF} - \nu_{LO} = 455\,\text{kHz}. \tag{8.3}$$

This becomes, in effect, a new carrier. The IF signal at $(\nu_{RF} - \nu_{LO})$ does *not* depend on ν_{RF} per se, but rather the RF signal frequency coming from the tuned circuit is shifted down to a *fixed* IF of ν_{IF} = 455 kHz. Then all subsequent amplifiers and filters can be tailored for optimal performance regardless of the carrier, leading to improved SNR.

There also appears a higher-frequency 'signal' at $(\nu_{RF} + \nu_{LO})$, but this is of no value and is filtered out.

New MRI machines employ direct detection, not an IF stage

Signal management in a modern MRI is much like that of an AM transmitter-receiver pair, but with several notable differences.

A simple AM radio transmitter utilizes direct audio modulation of the carrier amplitude (Figure 8.6a). An MRI *pulse programmer*, on the other hand, forms precisely sculpted (generally like a Sinc function) temporal pulses of RF (not just on-off!) for its transmission phase (Figures 5.10 and 7.6b), that deliver RF of uniform amplitude over a narrow band of frequencies (BW_{trans}). The 'carrier' is at about 63.87 MHz at 1.5 T, and pumps out instantaneous power of up to 25 kW, at least for a few microseconds at a time—enough to briefly energize 250 light bulbs. A *Faraday cage* of copper sheeting or mesh covers the walls, ceiling, floor, and even the windows of an imaging suite, providing about 100 dB of RF attenuation and insulation to keep the MRI power from getting loose and, similarly, to prevent any external RF electromagnetic noise from being picked up by the highly sensitive receiver coil.

EXERCISE 8.4 What would the RF power spectrum be for a simple on-off pulse?

The receive antenna of an MRI instrument detects a continuous RF signal coming from the patient, $S(t)$, and runs it through a low-noise RF pre-amplifier. You might reasonably expect from the above description of AM radios that an MRI machine mixes the incoming signal with that of a reference LO to create an intermediate frequency signal at, say, a few hundred kHz (Figure 8.6d). This $S_{IF}(t)$, in turn, is sampled at regular time intervals by an *Analog-to-Digital-Converter*, an electronic circuit that rapidly converts an analog input signal voltage into a sequence of digital numbers. (An ADC needs closely matched components, and usually comes as integrated circuits.) The newly created digital signal is now sent to memory in a reconstruction computer which demodulates and filters the signal to remove the intermediate carrier (all in the digital domain). Finally, the computer performs a discrete Fourier transform after all the data is collected, producing a digital image with a certain bit depth. This is, indeed, how earlier MR instruments operated when the electronics were relatively slow, causing a bottleneck that called for the creation of an IF and the use of heterodyning.

While this provided a useful first approach, several significant improvements soon followed to enhance the SNR. For one thing, modern sampling and ADC electronics have overcome the speed limitation, easily achieving a 32-bit ADC with sampling rates on the order of 100 to 200 MHz, thereby circumventing the need for heterodyning, which introduced some noise. Also, there is some random noise inherent in any analog signal, and more is introduced by the sampling and digitization processes themselves. The pre-amplifier and digitizer were a fair distance from the pickup coil, and the RF cable linking the receiver antenna to them was long and capable of acquiring other noise. Limiting the bandwidth of the data being conveyed by the carrier with narrow-pass filters cuts out much of the noise getting in to it.

The introduction of *direct digitization* of the MRI RF signal by way of so-called *digital coils* at the front end of the receive chain can resolve the speed and much of the noise problems [Hindorean et al. 1993]. The digitization of signal is pushed farther back toward the antenna itself, where a small preamp stage and a high-speed sampling/ADC are installed physically adjacent to the receiver coil (Figure 8.7). The signal is digitized and converted into digital *optical* pulses which are transmitted via nearly noiseless *fiber optic cables* to a computer at the imaging suite. Digitization at the front end and then replacement of meters of resistive copper RF cabling with optical cables can bring about a noise reduction of up to 40%, depending on the coil. Finally at the computer, the MR signal enjoys further digital processing (we're sweeping quite a bit under the rug here, as we'll see in the next section) and undergoes a Fourier transform into a real-space image.

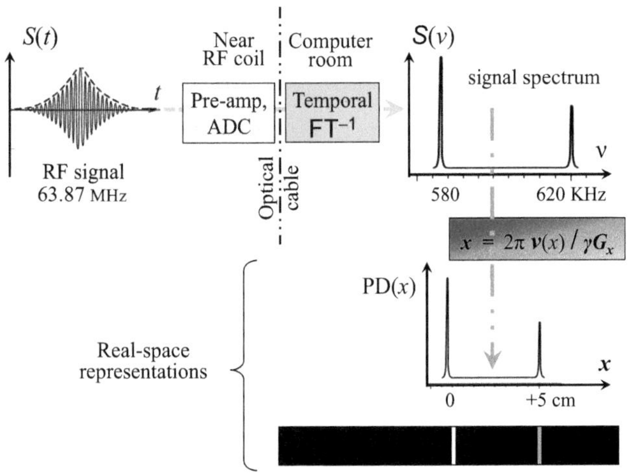

Figure 8.7 Produced simultaneously by large numbers of voxels, the incoming $S_{RF}(t)$ is a high-frequency echo signal created with a spin-echo or gradient-echo MRI pulse sequence. Low-noise pre-amplification, sampling, and digitizing occur in small electronic devices close to the receiver coil. Virtually noise-free optical cables then bring the signal into the computer room where, after further digital processing, it is subjected to a digital temporal Fourier transform to frequency-space and metamorphoses into a spectrum, $S_{RF}(\nu)$. With the simple change of variables, this, in turn, displays the distribution of proton density along the x-axis, PD(x), i.e., the 1D image. Of course, it's more complicated than that.

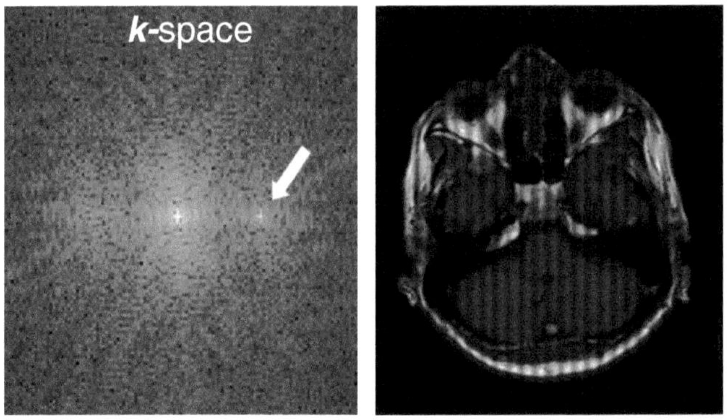

Figure 8.8 This herringbone or crisscross artifact arose because the nearby electronic equipment, or something within the MRI itself, produced a pulse of electrical noise that interfered with data processing. [Courtesy of Rao Gullapalli and J. Zhuo.]

EXERCISE 8.5 What kinds of 'further digital processing' steps might be needed here, and to do what?

During sampling, a real signal decays exponentially at a rate of 1/T2 or 1/T2*, as sin $(2\pi \nu t)\ e^{-t/T2}$, depending on the manner of data acquisition, because of spin relaxation. In general we want the duration of the readout window to be much smaller than the rate of signal loss so as not to compromise image quality. The relaxation times determine contrast, so it is necessary to wait through the appropriate TR and TE when playing out the pulse sequence; however, we desire the temporal window during which the receiver is turned on and the signal is being sampled to be somewhat shorter than the characteristic relaxation times of the tissue (at least that's a goal). Otherwise, the contrast of data will change during the readout window, as more and more data are read into a cell in ***k***-space.

Unfortunately, even Faraday screening cannot protect the instrument from misbehavior of electronic equipment in the imaging suite or within the MRI scanner itself. The MR electronics may occasionally hiccup, for example, or metal objects in the bore can vibrate, producing a voltage spike in the receiver. This may manifest as a narrow peak in ***k***-space, resulting in an image reconstruction anomaly referred to as a *herringbone artifact* (Figure 8.8).

Receiver bandwidth

The bandwidth of an apparatus commonly refers to the range of frequencies over which it can operate properly. It is necessary to distinguish between the RF bandwidths of the transmitter, BW_{trans}, and the receiver, BW_{rec}.

First, a short metaphor. Imagine that we are leisurely crabbing in the shallow waters of the Texas gulf coast, using a net that you lower into the water with a chicken neck tied to it (how coauthor NY misspent much of his youth). The longer you dip the net into the water—the longer it dwells on the bottom—the more crabs are bound to creep into the net. We'll call that a *crab-sampling time*, which is the duration of how long we trick crabs into the net. Or if we care about the reciprocal of the time, we can refer to a *crab-sampling rate*. You pull up the net gingerly and then dump the crabs into an ice chest (watch out for their claws!). Each chest corresponds to a position in ***k***-space, recording the signal strength (number of crabs). Then repeat the process: drop in the net, relax while you acquire crabs, then throw them into the next ice chest. We call the time it takes for the whole process to occur the *casting time* (or the inverse, a *casting rate*), or the time between casts of the net.

Ideally, we want to spend most of our time getting crabs and minimize the time that we are fussing with the net or the ice chests. That is, we would like for the crab-sampling rate to be as close (or equal) to the casting rate. The two rates are not necessarily equivalent, which could make for terrible fishing. We could dip the net in the water only for a split second (a super-fast crab-sampling rate) before putting the haul into a new chest. Then, to make matters worse, after this very limited haul, we grab a few brews before casting the net for the second time (a super-slow casting rate). We wouldn't get many crabs at all using this procedure. We might think that the number of crabs is related to how long we spend outside on such a day—but it's actually related instead to how long the net is in the water. The longer the *dwell time* of the net in water, the slower the *crab-sampling rate*, but the more crabs per chest. Thus it is the duration of the net dwell time—how long we sample crabs—that is what is important to maximizing the number of crabs (or signal, in MRI), not the duration of time we sit between tosses of the net into the water (related to the casting rate).

There is an equivalent difference in sampling rates for new scanners: the digitization rate of an ADC (super fast now) and the *data sampling rate* associated with dwell time (slower). Improvements in technology have allowed for the production of low-cost ADCs that can directly sample at rates comparable to the Larmor frequency, as noted earlier. But the sampling rate of a modern ADC is not directly relevant to dwell time. An ADC might chop up the signal into quick, discrete chunks, but at the end of the pipeline, data are integrated, in a sense, from multiple chunks into a slower flow of longer-duration samples. Data sampling rates (related to dwell time) are to a large degree defined in the digital processing pipeline downstream of the ADCs. In effect, data are *resampled* digitally according to a system bandwidth/rate that we choose. The details of how it works is somewhat mysterious, unless you dig into the digital processing pipeline.

For pedagogical purposes, we shall limit the discussion to older scanners, for the time being, given the caveats about data sampling rates on new scanners.

If the *time between samples* is t_{dwell}, then the *sample rate* is defined as (per our crab example):

$$\nu_s = 1/t_{\text{dwell}} \tag{8.4a}$$

$$BW_{rec} = 1/t_{\text{dwell}}. \tag{8.4b}$$

This is what is meant by the commonly seen, but not particularly obvious or self-explanatory, expression: 'The bandwidth is the inverse of dwell time.' In modern scanners, setting the receiver bandwidth defines the dwell time—the time during which we collect data for an individual datum in ***k***-space—regardless of what the ADC digitization rate is.

Meanwhile, the Nyquist-Shannon theorem, Equation (5.29), demands that if the MRI signal contains Fourier components up to ν_{MRI}, the maximum frequency contained within the bandwidth of the input signal, then the sampling frequency must be at least twice that,

$$\nu_s \geq 2\,\nu_{\text{MRI}} \tag{8.5a}$$

to preserve the signal's fidelity during digitization. The sample rate, $\nu_s = 1/t_{\text{dwell}}$, must be more than twice the maximum frequency contained within the bandwidth of the input signal, Equation (8.4b). It adds to the confusion that some vendors define BW_{rec} as bandwidth per image, while others mean bandwidth per voxel. Even more confusing is the fact that bandwidth per image may be reported using half of the sample rate, which is really ν_{MRI}. So, a BW_{rec} given as 12 kHz may be defined as $\pm\frac{1}{2}\nu_s = 24$ kHz! Things to keep track of....

For our 1D phantom, typically 2×512 or more samples of the RF echo signal would be taken over adjacent intervals, the duration of each of which is t_{dwell}. The x-gradient field is kept continuously *on* during all of this. To these samples there correspond values of $S(k_x)$ calculated at steps Δk_x from one another on the k_x-axis (Figure 7.9). In this way, a horizontal line is laid down in ***k***-space and, from that, a 1D Fourier Transform produces a 1D MR image of the phantom.

How long would it take to obtain 512 samples of an MR *echo* signal (only) during readout for a receiver with a fairly high receiver bandwidth of, say, BW_{rec} = 100 kHz? $1/BW_{\text{rec}} = t_{\text{dwell}}$ indicates that $t_{\text{dwell}} = 10^{-5}$ = 10 μsec dwell time for each sample, so about 512×0.01 msec or ~5 msec are needed for all of them to be taken. With a 32 kHz *BW*, by contrast, it would take about 16 ms. In other words, the *readout time* is given by the width of the sampling matrix divided by the receiver bandwidth.

In the discussion of Figure 7.9b and Table 7.1 above, it was argued that the length of a single echo may be of the order of tens of milliseconds long. Let's assume, for the moment, that an entire echo comes and goes in t_{echo} = 10 ms. For a carrier signal of 64 MHz, that would correspond to $(64\times10^6$ Hz$) \times (10\times10^{-3}$ sec$)$ or ~64,000 cycles. But remember, as the carrier oscillates at this high frequency, its amplitude is modulated at the much lower frequency relevant to BW_{rec}. So, after the demodulation and filtering process (for non-direct digitization systems), the digitized signal will be changing slowly, relative to the carrier wave.

One more step, involving the familiar change of variables, $x = 2\pi\nu / (\gamma G_x)$, reveals the useful linkage among the three important parameters BW_{rec}, FOV, and G_x:

$$\text{FOV}_x = (x_{max} - x_{min}) = 2\pi BW_{rec} / \pi G_x . \quad (8.6)$$

If a voxel is kept at a certain size, increasing the gradient values will cause greater frequency differences (or *BW* per pixel) between voxels; this makes sense because the Larmor frequency is changing more rapidly along $\boldsymbol{x}$. Signals from pixels outside of the field of view will be outside of the full bandwidth of frequencies to which the receiver is tuned. Those signals will still be sampled, unfortunately, but the sampling rate will be too slow to capture the signal accurately, and this will lead to aliasing.

Sampling is an approximate operation that can give rise to several forms of deviations from the incoming signal, and these may be apparent in the MR image as distortions. *Quantitation error* is the noise introduced by the ADC's inability to measure the signal voltage better than some $\pm\frac{1}{2}\Delta$V. Likewise, *aperture error* arises when the sample is obtained as a time average within a sampling period, rather than exactly at $t = n$T. *Oversampling* at even higher frequencies is a commonly employed method of reducing the potential effects of both these problems, making it possible to accommodate the field of view without risking an aliasing artifact.

8.6 Antennae/Probes/RF Coils and the SNR

Magnetic resonance takes place only if the magnetic field of the exciting RF signal is perpendicular to the main magnetic field. Since detection of an NMR signal comes *after* production of the excitation RF pulse, a single coil could be used for both the production and the subsequent detection of the resonance signal. But the primary objective of pulse generation differs from that of echo reception, so separate and differently designed coils are commonly employed.

MRI requires the creation of a sufficiently intense and uniform RF field in the non-radiative *electromagnetic near-zone field,* which dominates adjacent to the antenna and just within the patient. (By contrast, a commercial radio antenna is intended to radiate into the *far zone,* away from the antenna.) MRI attempts to bring about NMR in the near zone, with minimal energy dissipation and not much radiation away from it. Very little power is absorbed by spins during transmission, incidentally, and only a tiny amount of it is released from them later during reception. The MR receiver, on the other hand, must be highly sensitive to weak signals, as low as nanovolts (10^{-9} V) in amplitude, emanating from precessing protons within the patient's body, with low noise generation.

A typical transmitted pulse carries up to about 25 kW of power, corresponding to a field strength on the order of 10 to 50 mT. The field of the signal returning from a patient is about 0.1% of what is transmitted.

The transmission and reception pathways are generally segregated by way of a transmit-receive (TR) switch to avoid running the risk of leaving on the receiver while transmitting, resulting in a burnt-out receiver pre-amp, a frustrated radiologist, and an MR service engineer gnashing her teeth. TR circuitry detunes or decouples receive coils during transmission. RADAR and other high-power applications also employ TR switches.

In general, signal-to-noise can be enhanced significantly with RF detection coils that lie close to the body. Volume coils have been designed to provide uniform signal reception and reasonable signal-to-noise ratio (SNR) within the FOV [Brown et al. 2014]. A so-called *birdcage* coil surrounds the head almost completely and allows for uniform excitation within the head (Figure 8.9a). Surface reception coils, which may be as simple as a single loop, provide high SNR for tissues close to the coil, but they exhibit strong nonuniformity in their spatial sensitivity to signal, which decreases as a function of tissue depth (Figure 8.9b). This is part of the reason for the development of phased arrays of coil elements (Figure 8.9c), the subject of the next section of the chapter.

Signal-to-noise ratio, again

Let's return to the SNR of the receive antenna, discussed earlier in connection with Equations (7.3). Just after a 90° excitation pulse is switched on and off, $\boldsymbol{m}_{xy}(x,t) = \boldsymbol{m}_0(x) \sin 2\pi \nu_{\text{Larmor}}\, t$ in the voxel at position x, as the magnetization precesses in the transverse plane, carrying with it the magnetic flux $\Phi(x,t)$. This, in turn, crosses an n-turn pickup coil of area A situated on the y-axis, say, Faraday-inducing in it a voltage

$$d\Phi(x)/dt \propto \mathtt{n}Adm_{xy}(x,t)/dt \\ = \mathtt{n}A2\pi\nu_{Larmor}m_0(x)\sin 2\pi\nu_{Larmor}t. \quad (8.7a)$$

The Boltzmann equation, Equations (4.4), indicates that the magnitude of the magnetization in any voxel is linear in the strength of the main magnetic field, $m_0 \propto B_0$. So also is the Larmor frequency for it, $\nu_{\text{Larmor}} \propto B_0$ and, therefore, the voltage it induces in a pickup coil. So the net signal from all voxels, including the time-dependent component in Equation (8.3a), is

$$V_s(t) = \int c_{\text{coil}}(x)s(t)B_0^{\ 2}\text{PD}(x)\text{d}x. \quad (8.7b)$$

The factor $c_{\text{coil}}(x)$ is a messy function of the receiver antenna coil design, including its value of nA, their shape and position relative to the location, x, of the voxel, the strength of their coupling to the spins, etc.

The receive antenna can be viewed as a resonant LC-tuned circuit like the one marked as B in Figure 8.6c, and the noise from it can be considered similarly. The *fluctuation-dissipation* theorem, widely used in

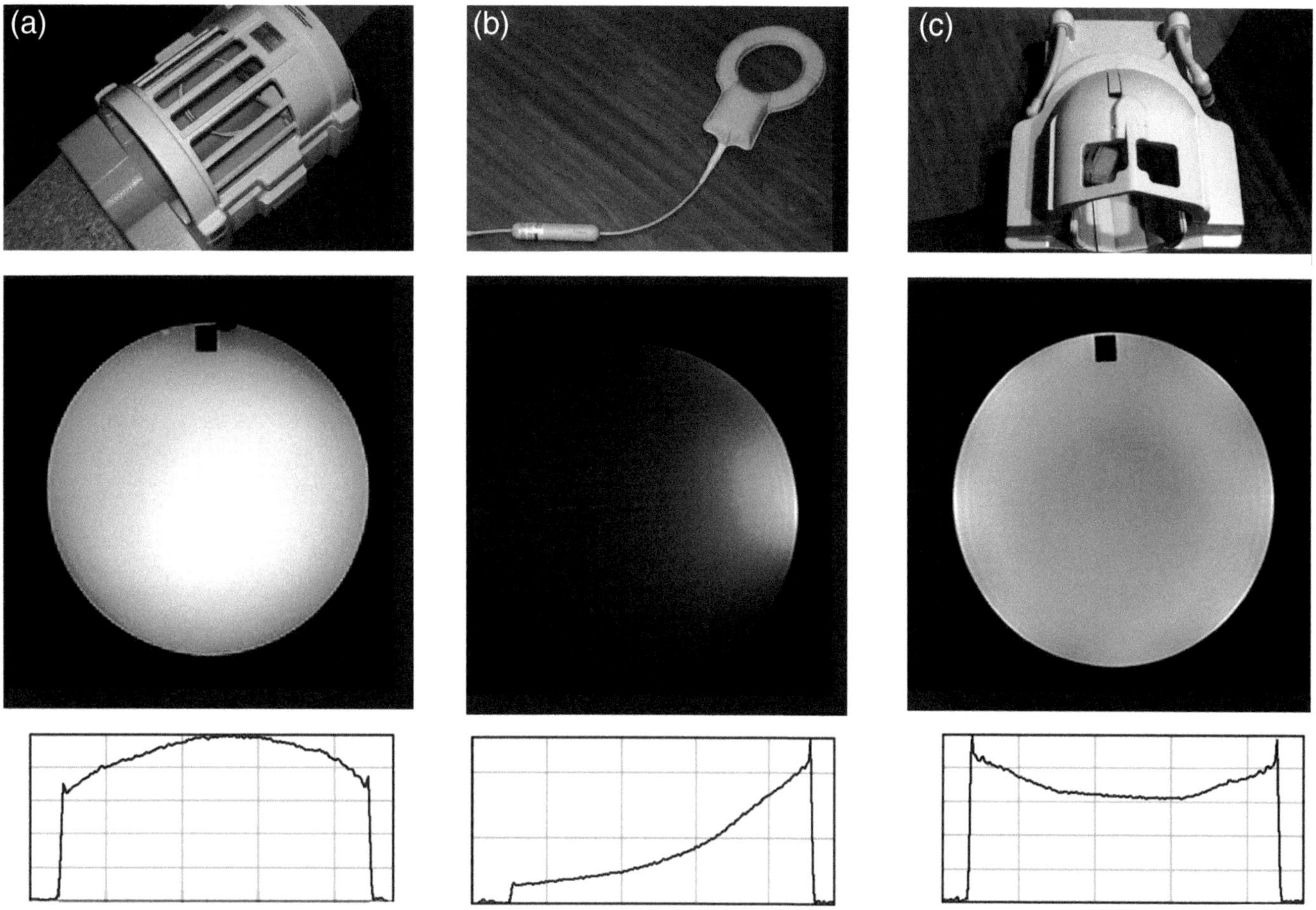

Figure 8.9 RF reception antennae for a modern scanner come in a variety of forms. (a) A *birdcage* head coil allows for localized transmission of RF energy to the head, and reception from it using the same device. Although it is losing favor to phased-array coils, it continues to serve for scanning of patients with implants, where confinement of RF to the head can minimize EM radiation to medical appliances elsewhere in the body. (b) Surface coil for the abdomen. (c) *Phased arrays* of multiple surface coil elements allow for multiple channels of readout during *parallel imaging*. Phased arrays of independent receiver coils have proved effective in improving both SNR and the speed of data acquisition. Middle row: typical image uniformity for each coil demonstrated with an American College of Radiology (ACR) QA phantom. Bottom row: intensity profiles along a horizontal line across each image; the PA coil has a characteristic central dip in image intensity, owing to its assembly from multiple surface coil elements.

electronics theory, tells us that the electronic noise voltage, V_N, from it is close to

$$V_N = (4k_B T R_c BW_{rec})^{1/2}, \quad (8.7c)$$

where k_B and T are Boltzmann's constant and the room temperature, R_c the electrical resistance of the circuit, and BW_{rec} its receiver bandwidth. This, however, is not the whole story. Dissipative body losses, the counterpart of those from R_c, increase with the volume of tissues inductively coupled to the coil. These come from dielectric losses arising from electric fields in tissues, electrolytic eddy currents induced by changing magnetic fields, and so on. The net result is that especially at higher fields, the patient contributes a relatively larger amount to the loss of SNR than does the antenna.

The SNR for a voxel can then be expressed approximately as

$$\text{SNR} = V_S/V_N \propto c_{coil} B_0{}^q \times \text{PD}(x) \times \Delta v\,(N_x \times NEX)^{1/2}\,(R_c BW_{rec})^{1/2}, \quad (8.8a)$$

where q is a real number somewhat less than 2. Here, Δv represents the volume of a voxel, and N_x is the dimension of the encoding matrix (i.e., how many voxels appear in an image). As we expect, SNR improves with field strength, voxel size, a narrower bandwidth, and NEX.

A simple, fairly intuitive formula is widely useful in MRI theory for estimating the SNR in 2D spin-echo imaging in terms of the imaging parameters, if one ignores relaxation:

$$\text{SNR} = \kappa \left\{ \frac{\text{FOV}_x \times \text{FOV}_y \times \Delta z \times \sqrt{\text{NEX}}}{\sqrt{N_x} \times \sqrt{N_y} \times \sqrt{BW_{rec}}} \right\} \times B_0. \quad (8.8b)$$

κ is a parameter that depends on details of the machine design, the size and shape of the RF coils, the pulse sequence, etc. Δz is the slice thickness, and $\Delta x \times \Delta y \times \Delta z = \Delta v$ is the volume of a voxel, and the meanings of N_x is evident from Figure 3.3. BW_{rec} is the RF bandwidth of the MR receiver, commonly of the order of 50 kHz, give or take a factor of 2 or 3. NEX is commonly 2.

From this, it appears that the SNR can be doubled by any one the following alone:

- increasing the dimensions Δx and Δy both by a factor of $2^{1/2}$, thereby doubling the number of protons in a voxel—meanwhile, the resolution along the ***x***- and ***y***-axes will decrease by $2^{1/2}$;
- doubling the slice thickness, Δz;
- quadrupling N_x or N_y;
- raising NEX by a factor of four;
- narrowing the BW by $\sqrt{2}$; or
- moving to a 3 T device, thereby doubling m_0.

Other factors may affect the SNR in more complicated ways. On some machines, for example, one cannot adjust Δx or Δy directly. Instead, Δx is determined by FOV_x/N_x. So, if resolution is decreased by increasing FOV_x by two, the SNR doubles. However, if resolution is diminished by decreasing N_x by two, the resulting SNR increase is only a factor of $\sqrt{2}$ greater.

EXERCISE 8.7 Confirm the physics behind each of these actions. What effect does each of them have on resolution in the x-y plane, resolution in the z-direction, data acquisition time, and apparent noise level? Any other concerns?

Discussion of the influences that degrade SNR becomes rather complex, but the above analysis should provide a taste of it. Going more deeply into it, such as analyzing the SNR characteristics of a low-signal image, can become a bit harrowing, drawing on the Rician statistical distribution with its modified zeroth order Bessel function of the first kind, and the like. Part of the problem is that while the noise is normally distributed in ***k***-space, for an image in real space, all voxel values of signal are ≥0.

8.5 Phased-Array of Receiver Coils

A *phased array* (PA) coil consists of a set of multiple loops or coils of wire, i.e., "elements," (Figure 8.10), each capable of functioning as a separate surface coil [Roemer et al. 1990]. A PA coil provides many advantages over a single surface coil or quadrature coil, and some disadvantages. Because surface coils are more sensitive than quad coils to neighboring tissue, an array of surface coils offers excellent SNR in adjacent tissue (e.g., brain cortex). This property unfortunately may increase the intrinsic nonuniformity of brightness in an image, but overall, SNR is generally better everywhere in tissue compared with a quad coil. PA receive antennae are employed in parallel imaging (pMRI), as discussed in chapter 13.

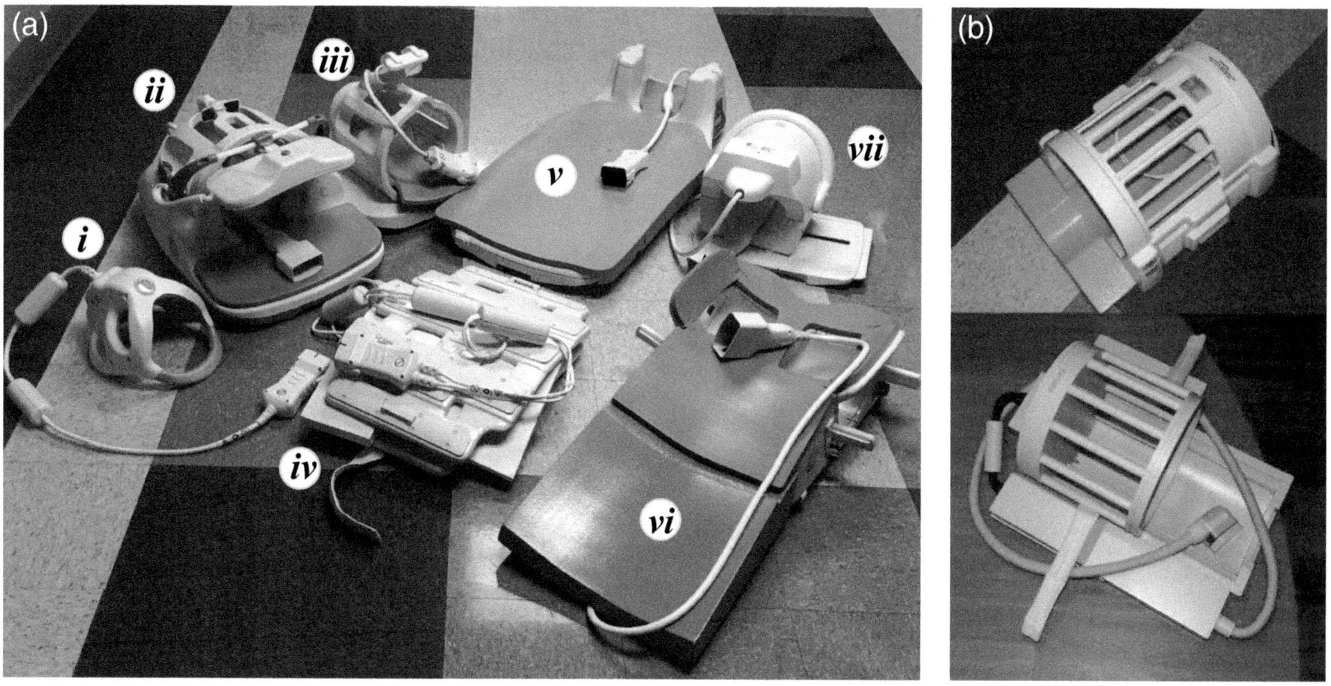

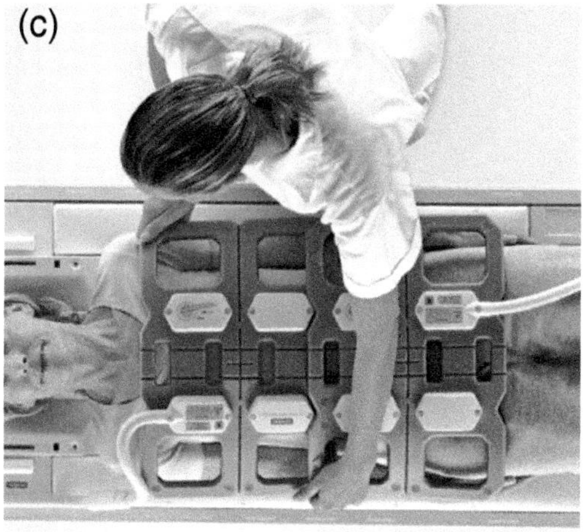

Figure 8.10 RF reception antennae for a modern scanner come in a variety of forms. (a) Five of these are receive-only coils and include: (i) shoulder coil; (ii) 8-channel neurovascular coil; (iii) 8-channel head coil; (iv) 8-channel torso coil; (v) multi-element PA spine coil; and (vi) 8-channel breast coil. Also appearing is (vii) an 8-channel transmit-receive knee coil. All of these are phased-array coils, and many allow for multiple channels of readout during parallel imaging. Phased arrays of independent receiver coils have proved effective in improving both SNR and the speed of data acquisition. (b) A bird-cage head coil. (c) Parallel imaging with phased arrays of independent receiver coils have proved effective in both improving SNR and decreasing image acquisition time. This 16-element coil provides coverage of the full torso, and two of these coils can be combined for full coverage from neck to toe. (Image courtesy of Siemens Healthcare.)

In conventional imaging, one can reconstruct a separate image for each element in the array. Large elements placed close together will overlap in their sensitivity to spatial features, while small elements placed farther apart will exhibit more discrete separation in spatial sensitivity. Poor performance in any one element can degrade image quality. However, image non-uniformity can be alleviated by optimizing how these images are recombined into a whole image [Roemer et al. 1990].

Finally, imaging information is shared between elements. In parallel imaging with PA, the redundancy of imaging information in its separate coil elements also leads to significant reduction in data acquisition time. Indeed, speeding up the image acquisition process with pMRI is known as *acceleration*, in the accepted parlance, chapter 13.

The above has dealt with multiple *receive*-element coils. Also of interest is the newer, complementary field of multi-*transmit*-coil MRI, which involves a coil with multiple elements, each capable of transmitting as a separate coil. Multi-transmit MRI allows for some interesting advantages, including SAR remediation, de-emphasis of dielectric effect artifacts, and selective excitation of small or oddly shaped volumes.

CHAPTER 9

T1 (Longitudinal, Spin-Lattice) Relaxation

As found in 1971 by Damadian (Figure 1.10a) the various soft tissues display different spin-relaxation behaviors (Table 9.1). The protons in a voxel inhabit a broad extent of environmental conditions, and they may take part in several kinds of relaxation process, each with its own characteristic time [Damadian 1971]. Two relaxation phenomena tend to dominate in a tissue, however, and normally each is close to being an exponential affair. It is, therefore, possible to correlate much of this activity with a single pair of distinctive, clinically meaningful parameters, T1 and T2. Fortunately, these two singular times, crude though they may be, are still sufficiently tissue-sensitive and tissue-specific for T1- and T2-based MRI maps to be highly informative in medicine.

In so-called *quantitative* MRI, an image is created as a reasonably precise, point-by-point record of the *absolute* value of either of these tissue parameters throughout a region of the body, and there can be clinical benefit to this.

In general, MRI studies do *not* directly display the distributions of T1 or T2 per se, but rather provide maps that reflect their spatial variations *in*directly. Almost always, the results of a standard clinical examination will reveal *relative* (rather than absolute) variations over space of the T1- or T2-values in tissue voxels. These sorts of images, said to be T1-*weighted* or T2-*w*, respectively, are, in effect, maps of spatial variations in the relative (rather than absolute) values of T1 and T2 tissues. With T1-*w* spin-echo MRI, for example, tissues

Table 9.1 For several tissues, typical values of proton density and also values of T1 at three field strengths, 1.0 T, 1.5 T, and 3.0 T, in milliseconds (ms). This demonstrates that T1 tends to increase with B_0. Values of T2 are typically in the tens of milliseconds, an order of magnitude *shorter* than those of T1, and largely independent of field strength. The numbers for the relaxation times from tabulated sources can vary considerably, in part because of the use of dissimilar data acquisition techniques. The ones listed here were selected by the authors as being representative.

Tissue	**PD** p^+/mm^3, rel.	T1, *1T* (ms)	**T1,** *1.5T* (ms)	T1, *3T* (ms)	**T2** (ms)
H_20	1		4000	4000	2000
Adipose	0.95	220	280	370	70
Blood, venous			1300		250
Brain					
white matter	0.7	600	700	1080	80
gray matter	0.8	700	1000	1300	100
CSF	0.95	2000	2400	4000	160
edema			1100		110
glioma		930	1000		110
Kidney			700	1200	60
Liver		320	600	800	50
hepatoma			1100		85
Lung			830		80
Muscle	0.9	600	1050	1400	40
myocardium			900		60
Spleen			1100	1300	80

with shorter T1-values show up brighter in the image, while T2-*w* makes shorter-T2 tissues appear darker—but in neither case is the degree of gray directly proportion to the value of the T-parameter. Yet this alternative MRI information can satisfy many medical needs, and—now wait for the punch-line —it can be obtained far more rapidly. In MRI, speed is of the essence, and quantitative imaging is like molasses.

The present chapter focuses, in particular, on the decay time T1, that of *longitudinal relaxation*, in a single voxel. It then turns to the creation of a T1-*w* image all along a 1D patient. The following chapter will address T2.

9.1 The Longitudinal Proton Spin-Relaxation Time, T1, in a Single Voxel

Equations (4.4) and the Figure 4.4d describe, from the quasi-QM, spin-up/spin-down perspective, the state of thermal equilibrium of a population of protons at body temperature in the field of an MRI main magnet. While protons in the voxel at any position $\boldsymbol{x}$ are continually being flipped up and down by thermal energy, Boltzmann statistics tells us that, on balance, an *average* of slightly (a few ppm of) more spins end up residing in the lower-energy state. This results in a time-averaged voxel *equilibrium* magnetization of $\boldsymbol{m}_0(x)$. Of primary interest in MRI is not the value of this thermal equilibrium-state magnetization itself, however, but rather the processes by which a disturbed population of spins moves back toward the equilibrium condition by way of nuclear spin relaxation, and the rates at which this occurs.

EXERCISE 9.1 How might thermal energy cause protons to undergo random spin-state transitions?

EXERCISE 9.2 Under what conditions can one replace $m_0(x)$ with m_0 in this discussion?

Relaxation is a common enough everyday phenomenon, whether it be the fading away of the peal of a church bell, the settling down of a swing after your child jumps off, or the decrease of current in a capacitor-coil-resistor electronic circuit. For all of these, the amplitude of resonance of the system diminishes nearly exponentially for the reasons discussed around Equation (5.1).

Likewise, after a tissue's nuclear spin system is somehow agitated, several distinct biophysical interactions occur at the molecular level among its protons. These encourage protons in the higher-lying energy level each to *relax*, and dissipate the right amount of energy for it to drop down into the lower level, as the system moves toward thermal equilibrium. The opposite is going on at the same time but, away from equilibrium, the net effect is to drive the system toward $\boldsymbol{m}_0(x)$.

EXERCISE 9.3 Why does the sound of a struck bell fall off exponentially? What other processes are exponential in time, distance, dose of radiation or drug, etc., and what physical phenomenon is common to them all?

The values of the various NMR proton relaxation times for a voxel depend on the local chemical environment and, therefore, on both the tissue type and its current state of health. It is the spatial variation in the relaxation times in various tissues that gives rise to the sometimes remarkable degree of contrast appearing in clinical MR images of soft tissue.

There are two specific kinds of spin relaxation that are not only readily detectable but also of special clinical importance. The first is the *longitudinal spin-relaxation* of the protons, which pushes $\boldsymbol{m}_z(t)$ back toward $\boldsymbol{m}_0$ at the rate 1/T1 after the system has been disturbed (e.g., by a 90° pulse). An MR image that emphasizes the variations in the value of T1 at the points throughout a slice of tissue is said to be T1-*weighted*. This can easily be described loosely in the semi-quantum spin-up, spin-down picture, so we shall tackle it first. The second, *transverse relaxation* with characteristic time T2, is emphasized in T2-*w* images. T2 relaxation incorporates T1-events, but also other happenings that are physically dissimilar, and will be the subject of the next chapter.

Imagine a collection of identical pocket compasses on a table that is being jostled lightly by a loudspeaker, as in Figure 4.3. In a magnetically shielded room, the needles point randomly in all directions. Take away the shield, however, and the needles all end up eventually pointing nearly north. They are still jiggling around a little because of the vibrational 'noise' being inputted, a macroscopic counterpart to thermal noise, but they do tend to pretty much align.

Now bang the table once hard so that the needles flop about wildly in all directions, but let the gentle vibrations continue. Each compass needle will settle back down exponentially, oscillating about north with an amplitude that diminishes over time. The length of time this settling down process takes, averaged over all the compass needles, is parameterized by their *mean needle relaxation time*, 'T.' More specifically, T is taken to be the average time it takes the net magnetization from all the needles together to drop 0.63 of the way down to background ($t = \infty$) level. For compasses, T is determined largely by the mass and shape of the needles, by the nature of the frictional forces occurring at the mechanical pivot points where the needles are supported, by the viscosity of any damping fluid, etc.

EXERCISE 9.4 Show that $(1 - e^{-t/T}) = 0.63$ at $t = T$. Does the amplitude of the speaker noise affect T?

T1 spin relaxation in a voxel of tissue protons is an analogous affair, but more nuanced, interesting, and, of course, clinically useful. The spins in a voxel of water with no magnetic field present would also point randomly in all directions, giving rise to zero net magnetization (Figure 4.4a). If we rapidly switch on a magnetic field, $\boldsymbol{B}_0$, however, quantum mechanics demands spatial quantization of the spins, and every proton will immediately snap into alignment either along or against the (vertical, *z*-directional) field, into an up- or down-state (Figure 4.4b). Roughly equal numbers fall into the lower- and higher-energy states in this fleeting moment of utter chaos, and the net magnetization is again briefly of zero magnitude.

Over a short time, some individual spins will begin a kind of seemingly spontaneous relaxation, with a modest majority of them tending downward toward the slightly more comfortable, lower-energy alignment *along* the field. The probability per unit time that any one proton in a population of them will flip over as the system thereafter equilibrates is independent of the time that it has already spent in the higher or lower energy state, of course, and also of what is happening to its fellow protons. Early on, $|\boldsymbol{m}_z(t)|$ appears to increase from a value of zero to 2μ at time 1, and continues on nearly linearly with time as 4μ, 6μ, 8μ, …, at times $t = 2, 3, 4,\ldots$ (Figure 9.1a). But over time, the system moves toward thermal equilibrium, so that the numbers of spins in the two quantized energy states approach a Boltzmann distribution.

By graphing this phenomenon over longer times, it becomes clear that what we have just seen is, in fact, only the beginning, linear portion of what is actually a curved graph of magnetization vs. time, the heavy blue line in Figure 9.1b. Over longer times, it grows as

$$\boxed{\boldsymbol{m}_z(t) = \boldsymbol{m}_0(1 - e^{-t/T})} \tag{9.1a}$$

from an initial value of zero,

$$\boldsymbol{m}_z(0) = 0, \tag{9.1b}$$

with equal numbers of spins pointing up and down. Thereafter, it grows toward a state of spin thermal equilibrium,

$$\boldsymbol{m}_z(\infty) = \boldsymbol{m}_0 . \tag{9.1c}$$

There is a slight excess of protons in the lower-energy state, which collectively produces the net mag-

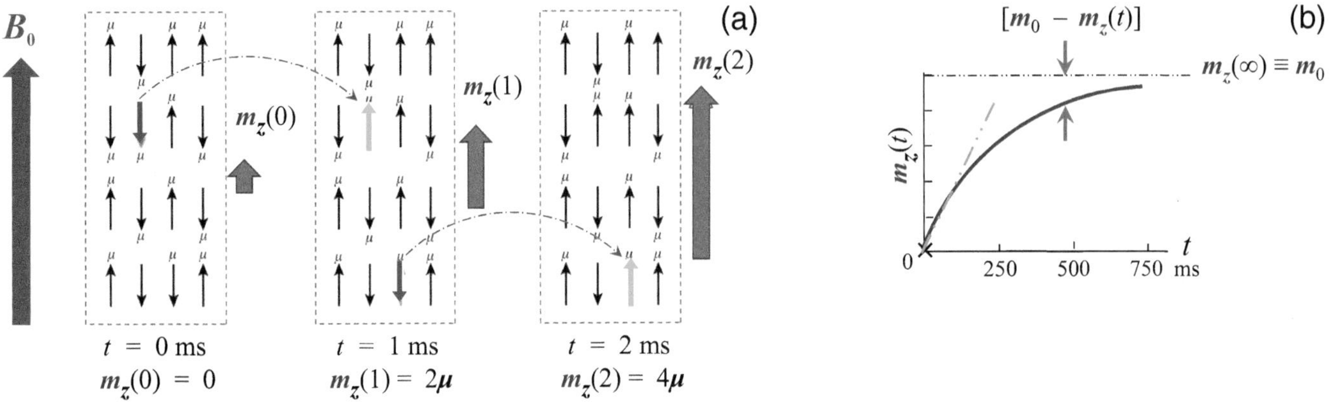

Figure 9.1 Proton spin re-equilibration following a disturbance. (a) Immediately after a magnetic field, $\mathbf{B_z}$, is switched on abruptly at time t = 0, equal numbers of protons will, for QM reasons, come to point up or down, and $\mathbf{m_z}(t{=}0) = 0$. But individual spins will flip over (both up and down) independently of one another, and the net number in the lower-energy level increases nearly linearly (at first) with the time. (b) Return of a voxel's $\mathbf{m_z}(t)$ toward its equilibrium value, $\mathbf{m_0} \equiv \mathbf{m_z}(\infty)$. Over longer periods of time, the net longitudinal magnetization curve grows back as $\mathbf{m_z}(t)/\mathbf{m_0} = (1 - e^{-t/T1})$ which, for small values of t, starts off as a straight line.

netization, $\boldsymbol{m_0}$. This resembles Equation (1.5) which, however, described a radically different system, an ensemble of compass needles.

At the same time, because of energy transfer from the thermal reservoir, some spins in the lower-energy level will be flipping over the to the opposite ('wrong') orientation, and into the higher-energy state. But the system *on average* moves toward an equilibrium configuration, and there is a general tendency for the system to evolve in this direction.

EXERCISE 9.5 There are n(0) radionuclei in a sample at t = 0 and n(t) remaining after time t. At the endpoint, or at 'equilibrium,' $n(\infty) = 0$. Derive and solve $dn/dt(t) = -\lambda\, n(t)$ to find $n(t) = n(0)\, e^{-\lambda t}$. Does $dn/dt = -\lambda\, [n(t) - n(\infty)]$ mean the same thing?

EXERCISE 9.6 Re-work the previous exercise but for spins, this time taking into account that spins are flipping up as well as down.

EXERCISE 9.7 Is it true that any well-behaved curve is approximately linear close enough to the origin?

9.2 Exponential Regrowth of $m_z(t)$ toward m_0 at the Rate 1/T1: $m_z(t) = m_0\,(1 - e^{-t/T1})$

It will be interesting to derive Equations (9.1a) from scratch.

The attenuation of a narrow beam of high- or low-energy photons passing into matter, or of an ultrasound beam, occurs nearly exponentially with depth in matter, Equations (5.1). Likewise for the decrease of stimulated fluorescence of a phosphor over time, or of the number of radionuclei remaining in a sample of radiopharmaceutical, or of tracer concentration in a biological compartment. A population of bacteria grows exponentially with time as long as its environment provides necessary nutrients and removes toxic wastes, and the rate of cell death remains constant. Cells are killed off exponentially as a function of dose of high-LET (linear energy transfer) ionizing radiation, such as a beam of alpha particles. These and many other disparate phenomena behave exponentially over time, distance, dose, etc., all for essentially the same reason as for radioactive decay (Equations 5.1a).

The difference in the numbers of proton spins in the lower- and upper-energy spin-states, from Equations (4.2), is

$$n(t) \equiv [N_-(t) - N_+(t)].$$

If the system is jolted out of thermal equilibrium, such as by a 90° saturation pulse, then $N_-(t) = N_+(t)$

briefly. After that, relaxation events occur, and the population evolves closer to its equilibrium condition. Over time, the system has to move through less and less remaining 'distance' to achieve this, and its rate of approach diminishes accordingly. That is, *dn/dt*, the rate of change of $n(t)$, is proportional to how far it still is from equilibrium, $n(\infty)$. Invoking Equations (4.3a), this notion can be expressed in terms of voxel magnetization, instead, as $[\mathbf{m}_-(t) - \mathbf{m}_+(t)] = [\mathbf{m}_z(t) - \mathbf{m}_0]$, and

$$d[\mathbf{m}_z(t) - \mathbf{m}_0]/dt = -(1/\mathrm{T1})[\mathbf{m}_z(t) - \mathbf{m}_0]. \quad (9.2a)$$

This the *Bloch Equation* for the longitudinal magnetization, where T1 relaxation is now being accounted for, and its solution is

$$[\mathbf{m}_z(t) - \mathbf{m}_0] = [\mathbf{m}_z(0) - \mathbf{m}_0]\mathrm{e}^{-t/\mathbf{T1}}. \quad (9.2b)$$

For those situations in which $\mathbf{m}_z(t) = 0$, this returns us to Equation (9.1a). Again, the constant of proportionality T1 serves as the parameter that characterizes how long this process takes, and 1/T1 is the rate at which it happens.

EXERCISE 9.8 Obtain Equation (9.2b) from Equation (9.2a).

For the particular instant at which t = T1, this becomes

$$\mathbf{m}_z(\mathrm{T1}) = \mathbf{m}_0[1 - \mathrm{e}^{-\mathrm{T1/T1}}] = 0.63\mathbf{m}_0. \quad (9.2c)$$

For this reason, T1 is also defined (somewhat arbitrarily) as the time required by the system to evolve a particular part (1/e = 63%) of the way from a configuration of zero magnetization to equilibrium (Figure 9.2). It will be 86% of the way to $\mathbf{m}_0$ after 2×T1, 95% after 3×T1, and so on.

1/T1 refers to the *rate* of regrowth of $\mathbf{m}_z(t)$ toward thermal equilibrium and $\mathbf{m}_0$ following complete or partial spin saturation. The rates at which relaxation processes occur and the associated *relaxation times* are inverses of one another: the faster the rate, the less time the process takes. Physicians and chemists tend to discuss the relaxations *times* that parameterize clinically interesting properties of a tissue; physicists and other researchers who are into the underlying biophysical events may be more inclined to speak of the *rates* at which they take place. These are but two sides of the same coin.

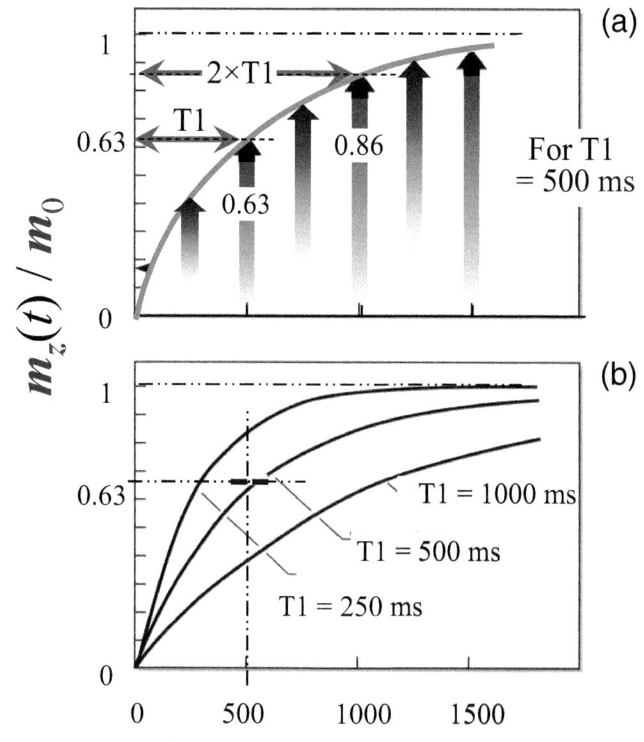

Figure 9.2 1/T1 parameterizes the rate of equilibration of $\mathbf{m}(t)$ after a disturbance. (a) At time t = T1, $\mathbf{m}_z(\mathrm{T1})/\mathbf{m}_0 = 0.63$, which provides a quantitative definition of T1. In addition, $\mathbf{m}_z(2\times\mathrm{T1})/\mathbf{m}_0 = 0.86$, and $\mathbf{m}_z(t)/\mathbf{m}_0 = 0.95$ at t = 3 T1. (b) Three different materials relax with T1 values of 250 milliseconds, 500 ms, and 1 second, respectively. The one with T1 = 500 ms is highlighted.

EXERCISE 9.9 Why is it meaningful and legitimate to refer to (1/T1) as a relaxation rate?

9.3 The Bloch Equations, Including T1 Relaxation

The most rudimentary Bloch equations—which describe precession and nutation, but not relaxation or other processes (Equations 6.4)—are a starting point for the interpretation of MRI data. These were amended in Equation (6.7) to account for the input of resonant RF and the production of nutation. It is necessary to modify them further to incorporate the effects of nuclear spin relaxation with the characteristic times T1 and T2. We shall deal with T1 here.

Guided by Equations (4.9) and (6.4), and assuming that T1 relaxation is the only such process at work in a spin system, the Bloch equations expand into

$$\begin{aligned} dm_x/dt &= 2\pi \nu_{Larmor}\, m_y \\ dm_y/dt &= -2\pi \nu_{Larmor}\, m_x \\ dm_z/dt &= -[m_z(t) - m_0]/\mathrm{T1}. \end{aligned} \tag{9.3}$$

These classical equations—which will be augmented further to cover additional processes like T2 relaxation, water diffusion, blood flow, and the like—delineate the time development of what the cohort of tissue protons in a voxel are actually going through during a specific pulse sequence. Quite simply, it is the Bloch equations that describe and can explain the dynamic physics of MRI.

The relaxation times T1 and T2 in the early days of Equation (9.3) were found empirically. Experiment indicated that for numerous tissues, a single value of T1 and one for T2 were adequate to describe both types of relaxation closely, as would be the case if each obeyed first-order kinetics, like that of simple diffusion or radioactive decay. The pathbreaking derivation of the two relaxation times [Bloembergen et al. 1948] served as the starting point for studies that continue to this day.

9.4 Biophysical Mechanism of Proton T1 Relaxation: Spontaneous, Random, ν_{Larmor} Fluctuations in the Local $B_z(t)$, Created by the Tissues Themselves, Can Induce Spin-State Transitions

So far, we have taken it on faith that thermal equilibrium, which maintains a small population difference between spin states in accord with the Boltzmann ratio, will just happen on its own. In the quasi-QM picture, this process involves spin flips—but, when the RF is not being applied (which is nearly all the time), something else must be making such transitions take place. That something is the interactions of the protons with their magnetic environments: these can lead to T1 *spin-lattice relaxation* events, so-named because they were first studied in connection with transfers of energy in the form of *phonons* (quanta of vibrational energy) between nuclear spins and crystal vibrations, molecular rotations, and other lattice degrees of freedom. As water protons move about, they feel changing magnetic fields due primarily to motions of their partner or other nearby protons. Some of these will occur at the Larmor frequency, and those can thereby induce spin-state transitions, both up and down. This is what allows a system of protons, after being disrupted, to evolve unaided toward thermal equilibrium at the rate 1/T1.

So there are two general categories of mutual influence between tissue protons and the outside world. Pulses of ν_{Larmor} radiation coming from the scanner's transmit antenna can cause spin-flips (both up and down) during those brief periods that the RF is present. Alternatively, the ν_{Larmor} components (and others, it turns out) of *background magnetic noise* are inducing *spontaneous* transitions at *all* times, again both up and down, in ways that differ significantly from the radiation-induced type. Chapter 4 explored this issue a little, and we shall do so a bit more here.

Analysis of the relevant physics reveals that NMR's RF photon-*nucleus* interactions differ in a fundamental way from the photon-*electron* interactions that play central roles in optical, gamma-, and x-ray imaging. With a photoelectric or Compton interaction, it is primarily the *electric* field of the incident photon that exerts a force on an atomic electron, resulting in a change in its orbital quantum state, along with absorption or scattering of the photon (Figure 2.2b, far right). With NMR and MRI, by contrast, it is the rapidly oscillating *magnetic* field component of the incident Larmor-frequency electromagnetic radiation that interacts with the magnetic moment of the nucleus, thereby bringing about a change in the nuclear spin orientation quantum state.

This suggests that *any* magnetic field that happens to be varying at the Larmor frequency (not only the one from a photon pumped in by an RF transmitter and coil) could couple with a proton to elevate it from its lower- into its higher-energy spin state (Figure 2.5). Just as likely, for reasons explained by quantum mechanics and indicated in Equations (4.6), it can tickle a proton sitting in the higher-energy state down into the lower one, with the release of the difference in their energies into the environment. That is, if there happen to be naturally occurring magnetic fields present with a component fluctuating at the Larmor frequency, then those fields are capable of connecting with a proton and stimulating a spin flip. And if the system is not already at thermal equilibrium, then the Larmor-frequency component of the naturally fluctuating local magnetic field that the protons experience will drive it over time preferentially toward that condition. There's nothing mysterious about this, just thermodynamics at work. The population dynamics of protons in a strong field is generally not static, but rather they are churning away in a sea of phonons, causing protons all over the place to continually flop up and down in a thermally induced Saint Vitus Dance. But during all this, there is an overwhelming tendency to drive the magnetization of a voxel relentlessly toward $\boldsymbol{m}_0$, over a period on the order of several times T1.

There are a number of ways that protons produce and are subjected to random fluctuations in the local magnetic field. The most obvious involves the magnetic interaction between the two hydrogen nuclei of the same water molecule as it rotates and vibrates and strikes other molecules. The spin axis of each proton, and the magnetic field it produces, will remain steadily pointing either along or against the external magnetic field. The strength of the weak dipole field from one proton overlapping its partner is determined, at any instant, by their instantaneous relative positions in the external field (Figure 9.3). The two are fixed at 1.5×10^{-10} m (1.5 angstrom, Å) apart, and they produce dipole fields that overlap one another by amounts that vary from near 0 T (when the two are exactly above one another, relative to the external field) to about 4×10^{-4} T = 0.4 mT = 400 μT (when they are side-by-side). A dipole field falls off rapidly with distance from the source, d, as d^{-3}, so that the randomly fluctuating magnetic fields produced at a proton by any nuclei on separate molecules are generally much weaker. The same sort of argument applies also to protons in lipids.

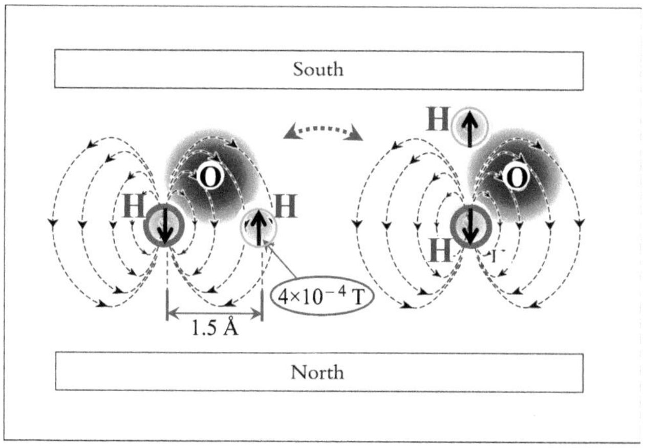

Figure 9.3 The principal T1-relaxation mechanism for water comes from inter-proton dipole-dipole interactions within the same molecule. Each hydrogen nucleus aligns either along or against the strong external magnetic field, and generally stays that way. But a proton also experiences the small, fluctuating magnetic dipole field created by its partner proton. The separation of the protons in water is about 1.5 angstroms, where 1 Å = 10^{-10} m, and the maximum field generated this way, 4×10^{-4} T = 0.4 mT = 400 μT, is actually greater than that of the RF typically produced by the system's transmitter and coil. If the molecule happens to be tumbling at the Larmor frequency, the resulting ν_{Larmor} field can induce proton spin transitions either up or down.

EXERCISE 9.10 A bit of topology: what sort of continuous motion will get one from the left-hand side of Figure 9.3 to the right?

EXERCISE 9.11 Must there be a component of the axis of symmetry of the water that lies perpendicular to the z-axis for the causation of spin flips? Parallel?

Whenever a water or lipid molecule happens to be tumbling, vibrating, etc. at the Larmor frequency, then the local magnetic dipole fields at each of the two protons will also cycle at that rate—and it is such naturally occurring Larmor-frequency fields that (somewhat like an RF photon supplied by an external RF coil) stimulate proton spin-state transitions. If the molecule is tumbling at or near half the Larmor frequency, moreover, then the time-varying magnetic field it produces may contain a small component at the Larmor frequency as well (the second harmonic of the tumbling frequency).

The above discussion drives straight to the heart of the matter: the only thing a proton is ever aware of, or reacts to, is the local magnetic field, $\boldsymbol{B}_{\text{local}}(t)$, that it experiences:

In MRI, the ***only*** thing a proton
is ***ever*** aware of, or reacts to,
is its ***local magnetic field***, $\boldsymbol{B}_{\text{local}}(t)$!

In addition to $\boldsymbol{B}_0$ and $G(t)$, the source
of $\boldsymbol{B}_{\text{local}}(t)$ can be either ***external*** (B_1)
or ***internal*** (e.g., partner-proton,
chemical-shift).

We can rephrase this as a question: how many water protons are experiencing any sort of ν_{Larmor} magnetic fields? Or, almost equivalently, how many of a cell's water molecules will be rotating, at any instant, at the Larmor frequency?

9.5 Tumbling Frequencies of Water Molecules in Three Aqueous Environments

The strongest MRI signals commonly come from intra- and inter-cellular water, so we shall focus on these for the moment.

How much of the water *is* tumbling at the Larmor frequency? The motions of the water molecules within a voxel of tissue are strongly influenced by how tightly they are embracing the various ions or biomolecules in solution, or the cellular membranes. It makes matters easier, and doesn't oversimplify things too much, to think of the fluid content of a tissue as comprising three distinct, separate general sub-populations of water and lipid molecules (Figure 9.4a):

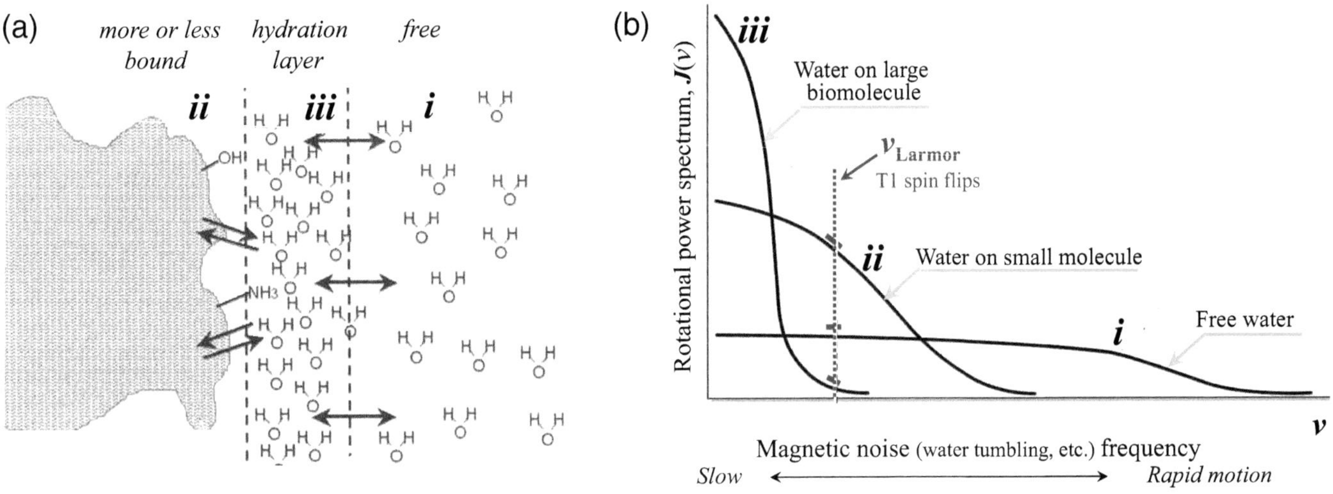

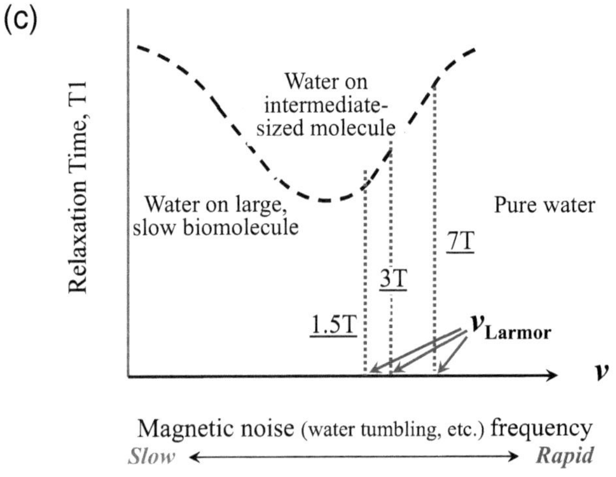

Figure 9.4 Consequences of restrictions on rotations and other movements of water molecules. (a) T1 depends on the tightness of the bonds between water and the cellular biomolecules of various sizes, and on the viscosity, etc. of the cellular fluid. Both factors affect the rate at which the water molecules move about and the likelihood that they will experience naturally occurring Larmor-frequency fields. One can view the behavior of water molecules as if they exist in three discrete local environments: (i) free water, in which most (but not all) molecules rotate much faster than ν_{Larmor}; (ii) water bound tightly to large, slow-moving macromolecules and membranes; and (iii) water in between these extremes, such as in a layer of hydration covering a cell membrane. In reality, of course, these zones form a continuum. The same kind of argument applies to tissue lipids. (b) The rate of proton longitudinal relaxation, 1/T1, is proportional to the number of molecules that are tumbling at and near the rate ν_{Larmor}, and are therefore experiencing magnetic noise with a significant component at that specific frequency. That, in turn, depends largely on how many are in each of the three categories, on the sizes of the biomolecules to which water tends to attach, and on the strength of those attachments. The rotational *power spectrum*, $J(\nu)$, can be viewed as the relative number in each category that are rotating or otherwise moving about at any particular frequency ν. For pure water, $J(\nu)$ is broad and flat, giving rise to only a small amount of it rotating at the Larmor frequency. Likewise, few of the water molecules that are bound firmly to the large, slowly moving macromolecules or membranes will experience Larmor-frequency field intensity. The major contribution to relaxation, where $J(\nu_{\text{Larmor}})$ achieves its greatest value, comes from water adhering to, or held loosely by, mid-size molecules, or by membranes. (c) It is primarily the waters on such intermediate-size molecules that determine the rate 1/T1, and for the measured relaxation time T1. As ν_{Larmor} moves upward from that at $\mathbf{B}_0$ = 1.5 T to higher main fields, incidentally, T1 values tend to lengthen, roughly as $\mathbf{B}_0^{1/3}$.

- one compartment or category consists of those that are essentially free;
- another, those bound fairly tightly to membranes or to massive molecules; and
- a third, those in between, such as in the *layer of hydration* loosely surrounding a large biomolecule.

The third of these is the important one, with water or small lipid molecules adhering insecurely to intermediate-sized biomolecules, some of which themselves rotate and vibrate at intermediate frequencies comparable to ν_{Larmor}.

Consider first a *free water* molecule surrounded only by pure water, (*i*) in Figure 9.4b. Many a free water molecule will briefly rotate about some axis in space, bang into another water molecule, start rotating at a different angular velocity about another axis, collide again, and so on. The span of the frequencies of its possible motions is displayed as the lowest of the three curves. Because they are small and light, many will be off far to the right in the figure, rotating at rates much higher than the Larmor frequency. It may be surprising, but there will also be a spread of motions at the low-ν end, some even at frequencies down near zero, to the left end of the figure. A plot of the likelihood of a molecule tumbling or vibrating at a certain frequency, $J(\nu)$, vs. that frequency, ν, serves as a loose definition of its *power spectrum.* For free water, the power spectrum of local 'magnetic noise' frequencies is flat, low, and broad, and the relative number of such molecules causing magnetic field fluctuation *at the Larmor frequency*, in particular, will be relatively small. For those few free water molecules that *do* happen by chance to be moving about near ν_{Larmor}, and thereby generating magnetic noise at that frequency, either of the partner protons may undergo a spin-state relaxation transition. Not too many will do this, however, and if that were the whole story, T1 for the tissue would be quite long (Figure 9.4c).

Likewise, for water connected firmly to membranes or to large macromolecules that tumble or vibrate only very slowly (*iii*), each proton again experiences an overlap field from its partner or others that varies at the slow tumble frequencies. But those tend to be much *lower* than the proton Larmor frequency at 1.5 T. There will be relatively few of them able to undergo relaxation, as well, so again T1 is long. (As chapter 10 will show, however, these slow motions *do* play an important role in determining T2!)

That leaves the third category of protons, those in water bound loosely and intermittently (typically for microseconds or less) in the hydration layers of macromolecules (*ii*). Likewise for those held to mid-sized biomolecules that are themselves rotating a good deal more slowly than free water, because of their greater mass and friction-like forces in the solution dragging on them. The tightness of the binding and the extent to which that will hinder rotation depend on the macromolecule's chemical makeup and three-dimensional structure, and on its tendency to hold on to water of hydration. A fair amount of such water may end up rotating at rates close to the Larmor frequency—and protons in this sub-population, unlike those in the free and bound groups, may undergo relaxation at a fast rate, causing the tissue to display relatively *short* T1.

So what is the connection between water tumbling rates and T1? As mentioned on several occasions, Bloembergen [1948] published the first rigorous explanation of the shape of Figure 9.4, and his derivation still forms the basis for much of the theory of relaxation in use today. It assumes that a molecule of water is undergoing certain rotational and diffusional motions for a brief period of time, then suffers a random collision with a neighbor and begins a new mode of movement. Because of these collisions, the positions and orientations of the water molecule will vary randomly, frequently, and sometimes significantly over time. The *correlation time* τ_c, the average duration between these interactions, is a measure of the correspondence of the orientations or locations of the molecule over short periods of time. In many situations, the correlation time may be described by the *Arrhenius equation*, which arises frequently in the study of chemical reaction rates, diffusion, radiation damage of matter, and other processes that depend on temperature:

$$\tau_c \sim e^{-E_a/kT}. \qquad (9.4a)$$

For the interacting molecules here, T is body temperature, the measure of the thermal energy and activity within the local environment, k is the Boltzmann constant, and E_a is an *activation energy* related to what is typically transferred during a collision. For a simple model of water at body temperature, $\tau_c \sim 10^{-10}$ sec, but in more sophisticated calculations E_a and τ_c are represented as continuous probability functions. The probability that a water molecule's motion does *not* change abruptly over time Δt is of significance, and it decreases rapidly as $e^{-\Delta t/\tau_c}$.

Proceeding from here gets a little hairy and involves QM time-dependent perturbation theory, Fourier transforms of autocorrelation functions, etc. We shall leave that to other treatments, but will reproduce one key result of it: the *noise spectral power density*, the magnitude of noise power as a function of noise frequency, commonly assumes the form

$$J(\nu) \sim \tau_c / [1 + 4\pi^2\nu^2\tau_c^2], \qquad (9.4b)$$

and three examples of it appear in Figure 9.4b. Of greatest interest here, naturally, is the component of the noise power that is available at the Larmor frequency, which is what causes T1 transitions, at a rate

$$\boxed{1/\mathrm{T1} \propto J(\nu_{Larmor}).} \qquad (9.4c)$$

We have focused attention so far on the protons of water molecules with high, middle, and low values of the tumbling frequency. Protons on other kinds of molecules can also undergo NMR, and this is particularly significant in cells with high aliphatic lipid content (in which the carbon atoms form chains, rather than aromatic rings). Such molecules undergo motions that are dissimilar from those of free water, and they display correspondingly different relaxation behavior. Proton T1 relaxation times for cerebral white matter, for example, tend to be shorter than those for gray matter because of the relatively greater number of fast-relaxing lipid protons (Table 9.1).

Numerous normal and abnormal physiologic processes affect the amounts of water residing within and between cells, the concentrations of certain specific molecules and organelle parts, the strength of binding of water to the cellular contents, the viscosity and ionicity of the solution, etc. The type and status of health of a tissue determine not only how much water there is in a cell, but also where in the cell it is located, what is dissolved or suspended in it, and what it has attached itself to and how tightly. All of that is reflected in the measured values of relaxation times. This is why spatial variations in the values of T1 and T2 are of such interest: because water proton spin relaxation times depend on cell type, MRI can distinguish one tissue or organ from another. Because they also reveal the physiological status of a tissue, MRI commonly indicates pathological conditions.

9.6 T1 Increases with the Strength of the Main Field, B_0

There is a fundamental trade-off in MRI among contrast/SNR, spatial resolution, and imaging time. Any of these three can be made better, but only at the expense of one of the other two. One way to improve on this balance is to switch to an MRI device with a greater main magnetic field strength, if possible.

We have discussed the effect of the strength of the main magnetic field on image quality through its impact on voxel magnetization, hence on signal strength, Equation (4.5). The magnitude of m_0 is proportional to B_0, and you might suspect that the strength, $S(t)$, of the MR signal coming from the patient goes up linearly with field strength. By Equations (8.3), it actually increases more closely with the square of the field strength, as B_0^2, but SNR increases only roughly with B_0. Nonetheless, a new and significantly better trade-off may be possible among tissue contrast, imaging time, and resolution at higher fields. Indeed, details that are not even suggested at 1.5 T can appear clearly in 7 T images (Figure 4.5). Greater field strengths enable higher SNR and smaller voxel size, thereby increasing resolution and decreasing the *partial volume effect* (overlapping information from adjacent voxels) and generally improving anatomical imaging, such as in MR microscopy (Figure 8.5).

Stronger B_0 has other effects as well. Since it leads to more pronounced chemical shifts, a higher field can separate overlapping resonance peaks somewhat, enhancing magnetic resonance spectroscopy studies. Both functional MRI (fMRI) and MR diffusion-tensor imaging (DTI) tend to do better at 3 T, moreover, and even more so at 7 T (Figures 1.5a,b). On the other hand, motion, blood flow, and chemical shift artifacts tend to worsen as field strength increases, along with geometrical distortions associated with external field inhomogeneities and patient susceptibility effects.

T1-values are found to lengthen with field strength approximately as

$$\mathrm{T1}(B_0)/\mathrm{T1}(1\,\mathrm{T}) = B_0^{1/3}. \qquad (9.5)$$

(T2 is much less sensitive to B_0). An increase in B_0, and the resulting shift of ν_{Larmor} to the right in Figure

9.4c, means that less water is now experiencing Larmor-frequency magnetic noise—hence slower relaxation and an increase in T1. There are two effects ongoing here. First, T1-*w* images tend to show a little less contrast among different tissues at higher fields, so people familiar with reading T1 images at 1.5 T may have to readjust their eyes a little at 3 T or 7 T. Second, they will find that a T1-*w* scanning protocol devised for 1.5 T is not necessarily optimal for higher-field-strength imaging. So it is not enough to state T1; one must know the field strength at which the measurement was made, as indicated by the several T1 columns in Table 9.1.

Greater safety issues and novel biological effects arise above 3 T, as will be discussed in chapter 15 on MR safety. Most obvious is the pull on magnetizable objects within the patient or the imaging suite (Figure 1.11). The active shielding of the main magnet makes the footprint of the fringe fields surrounding both 1.5 T and 3 T magnets about the same.

Also of concern are possible hazards from stronger rapidly switching gradient fields. In addition, the *specific absorption rate* (SAR), the rate at which RF power is absorbed per kilogram of tissue, increases roughly as the square of the field strength,

$$\mathrm{SAR}(B_0) \sim B_0^{\,2}, \tag{9.6}$$

and tissue heating may become more problematic. Indeed, most of the MRI incidents reported to the FDA involve local tissue RF burns. And cardiac electrophysiology (MRI pulse sequences that gate with the heart) and other techniques may require new methodologies at high fields.

While machines that operate at 1.5 T are still the norm, the sale of 3 T devices is increasing briskly. The FDA cleared the first 7 T machine for clinical use in late 2017, and there are now about 40 of these installed worldwide for both clinical and research purposes.

9.7 Determining T1 in a Voxel

Before addressing techniques for generating T1 contrast among different tissues in an image, it would be good to take a pace or two back and consider how one might actually go about assessing this relaxation time for a real material. Rather than describe details of how to do this in practice, here we shall just provide a simple, heuristic proof of concept. It involves making a number of *saturation-recovery* measurements with *pairs* of 90° pulses.

EXERCISE 9.12 Think of a simple way to use saturation-recovery or other pulse sequences to assess T1.

Recall the saturation-recovery event of Figures 7.1a and 7.1b. Suppose the voxel's spin system is left alone for a good deal longer than T1, during which the magnetization, $\boldsymbol{m}(t)$, moves asymptotically toward its thermal equilibrium state, $\boldsymbol{m}_0$. Just before time 0, at $t = 0_-$, $\boldsymbol{m}(t)$ lies almost entirely along the z-axis, becoming the fully longitudinal $\boldsymbol{m}_z(0_-) \sim \boldsymbol{m}_0$, the vertical green arrow in Figure 9.5a. Precisely at $t = 0$, the first of two Larmor-frequency 90° saturation pulses, red arrow, is applied. It drives (nutates) the longitudinal magnetization, $\boldsymbol{m}_z(0_-)$, rapidly down from the vertical into the $\boldsymbol{x}$-$\boldsymbol{y}$ plane. During this, $\boldsymbol{m}(t)$ morphs into the *transverse* magnetization $\boldsymbol{m}_{xy}(0_+)$, the white-dotted green arrow, still of magnitude $|\boldsymbol{m}_0|$. The direction in which the magnetization pointed changes, but not its magnitude, and the bottom-line take-home message for the event is: following the first 90° saturation pulse,

$$\boxed{|\boldsymbol{m}_{xy}(0_+)| = |\boldsymbol{m}_z(0_-)|.} \tag{9.7a}$$

This self-evident but important concept was seen earlier as Equations (6.8) and (7.1), and it will appear again in other guises. Its great significance is that it lets us assess the relative magnitude of $\boldsymbol{m}_z(0_-)$ from the voltage $\boldsymbol{m}_{xy}(0_+)$ induces in the pickup coil, $V(0_+)$.

EXERCISE 9.13 Convince yourself of the validity of Equations 9.7 and Figure 9.5 and of their physical meaning.

It is necessary to introduce an important new concept in this section, the *repetition time* (TR), so called for reasons that will become clear. This is usually defined as the interval between successive 90° pulses that excite the same tissue in a voxel, and it plays an essential role in all clinical sequences, and here, too.

After the events of Equation (9.7a), the spins begin to relax, causing $\boldsymbol{m}(t)$ to extend back upward along the z-axis (Figure 9.5b, green block), a phenomenon that

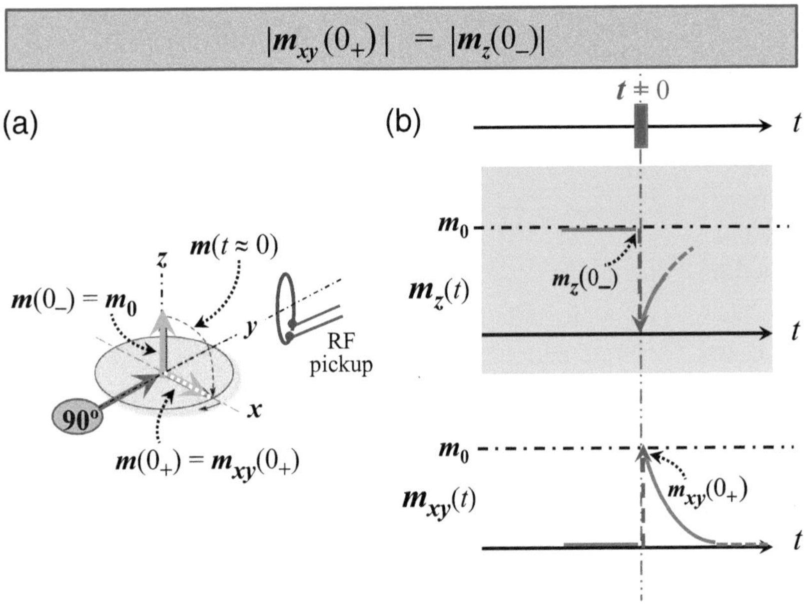

Figure 9.5 Equation (9.7a) illustrates the events occurring around $t = 0$ for a spin system originally at equilibrium. (a) With application of the first of two 90° RF pulses, the spin system snaps from $\boldsymbol{m}_0$ down into the horizontal plane, (b) where it becomes $|\boldsymbol{m}_{xy}(0_+)|$ and starts precessing at ν_{Larmor} and induces the signal $V(t)$ in the receiver antenna coil.

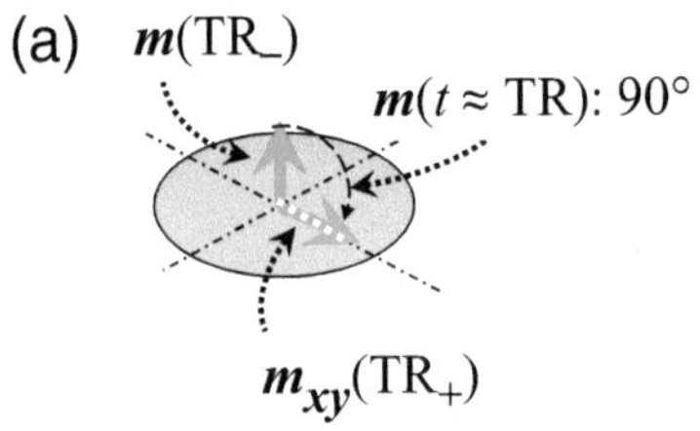

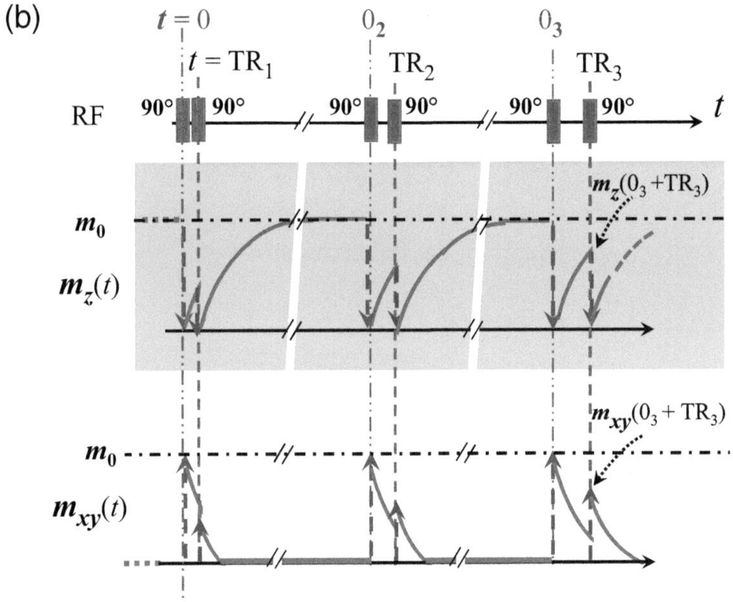

Figure 9.6 This figure is similar to Figure 9.5. Are you able to explain what is happening in this figure after reviewing Figure 9.5? Exercise 9.14 asks you to try to do just that.

occurs at the rate 1/T1 (don't ask; we're mixing the quasi-QM and classical pictures here). The value of the repetition time is specified to be a little less than T1 for the voxel material, and by the time t = TR, $\boldsymbol{m}(t)$ will have risen to $\boldsymbol{m}_z$(TR). Meanwhile, the transverse magnetization will have dropped to $\boldsymbol{m}_{xy}$(TR). Note that this almost always happens at a rate much faster than 1/T1.

At t = TR, the second of the two 90° pulses drives the remaining magnetization back into the $\boldsymbol{x}$-$\boldsymbol{y}$-plane (Figure 9.6a) where, at that instant,

$$|\boldsymbol{m}_{xy}(\mathrm{TR}_+)| = |\boldsymbol{m}_z(\mathrm{TR}_-)|, \qquad (9.7b)$$

and $\boldsymbol{m}_{xy}(t = \mathrm{TR})$ will be smaller than $\boldsymbol{m}_0$. $\boldsymbol{m}_{xy}(\sim\mathrm{TR})$ precesses and produces $V(\sim\mathrm{TR})$ in the pickup coil.

EXERCISE 9.14 Based on Figure 9.5, explain what is going on in Figure 9.6.

The top row of Figure 9.7a,b summarizes what has happened so far. The couplet [TR, V(TR)] constitutes our first data point in Figure 9.7c, and so we shall relabel it [TR_1, $V(\mathrm{TR}_1)$]. We are interested in how $V(\mathrm{TR}_j)$ depends on TR_j, so we go through this several more

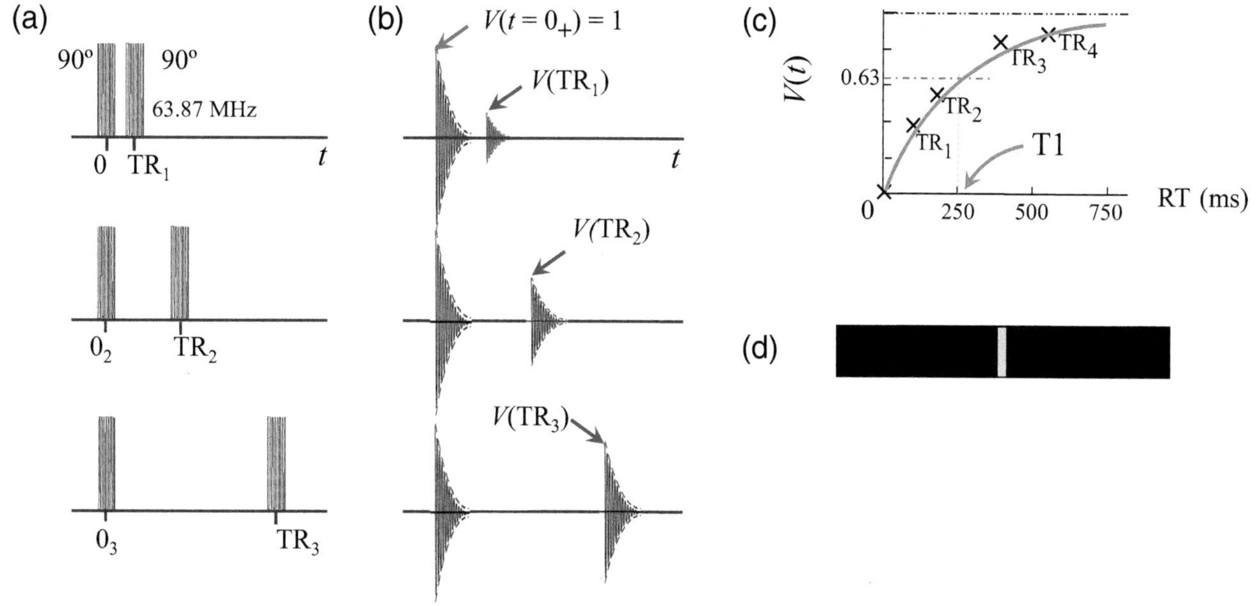

Figure 9.7 T1 of tissue in a voxel can be assessed with several pairs of 90° RF pulses, each pair separated by a different *repetition time* (TR). Shown here are the saturation-recovery cycles for three choices of RT. *Top row, first RT*: (a) Starting out with the system at equilibrium in 1.5 T, a 63.87 MHz *saturation* pulse at $t = 0$ drives $\boldsymbol{m}_z(0)$ down into the **xy**-plane, Equation (9.7a). It precesses there as $\boldsymbol{m}_{xy}(t\sim 0)$, with initial magnitude $|\boldsymbol{m}_0|$, and (b) induces the voltage $V(\sim 0)$ in the receiver antenna. The system is left alone for the short time TR_1, during which $\boldsymbol{m}_z(t)$ regrows to $\boldsymbol{m}_z(RT_1)$, only part of the way to $\boldsymbol{m}_0$. Then at $t = TR_1$, a second 90° pulse swings $\boldsymbol{m}_z(TR_1)$ back down into the **xy**-plane again, where it induces a smaller voltage $V(TR_1)$ in the pickup coil. (c) The ordered pair $[TR_1, V(TR_1)]$ constitutes the first data point in the creation of the spin recovery curve. For the next data point, the system is first left alone long enough to fully re-equilibrate. The above procedure is then repeated, and this set of events is replicated, but with a longer delay interval TR_2. Likewise for TR_3, TR_4,... The values of $V(TR_q)$ so acquired are graphed against $t = TR_q$ for a half dozen or so TR settings, and T1 is obtained by fitting these data to $V(TR_q) = (1 - e^{-TR_q/T1})$. (d) Given that the voxel is located at the center of our 1D phantom, it will be bright if the measured value of T1 is *short*.

times, each with a successively longer choice of TR. Finally, the resulting set of data pairs $\{TR_q, V(TR_q)\}$ is fitted to $V(TR) = [1 - e^{-TR/T1}]$, which allows the extraction of T1 from the information content of Figure 9.7c.

Specialized pulse sequences (e.g., the *Look-Locker* sequence) are available that allow for easier and more accurate T1 measurements. Generally, however, it is not precise values of $T1(\boldsymbol{x})$ that are needed, but rather *relative* values of that relaxation time that can be mapped out in a clinically meaningful way (Figure 9.7d).

9.8 T1 MRI in a 1D Patient

The above approach to finding tissue T1 in a voxel works just as well for a row of tissue voxels, each with its own value of T1, when we add a position-location gradient, like that of Figures 3.5a and b. We re-consider the 1D, 2-voxel phantom of Figure 7.3, with two changes. First, to aid in discussion of contrast, the two non-empty voxels will be adjacent, located at $x = 0$ and 0.1 cm (Figure 9.8a). Also, the chambers now contain lipid and cerebral spinal fluid (CSF), respectively, rather than only pure water. (CSF is a clear, colorless fluid comprised 99% of water with small amounts of salts and other materials; it surrounds and fills in the ventricles of the brain and the central canal of the spinal cord.) We arrange for the two voxels to contain equal numbers of protons, so that their values of $\boldsymbol{m}_0$ will also be the same. The applied $\boldsymbol{x}$-gradient will allow examination of each of the two chambers separately, including determination of their x-positions, and the RF pulse will be of wide-enough bandwidth to allow resonance in both.

EXERCISE 9.15 How would it affect our image if the values of PD(x) were not the same?

EXERCISE 9.16 In Figure 7.6b, G_z was activated as a narrow-band RF pulse was being applied. Compare that with what is going on in Figure 9.8.

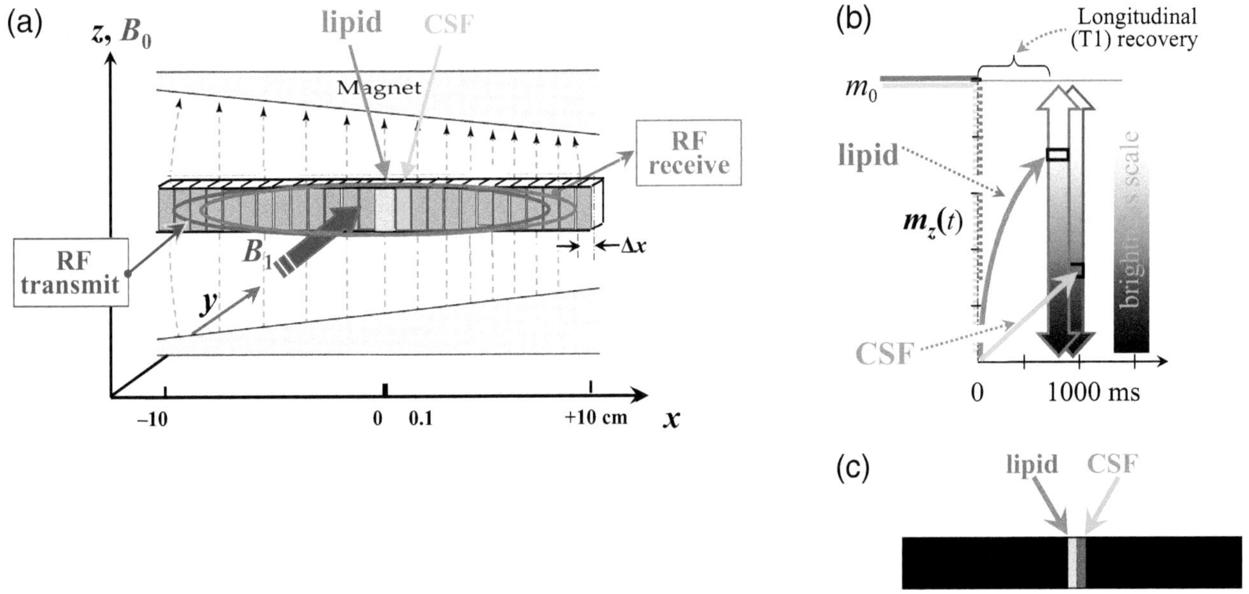

Figure 9.8 FID T1-weighted mapping in a 1D phantom with the two adjacent voxels, at $x = 0$ and 0.1 cm, containing lipid and CSF, respectively. (a) The two voxels are arranged to have the same PD, but they will display the different T1 values. An **x**-gradient allows separation of the contribution from each of the voxels, and the 90° saturation and recovery pulses contain a wide enough band of RF frequencies to cause resonance in both voxels. (b) A faster path to obtaining T1-*weighted* tissue contrast, employing only a single TR value, rather than a range of them. (c) T1-*w* MR image of the pair of voxels. In a T1-*w* examination, the material with shorter T1 shows up brighter.

Figures 9.6 and 9.7 indicate a way to create an MR image of a 1D patient, one voxel at a time, that shows T1 as a function of position. Most of the MR image is black, since there are no protons in the corresponding voxels. The voxel at $x = 0$ is bright because, in images designed to emphasize voxel T1-values, tissues in which protons that relax rapidly (*short*-T1) like lipids (¼ sec), are displayed as nearly white. T1 of CSF is close to 1 s, on the other hand, and its voxel is correspondingly darker. This choice of grayscale is a matter of convention, and does not come from any fundamental law of biophysics. This is an example of *quantitative MRI*, in which the absolute value of T1 in each voxel is obtained separately from a curve like that in Figure 9.7c. Such a study has enormous potential in the clinic, but currently it is not ready for prime time.

What is normally sought clinically is a more approximate anatomic map, known as a T1-*weighted* image (or a T2-*w* or PD-*w* image) that is created much more rapidly but that still somehow indicates the variations in relaxation times among the various healthy and pathologic tissues (Figures 9.9a and 9.9b). One swaps maximal contrast for swiftness, and nearly always it's a worthwhile exchange. It speeds things up significantly by assessing the magnitude of the MR signal at only *one* particular TR, rather than for a number of values of it (Figure 9.8b). Instead of fitting multiple data points to $(1 - e^{-TR/T1})$ for each voxel, a T1-*w* image is made from a *single* measurement, and for all the voxels in the region of interest (ROI) simultaneously. This yields only one magnetization amplitude in a voxel, at the selected TR, rather than a complete $m(t)$ curve-fitted value for each voxel. The measurement is made for all voxels at the same time and with the same TR value, and it generates a quick anatomic map of *relative* T1(x) values—but that is likely to be all that is needed to generate a clinically meaningful image. This usually does not lead to *optimal* contrast among tissues, so there is need for a trade-off here—typically one sets TR near the average of the various tissue T1 values for the region being examined. The topic of paramount importance—the optimization of contrast for T1-*w*, T2-*w*, PD-*w*, and other modes of MRI—will be discussed later.

This section has argued that an image of precise T1 values would be nice, but a simpler, much faster T1-*w* image that still allows the radiologist to discriminate among tissues, may well be fine for diagnostic purposes (Figure 9.9). Normally this is done with a spin-echo or gradient-echo pulse sequence, but for simplicity we have introduced the notion of weighting via FID.

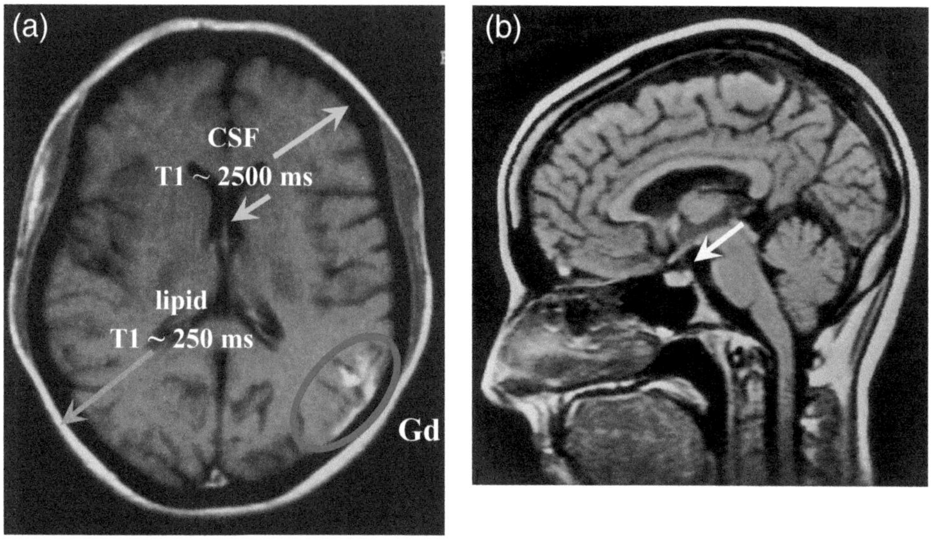

Figure 9.9 Two T1-w clinical images. (a) This transverse T1-*w* spin-echo scan reveals CSF surrounding the brain and in a ventricle, and the subcutaneous fat. Lipid relaxes more quickly than CSF, and has a shorter T1 value; this leads to a larger longitudinal magnetization at TR, and to a larger subsequent transverse magnetization vector following the pulse at t = TR, hence a brighter signal on the display. This also shows the take-up of gadolinium contrast agent by a so-called 'enhancing' lesion. (b) The indicated bright area is a lipoma, a tumor composed largely of adipose (fat) tissue. Do you see skull bones?

EXERCISE 9.17 How might one go about selecting a value of TR for optimal contrast? In fact, what does 'optimal contrast' really mean?

9.9 Speeding Things Up: Short TRs and Small Flip Angles

MR sequences such as spin-echo and gradient-echo typically comprise not single RF and gradient pulses, but rather sequences of a number of them.

In Figure 9.6b, the system has had ample time to nearly re-equilibrate between saturation pulses. As indicated by Table 9.1, T2 is normally an order of magnitude, a factor of 10, shorter than T1, and the transverse magnetization can decay away almost to nothing by the time a new pulse is applied. That way, the transverse magnetization from one excitation does not contribute to the transverse magnetization in the next one. (When this does not happen by itself, it can be made to occur by *spoiling* the spin system with special RF pulses or gradients, as will be discussed later.) For now, one might as well just assume that *transverse* magnetization is *not* preserved from excitation to excitation.

Such a situation is simple to analyze, but it is rarely the case in practice, largely because it is slow. Two changes can speed things up considerably.

The first is to reduce the separation between 90° pulses, so that time is not wasted in letting the system fully re-equilibrate. This strategy consists of applying a second 90° pulse before the T1 relaxation process from the previous one is completed. If TR is less than T1 or so, the system is made to remain in a *steady-state* dynamic, always somewhat removed from equilibrium, with the longitudinal magnetization able to rise only to some $\boldsymbol{m}_z(t)_{max}$ which is less than $\boldsymbol{m}_0$ (Figure 9.10).

A second method of moving things along faster is to not completely saturate the spin system with the saturation pulse. Until now, imaging sequences have begun with a 90° ν_{Larmor} pulse designed to drive the entirety of the magnetization (presumably lying along the z-axis) down into the $\boldsymbol{xy}$-plane. But as is apparent from Figure 9.10b, it is possible to obtain a relatively large $\boldsymbol{m}_{xy}(\text{TR}_+)$, and corresponding signal in the pickup coil, after the most recent such pulse even with a small *flip angle*, $\alpha°$, of the magnetization away from the vertical. Flip angles of 20° or less are not uncommon in practice.

We could address these two topics separately, of course, but shall combine them here, and leave their untangling as an exercise.

EXERCISE 9.18 Is it true that $|m_{xy}(TR_+)| = |m_z(TR_-)|$ *even if the system does not have time to re-equilibrate between pules? If the flip angle is less than 90°? If both?*

In Figure 9.10a, an excitation pulse is applied at t = 0, even though the *longitudinal* magnetization has not had time to fully relax. With a flip angle of α less than

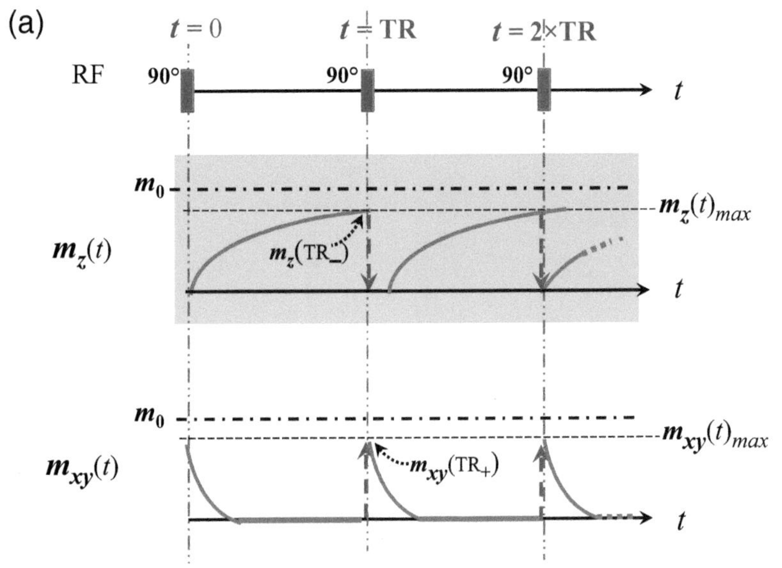

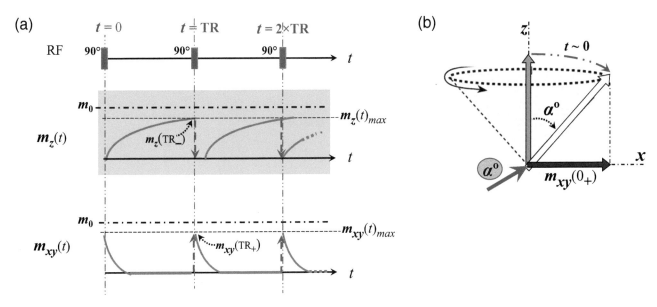

Figure 9.10 Ways to speed up data acquisition. (a) Series of equally-spaced multiple RF pulses. A sequence, beginning at $t = 0$, in which the repetition time is comparable to T1, and $\mathbf{m_z}(t)$ can return only to some value $\mathbf{m_z}(t)_{max}$ that is only part of the way to $\mathbf{m_0}$. (b) When the magnetization $\mathbf{m_z}(0_-)$ in a voxel is driven through only a small *flip angle* $\alpha°$, rather than 90°, the newly formed $\mathbf{m_{xy}}(0_+)$ can still be significant, even though the time taken for it to recover to near $\mathbf{m_0}$ is considerably less than for $\alpha° = 90°$.

90°, Equations (9.7) no longer apply as they stand, and the new longitudinal magnetization becomes

$$\mathbf{m}_z(0_+) = \mathbf{m}_z(0_-)\cos(\alpha). \tag{9.8a}$$

Transverse magnetization is given by an equivalent expression:

$$m_{xy}(0_+) = m_z(0_-)\sin(\alpha).$$

If this were the first excitation (Figure 9.10a), the expression would be

$$m_z(0_+) = m_0\cos(\alpha),$$

where, again, m_0 is the magnetization at thermal equilibrium. Similarly, following the next pulse in Figure 9.10a,

$$m_z(\text{TR}_+) = m_z(\text{TR}_-)\cos(\alpha). \tag{9.8b}$$

This is a general form, but if the voxel is excited only twice, it is easy to pursue the issue more thoroughly. The longitudinal magnetization just *before* the second excitation, $m_z(\text{TR}_-)$, is the residual longitudinal magnetization after the first excitation, summed in quadrature with the relaxed transverse magnetization:

$$\begin{aligned} m_z(\text{TR}_-) &= m_z(0_+)e^{-\text{TR}/\text{T1}} + m_{xy,relaxed}(\text{TR}_-) \quad (9.9) \\ &= m_z(0_-)\cos(\alpha)e^{-\text{TR}/\text{T1}} + m_0(1-e^{-\text{TR}/\text{T1}}). \end{aligned}$$

To follow what happens after a number of excitations, we adopt a simplifying short-hand notation. We'll refer to the nth longitudinal magnetization (i.e., after the nth pulse) as $m_{z,\,n}$. If the same sequence is applied enough times, the system will reach a condition of *steady-state magnetization* which means, for the longitudinal component, that $m_{z,\,n} = \mathbf{m}_{z,\,n-1}$ for a series of RF excitations in the future. Rewriting Equation (9.9) with $m_{z,\,n}$ notation for $m_z(t)$—that is, $t = \text{TR}_-$ for the nth excitation, and $t = 0_-$ for the $n{-}1^{\text{th}}$—and assuming a steady state leads to

$$m_{z,n}) = m_0(1-e^{-\text{TR}/\text{T1}})/(1-e^{-\text{TR}/\text{T1}}\cos\alpha). \tag{9.10a}$$

In other words, if RF pulses are applied repeatedly and TR < T1 apart, then after a while the longitudinal magnetization will asymptotically approach the value given by Equation (9.10a), as in the green box in Figure 9.10a. Longitudinal magnetization does not fully relax between excitations, but the signal can still be read out after each pulse. If different voxels have different T1 times, the longitudinal magnetizations will be different for all of them.

After the magnetization reaches a steady-state value, the voxels are all excited for the $(n+1)^{\text{st}}$ time, and a signal is read out. This next excitation results in a transverse magnetization

$$m_{xy,n+1} = m_{z,n}\sin(\alpha), \qquad (9.10b)$$

which is a function of T1, TR, and α. Once values of α and TR have been selected, then the signal strength, which is proportional to the value of $\boldsymbol{m}_{xy,n+1}$, will depend on T1 values. We've achieved steady-state T1-weighting!

Equations (9.10) depend not only on the T1 of the tissue, but also on the choices of flip angle and of TR. Clinically, the signal in T1-weighted images is determined both by the tissue T1 value in any voxel—known as an *intrinsic contrast parameter*—and the choice of TR and flip angle on the scanner (*extrinsic contrast parameters*).

But what value of flip angle will lead to the best signal? From Equations (9.10) and a specific value of TR, what value of $\alpha°$ leads to a maximum in $\boldsymbol{m}_{xy}(t)$? The maximum means that

$$d\boldsymbol{m}_{xy,n+1}/d\alpha = 0 = \left[m_0(1-e^{-TR/T1})/(1-e^{-TR/T1}\cos\alpha)\right]\times \left\{\cos\alpha - \left[e^{-TR/T1}\sin^2\alpha/(1-e^{-TR/T1}\cos\alpha)\right]\right\},$$

which reduces to

$$\cos(\alpha_E) = e^{-TR/T1}. \qquad (9.11)$$

The flip angle, α_E, that optimizes signal is called the *Ernst angle*. Operationally, for a specified TR value and tissues with different T1 values, Ernst angles can be calculated for each, and a mean angle will suffice that comes close to maximizing all signals.

CHAPTER 10

T2 (Transverse, Spin-Spin) Relaxation

You might imagine, after plowing through the previous chapter, that $\boldsymbol{m}_{xy}(t)$ of Figure 9.6 would decline at the same rate that $\boldsymbol{m}_z(t)$ regrows. While eminently reasonable from what has been said so far, such a supposition would be incorrect.

In fact, the FID signal arising from the precession of a voxel's protons in the ***x-y*** plane decays much faster than it would if only T1-events were involved! Almost always and for nearly all tissues, $\boldsymbol{m}_{xy}(t)$ is found to fall off 10 or so times faster than $\boldsymbol{m}_z(t)$ recovers (which is at the rate 1/T1) following a disruption of the spins. As it happens, longitudinal T1-events do partake in the decay of $\boldsymbol{m}_{xy}(t)$ as well, but there must be a second, highly effective process going on that helps to bring about a much more rapid decline of the FID signal. Working together, the two give rise to *transverse* or *spin-spin* relaxation, which takes place with a characteristic time known as T2. Such T2 relaxation times are an order of magnitude shorter for a given tissue than the measured values listed for T1 in Table 9.1. T1-type events are thus not only important in their own right, but they also contribute significantly to T2 relaxation, speeding it along and ensuring that $1/T2 \geq 1/T1$.

Something ***BIG***
must be going on
in addition to T1.

Another Kind of Relaxation...!!!

10.1 Transverse (Spin-Spin) Proton Relaxation: Spin De-Phasing in the *x-y* Plane with Characteristic Time T2

There are two separate and independent mechanisms that are responsible for transverse (T2) relaxation. *Lon-*

gitudinal relaxation at the rate 1/T1 comes from discrete spin-lattice proton flips alone and, as seen in the previous chapter, T1 is clinically important as it stands. But longitudinal relaxation happens to be only one of the *two* atomic-scale processes that together result in *transverse* relaxation, parameterized by 1/T2. While 1/T1 is the first of these, the secular contribution, $1/T2_{secular}$ is an order of magnitude faster. And the two rates 1/T1 and $1/T2_{secular}$ are additive, as seen in the bottom rows of Figure 10.1.

Ordinary T1 transitions are the first of the two mechanisms giving rise to T2 relaxation. T1 is a measure of the rate at which naturally occurring, Larmor-frequency magnetic field fluctuations cause $m_z(t)$ regrowth along the z-axis following a disruption, as brought about by *discrete* proton spin transitions from the spin-down state to spin-up, and vice versa. These involve the exchange of energy of the amount E_{Zeeman} between a proton and the local environment, as discussed in the previous chapter.

As just seen, the longitudinal relaxation portion of the rate 1/T2 is simply 1/T1. It may seem odd that changes of orientation along and against the z-direction can have an impact on spins presumably precessing in the x-y plane. This self-contradiction arises because we are shamelessly mixing the quasi-QM and the classical pictures, and such messiness does not arise in a rigorous, fully QM treatment. The bottom line is that T1-

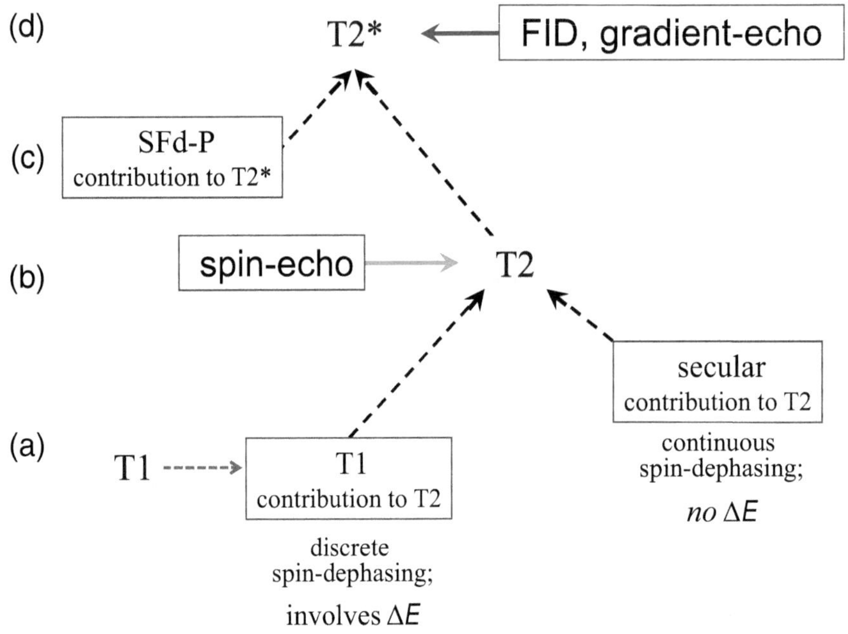

Figure 10.1 The complete proton spin-relaxation family tree. (a) T1 spin-state transitions are caused by random, Larmor-frequency (i.e., time-varying) magnetic field variations occurring naturally in tissue. Such erratic ν_{Larmor}-component field fluctuations 'tickle' protons from the higher-energy spin state down into the lower at the rate 1/T1, releasing Zeeman energy to the local environment as phonons or as other forms of energy. In addition to being responsible for T1 relaxation, these spin-flips along the **z**-axis also result in loss of phase in the **x-y** plane. (No, this may not really make much sense, but that's mainly because we're attempting to unify the spin-up/spin-down with the classical pictures, which is neither commonsensical nor quantum mechanically fully legal. Please forgive these minor problems—hopefully they may provide a little pedagogical relief.) (b) T2 de-phasing comes from both T1 events and the secular contribution to de-phasing. *Secular de-phasing* arises from relatively *slow*, temporally random field variations in the **z**-component of the local magnetic field. These *low*-frequency variations are brought about by proton-proton-dipole interactions in water bound to slowly moving large molecules; because of them, some proton packets briefly precess faster than average and others slower. These do *not* involve spin-flips, nor the exchange of energy, but they do provide the spins with sufficient time and motivation to de-phase a little from one another in the transverse plane. Because the magnetic fields underlying both contributions to 1/T2 occur *randomly* over time, they cannot be reversed or eliminated (by, say, the spin-echo pulse sequence). (c) On the other hand, $T_{SFd\text{-}P}$ de-phasing, which contributes to 1/T2* signal decay, *can* be reversed and eliminated as a relaxation mechanism with the spin-echo RF pulse sequence. The 'spin-echo' box indicates that a SE pulse sequence can carry out T2-*w* imaging by removing the SFd-P process from the picture. $T_{SFd\text{-}P}$ de-phasing is *not* removed by the *gradient-echo* sequence, however, and GE is clinically somewhat less useful than 1/T2 of spin-echo. It does have the advantage of generally being much faster than spin-echo. (d) 1/T2* is the rate of the fastest relaxation process of all, combining contributions from 1/T1, 1/T2, and $1/T_{SFd\text{-}P}$ processes. FID and gradient-echo (GE) sequences can generate T1-*w* images, as does S-E, but they fail to yield T2-*w* for almost all variants of the sequences. Their best efforts most frequently yield only the T2*-*w* study. Fortunately, sometimes that actually suffices.

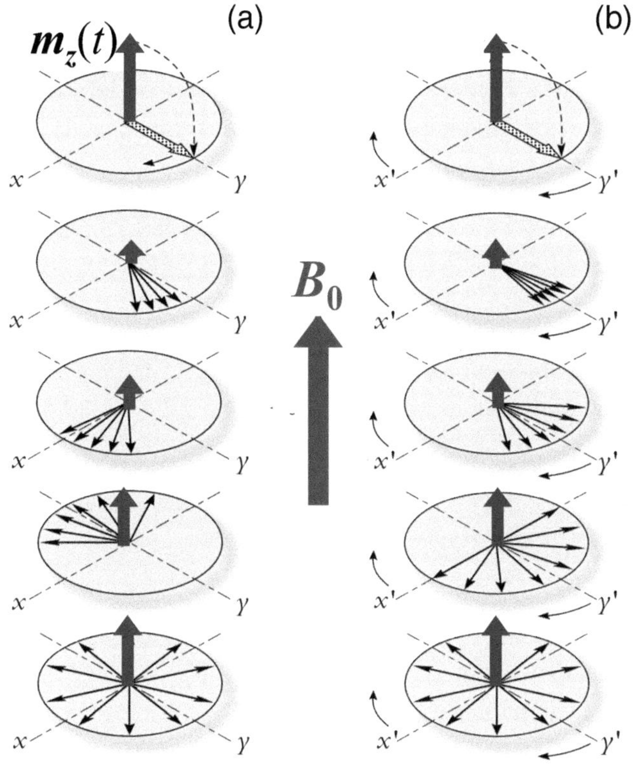

Figure 10.2 The secular component of T2 relaxation. The in-plane de-phasing of spin packets and spreading out of $\boldsymbol{m}_{xy}(t)$, and the associated regrowth of $\boldsymbol{m}_z(t)$, as viewed here from (a) the laboratory and (b) from a frame rotating at the local Larmor frequency. 1/T2 is the measure of the rate at which spin packets lose proton-spin phase-coherence in the **x-y** plane following, for example, a 90° saturation pulse. As suggested in Figure 10.1, two mechanisms give rise to this T2 de-phasing, both caused by spontaneous, random variations in the local $\boldsymbol{B}_z(t)$ fields. The first consists simply of longitudinal spin-flip events, which not only determine the value of the parameter T1, but also contribute to loss of phase in the **x-y** plane. The purple vertical arrows indicate the resulting regrowth of $\boldsymbol{m}_z(t)$. The other, so-called *secular* contribution to 1/T2 comes from small, very slow *low*-frequency random fluctuations in the local $\boldsymbol{B}_z(t)$. These cause tiny, brief variations in the processional ν_{Larmor} values of individual protons, hence a fanning out of their magnetic moment arrows in the transverse plane. Because the magnetic fields underlying both contributions to 1/T2 occur *randomly* over time, they cannot be reversed or eliminated, such as by the spin-echo pulse sequence.

type relaxations do, indeed, help to disrupt $\boldsymbol{m}_{xy}(t)$ and expedite the decline in its magnitude.

The second, *secular*, contributor to 1/T2 is the intermittently continuous de-phasing stemming from small, naturally occurring differences in proton precession frequencies, seen in Figure 10.2 from both the fixed and the rotating frames of reference.

Suppose that following a 90° saturation RF pulse, the two protons of a water molecule are precessing in the $\boldsymbol{x}$-$\boldsymbol{y}$ plane (Figure 10.3); the water is weakly bound to the hydrogen of some biomolecule, whose nucleus happens to be still aligned along the z-axis. (Perhaps it just underwent a T1 transition.) This affects the two water protons a little differently because of their relative positions; the local fields at the two are slowly varying, moreover, because of the rotations of the water itself and of the biomolecule. But both water protons can also be affected by other clinically relevant biophysical factors. So the two water protons will spend time in local environments that differ a little; these periods may be brief, but long enough to allow for some proton-proton mutual de-phasing to occur. That explains the *spin-spin* name for this type of relaxation. This secular component of 1/T2 does not involve

Figure 10.3 A heuristic, hand-waving explanation, mixing the quasi-quantum and classical pictures, of how stochastic de-phasing can occur. A water molecule lingers a while near a proton that is aligned along $\boldsymbol{B}_0$, and that is part of an organic macromolecule. The two water protons happen to be precessing in the **x-y** plane, but because one of these is closer to the hydrogen of the slowly moving macromolecule, their local fields differ a bit, as do their precession rates. They lose phase coherence with one another at a rate parameterized by $1/T2_{secular}$.

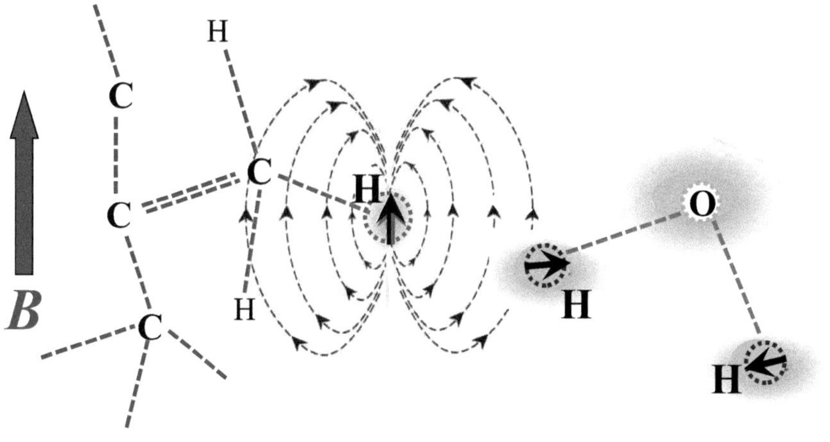

spin flips, nor the expenditure or release of energy (unlike the case for T1 transitions).

To summarize, 1/T2 is the rate at which a cohort of spins would spontaneously de-phase in the transverse plane, following a disruption such as a 90° pulse. The net relaxation rate 1/T2 can often be expressed as the sum

$$1/\mathrm{T2} = \tfrac{1}{2}/\mathrm{T1} + 1/\mathrm{T2}_{\mathrm{secular}}. \qquad (10.1\mathrm{a})$$

The ½ is there because $\mathrm{T2}_{\mathrm{secular}}$ events occur in the two-dimensional ***x-y*** plane, while T1 relaxation occurs only along one axis. That is, both 1/T1 and $1/\mathrm{T2}_{\mathrm{secular}}$ events contribute to the proton spin de-phasing within the ***x-y*** plane that determines the rate 1/T2, and

$$1/\mathrm{T2} > 1/\mathrm{T1}. \qquad (10.1\mathrm{b})$$

Indeed, T2 times are generally between 3 and 10 times shorter than T1 times for the same tissue (Table 9.1), and generally 1/T2 = 1/T1 only when all molecular rotations are totally inhibited, as in a solid.

EXERCISE 10.1 Why does T2 = T1 for ice? Will T2 and T1 be long or short?

T2 is a measurable parameter that happens to be of biologic interest, and images using T2 to generate contrast are clinically as important as the T1 variety. Like T1 relaxation, and for essentially the same reasons, T2 relaxation is approximately exponential,

$$|\boldsymbol{m}_{xy}(t)| \,/\, |\boldsymbol{m}_{xy}(0)| = e^{-t/\mathrm{T2}}. \qquad (10.1\mathrm{c})$$

This means that the transverse magnetization is likely to be long gone by the time that the longitudinal magnetization has regrown significantly. T2-*w* relaxation can even be, in fact, too fast to be imaged.

10.2 Biophysics of Proton T2 Relaxation

The amounts of either Larmor-frequency or very-low-frequency magnetic fields that will arise spontaneously and be experienced by a proton in a water or lipid molecule depend, of course, on the molecule's motions (Figure 10.4a). Like its T1 predecessor (Figure 9.4b), this displays the power spectrum, essentially the number of water molecules rotating around random axes at different frequencies, for three principal sub-populations of water molecules: free water, water that is tightly bound to something big and immobile, and the

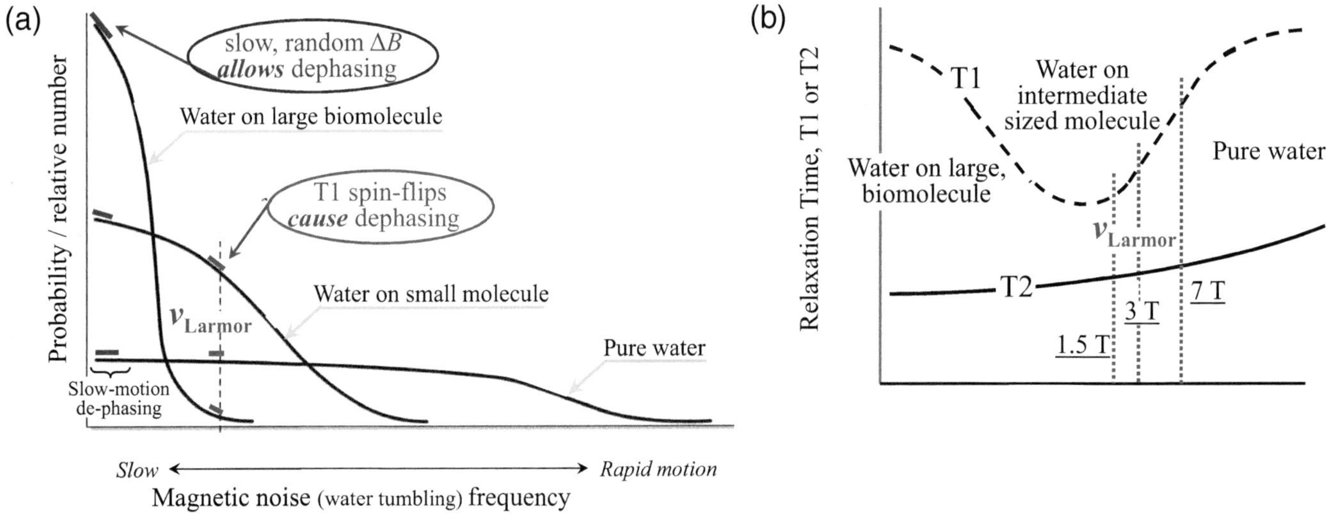

Figure 10.4 T2 relaxation. (a) One of the two contributions to the rate 1/T2 comes from the T1-relaxation process itself, which is caused by magnetic noise at the local Larmor frequency. Every time a spin flips up or down relative to the **z**-axis, triggered by a ν_{Larmor} local field fluctuation, the affected proton loses whatever precessional phase coherence in the ***x-y*** plane that it had with the others in the voxel. The second, or *secular*, mechanism involves slow, irregular molecular motions of water molecules. These entail neither spin flips nor the exchange of energy, but they do afford protons time to de-phase from other protons in the voxel. (b) As a result, water molecules in cellular environments that tend to provide plenty of low-frequency fields, as in Figure 10.3, undergo rapid T2 secular decay. On machines of different field strengths, T1-values can vary considerably, but T2 much less so.

loosely tied-up water, in between, that is of primary biological interest. The two regions of these spectra that are of especial concern are the ones near ν_{Larmor} and $\nu \sim 0$.

The impact of molecular motions on the relaxation times may be illustrated with one striking observation: when water freezes, its T2 drops from a few seconds to 10 μs = 10^{-5} s. In solids, there is relatively little motion of the nuclei, so that the slowly changing local fields give the spins ample time to de-phase. The secular relaxation mechanism is therefore fast, and the transverse relaxation time is much shorter than T1 alone.

Relaxation within living tissue is interestingly complex. Again, one can visualize three general categories of water mobility: a water molecule that is tightly bound to a very large macromolecule is like that in a solid. The paucity of Larmor frequency magnetic fields causes T1 to be long, but the abundance of low-frequency fields (since the molecular motions are generally slow) leads to *short* T2 (Figure 10.4b). Protons in solid-like environments undergo $T2_{\text{secular}}$ relaxation that is so fast, in fact, that their transverse magnetization may decay away even before it can be detected for construction of an image, in effect removing these protons from the image altogether.

At the other extreme, a relatively non-viscous liquid such as pure water will display a wide range of motions that the molecules are undergoing, and a broad spectrum of local magnetic fields and precession frequencies for all the protons in the voxel. For most of the water molecules in Figure 10.4a, the motions and the resulting variations in the local fields are far too rapid to be effective in causing secular spin de-phasing; the amplitudes of the fluctuations, both at the Larmor frequency and at low frequencies, in particular, will therefore be small, so both T1 and T2 will be *long*. For them, T2 is largely attributable only to T1-type transitions, so T2 ~ T1.

In between, some water will naturally bind, whether tightly or loosely, to medium-sized molecules, which may be either free-floating or part of a membrane or organelle. Water may be attached permanently or intermittently to such molecules, which are themselves rotating a good deal more slowly than free water. The noise magnetic fields that they create and experience may have significantly greater Larmor-frequency or low-frequency components than does free water; T1 and T2 may then both be much *shorter* than for pure water.

So the protons in any voxel of tissue inhabit a broad range of possible environmental conditions and may take part in a variety of spin relaxation processes, each with its own characteristic time. It is possible, nonetheless, to correlate much of spin behavior with the single *pair* of gross 'effective' relaxation parameters, T1 and T2. This is plausible when relaxation processes are at least nearly exponential, as would be appropriate for microscopic systems that can be described by so-called 'one-compartment' kinetic models.

Fortunately, the two relaxation times, crude as they may be, are still sufficiently tissue-sensitive and tissue-specific that T1- and T2-weighted MRI maps are clinically extremely useful.

With both T1 and T2 relaxation processes active, the Bloch equations become

$$\begin{aligned} dm_x/dt &= 2\pi\,\nu_{Larmor}\,m_y - m_x/\text{T2} \\ dm_y/dt &= -2\pi\,\nu_{Larmor}\,m_x - m_y/\text{T2} \\ dm_z/dt &= -\left[m(t) - m_0\right]/\text{T1}. \end{aligned} \tag{10.2}$$

During the brief times that a ν_{Larmor} RF pulse is being applied to induce nutation, on the other hand, T1 and T2 effects can generally be ignored.

10.3 Static Field De-Phasing at the Rate $1/T_{\text{SFd-P}}$; Spin-Echo Sequences Remove It

There is another influence that can speed up the decay of $\boldsymbol{m}_{xy}(t)$ considerably, above and beyond T2-relaxation, namely *Static Field de-Phasing* (SFd-P). It is not a genuine proton-spin relaxation mechanism, but rather it has to do with the fact that the short-range internal magnetic fields, even within a single voxel, are not perfectly uniform. It is accounted for as a modification of T2, which transforms it into the shorter time known as T2* ('tee-two-star'), unless the effect is somehow eliminated.

It happens that this predominantly instrumental defect can, indeed, be reversed through application of a *spin-echo* RF-and-gradient pulse sequence. This technique eliminates it from T2-*w* SE studies, leaving behind a purely biological T2 image. But this happy circumstance does *not* occur with the commonly much faster (and, therefore, in some situations preferable) *gradient-echo* sequences.

Suppose we apply a standard 90° pulse to the protons in a voxel, in the hypothetical situation where there are *no intentional gradient fields* present and *no relaxation* processes in play. Initially in phase, the spins begin precessing in the transverse plane at the Larmor frequency for $\boldsymbol{B}_0$. They give rise to the transverse magnetization that produces the FID signal, and we would expect the amplitude of the signal from $\boldsymbol{m}_{xy}(t)$ to remain constant, with no gradient or relaxation.

But while the main field is nominally uniform across the dimensions of a voxel, and everywhere across it nominally of strength B_0, in fact there are slight, fixed spatial variations within the voxel due to several factors. The field is supposed to be homogeneous, to within less than one part per million or so, within a field of view for standard MRI, but that still allows for some spatial unevenness on the near-microscopic level. This will cause some proton packets in the voxel to precess slightly faster or slower than average. This Static Field de-Phasing causes them to spread out in the $\boldsymbol{x}$-$\boldsymbol{y}$ plane, exactly as in Figure 10.2, albeit for a totally different reason than the factors that stimulate T2 relaxation! As a result, the length of the net magnetization vector of the voxel will diminish exponentially at the characteristic time $T_{SFd\text{-}P}$. When T1 and T2 processes are turned back on again, on the other hand, these will also separately contribute to the temporal decay of the voxel magnetization and of the MR signal strength.

The static field de-phasing arises from two sources. First, there are variations of the external field strength within the voxel, of average value ΔB_0, even in the absence of an applied gradient, caused by imperfections in magnet design and construction. Second, slight, short-range naturally occurring field gradients arise at abrupt changes in the susceptibilities, $\Delta\chi$, of certain tissues and other materials [Equation (2.4)]. These emerge at interfaces between soft tissue and cerebro-spinal fluid, for example, or at air, bone, or any non-ferromagnetic metal at interfaces among them. Indeed, when $\Delta\chi$ is large, as when caused by dental fillings, the resulting *susceptibility artifact* can be quite spectacular (Figure 3.19c).

EXERCISE 10.2 Does $\Delta\chi$ refer to electronic, e.g., paramagnetism as in Equation (2.3), or nuclear susceptibilities, or both?

The overall rate at which the SFd-P process occurs can be viewed as driven by the two factors parameterized by ΔB_0 and $\Delta\chi$, respectively,

$$1/T_{SFd\text{-}P} = \kappa\,\Delta B_0 + \kappa'\Delta\chi, \qquad (10.3a)$$

where κ and κ' are constants. $T_{SFd\text{-}P}$ is an instrumental effect and fortunately, as we shall see, the *spin-echo pulse sequence* can largely counteract such static field effects and remove them from the picture—literally. But the $\kappa'\Delta\chi$ term can be put to good (albeit limited) use in *susceptibility imaging*, which is like a clinically useful version of Figure 3.19c.

So, the availability of gradient- and spin-echo pulse sequences within a single exam gives us two tools to adjust the sensitivity of tissue to transverse relaxation—the use of spin echo to decrease sensitivity to susceptibility (e.g., avoiding artifacts from dental hardware), and the application of gradient echo to increase speed and sensitivity (e.g., in searching for micro-hemorrhages).

A point that is crucially important, as we shall find in the upcoming discussion of the spin-echo pulse sequences in the next chapter, is that whatever the cause of the spatial variations in local field strength, the static field de-phasing process is *fixed in time* and does not randomly fluctuate temporally. Hold on to that thought.

The point of the previous paragraphs is that individual protons experience *static* local magnetic fields that are slightly different from one another, and that such irregularities will bring about tiny but constant spatial differences in the precessional frequencies of the proton clusters. At any instant, half the protons will be precessing a little faster than the median proton because of their spatial positions, and the others will go around more slowly. So after a 90° pulse, the spins will lose their initial phase coherence, over time, and their spin orientations will fan out evenly in the transverse plane. This appears to be the same phenomenon observed in Figure 10.2, but the cause is altogether different. Either way, if you took a snapshot of all of the spin packets in a voxel, the individual packets will spread out to point in all directions over time, ending up like the spokes of a bicycle wheel, and their magnetic moments will cancel one another out. As a result, the net magnetic field they produce at the pickup coil drops to zero, as

$$m_{xy}(t)_{\text{SFd-P}} / m_{xy}(0) = e^{-t/T_{\text{SFd-P}}}, \quad (10.3b)$$

with characteristic decay time $T_{SFd\text{-}P}$, assuming for the moment no spin relaxation. As with other exponential processes, $T_{SFd\text{-}P}$ represents how long it takes for static field inhomogeneities to cause the magnitude of $m_{xy}(t)$ to de-phase to 0.37 of its original value. A short $T_{SFd\text{-}P}$ corresponds to a fast *rate* of exponential decline.

EXERCISE 10.3 How great is the effect of a typical applied x-gradient across a single voxel during readout?

EXERCISE 10.4 What is rate of dephasing in a voxel caused by normal magnet inhomogeneity with no gradient on? Is it the same as for the entire FOV?

10.4 1/T2* is 1/T2 Speeded Up by $1/T_{SFd\text{-}P}$

Let us turn now to a realistic situation in which $\boldsymbol{m}_{xy}(t)$ decays both because of T2 relaxation and from the static field de-phasing associated with ΔB_0 and $\Delta\chi$ of Equations (10.3). Following a 90° pulse, an FID signal falls off exponentially at the rate

$$1/\text{T2*} \equiv 1/\text{T2} + 1/T_{\text{SFd-P}} = 1/\text{T2} + [\kappa\Delta B_0 + \kappa'\Delta\chi]. \quad (10.4)$$

T2* is called "tee two star" and, as suggested by Figure 10.1, it has the greatest accumulation of physical effects contributing to it. It is, therefore, the fastest member of the 'relaxation' or, more correctly, MR signal-decay family.

Much of magnetic resonance imaging employs variations on the spin-echo sequence because signal strength is relatively high and $T_{SFd\text{-}P}$ effects are minimized. But other approaches, in particular sequences in the gradient-echo (GE) family, also generate T1-*w* maps of T1, like S-E, but of T2*-*w* rather than of T2-*w*. This is tolerable because while such T2* MRI images are only partially biologically based, they are still of some clinical use. A GE T2* examination, for example, may reveal susceptibility effects, such as for old bleeds, slow leaks, calcifications, and vascular malformations. Iron loading in liver or the heart can alter the T2* relaxation rate (in proportion to Fe concentration in the tissue) and susceptibility-based contrast in patients with these conditions (e.g., sickle-cell patients undergoing blood transfusions and requiring iron-chelation therapy). Also importantly, GE imaging is generally very, very fast—and time, in MRI, is of the essence.

10.5 Shape of an NMR Resonance Line

As apparent in Figure 9.4a, water in the hydration layers of macromolecules experiences a broad range of local environments and possible motions. These include those of the vigorous and rapidly varying protons in free water or loosely attached to small molecules, at one extreme, down to the practically immobile ones that are bound tightly to membranes and huge biomolecules. Figure 9.4b indicates, in the form of *probability distribution functions* (PDF), the relative number of water molecules that are rotating and otherwise moving at various rates across this range. In particular, it reveals that the number of protons that are rotating at or sufficiently near the Larmor frequency, and capable of undergoing resonance, is quite small. This kind of statistical information finds use in *ab initio* calculations of T1 and other relaxation parameters.

Figure 10.5a offers something related but different. For each of two populations of water protons shown, which inhabit distinctly different local environments, it plots the relative numbers of water protons undergoing resonance as a function of the *offset* distance, $\Delta\nu_{Larmor}$, from their central resonance frequency—in effect, their NMR resonance line-shapes.

For each population, as an applied ν_{RF} approaches and passes through its central resonance frequency ν_{Larmor}, the NMR power absorption in solutions most commonly follows the bell-shaped *Lorentz* line-shape,

$$dE/dt(\nu) \sim 1/\text{T2} / \left[1/\text{T2}^2 + 4\pi^2(\nu - \nu_{Larmor})^2\right], \quad (10.5a)$$

introduced in Equation (6.3) and Figure 6.5. The absorption line-shape is not an infinitesimally thin delta-function (Figure 5.5a), of course, but rather it is fattened up a little by friction-like relaxation processes. Comparison with Equation (6.3) leads to the correct supposition that

$$1/\text{T2} = \text{linewidth} \quad (10.5b)$$

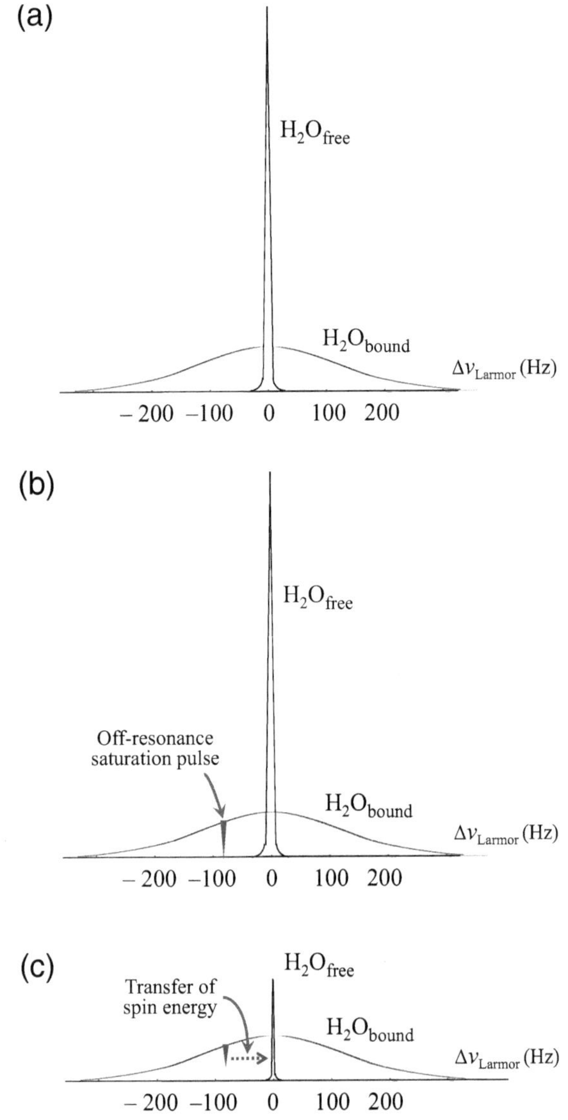

Figure 10.5 NMR line-shapes/spectra for two sub-populations of water, expressed as frequency shifts/offsets, $\Delta\nu_{Larmor}$ (Hz), from the population mean proton resonant frequency. (a) One sub-population consists of free/bulk water, with the much narrower line, and the other of loosely bound hydration water. These two probability distribution functions (PDFs) also serve as resonance spectra. (b) In *magnetization transfer* (MT) imaging, RF power saturates the polarization of water of a relatively slender sub-population of the bound protons. (c) The energy released in the process can be transferred to free water protons in this MT exchange, reducing the height of their PDF, which is a detectable phenomenon.

The tall curve in Figure 10.5a is very narrow because the local surroundings of all of the *free* water molecules are virtually identical, so that NMR occurs at nearly the same frequency. Because only a few of these molecules are rotating either at ν_{Larmor} or very slowly, moreover, spin-spin events are relatively rare. T2 is long (~4000 ms), de-phasing is slow, and, by Equations (10.5), the resonance peak is slender—the same as with the peal from a struck crystal glass.

The broader curve provides the same kind of information about the water molecules that are interacting with macromolecules. It shows the relative numbers of protons in hydration water as a function of their offset resonant frequencies, $\Delta\nu_{Larmor}$, away from the central resonance peak. As such, it corresponds to a broad span of molecular environments, hence over a significant spread of local magnetic fields and resonance frequencies.

Now we shall make some practical use of this discussion of absorption lines and linewidths.

EXERCISE 10.5 T2 ~ T1 for free water. What does that imply about the motion of water for the two relaxation processes?

10.6 Magnetization Transfer (MT) Imaging

Again, and as suggested by Figure 9.4, there are three general categories of bioenvironmental conditions for water molecules. Some are tightly bound to macromolecules and membranes and are largely immobile. Bulk or free water, at the other extreme, is surrounded only by more water, and can rotate rapidly around any axis. And the hydration layer comprises those water molecules in between, those that are loosely held or are moving about with small biomolecules.

Protons on large biomolecules, or on water attached firmly to them, tend to have T2 values that are too short (microseconds) to enable most pulse sequences to carry on to completion before the signal decays away to nothing. As a result, resonances from the protons of bound water are not commonly seen in MRI, although the development of *ultrashort-TE* (UTE) sequences has partially overcome this problem [Robson et al. 2006].

Which brings us to the subject of *magnetization transfer* (MT) [Henkelman et al. 2001; Symms et al. 2004]. Consider first the two resonance spectra of bound and bulk water (Figure 10.5a). Because the protons on biomolecules will experience a wide range of magnetic environments, their NMR spectrum will be broad, quite unlike that for free water. Suppose that an RF pulse at an 'off-resonance' frequency somewhat distant (several hundred Hz) from the ν_{Larmor} of bulk water, now jolts a small sub-population of the bound proton population (Figure 10.5b). If the RF is intense enough in this off-resonance part of the spectrum to partially *saturate* the spins there, driving their polarization and net magnetization to zero and 'burning a hole' in the spectrum, the ratio of the numbers of spin-up to spin-down protons there, N_{up}/N_{down}, shifts nearly to 1.

This can, quite remarkably, result in a detectable impact on the narrow resonance peak of the *free* water (Figure 10.5c). Some of the bound protons cross-relax with protons in the free water pool. They thereby create a kind of contrast by transferring some of their (formerly bound) proton spin polarization to protons in the bulk water. This transfer of energy or hydrogen nuclei between the populations is brought about by several physical mechanisms, mainly proton dipole-dipole interactions, Equations (2.6), and chemical exchange, and it provides a novel form of MR contrast.

Magnetization transfer (MT) imaging has proven particularly beneficial in time-of-flight (TOF) MR angiography (MRA). When the MT protocol is employed, the lower concentration of macromolecules in fine blood vessels, which have few protons at off-resonance that can be saturated, allows for a contrast enhancement compared with macromolecule-rich brain parenchyma. Other less widely spread uses of MT include the detection of multiple sclerosis, neoplasms that elude standard MRI, and irregularities in the breast, heart, eye, and elsewhere.

10.7 Contrast Agents Such as Gadolinium Chelates

Most imaging modalities make some use of chemical or other agents to enhance contrast among tissues. The ideal contrast agent is tissue-specific and safe at concentrations that are high enough for it to have enough effect to be visible. In radiography, iodine and barium atoms are particularly adept at absorbing x-ray photons. Gas-filled micro-bubbles increase the reflectivity of blood vessels in ultrasound. And the radiopharmaceutical tracers in nuclear medicine themselves play roles analogous to contrast agents.

MRI contrast agents work indirectly, by altering the local magnetic environments of some of the nearby tissue protons, thereby affecting their relaxation rates and T1, T2, and T2* values. While some such electron paramagnetic materials occur naturally in the body, such as deoxyhemoglobin, others are intentionally introduced from outside as MR contrast agents. The substances employed usually involve strongly electron-paramagnetic transition metals or rare earth elements. (We write here about electron-paramagnetism, as in Table 2.4, which can have effects on MRI images, as opposed to nuclear paramagnetism, which is the fundamental phenomenon that underlies NMR and MRI.)

Unfortunately, many of the strong paramagnetic elements are transition metals or rare earth elements, which are frequently quite toxic in raw form. MR contrast agents, therefore, commonly consist of a strongly paramagnetic ion trapped within a *chelating* molecular cage, the *ligand*, which ties them up with multiple bonds and greatly diminishes their toxicity (Figure 10.6a). Its effectiveness depends on whether its ligand is hydrophilic or hydrophobic and other physicochemical factors.

Chelating also helps with purging the paramagnetic ion from the patient's system (e.g., via the kidneys), as the lifetimes for many chelate complexes in the body are fairly short, commonly less than 24 hours.

In most atoms, the inner orbital electrons of atoms are spin-paired, in the sense that the magnetic field generated by one is effectively canceled out by that from another oriented in the opposite direction. As a result, water, many organic compounds, soft tissues, oxyhemoglobin, and some other materials are slightly diamagnetic, $\chi < 0$ in Table 2.4. A few atoms, however, find it energetically favorable for one (or more) of their electrons to go against the grain and remain unpaired. The magnetic moment of one unpaired orbital electron is 657 times that of a proton, so the field it produces at a nearby water proton will be far stronger than even that of its own partner.

The most widely explored agents contain chelated ions of iron, manganese, dysprosium, and especially gadolinium, which is commonly complexed with diethelenetriaminepenta-acetic acid (Gd-DTPA) (Figure 10.6). A Gd atom is an extreme-case actor, carrying seven unpaired electrons in the 4f subshell. As it rotates, it gives rise to a fluctuating magnetic field thousands of times stronger than that from any proton, and this can radically affect the local environments of nearby water or lipid protons. Clinical trials have already demonstrated the efficacy and general safety of Gd-DTPA in imaging renal lesions, malignancies of the brain that cause breakdown of the blood-brain bar-

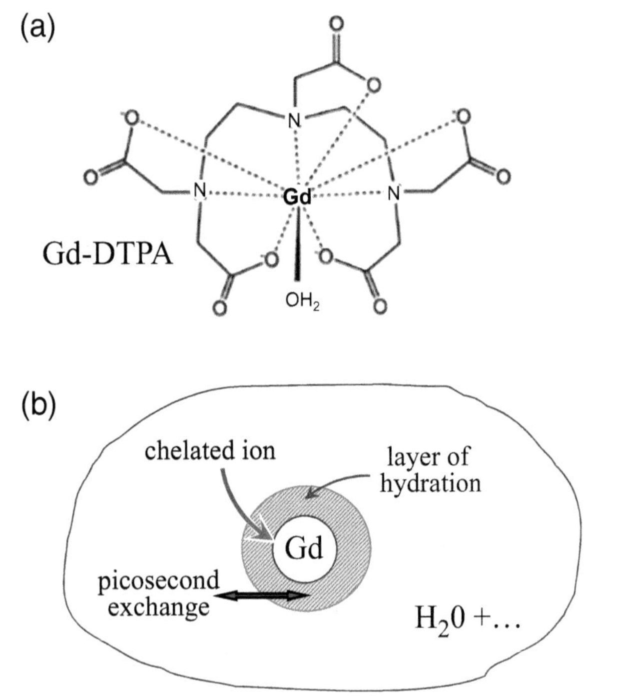

Figure 10.6 Most MRI contrast agents are electron paramagnetic (gadolinium, dysprosium, manganese) ions, or superparamagnetic (iron oxide) nanoparticles, *chelated* (encaged by an organic exo-structure) to reduce their toxicity. (a) The structure of Gd^{7+}, with seven unpaired electrons attached, chelated within DTPA. Its fields can be thousands of times stronger than those from a proton, since the magnetic moment of an unpaired electron is hundreds of times greater than that of a proton. Interactions between contrast agents and protons can speed up proton relaxation rates 1/T1 considerably and, to a much lesser extent, 1/T2, thereby shortening T1 and T2. (b) A gadolinium-based contrast agent, whose surrounding layer of hydration separates the chelated Gd ion from the free water, as in Figure 9.4a.

rier (Figure 1.6b), etc. It is cleared rapidly from the liver by glomerular filtration but, as will be discussed in chapter 15 on MRI safety, there are certain serious drug-reaction hazards, in particular with the possible causation of nephrogenic systemic fibrosis (NSF).

EXERCISE 10.6 Is Gd a highly chemically reactive free radical?

Non-chelated materials, such as manganese chloride, have higher levels of toxicity in part due to the fact that they do not evacuate the body rapidly; they can probe some internal cellular properties more readily, however, due to the small size, properties, and chemical function of the free ion. Manganese chloride is currently approved for use in Europe, but not in the United States. Nanoparticles of ferric oxide and a few other *superparamagnetics* have magnetic susceptibilities hundreds or thousands of times stronger even than those of paramagnetic materials, but they find limited use as agents.

The overall 1/T1 relaxation rate for a tissue depends both on the inherent pure tissue rate, $1/T1_{tissue}$, and on the concentration of agent within it. Because the two processes are random and independent, their rates are additive:

$$1/T1 = 1/T1_{tissue} + r1[Gd], \tag{10.6}$$

with the agent- and tissue-specific rate constant $r1$, which may be dependent on pH, etc. The same equation applies for 1/T2 (with parameter $r2$) and for 1/T2* (with parameter r2*). This suggests, of course, the involvement of several distinct populations: free water molecules in tissue away from the chelated ions, and those in the contrast agent hydration layers, which exchange rapidly with the pool of tissue water, somewhat as in Figure 10.6b.

Contrast agents are much more effective in speeding up the T1 relaxation rate than that for T2. Typically a clinical dose of Gd-DTPH will hasten T1 relaxation by a factor of 25%, with an effect of only about 4% on T2. Contrast agents tend to increase signal intensity in T1-*w* imaging, except in the case of extremely high concentrations (e.g., kidney and bladder post-contrast scans), where T2 or T2* effects destroy the transverse magnetization quickly, even on a T1-w image.

It usually would require high doses of gadolinium to affect T2, and it is not routinely used in T2-*w* imaging. However, a bolus—which for all purposes is a temporary locally high concentration of gadolinium-based contrast agent passing through blood vessels and capillary beds—dramatically increases T2* relaxation in that tissue for a short period of time. With a fast gradient-echo type of sequence called an *echo-planar imaging* (EPI) sequence, one can acquire several images within a second to visualize the passing bolus as tissue darkening and brightening.

Dynamic contrast-agent enhanced (DCE) MRI is another *perfusion* modality that employs rapid sequential acquisition of multiple T1-w MR images to follow the flow and migration of contrast agents into certain tissues, and the subsequent washout. DCE-MRI will be described in greater detail in chapter 14 on fluid flow.

CHAPTER 11

Spin-Echo (SE) Pulse Sequences in 1D

The specific study chosen for a given patient and medical problem is determined by the kind of information felt to be clinically needed, and also by the capabilities of the diagnostic equipment available. The examination, in turn, is then made with optimal settings of the instrument's operational controls, the selection of which may depend upon complex and interrelated trade-offs that influence contrast, resolution, noise level, artifacts, and image acquisition time. While some of the parameters are subject to established guidelines or instrumental restrictions, for MRI the settling on others is a subtle and still evolving science and art. Indeed, from the perspective of the novice physician, radiographer, engineer, or physicist, what most distinguishes learning to run an MRI device, in particular, may well be the number of possible combinations of parameter choices available, and the extent to which the quality of the resulting images is sensitive to the set actually adopted.

Some parameters are difficult or impossible to alter, such as the strength of the main magnet, the general design of commercially available pulse sequences, the scanner's inherent sensitivity and selectivity, and, of course, characteristics of the tissues being observed. Others are under the immediate control of the operator (Table 11.1). Among them are the dimensions of the voxel matrix and of the field of view (which together limit voxel size and image resolution), tissue slice thickness and inter-slice separation, the receiver RF bandwidth, the coils employed to detect the RF echo

Table 11.1 Some operator-controlled parameter settings for the various forms of MR imaging

Sequence type (SE, GE, IR, FLAIR, DTI, fMRI, MRS, MRA, et al.)
Weighting (T1-*w*, T2-*w*, PD-*w*)
Pulse timing (TE, TR, TI)
Voxel matrix, FOV
Slice thickness, separation
Multi-2D vs. 3D
NEX
RF coil, RF bandwidth ($BW_{receive}$)
Parallel coils
Contrast agent, Gating
Display matrix
etc.

signals, NEX, the use of contrast agents or physiologic (such as electrocardiogram or breathing) gating, and other factors that we shall discuss.

But of all these, the fundamental choice is of the MR sequence to be applied, and of the timing of its RF and gradient pulses. Free Induction Decay/Saturation Recovery, discussed in the past few chapters, was fine for introducing the general idea of pulse MRI, but it is rarely called upon in practice. Certain extensions of FID are far more rapid, flexible, and effective. *Spin-echo* (SE) and *gradient-echo* (GE) are the two most important families of sequences, and in this chapter we shall begin to explore the first of them.

11.1 Spin-Echo (SE) RF Pulse Sequence in a Single Voxel: 90°–180° – read

Figure 9.6 present a conceptually simple but impractical way to obtain values of T1 for biologic materials in a voxel with FID. That *quantitative* approach requires driving the longitudinal magnetization, $\boldsymbol{m}_z(t)$, down into the transverse plane and then following its regrowth for several different periods of *repetition time*, TR_i. Fitting the recovery data to $(1 - e^{-TRi/T1})$ for a set of TR_i times can yield a reasonably close estimate of the actual value of T1. Unfortunately, this approach would require an unacceptably long scan time.

It was then suggested that it is much faster, yet still clinically tolerable, to produce T1-*w* images with FID and a *single* value of TR. In fact, it is such weighted images that are commonly used for clinical diagnoses. This was demonstrated with one measurement of signal strength of tissues in adjacent voxels (Figure 9.8). It did not lead to precise values of T1 for each of the various tissues of interest, but rather to a weighted parameter that is roughly proportional to T1 itself, and with which one can map spatial *variations* in T1, generating a T1-*w* image.

The standard *spin-echo (SE) RF pulse sequence*, however, goes way beyond that, and it provides two immediate, great advantages over FID in generating relaxation time-weighted or proton density-weighted images.

First, as the name implies, SE leads to the creation of one or more RF *echo signals*, which provides a solution for the problem of the rapid FID signal decay due to T2* relaxation. These echoes are regenerated at a much later time than T2* would appear to allow (they are governed instead by T2), and can be harnessed in image creation long after an FID signal would have fallen away to nothing.

Second, T2* relaxation following a Larmor-frequency pulse is brought about by two separate processes. The first is the biologically significant spin-spin relaxation rate (1/T2). The second, at the rate $1/T_{SFd\text{-}P}$, is effected by spin de-phasing from field gradients that result from nonuniformities of sub-voxel size in the external field, and by diagnostically useless (for the most part) susceptibility effects at tissue interfaces, Equations (10.3) and (10.4). Fortunately, as we shall now show, the spin-echo pulse sequence allows us to get rid of this effect and to turn from the assessment of T2* to the purely biologically meaningful T2. SE can reverse and thereby eliminate the static field de-phasing (SFd-P) effects because, and only because, such disruptions in the local fields are *static*, not varying appreciably over time like random magnetic noise. While physiologically based susceptibility effects may be useful to visualize in a clinical context (e.g., hemorrhage), they may also lead to image distortion and other artifacts that hinder diagnosis—which is why SE is such a fine tool to have at hand.

We can illustrate this remarkable immunity to T2* effects by analogy to Kentucky's glorious and world-renowned Derby: spin-echo works just like the First Shall Be Last Kentucky Turtle Festival, held annually near the home town of one of the authors in Clover Bottom, KY (Figure 11.1). The progress of the turtles

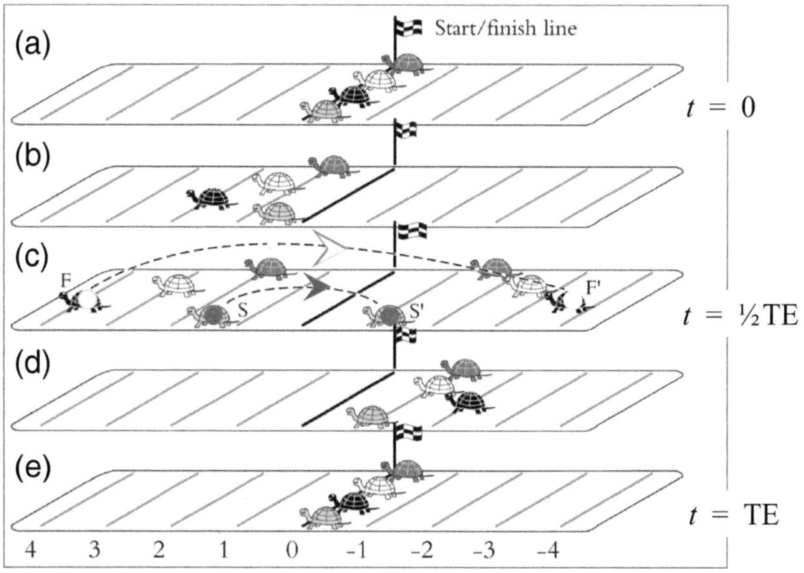

Figure 11.1 The First (Annual) Shall Be Last Kentucky Turtle Derby race (a) begins at time $t = 0$, and (b) the contestants begin spreading out along the course. (c) At exactly $t = ½$TE, each turtle is snapped to the other side of the start/finish line, the same distance from it as just before this teleporting, but continues racing at the same speed and facing in the same direction. (d) Now, somewhat after $t = ½$TE, they converge and (e) all cross the line together at TE. They can repeat the race again and again, finishing in unison each time.

is analogous to what is going on with the phases of spin packets and the Static Field de-Phasing (SFd-P) business.

(a) At $t = 0\_$, just before the race begins, all the competitors are aligned at the start/finish line. With the initiating Larmor-frequency **90°** *excitation* or *saturation pulse*, the noble beasts bolt from the line in a fury of dust and thundering hooves.

(b) Because of inherent differences in ability among them, some gallop faster than others, and the field of competitors begins to spread out. In MRI-speak, modest, static magnetic field inhomogeneities drive the protons to precess at slightly different rates.

(c) At a certain prespecified time, exactly *one-half* the *echo-time* (½TE), the fleetest steed (F, yellow dot) will have traveled 4 meters along the track, say, and the slowest (S, red), 1½ m, with the others in between. In a flash, precisely at $t = ½$TE, a 180° Larmor-frequency *refocusing pulse* teleports the lead turtle to a point 4 m on the *other* side of the start/finish line, F', but leaves her streaking full tilt ahead in the original direction. At the same time the slowest one is similarly re-positioned to 1½ m across the line, S', and likewise for the others.

(d) Over the next period of length ½TE, each turtle has the exact amount of time needed to reach the start/finish line...

(e) ...so they all cross it together exactly at t = TE. Nose-to-nose photo finish!

(f) Not shown here is a second lap around the course, which is just like the first. That is, we let them continue racing forward, applying a second refocusing pulse another ½TE later; then they'll all re-cross together once more at t = 2TE. And again and again, ad

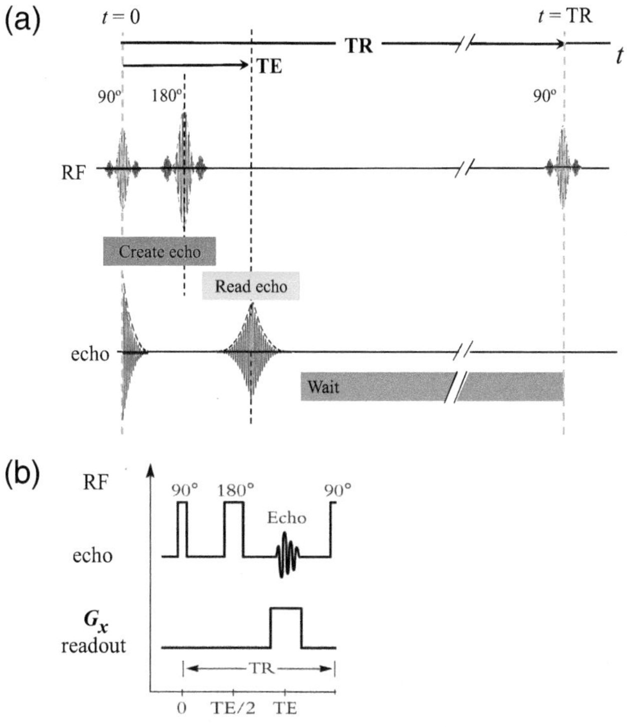

Figure 11.2 The SE pulse sequence. (a) With a single voxel, no ***x***-gradient is needed for spatial localization. The echo appears TE milliseconds after the initiating 90° RF. Following the arrival of the echo, the system is left to recover for a further (TR – TE) ms, where generally the *repetition time*, TR, is greater, or much greater, than TE. The next sequence begins at t = TR, with a new 90° pulse. (b) For a 1D row phantom, G_x is on briefly during readout.

infinitum, where it is assumed that they don't tire or begin to *relax*.

With spin-echo NMR, or with MRI in a single voxel, the same sort of thing happens to clusters of protons—except that it is the *spin orientation* of each subgroup of protons in the voxel that is transformed at ½TE rather than its *spatial location*.

The general form of the SE pulse sequence is 90°–180°–*echo/read*. The Larmor-frequency 90° *excitation* or *saturation* pulse occurs at time 0, and the 180° *refocusing* or *echo-generation* pulse at t = ½TE; the echo signal appears at time t = TE (Figure 11.2a). For now we are concentrating on SE in a single voxel, so there is no need here of an $\boldsymbol{x}$-gradient for spatial localization. (For examination of a row phantom, Figure 9.8, $\boldsymbol{G_x}$ is activated during readout of the echo, Figure 11.2b.)

EXERCISE 11.1 Why is G_x not needed for the single-voxel case?

EXERCISE 11.2 Why are G_z gradients (not shown here) activated during application of the narrow-band 90° saturation and 180° refocusing (echo-generating) ν_{Larmor} pulses?

EXERCISE 11.3 Why does an echo-like signal appear at t = 0?

A signal somewhat like a half-echo is generated during transmission of the initial saturation pulse at t = 0, in the lower left corner of Figure 11.2a. The true echo signal appears at t = TE, and both it and the one at t = 0 are sensed by the receiver antenna. Following a quiescent period of duration TR, the *repetition time* (the time from the prior 90° RF pulse, usually a good deal longer than TE), the whole SE sequence can be repeated. During the period following each saturation pulse, the longitudinal magnetization, $\boldsymbol{m_z}(t)$, is undergoing recovery. TR can be long enough to reestablish spin equilibrium, but in practice it is generally a good deal less than that, to cut down on imaging time. It also emphasizes tissue contrast differences in a T1- or T2-weighted image.

Proton spin-echo works like this (Figure 11.2b):

(a) Suppose that at $t = 0_-$, just before anything interesting happens, the magnetization is aligned completely along the main field and the z-axis, and $\boldsymbol{m}(t = 0_-) = \boldsymbol{m_z}(0_-)$. The green arrow in Figure 11.3a represents the *longitudinal* magnetization at that moment. If the system had been left undisturbed for a good while before, and had time to establish thermal equilibrium, then $\boldsymbol{m_z}(0_-) = \boldsymbol{m_0}$, but that is usually *not* the case.

The sequence begins at t = 0 with an initial 90° excitation/saturation pulse propagating along the $\boldsymbol{y}$-axis, red arrow. Its magnetic component, $\boldsymbol{B_1}(t)$, points perpendicular to both the $\boldsymbol{y}$- and z-axes (hence to $\boldsymbol{B_0}$). Over a very brief period just after t = 0, $\boldsymbol{m}(t)$ nutates down into the $\boldsymbol{x}$-$\boldsymbol{y}$ plane; it is here renamed the *transverse* magnetization (orange arrow), and designated $\boldsymbol{m_{xy}}(t)$. All the spins are now lying parallel to one another and along the $\boldsymbol{x}$-axis, and nothing remains of the longitudinal magnetization, $\boldsymbol{m_z}(0_+) = 0$. The magnitude of the new transverse magnetization is equal to that of the just-vanished longitudinal magnetization,

$$\left|\boldsymbol{m_{xy}}(0_+)\right| = \left|\boldsymbol{m_z}(0_-)\right|. \qquad (11.1)$$

We ran into this important concept in Equations (6.8), (7.1), and (9.7b), and we shall soon see it again.

(b) Immediately after the spin magnetization has been relocated in the $\boldsymbol{x}$-$\boldsymbol{y}$ plane, two closely related but separate things happen—one within the $\boldsymbol{x}$-$\boldsymbol{y}$ plane itself and the other along the z-axis.

First, as $\boldsymbol{m_{xy}}(t)$ precesses in the transverse plane (Figure 11.3b), the spins of those protons that happen to reside in a slightly above-average static local field will precess a little faster and farther than most, and conversely for the slower protons in lower fields, as with T2 relaxation. Both T2-relaxation and SFd-P effects cause spin packages to de-phase, which leads to a spreading out of spin-axes and a corresponding decrease in the length of $\boldsymbol{m_{xy}}(t)$.

Second, also in Figure 11.3b, the longitudinal component of the magnetization, $\boldsymbol{m_z}(t>0)$, the shortish vertical green arrow, begins to regrow upward at the rate

$$\boldsymbol{m_z}(t) = \boldsymbol{m_z}(0_-) \times [1 - e^{-t/T1}], \qquad (11.2)$$

until t = ½TE, at which point it has climbed to $\boldsymbol{m_z}(t$ = ½TE). If the spin system happens to have sufficient time to fully relax back to thermal equilibrium (which most likely it will not), a slightly different version of this applies.

(c) Exactly at t = ½TE, a 180° refocusing/echo-generating RF pulse, either twice the amplitude or duration (or some combination thereof) of a 90° pulse is applied (Figure 11.3c). It snaps every spin in the transverse plane over into its mirror-image orientation relative to the ***x-z*** plane, but it leaves the protons precessing in the same direction as before. Now they tend to head back to their origin.

Also, and separately, this 180° pulse forces the green longitudinal magnetization to flip rapidly *down* through the ***x-y*** plane; it ends up with the same magnitude that it had just before $t = ½$TE, namely $|\boldsymbol{m}_z(½\text{TE}\_)|$, but now it is aligned south, along the *negative* z-axis.

(d) Over a second period of duration ½TE, spins keep on precessing. But the 180° pulse has caused the spreading-out of spin orientations that came from SFd-P de-phasing to be *reversed*, and the 'fanning out' of spins brought about by SFd-P starts to collapse inward (Figure 11.3d). At the same time, the secular T2 relaxation does continue to spread the vectors out randomly, as those *stochastic effects cannot be reversed.* This process is generally far slower and less evident than the T2* relaxation.

Meanwhile, $\boldsymbol{m}_z(t)$ again regrows upward as $(1 - e^{-t/T1})$. After passing again through the transverse plane, $\boldsymbol{m}_z(t)$ keeps on lengthening upward toward $\boldsymbol{m}_0$, but it probably will not have enough time to reach it. Meanwhile, spins continue to precess in the transverse plane.

(e) The spins remaining in the transverse plane converge and reach a maximal length of $\boldsymbol{m}_{xy}(t)$, at (and

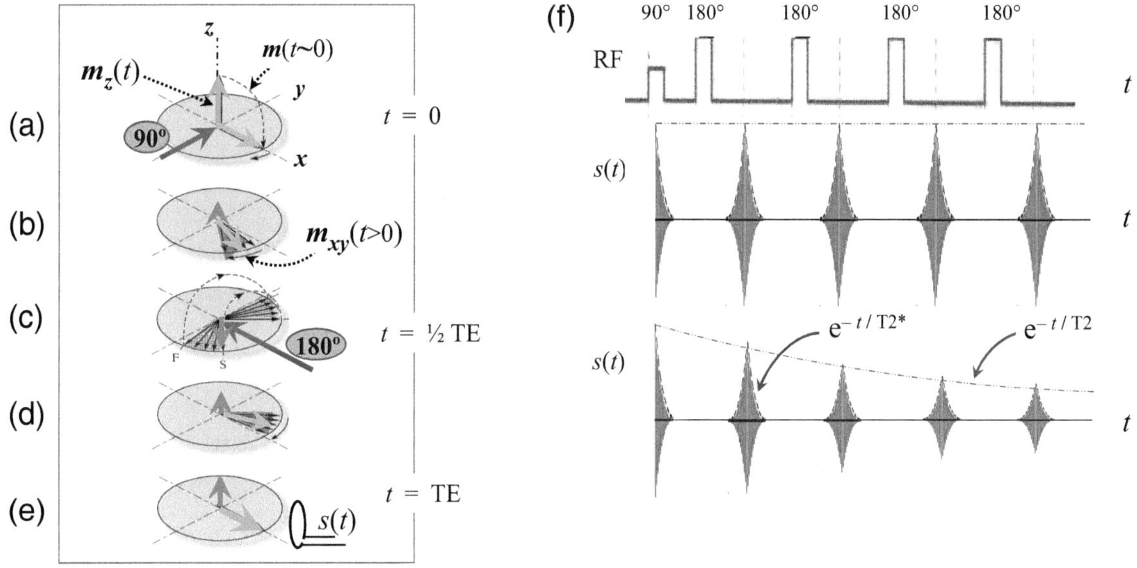

Figure 11.3 Spin echo in a single voxel operates somewhat like the Turtle Derby. Since we are dealing with only one voxel here, there is no need for a field gradient. Local fields fluctuate randomly in time, some at ν_{Larmor} and at $\nu \approx 0$, that can induce T1- or T2-type relaxation. This diagram displays the lab frame perspective. (a) Just before $t = 0$, the overall magnetization of the voxel (green arrow) initially points straight up, lying parallel to the static magnetic field along the **z**-axis, and comprises the purely *longitudinal* magnetization, $\boldsymbol{m}_z(0\_)$. The 90° *excitation* pulse at $t = 0$ (red arrow) causes it to nutate very rapidly down into the **x-y** plane, during which it morphs into the new *transverse* magnetization, $\boldsymbol{m}_{xy}(0_+)$. The magnitude of $\boldsymbol{m}_z(t)$ is now zero. (b) The newly created *transverse* magnetization $\boldsymbol{m}_{xy}(0_+)$ precesses about the **z**-axis at the nominal Larmor frequency. There are millions of protons in the voxel, and various clusters of them experience fixed, *static* local fields that are slightly higher or lower than average. Some, therefore, precess slower and others faster, and their orientations fan out. In addition to this SFd-P effect, the random fields also produce the secular T2 effects, and these also contribute to the de-phasing. Meanwhile, because of T1 and the T1 component of T2 relaxation, $\boldsymbol{m}_z(t)$ regrows by a small amount, green arrow. (c) At $t = ½$TE, a 180° *echo-generating* pulse (twice as long or strong as its predecessor) is applied along the **y**-axis. This flips all the spin packets precessing in the **x-y** plane back across their 'starting' line, the **x**-axis, after which they continue to rotate in the same direction as before. Simultaneously, $\boldsymbol{m}_z(½\text{TE})$ nutates down thru 180°, ending up aligned in the opposite direction, along the *negative* **z**-axis. (d) Each packet will precess at its own fixed speed, so that the spin axes steadily close up together again. However, the SFd-P effects have reversed direction. The random de-phasing associated with the secular T2 relaxation continues to spread out the vectors according to their local T2 relaxation rates. $\boldsymbol{m}_z(t)$ grows upward, crossing the transverse plane, and keeps on going. (e) The spin packets remaining in the **x-y** plane complete their recombination exactly at TE, and together they create an echo signal, $s(t)$, which is detected by the pickup/receive coil/antenna. Its amplitude is diminished primarily by secular T2 now. (If there were no spin relaxation, the amplitude of the echo at TE would be the same as that of the initial FID-signal at $t = 0$.) SE thus removes from the picture the effect of small *static* ($T_{SFd\text{-}P}$) fields, which would otherwise degrade the image, and are biologically uninteresting anyway. (f) *Spin multiple-echo* is generated by following the 90° excitation pulse with n 180° re-focusing pulses, Equation (11.2b). The envelope of each separate echo falls off as $e^{-t/T2^*}$, while the echo peak amplitudes decline as $e^{-t/T2}$. This pulse train is typically 8 or 16 echoes long.

only at) $t \approx$ TE (Figure 11.3e). The protons cross the $\boldsymbol{x}$-axis in unison, orange arrow again, and $\boldsymbol{m_{xy}}(t \approx \mathrm{TE})$ Faraday-induces an MR *echo signal* voltage, $s(t \approx \mathrm{TE})$, in the receive antenna.

Because of T1 relaxation events from $t = 0$ on, some protons will have escaped the transverse magnetization and are now part of $\boldsymbol{m_z}(t)$. This leaves behind a diminished $\boldsymbol{m_{xy}}(t)$, so the pickup coil generates a weaker signal than it did at $t = 0$: $< s(\mathrm{TE}) < s(0)$.

(f) In the commonly seen *Spin Multiple-Echo* sequence, the single 90° excitation pulse is followed with more than one 180° re-focusing pulses (Figure 11.3f). (This is commonly referred to as *Fast Spin Echo* because of a savings in image acquisition time over conventional spin echo, as will be discussed in chapter 13). The resulting echoes appear at times $t = \boldsymbol{n} \times \mathrm{TE}$ for integer $\boldsymbol{n}$. If there were no decay processes at play, then the amplitudes of the echoes would be the same. MRI is more interesting and clinically useful than that, of course, and there *are* spin-relaxation processes ongoing. The amplitude of the signals will fall off exponentially with the characteristic time T2 of the material in the voxel, as

$$s(n \times \mathrm{TE}) = s(0)\mathrm{e}^{-n \times \mathrm{TE/T2}}. \quad (11.3)$$

EXERCISE 11.4 Explain the differences between Figures 11.3 and 10.2.

EXERCISE 11.5 $m_{xy}(t)$ *produces a signal near* $t = 0_+$ *before the spins have time to de-phase much. What's going on?*

EXERCISE 11.6 How much of the above is different for SE on a linear phantom (Figure 11.2b)?

EXERCISE 11.7 How can the same 180° pulse flip $m_{xy}(t)$ *over in the x-y plane, and* $m_z(t)$ *through it?*

11.2 Put Another Way...

There is a somewhat different but complementary way to view SE, and we shall describe it now for a *train* of *multiple spin-echo* RF pulse sequences (*not* spin multi-echo!) in a single voxel. Three repetitions in this multi-SE sequence are labeled n–1, n, and n+1 in Figure 11.4a, which are of the form of Figure 9.6. The green (upper) and pale ochre regions highlight what is going on with $m_z(t)$ and $m_{xy}(t)$, respectively. The two heavy, dashed vertical arrows indicate what happens at the 90° excitation pulse at $t = 0$: $m_z(t)$ free-falls to 0, and $m_{xy}(t)$, of exactly the same magnitude, comes into existence. The horizontal dashed line indicates m_0; the dotted one just below shows, for a repetition of multiple, sequential SE runs, the steady-state maximum values, $\boldsymbol{m}_{\boldsymbol{xy}\text{-max}}$ and $\boldsymbol{m}_{\boldsymbol{z}\text{-max}}$ achieved by the transverse and longitudinal magnetizations when TR is not much longer than the tissue T1 value in the voxel.

EXERCISE 11.8 Produce the most succinct, bare-bones, but still complete caption possible for Figure 11.4. See Figure 9.5.

EXERCISE 11.9 Does $m_{xy\text{-}max} = m_{z\text{-}max}$*?*

The earliest segment shown is the n–1st. Following a previous disturbance, the recovering spin population (brown line) will have reacquired the longitudinal magnetization $\boldsymbol{m_z}(0_-) = \boldsymbol{m}_{\boldsymbol{z}\text{-max}}$ by $t = 0_-$, just before another 90° pulse. If the protons have been resting undisturbed for a long time (not the situation here), $\boldsymbol{m_z}(0_-) \sim \boldsymbol{m_0}$. Otherwise, as in Figure 11.4a, $\boldsymbol{m_z}(0_-) < \boldsymbol{m_0}$.

At $t = 0$, a 90° excitation pulse initiates repetition $\boldsymbol{n}$ by driving the extant purely *longitudinal* $\boldsymbol{m_z}(t{=}0_-) = \boldsymbol{m_z}(t)_{\max}$ down into the $\boldsymbol{x}$-$\boldsymbol{y}$ plane—the *down*ward-pointing heavily dashed black arrow at the end of the green area. As a result, and at the same time, $\boldsymbol{m_z}(t)$ instantaneously transforms into, and reappears as, the purely *transverse* magnetization, $\boldsymbol{m_{xy}}(0_+)$, the *up*ward dashed arrow at the end of the pale ochre region. The spin packets then begin precessing there at the nominal Larmor frequency, and again

$$|\boldsymbol{m_{xy}}(0_+)| = |\boldsymbol{m_z}(0_-)|. \quad (11.1')$$

EXERCISE 11.10 What happens to $m_{xy}(t)$ *at* $t = 0$*?*

After the 90° pulse, and during the subsequent fanning out of the spins via SFd-P and secular T2 relaxation (Figures 11.3b and 11.4b), the magnitude of $\boldsymbol{m_{xy}}(t)$ falls exponentially from $|\boldsymbol{m_{xy}}(0_+)|$ as

$$|\boldsymbol{m_{xy}}(t)| = |\boldsymbol{m_{xy}}(0_+)| \times \mathrm{e}^{-t/\mathrm{T2}^*}, \quad (11.4a)$$

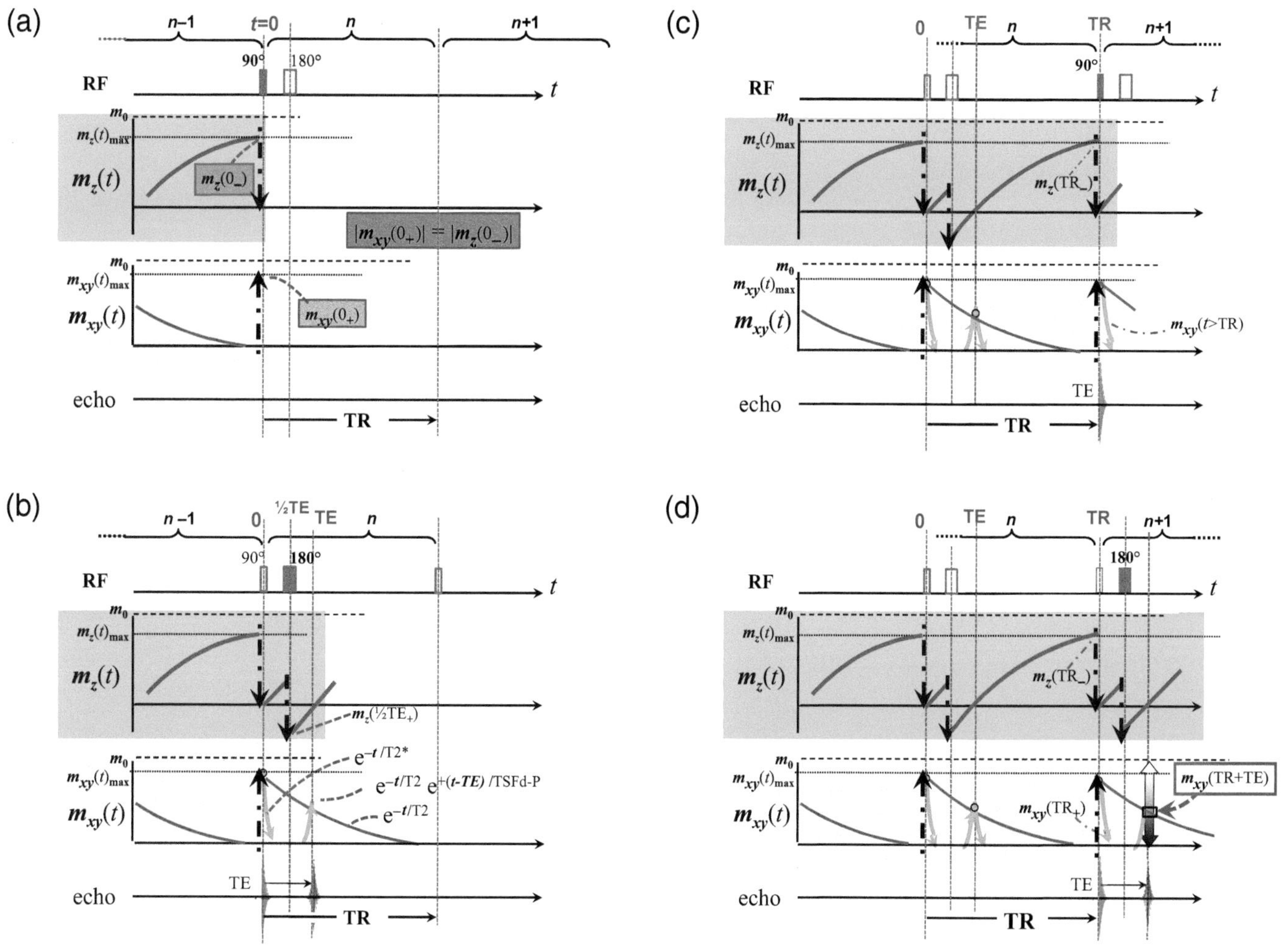

Figure 11.4 See Exercise 11.8.

indicated by the downward-pointing light blue arrow. As defined in Equation (10.4) and repeated here,

$$1/T2^* \equiv 1/T2 + 1/T_{SFd\text{-}P}\,. \qquad (10.4)$$

The rate 1/T2* is the spin-spin relaxation rate speeded up by the Static Field de-Phasing process of Equations (10.3) and, as such, T2* tends to be much shorter than T2. We can rewrite Equation (11.4a) explicitly with both terms, showing that the downward blue curve in Figure 11.4b is the product of two exponentials:

$$|\boldsymbol{m}_{xy}(t)| = |\boldsymbol{m}_{xy}(0_+)| \times e^{-t/\mathbf{T2}} \times e^{-t/\mathbf{T}_{SFd\text{-}P}}. \qquad (11.4b)$$

The purple curve on the plot of transverse magnetization $\boldsymbol{m}_{xy}(t)$ in Figure 11.4b shows the T2 relaxation exponential, which is much slower than the static field relaxation.

At the same time, $\boldsymbol{m}_z(t)$ begins to regrow along the *z*-axis from $\boldsymbol{m}_z(0_+) = 0$, as with Equation (11.2), the short solid brown *diagonal* line springing up from the abscissa at $t = 0$. At $t = ½$TE, however, everything turns upside-down: a 180° re-phasing pulse is applied, and this has two important and linked consequences.

First, the 180° pulse flips the spins precessing in the transverse plane back over across the $\boldsymbol{x}$-axis in such a manner that each spin packet continues precessing in the original direction, but with the new phase that corresponds to the mirror orientation on the other side of the 'starting' line, as in Figure 11.3c. This reverses the previously ongoing de-phasing process of the spin packets within a voxel, and now they start coming together again.

The flipping of spins with the 180° pulse directly reverses the decay arising from the Static Field de-Phasing process. T2 relaxation, however, continues to occur. A description of the transverse magnetization value would be similar to Equation (11.4b), except that the negative sign on the SFd-P term would now be positive, indicating regrowth. The transverse magnetization for $t > ½$TE may be viewed as its value at time $t = ½$TE, with continued T2 relaxation but with exponential regrowth of the SFd-P component.

Immediately after $t = ½$TE, the transverse magnetization would be

$$|\boldsymbol{m}_{xy}(t > \tfrac{1}{2}\mathrm{TE})| = |\boldsymbol{m}_{xy}(\tfrac{1}{2}\mathrm{TE}_-)| \times e^{-(t-\frac{1}{2}\mathrm{TE})/\mathrm{T2}} \times e^{+(t-\frac{1}{2}\mathrm{TE})/\mathrm{T}_{\mathrm{SFd\text{-}P}}}.$$

Here, the 180-degree pulse 'resets the clock' on the exponential terms, and subsequent exponential decay refers to a time after $t = ½$TE. It follows from Equation (11.4b) that

$$\begin{aligned}|\boldsymbol{m}_{xy}(t > \tfrac{1}{2}\mathrm{TE})| &= \left[|\boldsymbol{m}_{xy}(0_+)| \times e^{-\frac{1}{2}\mathrm{TE})/\mathrm{T2}}\right] \\ &\quad \times e^{-\frac{1}{2}\mathrm{TE})/\mathrm{T}_{\mathrm{SFd\text{-}P}}} \times e^{-(t-\frac{1}{2}\mathrm{TE})/\mathrm{T2}} \times e^{+(t-\frac{1}{2}\mathrm{TE})/\mathrm{T}_{\mathrm{SFd\text{-}P}}} \\ &= |\boldsymbol{m}_{xy}(0_+)| \times e^{-t/\mathbf{T2}} \times e^{(t-\mathrm{TE})/\mathrm{T}_{\mathrm{SFd\text{-}P}}}.\end{aligned} \quad (11.4c)$$

The last term grows as t increases beyond ½TE as indicated by the *up*turned blue arrow shortly before TE. By the time $t =$ TE, those spins remaining in the transverse plane (i.e., those that have not undergone a T1-type relaxation event and left it) will have fully coalesced, and over that one brief instant, they together generate an echo. The SFd-P decay will have been counteracted, and the transverse magnetization at $t =$ TE is equal to the transverse magnetization at $t = 0_+$ with the added T2 relaxation only. In other words, the last exponential factor in Equation (11.4c) goes to 1.

Second and simultaneously, the newly resuscitated small z-component of the magnetization (which did regrow between $t = 0$ and ½TE) is forced by the same 180° pulse to flip over, the green arrow in Figure 11.3c and the second downward heavy black arrow in the green area of Figure 11.4b. The longitudinal magnetization ends up aligned in the opposite, 'wrong' direction, along the *minus* z-axis:

$$\boldsymbol{m}_z(\tfrac{1}{2}\mathrm{TE}_+) = -\boldsymbol{m}_z(\tfrac{1}{2}\mathrm{TE}_-). \quad (11.5)$$

$\boldsymbol{m}_z(t)$ now starts anew to grow upward exponentially from *below* the time axis, the second short diagonal solid brown line, to the right of ½TE. As indicated in Equation (11.5), what we find just after the 180° flip is the negative of the situation that existed just prior to it. What extends upward along the z-axis after ½TE is similar to but not the same as Equation (11.2). All this starts to get a little bit hairy, so hang on.

The immediate objective is to represent the *long brown curve* that arches up toward $\boldsymbol{m}_z$(TR) in Figure 11.4c and to find its *value at* $t =$ TR. There are several ways to do this. One is to ignore what goes on before TE, in effect, and start counting at TE. From Equation (11.2),

$$\boldsymbol{m}_z(t > \mathrm{TE}) = \boldsymbol{m}_z(0_-) \times \left[1 - e^{-(t-\mathrm{TE})/\mathrm{T1}}\right], \quad (11.6a)$$

hence at TR

$$\boldsymbol{m}_z(\mathrm{TR}) = \boldsymbol{m}_z(0_-) \times \left[1 - e^{-(\mathbf{TR}-\mathrm{TE})/\mathrm{T1}}\right]. \quad (11.6b)$$

When TE ≪ TR, which is not uncommon but not always strictly true, this expression reduces to the now-familiar form:

$$\boldsymbol{m}_z(\mathrm{TR}) \approx \boldsymbol{m}_z(0_-) \times \left[1 - e^{-\mathbf{TR}/\mathrm{T1}}\right]. \quad (11.6c)$$

Alternatively, at $t = ½$TE the curve can be shifted upward by the amount $2 \times \boldsymbol{m}_z$(½TE), so that a newly created single continuous function begins at $t = 0$ and grows monotonically up to TR. Then, at the end, subtract $2 \times \boldsymbol{m}_z$(½TE_) from that value to shift everything back down again:

$$\boldsymbol{m}_z(\geq \tfrac{1}{2}\mathrm{TE}) = \boldsymbol{m}_z(0_-) \times \left[1 - e^{-t/\mathrm{T1}}\right] - 2 \times \boldsymbol{m}_z(\tfrac{1}{2}\mathrm{TE}), \quad (11.7a)$$

$$\boldsymbol{m}_z(\mathrm{TR}) = \boldsymbol{m}_z(0_-) \times \left[1 - e^{-\mathrm{TR}/\mathrm{T1}}\right] - 2 \times \boldsymbol{m}_z(\tfrac{1}{2}\mathrm{TE}). \quad (11.7b)$$

This expression is similar to Equation (11.6c), but with a correction factor. If you reference the brown curve describing $\boldsymbol{m}_z$ in Figure 11.4b, the second term in Equation (11.7b) corresponds to the regrowth of longitudinal magnetization around $t = ½$TE. For short values of TE, the amount of regrowth is minimal, going to zero as TE goes to zero. In this case, Equation (11.7b) reduces to Equation (11.6c).

The derivations of Equations (11.6c) and (11.7b) were based on heuristic, intuitive descriptions of the longitudinal magnetization regrowth. A better approach is to build directly upon a rigorous solution for the longitudinal component of the Bloch equations, seen earlier as Equations (9.2b) and repeated here:

$$[\boldsymbol{m}_z(t) - \boldsymbol{m}_0] = [\boldsymbol{m}_z(0) - \boldsymbol{m}_0]e^{-t/\mathbf{T1}}. \quad (9.2b)$$

Moving $\boldsymbol{m}_0$ to the right,

$$\boldsymbol{m}_z(t) = \boldsymbol{m}_0 + \left[\boldsymbol{m}_z(0_+) - \boldsymbol{m}_0\right] \mathrm{e}^{-t/\mathbf{T1}} \qquad (11.8a)$$
$$= \boldsymbol{m}_0 \times (1 - \mathrm{e}^{-t/\mathbf{T1}}) + \boldsymbol{m}_z\ (0_+)\ \mathrm{e}^{-t/\mathbf{T1}},$$

describes the regrowth of the longitudinal magnetization for $t>0$, i.e., for $\boldsymbol{m}_z(t>0)$, after a pulse occurs at t = 0 (Figure 11.4b). $\boldsymbol{m}_0$ is the long-term equilibrium value achieved if the system is left alone long enough for the longitudinal magnetization to relax fully.

11.3 And More Realistically....

In practice, systems of interest often do not have time to re-equilibrate between repetitions, and the magnetization does not have time to recover completely to $\boldsymbol{m}_0$. The above formalism must therefore be modified so as to reflect the reference time at which a new pulse is applied—in effect, resetting the relaxation process, indicated by the rallying of $m_z(t)$ in Figures 11.4b and 11.4c after t = ½TE. In other words, $m_z(t)$ relaxes upward after t = 0, and then continues doing so after t = ½TE, but with a new starting value.

In the trivial case in which the reference time is set at t = 0, Equation (11.8a) can be re-written in the silly but suggestive manner:

$$\boldsymbol{m}_z(t > 0) = \boldsymbol{m}_0 \times (1 - \mathrm{e}^{-(t-0)/\mathrm{T1}})$$
$$+ \boldsymbol{m}_z(0_+) \times \mathrm{e}^{-(t-0)/\mathbf{T1}}.$$

Near t = 0, this reduces to $\boldsymbol{m}_z(t=0_+)$, as expected.

For a more general future reference time, 0 is replaced with $t = t_{fut}$ at which the pulse occurs, resulting in a Bloch equation solution for $\boldsymbol{m}_z(t)$:

$$\boldsymbol{m}_z(t > t_{fut}) = \boldsymbol{m}_0 \times (1 - \mathrm{e}^{-(t-t_{fut})/\mathbf{T1}}) \qquad (11.8b)$$
$$+ \boldsymbol{m}_z\ (t_{fut+}) \times \mathrm{e}^{-(t-t_{fut})/\mathbf{T1}}.$$

To be more specific, suppose a 90° pulse is applied at t = 0, to be followed by a 180° pulse at t = ½TE. After the 90° excitation but before the 180°, the longitudinal magnetization, from Equation (11.8a), is

$$\boldsymbol{m}_z(t > 0) = \boldsymbol{m}_0 \times (1 - \mathrm{e}^{-t/\mathbf{T1}}) + \boldsymbol{m}_z\ (0_+) \times \mathrm{e}^{-t/\mathbf{T1}}. \qquad (11.9a)$$

If there had been no longitudinal magnetization present when the 90° pulse was applied, i.e., $\boldsymbol{m}_z(0_+) = 0$, then immediately before the 180° pulse this equation would reduce to

$$\boldsymbol{m}_z(\tfrac{1}{2}\mathrm{TE}_-) = \boldsymbol{m}_0 \times (1 - \mathrm{e}^{-\frac{1}{2}\mathbf{TE}/\mathbf{T1}}). \qquad (11.9b)$$

Now comes the 180° pulse at t = ½TE. After that, from Equation (11.8b),

$$\boldsymbol{m}_z(t > \tfrac{1}{2}\mathrm{TE}) = \boldsymbol{m}_0 \times (1 - \mathrm{e}^{-(t-\frac{1}{2}\mathrm{TE})/\mathbf{T1}}) \qquad (11.9c)$$
$$+ \boldsymbol{m}_z(\tfrac{1}{2}\mathrm{TE}_+) \times \mathrm{e}^{-(t-\frac{1}{2}\mathrm{TE})/\mathbf{T1}}).$$

The 180° pulse inverts the longitudinal magnetization at t = ½TE and, with a little help from Equation (11.5) and (11.9b),

$$\boldsymbol{m}_z(\tfrac{1}{2}\mathrm{TE}_+) = -\boldsymbol{m}_z(\tfrac{1}{2}\mathrm{TE}_-) \qquad (11.9d)$$
$$= -\boldsymbol{m}_0 \times (1 - \mathrm{e}^{-\frac{1}{2}\mathrm{TE}/\mathbf{T1}}),$$

So, the longitudinal magnetization after the second pulse takes the final form:

$$\boldsymbol{m}_z(t > \tfrac{1}{2}\mathrm{TE}) = \boldsymbol{m}_0 \times \left[(1 - \mathrm{e}^{-(t-\frac{1}{2}\mathrm{TE})/\mathrm{T1}})\right]$$
$$- \boldsymbol{m}_0 \times \left[(1 - \mathrm{e}^{-\frac{1}{2}\mathrm{TE})/\mathrm{T1}})\right] \qquad (11.9e)$$
$$\times \left[\mathrm{e}^{-(t-\frac{1}{2}\mathrm{TE})/\mathrm{T1}})\right].$$

This achieves a peak magnitude, at t = TR, of

$$m_z(\mathrm{TR})_{\max} = \boldsymbol{m}_0 \times (1 - 2\mathrm{e}^{-(\mathrm{TR}-\frac{1}{2}\mathrm{TE})/\mathrm{T1}} + \mathrm{e}^{-\mathrm{TR/T1}}) \qquad (11.9f)$$

(which most likely will be somewhat less than $\boldsymbol{m}_0$). This expression is similar to but more rigorous than Equation (11.6b).

To summarize the plan: a 90° pulse is applied at t = 0, a 180° pulse ½TE later, another 90° pulse at t = TR, and a 180° pulse a time ½TE after that, and so on. The effect of this is that, after many TR periods, the longitudinal magnetization at the end of any period will try to relax back toward the original value $\boldsymbol{m}_0$, but will actually reach only to $\boldsymbol{m}_{z\text{-max}}$. For most clinical applications of a spin-echo sequence, TR is somewhat longer than TE, which means that the middle term in Equation (11.9f) disappears, leading back to the approximate expression at t = TR that we saw earlier as Equation (11.6c).

$$\boldsymbol{m}_z(\mathrm{TR}) \approx \boldsymbol{m}_0 \times (1 - \mathrm{e}^{-\mathbf{TR}/\mathrm{T1}}) \sim \boldsymbol{m}_{z\text{-max}}. \qquad (11.9g)$$

At this point, the second 90° excitation pulse initiates the $\boldsymbol{n+1}^{\text{th}}$ repetition. It swings all the spin packets currently aligned along the z-axis down into the transverse plane, and the actions of Figures 11.4a and 11.4b are repeated. The magnitude of the newly precessing $\boldsymbol{m}_{xy}(\mathrm{TR}_+)$ is the same as that of $\boldsymbol{m}_z(t)$ before TR. This transformation can be described by an expression equivalent to Equation (11.1), but shifted TR forward:

$$|\boldsymbol{m}_{xy}(\mathrm{TR}_+)| = |\boldsymbol{m}_z(\mathrm{TR}_-)|. \qquad (11.10a)$$

The newly created $\boldsymbol{m}_{xy}(t > \mathrm{TR})$ is precessing in the transverse plane, but undergoes de-phasing at the rate 1/T2* (much faster than 1/T1 or 1/T2):

$$m_{xy}(t > \mathrm{TR}) = m_{xy}(\mathrm{TR}) \times (\mathrm{e}^{-(t-\mathrm{TR})/\mathrm{T2}^*}), \quad (11.10b)$$

as indicated by the *descending* blue curve to the right in the pale ochre block of Figure 11.4c.

$$t = \mathrm{TR} + \tfrac{1}{2}\mathrm{TE} \qquad (11.11a)$$

causes $\boldsymbol{m}_{xy}(t)$ to begin precessing in the opposite direction in the $\boldsymbol{x}$-$\boldsymbol{y}$-plane, and to start to *re*-phase (Figure 11.4d). Exactly at TR + TE, all the spin packets that comprise $\boldsymbol{m}_{xy}(t)$ have briefly come back into phase with one another and are described similarly to Equation (11.4b):

$$|\boldsymbol{m}_{xy}(t = \mathrm{TR+TE})| = |\boldsymbol{m}_{xy}(\mathrm{TR}_+)|\, \mathrm{e}^{-(t-\mathrm{TR})/\mathrm{T2}}. \qquad (11.11b)$$

which, with Equation (11.11a), becomes

$$= |\boldsymbol{m}_{xy}(\mathrm{TR}_+)|\, \mathrm{e}^{-\mathrm{TE}/\mathrm{T2}}. \qquad (11.11c)$$

This induces the echo signal in the receiver antenna,

$$s(t_{readout}) \sim \mathrm{PD}(x) \times (\mathrm{e}^{-2\pi i \nu_L t}) \times \left[(1 - \mathrm{e}^{-\mathrm{TR}/\mathrm{T1}}) \times (\mathrm{e}^{-\mathrm{TE}/\mathrm{T2}})\right], \qquad (11.11d)$$

which is included in Table 11.2. PD, T1, T2, and the Larmor frequency, ν_L are all functions of the voxel position, $\boldsymbol{x}$, of course, although not explicitly shown as such here. After we've excited tissues several times and reached a steady-state, $t_{readout}$ occurs TE after any 180° pulse.

The above critically important equation describes and codifies a fundamental and essential concept in standard spin-echo MRI. It also makes clear that at the time that an echo appears, $s(x,\mathrm{TE})$ depends not only on the inherent values of the parameters T1, T2, and PD of the tissue being examined in the voxel at $\boldsymbol{x}$, but also on the values of the operational parameters, TR and TE, selected for the study. The choice of these last two parameters determines whether an SE image will be T1-*w*, emphasizing spatial variations in T1, or T2-*w* or PD-*w*, as we shall see.

Table 11.2 For simple spin-echo imaging, tissue parameters and settings for the standard operational parameters, TR and TE, can be adjusted to select for display of specific tissue characteristics such as T1, T2, or PD. The equation is widely useful with spin-echo, and Equation (11.11d) and this table demonstrate how to select T1-weighting, for example: a choice of TE ≪ T2 reduces $\mathrm{e}^{-\mathrm{TE}/\mathrm{T2}(x)}$ to near unity, so that the value of T2 has no impact on signal strength. Only T1 and, to a lesser extent PD, remain to do that. The color and gray-shade coding convention is that for those voxels where $m_{xy}(\mathrm{TE})$ is large, the corresponding pixel on display is bright, such as with the image to the left in Table 11.3.

Tissue Parameters	Operator Settings	
T1	TR	← time since previous 90° pulse
T2	TE	
PD		
flow (blood, water),		

$$s(t \sim \mathrm{TE}) \sim \mathrm{PD}\,[1 - \mathrm{e}^{-(\mathrm{TR})/\mathrm{T1}}]\, \mathrm{e}^{-\mathrm{TE}/\mathrm{T2}}$$

proton density — prior regrowth along *z*-axis — current dephasing in *x*-*y* plane

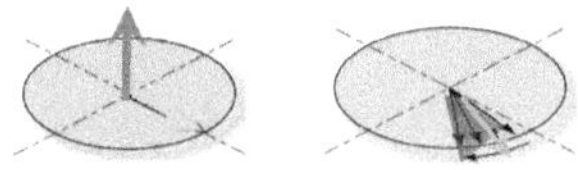

EXERCISE 11.11 What might be the meaning of the hollow vertical arrow in Figure 11.4d?

11.4 The Spin-Echo RF Pulse Sequence Eliminates the $T_{SFd\text{-}P}$ Effect

You might expect the *Static Field de-Phasing* effect of Equations (10.4) to alter everything, seeing how it affects the local fields at all the protons. SFd-P occurs because of two kinds of *static* irregularities in the local magnetic field. One arises from sub-voxel-sized, unchanging local inhomogeneities, ΔB_0, in the external main field, even when the gradients are off, that are created by imperfections in the construction of the high-field external magnet. The other comes from variations in the tissue susceptibilities, $\Delta\chi$, in nearby tissues. Surely these two also contribute to the de-phasing, and act only to degrade the image! The first of these two statements is true, but the second is not necessarily so.

Amazingly, these SFd-P problems are removed entirely from an MR image by the SE sequence! Equation (10.4b) indicates that the magnet imperfections and tissue susceptibilities together create a small field of magnitude $[\Delta B_0 + (\kappa'/\kappa)\,\Delta\chi]$, but that field is *fixed* in time and simply adds a time-invariant, small but constant amount to $\boldsymbol{B}_0$. This may shift the Larmor frequency slightly from place to place within a voxel, leading to de-phasing over time if nothing else is done to correct matters, but it should have no other effect on the events of Figures 11.3 and 11.4 for the SE sequence. In particular, just like $\boldsymbol{B}_0$ itself, it causes no enduring de-phasing of $\boldsymbol{m}_{xy}(t)$ by the time that the echo is being created at t = TE. With spin-echo, rather, any excess (induced by SFd-P) in the angles that the spins precess through between times 0 and ½TE is totally *un*done and reversed between ½TE and TE, so the SFd-P process does nothing permanent to corrupt an image. So at the time near the center of the echo, the loss of phase coherence caused by the static field de-phasing effect is completely eliminated from the echo signal, as if no SFd-P had ever occurred! The same cannot be said for the gradient-echo pulse sequence, addressed below.

Meanwhile, magnetic noise is causing the local proton fields to fluctuate continuously and *randomly*, as opposed to being static. Therefore, these spin relaxation events can*not* be reversed and undone. The noise-induced fanning out of the individual contributors to the net $\boldsymbol{m}_{xy}(t)$ vector within a voxel, and the resulting loss of echo signal are, therefore, attributable entirely to T2-type relaxation—and not at all to imperfect-magnet or susceptibility $T_{SFd\text{-}P}$ effects!

EXERCISE 11.12 Why does the half-echo-like signal following a 90° pulse fall off with characteristic time T2 in Figure 11.4?*

Once again: spin-echo displays the wonderful capacity to reverse and cancel out that part, but only that part, of the spin de-phasing that is caused by the frozen-in micro-inhomogeneities of the external magnet's field and by minuscule static variations in tissue susceptibility. T1- and T2-weighted SE images produce studies that emphasize spatial variations in those two clinically meaningful tissue parameters. At the conclusion of an SE run, the echo RF signal is detected by the pick-up coil, amplified, sampled, digitized, and sent to the computer for Fourier untangling, as was discussed in chapter 8.

The gradient-echo sequence, on the other hand, maps out variations in T1 and T2*, the latter of which may be contaminated with the clinically misleading instrumental $T_{SFd\text{-}P}$ bias (but it also might provide more sensitivity to certain pathologies such as micro-hemorrhage).

11.5 Selecting TE and TR for T1-Weighted SE

Multiple application of 180° refocusing pulses to a voxel following a single 90° excitation pulse will give rise to a sequence of echoes (Figure 11.3f). This approach is useful for 'accelerating' (i.e., speeding up the imaging process), as discussed in chapter 13. Alternatively, one can repeat a complete 90°–½TE–180°–TE sequence a number of times. An ultimate goal of either process to produce images *weighted* to emphasize spatial variations in T1, T2, or protons density. The nature of the information obtained in spin-echo MRI is a function primarily of the *echo time*, TE, and *repetition time*, TR, in the pulse sequence.

Let us re-visit a phantom like that of Figure 9.7, which contains only two filled voxels, but with one change: the one at the center contains *white matter* of the brain (comprising vast numbers of fatty, myelinated axons), while that which holds CSF is the immediately adjacent voxel at x = 0.1 cm. There is an x-gradient on during readout, and that will enable us to distinguish the echo signals from the two. For the moment, we shall consider each voxel one at a time.

A good strategy to create T1 weighting consists of two independent decisions. The operator selects a TR value that is comparable to T1 values for the materials of interest (see Table 9.1; white matter has T1 of ~700 ms at 1.5 T, while that of CSF is 2500 ms). A TR of 500 ms, for example, is long enough for $m_z(t)$ in white matter to regrow fairly far above $m_z(0) = 0$, but they will not yet have come close to merging together by t = TR. On the other hand, CSF will experience much less regrowth than white matter. Their equilibrium values, $\boldsymbol{m}_0(x{=}0)$ and $\boldsymbol{m}_0$(at 0.1 mm), are determined nearly entirely by their respective proton densities, Equation (4.4c.) A comparison of Figures 11.5a and 11.5b reveals that this setting for TR enhances the T1 contrast between the voxels, and by the time the two $m_z(x,t)$ approach their peaks, beyond t = TR, the curves will be widely separated (Figure 11.5c), revealing a significant T1 contrast between them.

Also, the radiographer sets TE to a value (e.g., 20 ms) that is short relative to the T2 values of both white matter (T2 = 60 ms) and CSF (200 ms), so that neither voxel will have sufficient time to undergo much T2 relaxation. By the time the system reaches mid-echo, $m_{xy}(t)$ will have shrunk from m_{xy}(TR) by only the small factor ($e^{-TE/T2}$) ~ e^{-0} = 1, and Equation (11.4) reduces approximately to

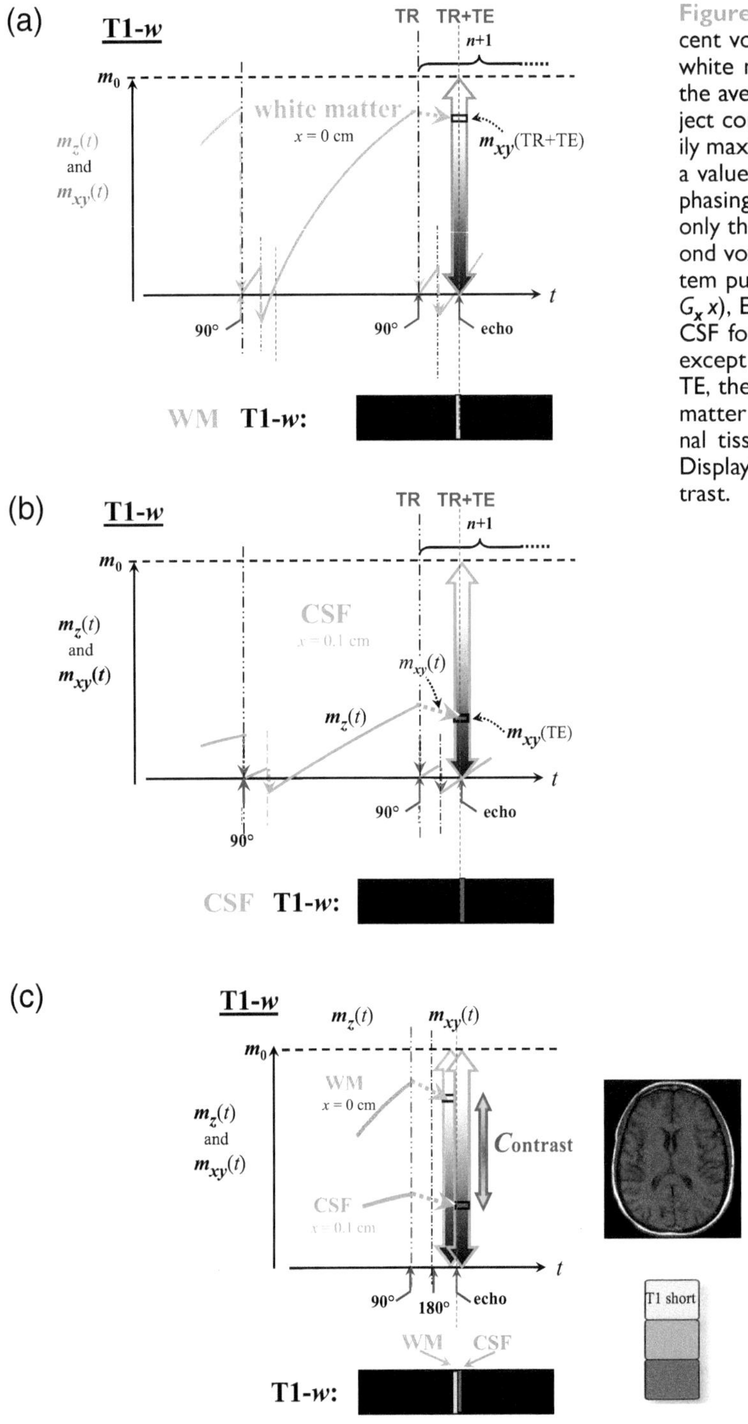

Figure 11.5 Display of the difference in T1 between two adjacent voxels in a T1-*w* study. (a) The voxel at x = 0 contains brain white matter. The T1 emphasis is accomplished with a TR near the average T1 values of the tissues of interest, to maximize subject contrast between them (even though this does not necessarily maximize signal strength). Also, a short TE will force $e^{-TE/T2}$ to a value of about unity, minimizing the effects of T2-type spin dephasing in the *x-y* plane. What remains is an image that reflects only the T1 (and PD, to some extent) for the voxels. (b) The second voxel in the phantom at x = 0.1 cm is filled with CSF. The system puts out RF pulses resonant for the local field, $B_z(x) = (B_0 + G_x x)$, Equation (3.1). Not surprisingly, the behavior of spins in the CSF for a T1-*w* study closely mirrors that for the white matter, except that its T1 value is one second, four times longer. Just after TE, the CSF signal at x = 0.1 cm will be far weaker than the white matter because of its much longer T1; by convention, strong-signal tissues (white matter, here) are made to appear bright. (c) Displaying both voxels together illustrates the origin of T1-*w* contrast.

$$s(t \sim \mathrm{TR} + \mathrm{TE}) \;\sim\; \mathrm{PD}(x)(\mathrm{e}^{-2\pi i\,\nu_L t}) \times (1 - \mathrm{e}^{-\mathrm{TR}/\mathrm{T1}}) \tag{11.12}$$

That is, while there is ample time for some T1-relaxation to take place, there is practically none for the much faster T2 events, and the effect of T2 is thereby largely removed from the picture (Table 11.2).

This combination of TR and TE settings yields an image that is heavily weighted toward T1 spatial variations, so that $s(x,\mathrm{TE})$ becomes a function only of T1, with some inevitable PD influence but virtually no impact of TE relaxation on the image. So much for the short-TE requirement in Table 11.3. It is with these two selection decisions, for TR and TE, that the operator is calling for a T1-weighting for the image, the left-hand column in Table 11.4.

Within the system's computer there is a pre-set *grayscale* that associates any value of $m_{xy}(\mathrm{TE})$ with a shade of gray on the display in a one-to-one fashion. The standard *grayscale convention* is that a voxel with a strong signal, a *large* value of $s(t = \mathrm{TE})$, shows up as a *bright* pixel in an MR image. The amplitude of $s(x,\mathrm{TE})$ decreases with longer T1, so that a *bright* pixel in a T1-*w* image corresponds to *short* T1.

Table 11.3 Representative TR and TE settings to generate T1-*w*, T2-*w*, and PD-*w* SE images, respectively, and standard clinical meaning of bright or dark voxels. At the top, typical scans with the three weightings.

	T1-*w*	T2-*w*	PD-*w*

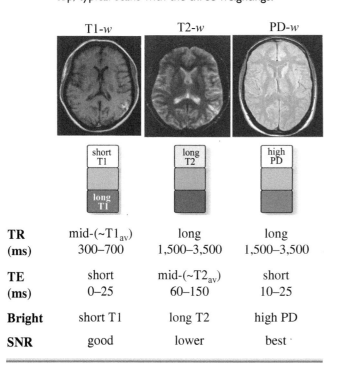

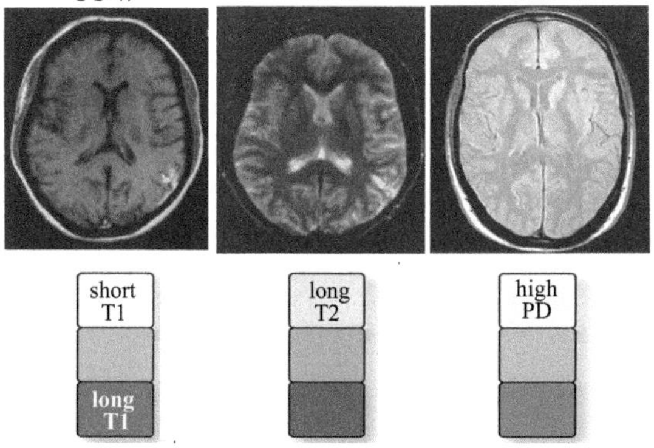

	T1-*w*	T2-*w*	PD-*w*
TR (ms)	mid-(~$\mathrm{T1_{av}}$) 300–700	long 1,500–3,500	long 1,500–3,500
TE (ms)	short 0–25	mid-(~$\mathrm{T2_{av}}$) 60–150	short 10–25
Bright	short T1	long T2	high PD
SNR	good	lower	best

EXERCISE 11.13 Show that the assignment of pixel brightness for T1-w is in accord with the grayscale convention.

11.6 Optimizing the T1 Contrast for a Pair of Nearby Voxels

Suppose there is need to create a significant difference in the amplitudes of the signals from the two voxels, reflecting the spatial distribution of their T1 values. To achieve high contrast in T1-weighted imaging, the operator chooses TE to be *short* relative to their T2 values and a TR that is comparable to *average* of the T1 values of the tissues in the region. Now to explain this last claim....

Guided by Equations (3.5a) and (3.7a), one might be inclined to define the MR *contrast* between adjacent voxels, and their pixels, that contain white matter and CSF, respectively, as

$$\begin{aligned} C_{\mathrm{ontrast}} \;\equiv\; & \left[m_{xy\text{-WM}}(\mathrm{TE}) - m_{xy\text{-CSF}}(\mathrm{TE})\right] \\ & /\left[m_{xy\text{-WM}}(\mathrm{TE}) + m_{xy\text{-CSF}}(\mathrm{TE})\right] \\ \sim\; & \left[s_{\mathrm{WM}}(\mathrm{TE}) - s\mathrm{CSF}(\mathrm{TE})\right] \\ & /\left[s_{\mathrm{WM}}(\mathrm{TE}) + s_{\mathrm{CSF}}(\mathrm{TE})\right]. \end{aligned} \tag{11.13}$$

Most importantly, for T1-*w* SE imaging (Figure 11.5c) the regimen of a short TE and longish TR comes close to *maximizing the contrast* between pixels. Although we have not explicitly demonstrated that the short TE, median TR regimen brings C_{ontrast} to its peak value, hopefully the idea is plausibly self-evident.

To summarize: in T1-*w* imaging, the three principal objectives are to achieve enough signal strength for an adequate SNR, to emphasize the difference in T1 between voxels by selecting a TR midway among the tissue T1 values, and to eliminate the impact of T2 on the image by selecting a very short TE (Table 11.4).

T1-*w*

Enough SNR
Emphasize T1 (TR ≈ T1)
Eliminate T2 (short TE)

Regardless of how good the contrast and resolution are, an image can be clinically worthless if the amount

Table 11.4 T1-*w*, from selecting an average TR and a short TE for the equation

T1-*w*

TR ~ $T1_{mean}$: maximize T1 contrast — TE short: remove T2 effect

$$s(t{\sim}TE) \sim PD\,[1 - e^{-TR/T1_{av}}]\,\cancel{e^{-TE/T2}}\;1$$

prior regrowth along *z*-axis — current de-phasing in *x*-*y* plane

Short-T1 tissues bright

Table 11.5 TR and TE choices for T2-*w*

T2-*w*

Long TR: eliminate T1 impact — TE ~ $T2_{mean}$: maximize T2 contrast

$$s(t{\sim}TE) \sim PD\,[1 - \cancel{e^{-TR/T1}}]\,e^{-TE/T2_{av}}$$

prior regrowth along *z*-axis — current de-phasing in *x*-*y* plane

Long-T2 tissues bright

Table 11.6 A sequence to emphasize PD(***x***), displaying minimal relaxation effects of either kind. A long TR and short TE to minimize T1 and T2 effects, leaving PD-*w*.

PD-*w*

Long TR: eliminate T1 impact — Short TE: eliminate T2 impact

$$s(t{\sim}TE) \sim PD\,(1 - \cancel{e^{-TR/T1}})\,\cancel{e^{-TE/T2_{av}}}$$

prior regrowth along *z*-axis — current de-phasing in *x*-*y* plane

of noise present overwhelms them. Noise is where clinical studies go to die. The SNR in a region of an image compares signal strength there with average background noise, Equation (3.10). It is defined as the ratio of signal strength to noise power, and it is commonly expressed in decibels. Equation (8.4) is a simple, fairly intuitive formula that, along with variations on it, is widely applicable in MRI theory.

11.7 T2-Weighting and PD-*w* for SE

The argument is much the same for T2-*w* images, just turned around a little, Table 11.5 and Figure 11.6. Here, apart from good SNR, the objectives are to emphasize the difference in T2 between voxels and to eliminate nearly all signs of T1 in the image. Because we now set TR quite *long* relative to the T1s of the two voxel materials, their magnetizations have sufficient time to reach a maximum near their respective m_0 values, which differ only by the amounts of their proton densities.

The right-hand side of Figure 11.6 indicates that as the operator lengthens TE, two important but counterbalancing processes occur. First, T2 contrast is enhanced, up to a point, because of the difference in T2(x) values and, therefore, in factors of $e^{-TE/T2(x)}$ for the two. But at the same time, signal strength decreases over time for each, again because of the $e^{-TE/T2(x)}$ terms. Selecting an optimal TE once again presents a

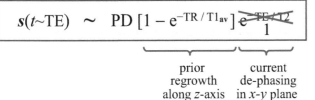

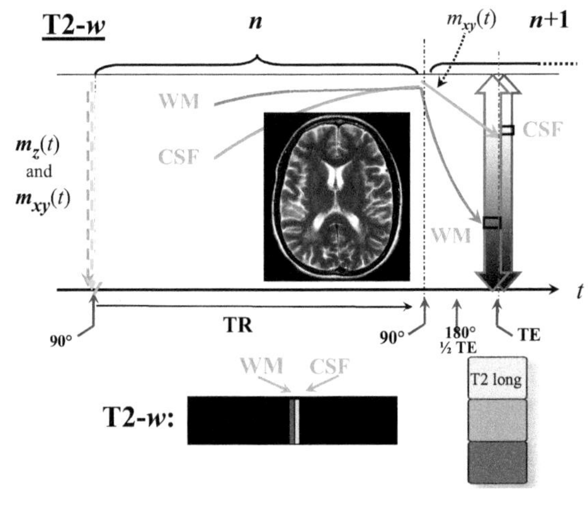

Figure 11.6 T2-weighted SE, produced with long TR, to null out the T1(x) effect, and longish TE to emphasize T2(***x***). Regrowth of $\boldsymbol{m_z}(t)$ values in two adjacent voxels containing white matter and CSF, respectively, before a 90° pulse, and decay of the associated values of $\boldsymbol{m_{xy}}(t)$ during the production of a T2-*w* SE MR image.

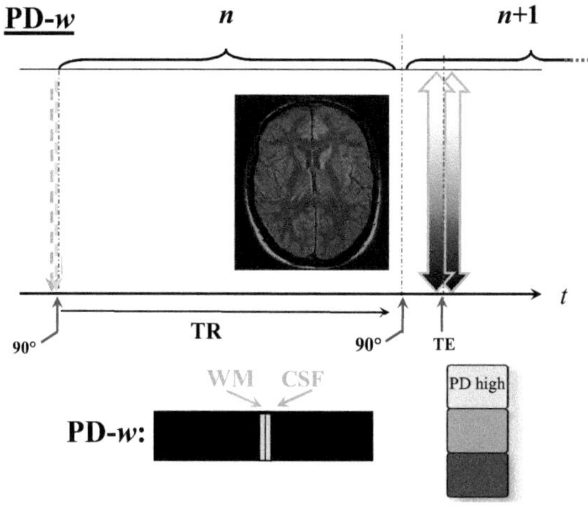

Figure 11.7 PD-weighted SE, produced with long TR and short TE.

trade-off decision, the resolution of which is commonly to choose TE midway among the T2 times of the tissues in the vicinity of interest. The transverse magnetization of a tissue in a voxel with a longer T2 will have decayed less at the time that the echo signal materializes, and shows up brighter.

Finally, a long TR and short TE together produce images that reveal little about either T1 or T2 processes (Table 11.6). What is left is primarily proton density (Figure 11.7), and PD-*w* images do have a few specialized applications. Since they involve a long TR, similar to that for T2-*w*, PD-*w* images can be collected together from sequences that acquire two echoes following the same 90° excitation—one at short TE to provide a PD-*w* image, and another with long TE for the T2-*w*. That's twice the bang for the buck at little more time, so why not obtain both?

EXERCISE 11.14 Fill in the missing information in Figure 11.7.

The connections between the imaging parameters TE and TR and the T1/T2/PD weighting of the resultant images are indicated in Table 11.3, along with representative images for a transverse slice through the cranium. Also indicated are the standard meaning of a bright region and values of the signal-to-noise ratio.

The big problem with a long TE, or even more so with a long TR, is the extended data acquisition time. This can slow patient throughput and sometimes leads to motional blurring. Anyone who's had an MRI scan knows how little effort it takes to wiggle and mess it up. Efforts to minimize these times, on the other hand—by reducing the gradient switching time, for example, or by shortening gradient field and RF pulse durations (but increasing their amplitudes correspondingly)—eventually run into a number of other technical problems. Some fast-imaging techniques have been developed to circumvent these difficulties, including the gradient-echo pulse sequence, as will be addressed in chapter 13 on fast imaging.

11.8 SE Reconstruction of the Image of a 1D Patient from a Single Line in *k*-Space

To image an entire 1D or 2D phantom, the near-monochromatic 90° and 180° pulses suitable for a single voxel are replaced with a more general pair that can affect all the voxels in a row or plane at the same time. This is involved, at the same time, with the choice of a thin transverse slice of tissue at some position $z = z_0$ along the phantom, as in Equation (7.7).

We have been referring to the MR signal from a single voxel at position x as $s(x,t)$. As discussed for 1D in the chapter on FID, the detected *multi-voxel* temporal MRI signal, $S(t)$, can be expressed as a superpositioning of *spatial* sine waves, in effect, each of its own spatial frequency k, in cycles/cm, Equations (7.16). At any instant, $S(t)$ is proportional to $M_{xy}(t)$, or

$$S(t) \propto M_{xy}(t) = \Sigma m(x)\, e^{-i(\gamma G x)t}, \quad (11.14a)$$

with a sum over the separate voxels. Interchange the x and the t in the exponent, so that

$$S(t) = \Sigma m(x)\, e^{-ikx}, \quad (11.14b)$$

where k was defined in Equations (7.16) as the rather odd duck

$$k(t) \equiv (\gamma/2\pi)\, G_x(t)\, t. \quad (11.14c)$$

Here this expression is meaningful only during the readout period, when $\boldsymbol{G_x}(t)$ is turned on. In chapter 12 on 2D imaging, we will look at a more general form of $\boldsymbol{k}(t)$ that will allow us more flexibility in acquiring an image. Stick around.

This change of variables, Equation (11.14c), transforms the incoming signal, $S(t)$, into its k-space counterpart, $S(\mathbf{k})$. As with 1D FID imaging (although we didn't mention it then), image reconstruction again takes place in $\mathbf{k}$-space. The time variable t is now embedded in $k(t)$ but with the difference that $k(t)$ is non-zero and varies only when $\mathbf{G}_x(t)$ is being applied. But during that on-time, it is *the* measure of the global impact of $\mathbf{G}_x$. One might think of $\mathbf{k}$ as just t gussied up for readout in its Kentucky Turtle Derby finery—but with the proviso that (unlike t) it can be viewed also as a vector in a (1D, for now) vector space. This is suggested in Figure 11.8, seen earlier (with a two-voxel beat signal, rather than an echo) in Figure 7.8. There we reversed the flow of actions of Figure 7.9 and reached Figure 7.10. Likewise, beginning with $S(t)$ and the change of variables from t to $\mathbf{k}(t)$, we get to PD(x) by way of $\mathbf{k}$-space (Figure 11.8).

The echo signal $S(t)$ is sampled a large enough number of times to satisfy the Nyquist criterion (such as 2×256) Δt apart (Figure 11.9a). Meanwhile, in accord with Equation (11.14c), the corresponding values of $\mathbf{k}_x(t)$ increase linearly with time as the echo is being read, and are likewise separated by Δk_x(Figure 11.9b). In effect, we are taking many little steps along the $\mathbf{k}_x$-axis, from $-k_{\max}$ through $\mathbf{k}_x = 0$ to $+k_{\max}$ (Figure 11.9c). After every step we note both the current $\mathbf{k}_x$-value and magnitude of $S(\mathbf{k})$, bearing in mind that larger-magnitude $\mathbf{k}$-values correspond to greater spatial frequencies. And that throughout this readout, at any specific instant of time t or any specific value of $\mathbf{k}(t)$, there are contributions to $S(\mathbf{k})$ from *all* discrete voxel *spatial* positions, $\mathbf{x}$, hence at all corresponding Larmor frequencies.

These data points can be plotted on a graph in $\mathbf{k}_x$-space of $S(k_x)$ versus $\mathbf{k}_x$ (Figure 11.9c). A spatial Fourier transform of these data from $\mathbf{k}$-space to real-space—which is to say a spectral decomposition of $S(\mathbf{k})$ into spatial frequencies—then makes it possible to complete the adoption of the $\mathbf{k}_x$-space to $\mathbf{x}$-space formalism of Equation (7.14). That yields PD(x) along the $\mathbf{x}$-axis, direct from the input signal, $S(t)$—that is to say, a PD-*w* MR image!

One should be forgiven for thinking this is all very weird and unpleasantly complicated; its only saving grace is that it works in 2D and 3D, while the earlier t-space formalism does not. The example here is for a 1D phantom, but the approach will be generalized into 2D and 3D by replacing the 1D, scalar $k(t)$ and with a vector $\mathbf{k}(t)$ in a two- or three-dimensional k-space.

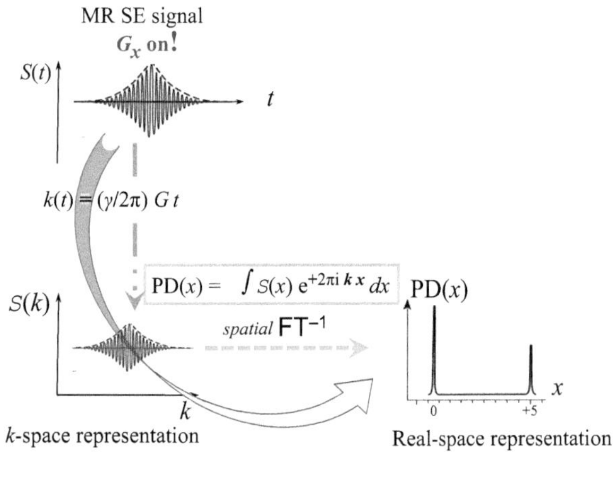

Figure 11.8 The $\mathbf{k}$-space SE reconstruction for the 1D patient, largely a spin-echo version of Figure 7.10, results in an SE echo rather than an FID RF beat signal. Reconstruction of the MR image, PD(x), from the RF signal, $S(t)$, is by way of the corresponding function, $S(k)$, in $\mathbf{k}_x$-space. The time variable in the signal $S(t)$ undergoes the change of variable $k(t) = (\gamma/2\pi)\, \mathbf{G}_x\, t$, Equation (11.14c), as $S(t)$ assumes the new form $S(\mathbf{k})$. A Fourier transform from $\mathbf{k}$-space to $\mathbf{x}$-space then yields the spatial distribution of the proton density, PD($\mathbf{x}$).

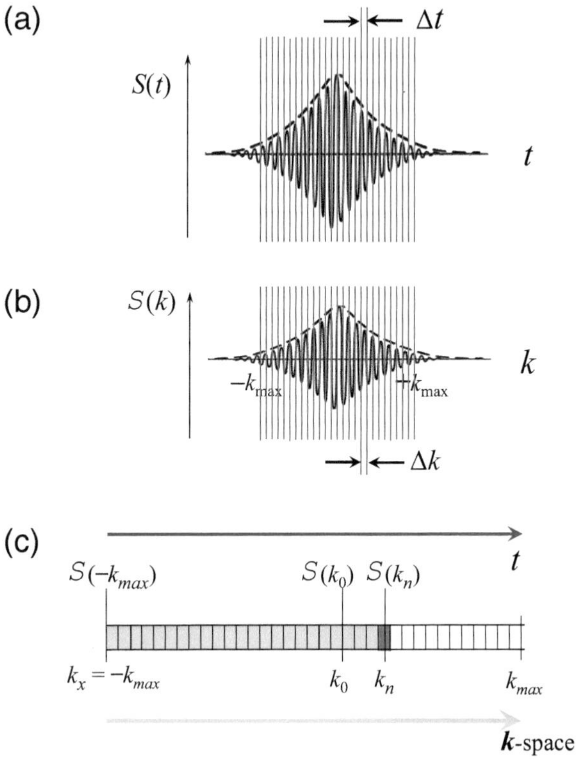

Figure 11.9 Sampling and $\mathbf{k}$-space. (a) Hundreds of samples of $S(t)$ for the echo are obtained Δt apart during the readout time. (b) With every sample, the $\mathbf{k}$-value steps a distance Δk along the $\mathbf{k}$-axis. (c) Completion of the 'filling' of a single line in $\mathbf{k}$-space. For a 1D patient, there is only one such line, but for a 2D image, there can be many (e.g., 192).

11.9 Dealing with Unwanted Phase Changes in the Echo Signal Caused by Gradients

By the time the middle of an echo signal arrives at the receiver coil, the x-gradient itself will have inadvertently introduced an additional voxel position-dependent phase shift of the transverse magnetization by the amount $\frac{1}{2}\Delta\phi$, as in the red portion of Figure 11.10. According to Equation (11.14.c), if we did nothing else, we would start with a $\boldsymbol{k}$-value of 0, and it would increase over time.

The solution involves counteracting the first $+\frac{1}{2}\Delta\varphi$ phase distortion with an intentional de-phasing by the amount $-\frac{1}{2}\Delta\varphi$ in the *opposite direction* before the echo arrives. The pre-emptive blue de-phasing (before t = TE) will cancel the red portion of the gradient near the middle of the reading process, so that maximal overall *re*-phasing will occur there at the peak of the echo. By centering the smallest values of $\boldsymbol{k}$ at exactly t = TE, we can maximize the echo signal in the center of $\boldsymbol{k}$-space, where contrast is all important.

At first glance, it might seem that the blue de-phasing gradient would add to the red, exacerbating the situation rather than countermanding it; that would, indeed, be so if it had been inserted *after* the 180° RF pulse, rather than *before* it.

To reduce complexity, we shall ignore this sort of correction in most future diagrams.

EXERCISE 11.15 Why are both applications of the gradient in the same direction, along the positive x-axis? Would the sequence of Figure 11.10 be valid if TE were longer, so that the anti-de-phasing gradient could fit in just before the echo, rather than in front of the 180° pulse?

11.10 Following a Simple Trajectory through 1D *k*-Space

We can demonstrate much of the 1D FID sequence of RF and $\boldsymbol{x}$-gradient pulses described above by explicitly following a trajectory in 1D $\boldsymbol{k}$-space during readout. A real 2D MR scan involves several hundred such sequences, one for each horizontal line in $\boldsymbol{k}$-space, i.e., for every value of $\boldsymbol{k_y}$, the meaning of which will be explained in the next chapter on 2D imaging. It is important that at the conclusion of any sequence, the spin magnetization in every voxel is left ready to begin the next one. But hold on there! We're getting a little ahead of ourselves, and we need to backtrack.

Figure 11.11a displays a pair of negative, $-\boldsymbol{G_x}(t)$ blocks, each half as long as the positive-polarity readout $\boldsymbol{x}$-gradient. Their purpose is simply to undo the disruption, in advance, to the phase relationships that the readout gradient itself causes among voxels, as just discussed. By the end of this set of multiple gradient triads at Figure 11.1i, ideally both $k_x(t)$ and $\phi_x(t)$ will have been re-wound to their initial configurations, and to

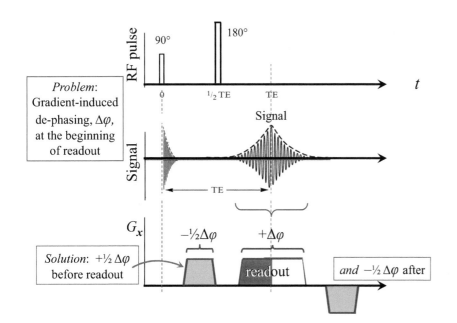

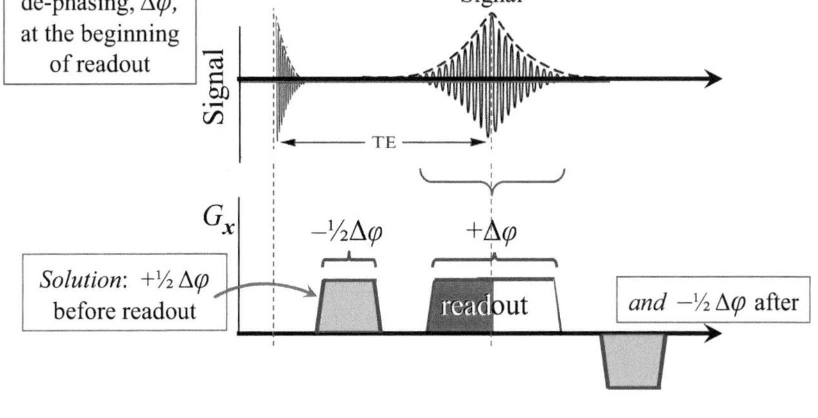

Figure 11.10 An undesirable aspect of applying the x-gradient, G_x, in spin-echo. A more refined version of the 1D SE sequence consists of two RF pulses and *three* applications of the x-gradient. The first and second halves of the *readout* gradient are red and white, respectively, but there is a blue *anti*-de-phasing pulse that precedes it, half as long as the readout gradient on-time. It is intended to counteract, in advance, the first half of the de-phasing ($\Delta\varphi$) caused by the readout gradient itself; that way, maximal overall *re*-phasing will occur at the center of the readout process, at the peak of the echo. The second blue gradient counteracts the effect of the first.

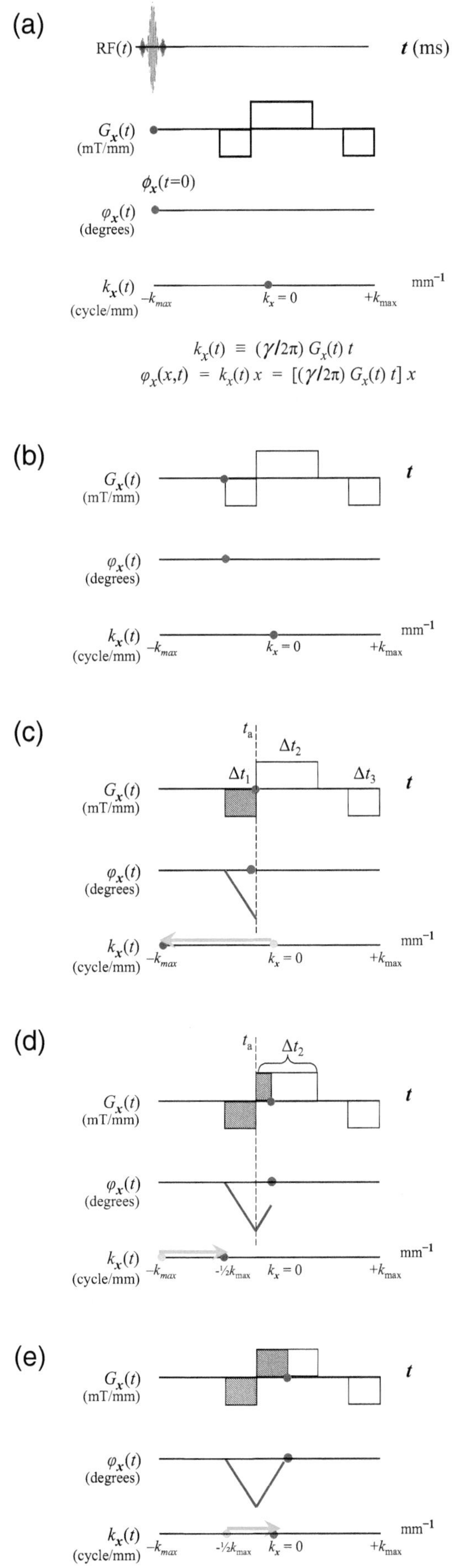

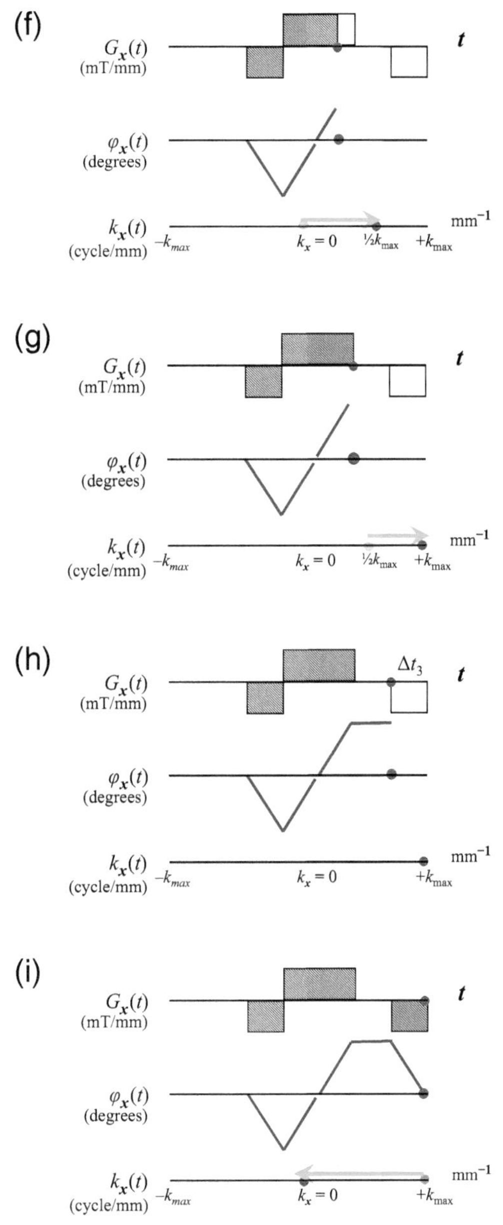

Figure 11.11 Two time lines and one 'sort-of' time line (showing $k_x(t)$) associated with a simple nine-step path (a)–(i) within 1D ***k***-space. $\boldsymbol{G_x}(t)$ traces the strength of the ***x***-gradient over time following the 90° saturation pulse. In the rotating frame, accumulated 'phase' $\varphi_x(t)$ refers to that of the precessing $\boldsymbol{m_{xy}}(x, t)$ in the voxel at any particular x, and it too glides along the time axis. $k_x(t)$ indicates how the magnitude of k_x changes over time. A green arrow represents the motion within k-space from the time at the end of the previous step to the current time, which is marked by the red dot along the $\boldsymbol{G_x}(t)$ and $\varphi_x(t)$ axes. As $k_x(t)$ progresses between $-k_{x\,max}$ and $+k_{x\,max}$, it doesn't move strictly from left to right, since $\boldsymbol{G_x}(t)$ is sometimes negative. It is assumed that $\boldsymbol{G_x}(t)$ is turned on and off instantaneously. [Idea for this figure from Jason Stafford.]

their original values, ready to start all over again. We assume here, unrealistically but for mathematical simplicity, that $G_x(t)$ is turned on and off abruptly, and has only the values 0 and 1.

Begin a short time period after a 90° Larmor-frequency pulse excites the spins (Figure 11.11a). The *z*-gradient, $G_z(t)$, is switched on briefly for slice selection (not shown). In this particular trajectory through 1D ***k***-space, the frequency-encoding ***x***-gradient, $G_x(t)$, has not been turned on yet, and $k_x(t)$ remains at zero, in the middle of 1D ***k***-space, by its definition, Equation (7.16b).

Likewise, no gradient-induced phase $\varphi_x(t)$ in the transverse magnetization has accumulated for the voxel at *x,* in the rotating frame up, until the time that $G_x(t)$ is activated (Figure 11.11b). $k_x(t)$, which changes only while the ***x***-gradient is on, remains at its origin.

The story gets more interesting when the $G_x(t)$ is turned on for an interval Δt_1 (Figure 11.11c). Since the gradient is *negative* for now, $k_x(t)$ slides the distance $\Delta k_x(t) = -(\gamma/2\pi)\,|G_x(t)|\,\Delta t_1$ to the left through ***k***-space, arriving at $-k_{max}$ just as the ***x***-gradient switches polarity. $k_x(t)$ is proportional to the product $G_x(t) \times t$, so that at any given time, its value depends on the total area under the gradient curve, from before it was turned on until the current moment. During the application of $G_x(t)$, the magnetizations will be precessing at different frequencies, depending on the voxel location, *x*. As a result, the magnetizations will fall out of phase with one another over time. The exact phase change accumulated by the magnetization in the voxel at position *x*, $\varphi_x(t)$, is specified by Equation (7.17c). The change in phase within the phantom during Δt_1 is negative. In other words, the phases of the magnetizations begin to lag, or fall behind, where they would be if there were no gradient. The amount of lagging depends on position. Magnetization in voxels at more positive values of *x* have greater amounts of lagging in phase than those located at smaller values of *x*, because of the negative gradient.

The polarity of the gradient is reversed at time t_a, at the end of the first *x*-gradient block, and becomes positive (Figure 11.11d). At this time, readout begins. The incoming MR signal may be sampled hundreds of times during readout, each sample corresponding to a unique point in ***k***-space, during application of the readout gradient of width Δt_2. Reversing the gradient polarity has the effect of shifting k_x toward 0. Now, the previously lagging magnetizations begin precessing faster because of the positive gradient. Indeed, magnetizations in the voxels at larger values of *x*—those that lagged more earlier—are now precessing faster than those at smaller values of *x*. The trend is that $\varphi_x(t)$ will begin to decrease, reversed by the quicker precession rates. The negative value of $k_x(t)$ at the gradient reversal demonstrates that ***k*** depends not only on its current circumstance, but also on its history of what has gone on before (i.e., shifting to the left because of the negative $G_x(t)$ acting over Δt_1). In other words, ***k*** begins to increase as the gradient is reversed, but it starts off at a negative value because of the previous gradient during Δt_1. The figure shows the situation at the time $\frac{1}{4}\Delta t_2$ after t_a, a quarter of the way through readout. After running the gradient for half the time it took to get to the edge of ***k***-space, but at reverse polarity, the area under the gradient curve is one-half that of the first negative application; we have swept out enough area under the curve to get us halfway back home to $k_x = 0$.

EXERCISE 11.16 What is the value of $\varphi_x(t)$ at time ($t_a + \frac{1}{4}\Delta t_2$), and why? Why does $k_x(t) = -\frac{1}{2} k_{xmax}$ at that point?

By leaving the gradient on longer (Figure 11.11e), until ($t_a + \frac{1}{2}\Delta t_2$), we sweep out the same area as for the prior negative gradient, and $k_x(t)$ continues on to the center of ***k***-space. All the gradient-induced phase change has been undone, and $\varphi_x(t)$ is returned to zero. This results from the refocusing of the transverse magnetization along the encoding axis, so that the protons in each voxel are singing in-phase at this moment. From the beginning, the negative gradient-introduced phase differences between the magnetizations in different voxels, *warps* the original phases over space. (Spin-warp is discussed later in connection with 2D imaging.) In other words, the total signal received by the scanner at any moment is produced by the sum of the magnetizations at all ***x***; these phase differences led to destructive interference and loss of signal. Now that we have fully reversed this phase dispersion in the transverse magnetizations, they are now precessing in-phase, for every voxel, as we sit in the center of ***k***-space.

With the gradient still on, the system passes through the center of *k*-space *en route* to $+k_{max}$ (Figure 11.11f). Positive phase continues to accumulate for each voxel, until G_x is turned off (Figure 11.11g).

There should be no further changes in either phase or location in $\boldsymbol{k}$-space in the absence of a gradient (Figure 11.11h).

Lastly, a negative gradient pulse of the same area as the first one navigates the spins back to the center of $\boldsymbol{k}$-space (Figure 11.11i). We may or may not choose to perform this last step, depending on the details of the pulse sequence.

If we want to use the residual transverse magnetization in the next imaging step, we would include the negative pulse. (Sometimes, it is best to obliterate transverse magnetization at the end of this sequence; in that case, an extremely strong gradient pulse known as a *crusher* is applied that has the opposite effect of Figure 11.11i: it greatly increases the amount of phase as a function of position.)

As pointed out at the beginning of this section, it takes multiple sequences like that of Figure 11.11 to create a real 2D MR image. A small complication: if we repeat the process before the longitudinal magnetization in a voxel has had a chance to relax fully, its z-magnetization will no longer start off in thermal equilibrium, but after several repetitions it will approach a 'quasi-equilibrium' *steady state* condition away from $\boldsymbol{m}_0$. Similarly, the transverse magnetization also may not have relaxed fully after one sequence. The timing and magnitudes of gradients and RF pulses prior to the next repetition will strongly influence the nature of this steady state situation, as we shall see, and on the achievable MRI contrast.

By adjusting the amplitudes of the gradients, the rate at which $\boldsymbol{k}$-space is covered can be modified.

Varying this rate can be helpful in controlling errors from flowing spins, as in diffusion imaging (Figure 1.7b) or in moving tissue. $\boldsymbol{k}_x$ is proportional to the product $\boldsymbol{G}_x \times t$, so to traverse $\boldsymbol{k}$-space more quickly, one could apply stronger gradients. The readout would take a correspondingly shorter time, which means that the SNR would generally be lower. And the readout time is intimately related also to receiver bandwidth, which we will discuss further in the next chapter on MR instrumentation.

11.11 Inversion Recovery (IR) in a Voxel: 180°–90° (TI)–180° (½TE) –*read*

Inversion recovery is a sequence employed originally as an alternative way to generate images with strong T1 weighting. Most current applications of IR incorporate a spin-echo aspect as well, which allows the creation of an echo and the suppression of some signal patterns that happen to obscure others of clinical interest. Let's turn again to our two-voxel phantom containing one voxel of white matter and another of CSF. The objective of this little exercise will ultimately be to eliminate all traces the CSF signal from the image, yielding a clearer image of the white matter.

But first.... Any IR sequence begins with a 180° RF *inversion* pulse, with the slice-selection gradient $\boldsymbol{G}_z$ on. This turns the longitudinal magnetizations in both voxels not through 90°, but rather completely upside down (Figure 11.12). If a voxel of spins had been sitting undisturbed for a while and was at equilibrium, so that $\boldsymbol{m}_z(t)$ was of magnitude $\boldsymbol{m}_0$, then immediately fol-

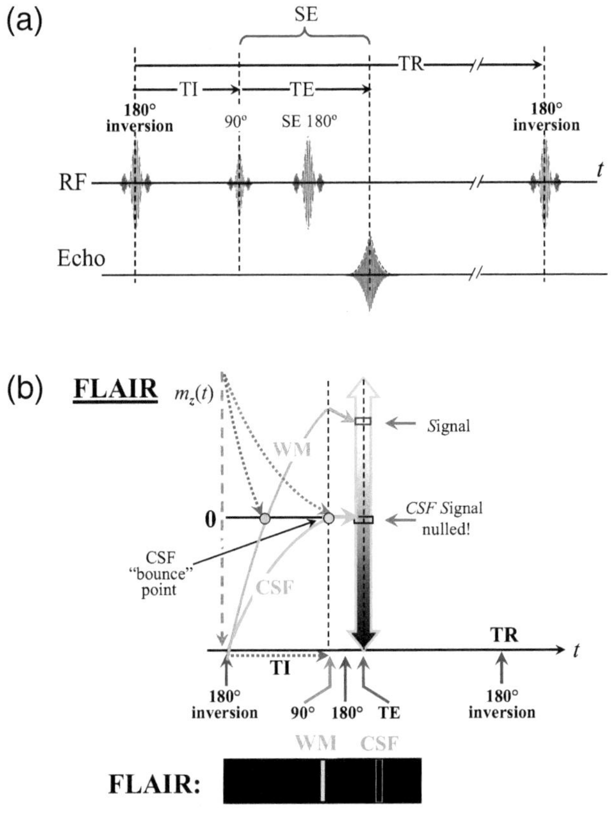

Figure 11.12 Inversion-recovery. (a) This IR sequence begins with a 180° *inversion pulse*, and the echo is read out later with a SE sequence. The sequence repeats after time, TR. (b) This FLAIR study is designed to remove CSF from the image. The transverse magnetization of CSF, $\boldsymbol{m}_{xy}(\text{TI})_{\text{CSF}}$, will be precessing, but at the exact moment of its *bounce point*, where the CSF longitudinal magnetization crosses the **x-y** plane, it is briefly of zero magnitude. It produces no signal and, for this reason, any voxel containing CSF does not appear in the FLAIR MR image (Figure 1.6b), which is compared there to a standard T1-*w* image.

lowing the inversion pulse the magnetization will point downward and be of magnitude $-\boldsymbol{m}_0$.

During the *inversion time* period, TI (note that this is TI, not T1), $\boldsymbol{m}_z(t)$ recovers from $-\boldsymbol{m}_0$ at the rate 1/T1,

$$m_z(t) \;=\; \boldsymbol{m}_0(1-2\mathrm{e}^{-t/\mathrm{T1}}). \qquad (11.15a)$$

(assuming a long TR). By t = TI, $\boldsymbol{m}_z(t)$ for each material will have separately regrown to its magnitude m_z(TI). At precisely that instant, a 90° pulse tips the longitudinal magnetizations into the ***x-y*** plane, where they transform instantaneously into $\boldsymbol{m}_{xy}$(TI). The two new transverse magnetizations precess freely there for ½TE, where TE is usually chosen to be short. After that a (spin-echo) 180° refocusing pulse is applied. Two echoes materialize ½TE later,

$$m_{xy}(\mathrm{TI{+}TE}) \;=\; \boldsymbol{m}_0(1-2\mathrm{e}^{-\mathrm{TI/T1}})\mathrm{e}^{-\mathrm{TE/T2}}. \qquad (11.15b)$$

and are read out with the x-gradient on. The impact of TE spin relaxation will be small for a short TE, so the image is T1-weighted. For our two-voxel phantom, both voxels will, in general, show up in the MR image.

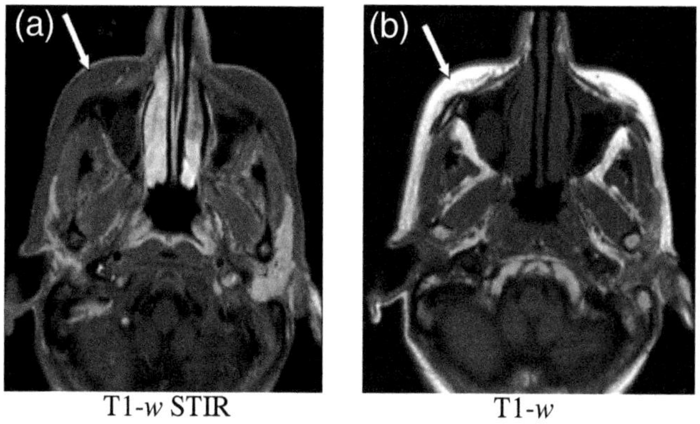

Figure 11.13 STIR inversion recovery is one type of fat suppression ("fat-sat"). (a) STIR nulls the fat signal, here rendering bone marrow edema more visible. (b) Standard T1-*w* image at same slice level.

A frequently employed variation on the IR technique is FLAIR (*FL*uid *A*ttenuated *I*nversion *R*ecovery). Damaged brain tissues may locally take up excess fluid, but the extent of this edema may be obscured by adjacent cerebrospinal fluid. FLAIR can largely null out the presence of the CSF and render more visible a region that is edematous but happens to lie near an interface with CSF (Figure 1.6). FLAIR also helps to reveal periventricular lesions like multiple sclerosis plaques.

What an IR image reveals depends on the selection of TI, hence on the timing of the *bounce point*. To illustrate how this works, a voxel containing CSF is left alone to equilibrate for a while, after which the IR sequence is initiated with application of a 180° inversion pulse. By selecting TI to be the exact, specific time of the bounce point, defined as

$$t_{\mathrm{bounce}} \;\equiv\; \ln 2 \times \mathrm{T1} \;=\; 0.693 \times \mathrm{T1}, \quad (11.15c)$$

$\boldsymbol{m}_z(t)$ will have recovered half-way to its equilibrium level—that is, up to the time when $\boldsymbol{m}_z$(0.693 T1) = 0 (Figure 11.12b). At the moment of CSF's bounce point, a 90° pulse swings the magnetizations in both voxels from the vertical down into the x-y plane, and leaves the resulting new $\boldsymbol{m}_{xy}(t)$ for each precessing there. But since the magnetization in the CSF voxel is now of zero magnitude at (and only at) that instant,

$$\boldsymbol{m}_{xy}(t_{\mathrm{bounce-CSF}}) \;\equiv\; 0, \qquad (11.15d)$$

and the CSF creates *no* signal that can be detected in the pickup coil. The CSF has, in effect, been removed from the picture, literally, and its pixel appears as black, as if nothing were there, as in the phantom MRI image.

A similar specialized IR sequence known as STIR (Short TI Recovery) works much the same way as FLAIR. But because T1 for white matter is a good deal shorter than for CSF, the bounce point occurs much earlier, hence the 'short TI.' STIR sequences, which can be T1-*w* or T2-*w*, find use in musculoskeletal studies by attenuating the signal from white matter, such as in marrow (Figure 11.13ab). STIR sequences can also counteract the presence of metal within the patient, and chemical shift artifacts.

Another form of *fat suppression*, unrelated to STIR, takes advantage of the chemical shift. There is a 3.5-ppm difference in the resonance frequencies of the protons in water and the methylene groups in white matter. Because of this separation, a narrow-band 90° pulse at the WM Larmor frequency can selectively drive the fat protons down into the x-y plane. Brief application of a 'spoiler' gradient field rapidly dephases them—and that leaves a window of opportunity to carry out a normal imaging study of the water-protons alone.

T1-*w* imaging can proceed with multiple RF pulses, where the combination of TR and flip angle do not allow for total relaxation of the magnetization back to its original, longitudinal orientation and magnitude between 90° or α° pulses. Since imaging time is costly (and because it depends on TR), most clinical scans have some marginal degree of T1 weighting arising purely from the very long T1 value. Chapter 13 will turn to the critically important topic of fast imaging.

CHAPTER 12

Image Reconstruction in Two Dimensions

Much of the excitement and success of the CT revolution stemmed from the ability of its x-rays to isolate and represent thin transverse slices of tissue that are not obscured by over- and underlying layers. This is even more so the case for MRI, which can display planes with *any* orientation—exhibiting transverse, coronal, sagittal, or other thin slices of tissue equally well. Some MR reconstruction techniques are even inherently 3D in nature, moreover, acquiring and displaying entire blocks of tissue at a time. And MRI can display sensitivity not only to spatial variations in tissue density and thickness, but, more importantly, to tissue physiology and pathologies.

Many MRI methods are planar, mapping out contrast in a tissue volume as a stack of separate, thin slices. Of the numerous 2D signal generation and reconstruction techniques, one of the most widely used is the *two-dimensional Fourier* method, of which *spin-echo/spin-warp* (SE/SW) is the most common version. We shall discuss it in the next chapter in some detail. After selecting a single slice of tissue, the approach captures a sequence of MR echo signals, filling in, line by line, the corresponding 2D ***k***-space representation. Further computation then reconstructs it as a clinical image in real-space (Figure 11.8). The first task is to store the relevant raw data in ***k***-space.

12.1 Selection of a Two-Dimensional Slice

Up until now, with 1D MR phantoms and images, the magnetic field pointed vertically, as in an open-magnet system, and the position of a voxel was expressed as its *x*-value (Figure 12.1a). In moving on to two and three

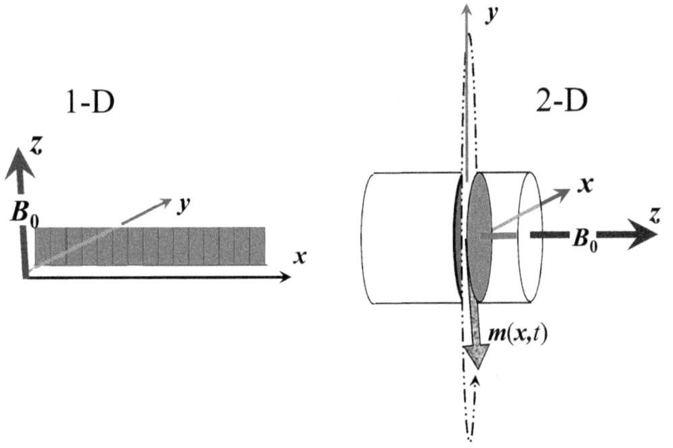

Figure 12.1 Coordinate systems and field geometries for (a) an 'open' magnet with a vertical **z**-direction and field orientation, which we have used until now for the 1D patient. (b) A standard superconducting magnet with horizontal **z**-axis, main field, and patient, as employed hereafter for 2D imaging.

dimensions, it simplifies matters to adopt another geometry—that of a standard superconducting magnet, with $\boldsymbol{B}_0$, the z-axis, and the patient lying horizontally and co-linear (Figures 8.1 and 12.1b). We can then select and focus on a single thin transverse (for simplicity here) slice of tissue, at some specific position along the body, z' say. This makes possible the mapping out of PD, T1, T2, etc. throughout the $\boldsymbol{x}$-$\boldsymbol{y}$ plane that crosses the z-axis there. Similar arguments pertain for sagittal, coronal, or arbitrarily oriented display planes.

Selection of a thin transverse slice of protons at a particular z-position, namely $\boldsymbol{z} = z'$, entails applying a *narrow-band* 90° pulse centered at $\nu_{\text{Larmor}}(z')$, with the z-gradient on, Equation (7.7), repeated here as

$$\nu_{\text{Larmor}}(z') = \nu_{\text{Larmor}}(B_0 + G_z z'). \quad (12.1a)$$

The z-gradient $G_z \equiv dB_z/dz$ (not the $\boldsymbol{x}$-gradient!) and the RF pulse mid-frequency together determine the z-position of the slice of body tissues. This combination drives the spins in the slice at the currently selected z'-position, and *only* them, over into the $\boldsymbol{x}$-$\boldsymbol{y}$ plane. Again, of all the protons in the phantom, it is only these in this single thin slice that will be imaged at present.

The narrow bandwidth of the RF pulse and the steepness of the gradient, $G_z = \mathrm{d}B_z/\mathrm{d}z$, control the *slice thickness* (Figure 12.2a). The band comprises a continuum of RF frequencies of equal power over the span from $\nu(z' - ½\Delta z)$ to $\nu(z' + ½\Delta z)$. That is, the RF pulse should contain a *transmission bandwidth* of frequencies about as wide as a small sliver of the z-gradient-induced spread in Larmor frequencies:

$$\begin{aligned} BW_{\text{trans}} &= \Delta\nu_{\text{trans}} \equiv \left[\nu(z' - \tfrac{1}{2}\Delta z) - \nu(z' + \tfrac{1}{2}\Delta z)\right] \\ &= (\gamma/2\pi) G_z \Delta z. \end{aligned} \quad (12.1b)$$

After that, with the z-gradient turned off, these protons precess in synchrony at and near $\nu_{\text{Larmor}}(z')$.

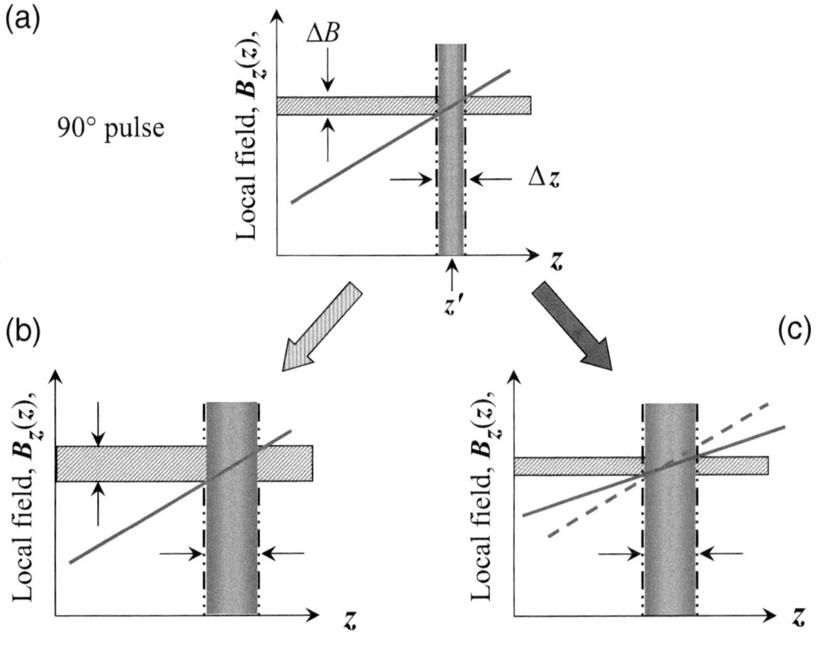

Figure 12.2 Transmitter RF bandwidth, **z**-gradient strength, and slice thickness, Δ**z**. Keep an eye on the separations of both the horizontal and vertical bands, and on the slope of the diagonal. (a) One can produce a thicker (in the **z**-direction) tissue slice either (b) by increasing the bandwidth of the RF pulse, $BW_{\text{trans}} = \Delta\nu_{\text{trans}}$, or (c) by reducing the strength of the **z**-gradient, $\boldsymbol{G_z}$. A third approach, familiar from multi-slice CT, is to combine the data from several adjacent tissue slices. However accomplished, acquiring data from more protons will improve signal-to-noise, but perhaps at the expense of reduced spatial resolution or contrast because of more overlapping of tissues and greater *partial volume* effects.

EXERCISE 12.1 Are the bandwidths of Equations (12.1b) and in the caption of Figure 3.3 the same? Explain.

We will normally be dealing with only the protons that lie within a single slice of small but finite thickness Δz. The BW_{trans} of the RF pulse should be to span the frequencies ranging from the top to the bottom of a voxel:

$$BW_{trans} = \Delta\nu_z \equiv 42.58\, G_z \Delta z, \qquad (12.1c)$$

which typically amounts to about 1 kHz per 1-mm thickness of slice. Here, the electronics must be adjusted to handle a bandwidth of only a few kHz. Bear it in mind that the term 'bandwidth' has several different applications and meanings in MRI, of which this is but one, and the temporal frequencies in Hz can differ considerably among them.

For the Δz-plane to be thin, which leads to good z-direction resolution, $\Delta\nu_{trans}$ must be narrow—but the narrower the frequency band and the thinner Δz, the fewer the number of protons in any voxel, and the worse the SNR for the slice. The transmitter bandwidth has virtually nothing to do with the *receiver bandwidth*, which has been discussed earlier and will be addressed again soon. Similar names, but largely unrelated attributes.

Either a broader band of RF frequencies in the pulse or a flatter field gradient will result in a thicker tissue slice (Figure 12.2). In practice, thickness is altered by changing the gradient strength, while keeping the transmit bandwidth constant. Current MRI machines can provide a choice of slice thicknesses ranging typically from 0.5 to 10 mm. As with CT, a thicker slice means better signal-to-noise and perhaps contrast (or not), but reduced spatial resolution because of tissue averaging along the slice-selection direction.

It is desirable that the two surfaces of the slice be flat and parallel—that is, the proton density, PD(z) should appear to be a step function of z-position, Rect(z) (Figure 5.10). This implies that the RF power spectrum of the narrow-band 90° pulse should be a rectangular function of *frequency*, Rect(ν), centered at $\nu_{Larmor}(z')$. A 90° pulse of RF power is required with contributions of *equal amplitudes* throughout that band, $P_{RF}(\nu) = \text{Rect}(\nu)$.

But what function of *time* can we provide to the transmit coils, $P_{RF}(t)$, that has Rect(ν) as its spectrum, and with all frequency contributions of the same amplitude? One can obtain it by way of a simple Fourier transform of Rect(ν) into the time domain, Equations (5.19). This yields **FT**[Rect(ν)] = Sinc(t) (Figure 5.10g) for the rather weird *envelope* shape of the RF power, $P_{RF}(t)$, of the 90° RF pulse. To generate a pulse with the characteristic RF spectrum Rect(ν), one must modulate the power of the relevant band of RF carrier frequencies according to $P_{RF}(t) = \text{Sinc}(t) \equiv \sin(t)/t$. To summarize, when all the waves included in Rect(ν) interfere with one another, they produce RF with an outline pattern of Sinc(t).

One doesn't need to manipulate the entirety of Sinc(t), which is fortunate because its domain (range of t-values) is infinite. Commonly, the central peak along with one or two periods on either side are retained, and the rest truncated away. This usually provides an acceptable compromise between a good slice profile and a reasonably short pulse length for faster operation.

A single transverse slice is selected by means of the 90° pulse (with the z-gradient on), and the system immediately turns off the z-gradient. With spin-echo, the 180° echo-generating pulse occurs at t = ½TE. The objective is now to determine the magnitudes of the magnetization vectors for the various voxels precisely at t = TE, when the echo signal appears. The spin-echo/spin-warp technique achieves this in 2D through a specific application of x-gradient and y-gradient magnetic fields and RF pulses.

EXERCISE 12.2 The z-gradient is activated again for a brief moment while the 180° echo-formation RF pulse is being delivered. Does it have to be?

After completion of the mapping throughout the slice (for some pulse sequences such as gradient echo, at least), the next one is selected. With the z-gradient on briefly again, the center frequency of the new 90° RF pulse is shifted slightly to $\nu_{Larmor}(z'')$. A new cohort of protons is thereby made ready for mapping throughout its x-y-plane.

12.2 Waves in 2D Real Space and Vectors in 2D *k*-space

For a digitized photograph, the data are readily manageable when housed in a 2D matrix, as with Figure

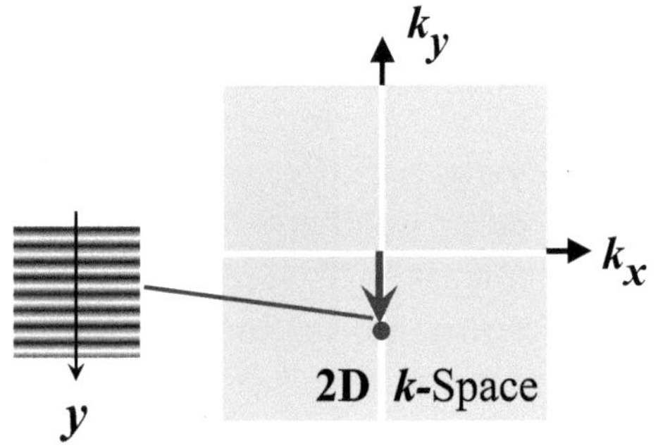

Figure 12.3 Relationship between points in 2D ***k***-space along the $\boldsymbol{k}_y$-axis and waves in 2D real-space. The farther a point is from the origin, the higher the spatial frequency of its associated wave.

5.15. But in what format should we partition the set of hundreds of thousands of digitized signal-valued data points in an MR image into a most meaningful and tractable formation of rows and columns?

We know that for digital radiography, the matrix of the final image in real space exactly mimics the x-ray transmission-data matrix, pixel-by-voxel (Figures 5.15 and 5.16). Also, we have already found that a 1D MR image can be represented as a line in a 1D $\boldsymbol{k}_x$-space (Figures 7.10, 11.8, and 11.11). But does that remain the case for ***k***-space data and 2D MR images? And once we have arranged the data properly, how do we then convert it into a clinical 2D pixel display of MR tissue PD, or T1, or T2, etc.?

Figure 5.9 demonstrated how a finite number of spatial sine waves of the correct wavelengths (i.e., ***k***-values) and amplitudes can combine to produce a rendering of a real-space entity—namely a periodic square wave—in effect, a 1D image. Figure 5.14 then redisplays the same idea in a slightly (not much!) different format. The two presentations are really saying very much the same thing. The k_x-axis is presented as a white line in the latter, and loci along it correlate to spatial frequencies. The associated α-values record the amplitudes of the several waves.

It's not a problem to extend the 1D approach to real 2D images. It requires only having the wave-vectors reside in a two-dimensional ***k***-space. A $\boldsymbol{k}_y$-wave can be viewed in the same manner as the $\boldsymbol{k}_x$-vectors of Figure 5.14 by just rotating everything 90 degrees (Figure 12.3). As before, the farther a point is from the origin, and the longer its ***k***-vector, the higher the spatial frequency of the wave.

A real-space wave of any alignment and wavelength can be specified by means of a ***k***-vector with the right choice of $\boldsymbol{k}_x$- and $\boldsymbol{k}_y$-components. Real waves associated with $\boldsymbol{k}_x$- and $\boldsymbol{k}_y$-vectors (Figure 12.4a) have a certain wavelength along the ***x***- or *y*-axis in the image. One can add $\boldsymbol{k}_x$- and $\boldsymbol{k}_y$-components to generate any wave in real-space. The direction and wavelength of the resultant diagonal wave are indicated by the larger purple dot—and its position in ***k***-space fully determines its wave-like properties. Figure 12.4b pro-

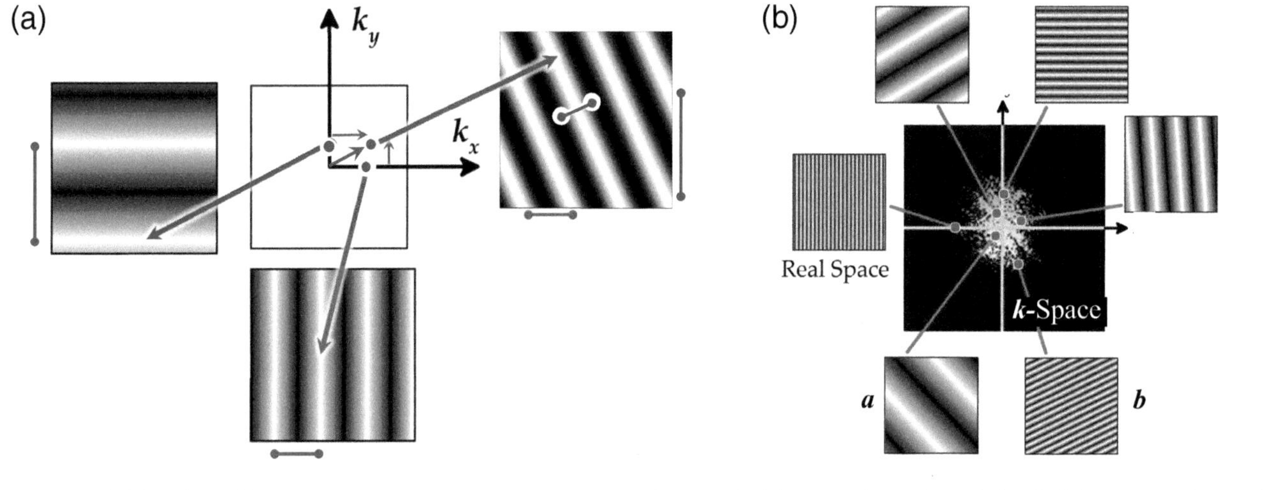

Figure 12.4 Adding $\boldsymbol{k}_x$ and $\boldsymbol{k}_y$-components to generate a wave in real space. (a) The spatial wavelength (barbells) associated with the wave specified by the $\boldsymbol{k}_x$- (blue) or $\boldsymbol{k}_y$- (red) component is preserved along both axes of the wave with both $\boldsymbol{k}_x$ and $\boldsymbol{k}_y$-components. (b) Various pairs of $\boldsymbol{k}_x$- and $\boldsymbol{k}_y$-components can combine (red dots) to give rise to real-space waves each of unique direction and wavelength. It is possible to re-create any real 2D image out of contributions, of proper amplitudes, of distinct points in ***k***-space. What role does wave amplitude play in this?

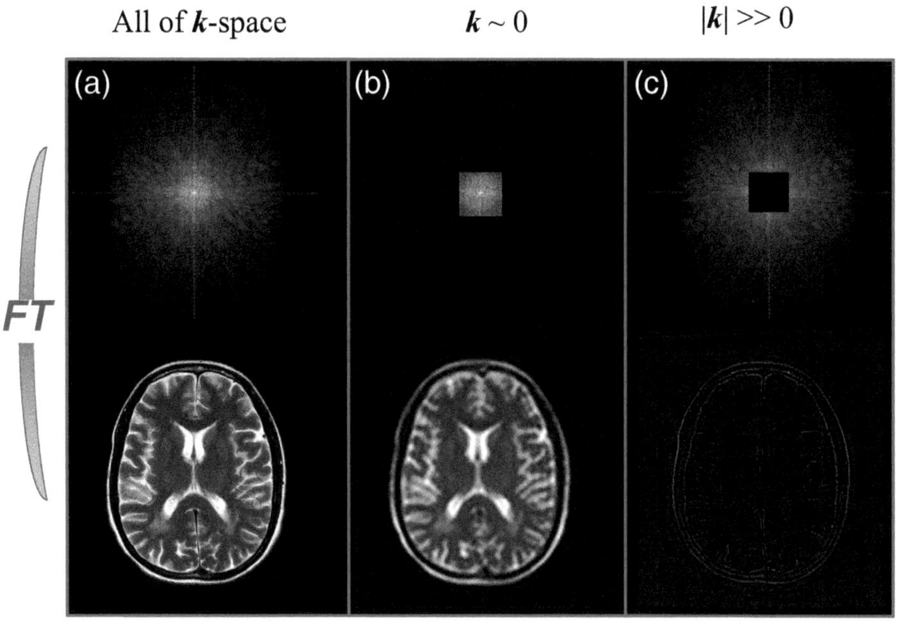

Figure 12.5 The quality of an image reconstructed from a representation in **k**-space depends on, among other things, how much and what part of **k**-space is being Fourier transformed to real-space. (a) All vectors from **k**-space. (b) Long-wavelength vectors near the center of **k**-space provide coarse contrast. (c) **k**-vectors of shorter spatial frequency, far from the origin, give the image its detail.

vides examples of real-space waves and their points in $\boldsymbol{k}$-space, both along and off the $\boldsymbol{k}_x$- and $\boldsymbol{k}_y$-axes.

Reminiscent of Figure 7.12, Figure 12.5 shows three $\boldsymbol{k}$-space, and corresponding real-space, representations of the same MR slice, with each in the lower row the inverse Fourier transform of the one above. The brighter a point in $\boldsymbol{k}$-space, the greater the amplitude of its wave contribution to the real-space image. These reiterate an important fact of life: to represent a real object as closely as possible, it is necessary to utilize *all* the points in $\boldsymbol{k}$-space in reconstruction (Figure 12.5a). If only the data points near the center of $\boldsymbol{k}$-space are incorporated, where wavelengths are long, the overall contrast comes out, but not the (short $\boldsymbol{k}$, high-1/$\boldsymbol{k}$) fine detail (Figure 12.5b). If only the shorter wavelength vectors from the outer reaches of $\boldsymbol{k}$-space are involved, conversely, there is plenty of sharpness, but the lack of contrast makes it hard to visualize what's actually there (Figure 12.5c).

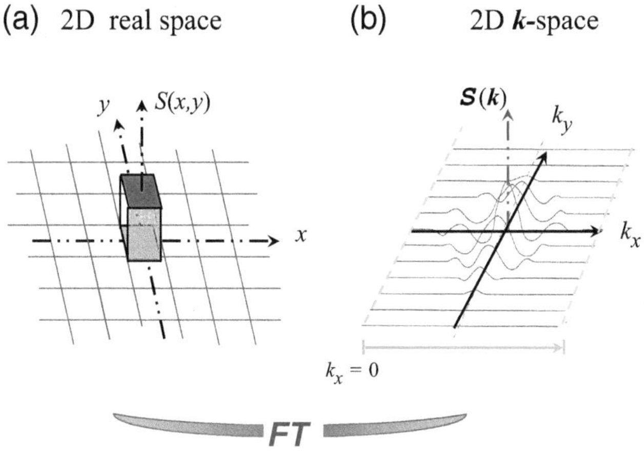

Figure 12.6 A simple 2D function, $S(x,y)$, and one possible representation of it, $S(\boldsymbol{k}_x, \boldsymbol{k}_y)$, in **k**-space.

Similarly, a function of two real variables in three-space, $S(x,y)$ (Figure 12.6a) can be re-expressed as its Fourier transform and displayed as a set of closely spaced 2D curves, $\{\boldsymbol{k}_x, S(k_x)\}$ (Figure 12.6b) with one line plotted for each $\boldsymbol{k}_y$-value. MRI works in the converse fashion, first acquiring and storing data along every row of a set (with 512 samples of each one, say) of parallel lines (perhaps 192 of them) parallel to the $\boldsymbol{k}_x$-axis, and then Fourier transforming that information to real-space.

EXERCISE 12.3 Figure 12.6 presents S(k) on a set of parallel vertical planes, each containing the k_x-axis. How else might this information be presented?

12.3 Phase Encoding of Voxel y-Position

Up until now, we have been dealing with 1D frequency encoding of voxel position along the x-axis alone. A spin-echo pulse sequence, for example, causes an echo signal, $S(t)$, to appear over a brief interval centered on TE. Simplistically, we say that $S(t)$ is comprised of contributions at N_x slightly different spatial frequencies, $\nu(x_n)$, one for each voxel position, x_n (it's a little more complicated than this because frequency varies continuously along x—there's a spread of frequencies in each voxel. However, read on…). A *discrete* Fourier analysis then separated the 1D $S(t)$ into its constituent N_x contributions exactly, providing the frequency spectrum of the signal, with the amplitude of each of the N_x spectral peaks proportional to the signal strength from its voxel (Figure 7.6). Essentially the same thing happened when image creation was by way of $\boldsymbol{k}_x$-space (Figures 7.10 and 11.8). This approach has been fine for imaging one-dimensional patients, but it simply cannot be extended directly into 2D or 3D.

Fortunately, there is a second, and independent, way to track a voxel's position throughout 2D space. It involves modulating and detecting the *phases* of various components that make up the RF echo signal, rather than their frequencies, and it serves nicely in introducing the $\boldsymbol{y}$-axis, along the vertical direction, into the picture (Figure 12.3). So it is necessary here to say a word or two more about the phase of a wave. For 2D MR imaging, it is 100% as important as its frequency or amplitude.

Phase don't get much respect! The concept of the phase of a wave is seems usually to be an afterthought and, at best, it gets second billing. We routinely adjust the frequency setting and amplitude of the radio, thinking not at all of the phase. But it can be just as significant as frequency in, for example, trying to excite a resonant system, such as a child on a swing. If the impulses are applied at the right frequency but 180° out of phase with the motion, the oscillations will be suppressed, rather than enhanced—and this is true in MRI as well.

The two dashed sine curves in Figure 5.5 are of the same frequency and amplitude, but differ in phase by 90° = π/2. When each is separately added to the solid sine wave, which in this example is one third of the frequency but three times the amplitude of the others, the two resulting composite signals are notably dissimilar. A sophisticated Fourier analysis program can tell the difference, and even determine the relative magnitudes of the phases. So the relative phases of waves, which can range between –180° and 180°, have an obvious importance when they are superimposed. That is central to how phase-encoding localizes k_y-position in $\boldsymbol{k}$-space.

In the past, we have drawn upon frequency encoding to distinguish voxels along the x-axis in real space. Now we introduce row-by-row phase shifts to keep track of voxel position on the y-axis as well.

12.4 A Simple 4-Voxel, 2D Spin-Echo, Spin-Warp Example

To make things a little more concrete and interesting, let's partition a small, square portion of our thin slice into four voxels that contain different water densities. This 2D toy model is simple enough to be solvable algebraically, and there is no need for the heavy machinery of Fourier transforms, etc., here. It is a little different from what actually occurs clinically, but it is useful to demonstrate 2D spatial encoding from an intuitive perspective—it really works! The labeling of the voxels and of the field gradient directions (and of the $\boldsymbol{x}$- and $\boldsymbol{y}$-axes) is indicated in Figure 12.7. Our mission is to use MRI to estimate PD($\boldsymbol{x,y}$) in each voxel by determining the magnitude of the net magnetization in each, $\boldsymbol{m}_i$, for i = 1 through 4, by carrying out the *spin-echo, spin-warp* procedure.

We'll accomplish that by detecting (with the pickup coil and receiver) the signal, $S(t)$, from the entire slice, from the protons in the four voxels acting together. We shall do this for three applications of the SE pulse sequence, with three different, specific sets of timing for the gradients G_x and G_y (Figure 12.8a). We

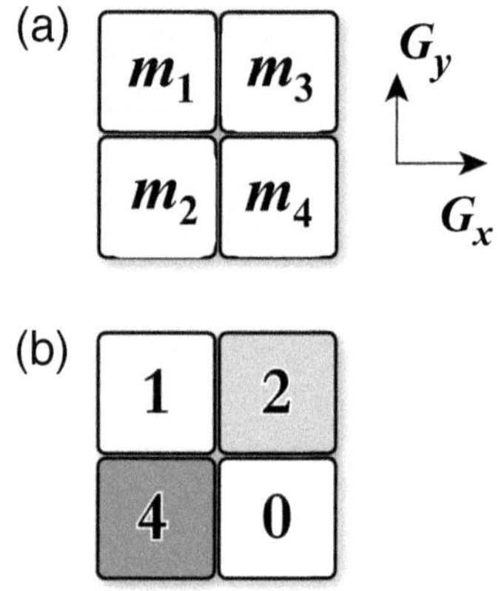

Figure 12.7 The 2×2, 4-voxel matrix of our phantom. (a) The patient consists of four thin, labeled voxels lying in the plane to which $\boldsymbol{B_0}$ and the $\boldsymbol{z}$-axis are perpendicular. (b) The solution to the simple example described in the text.

shall label these three the *preliminary* study, the *first iteration*, and the *second iteration*. We then untangle all the signal data to extract the part contributed by each voxel individually. In the clinic, we'd use Fourier analysis to carry out this last step, but here just a little algebra will suffice. Furthermore, our so-called *preliminary* study is not necessary in a clinical scan, but it can serve to provide a check on our arithmetic. Either way, the answer (spoiler alert) lies in Figure 12.7b, so now let's get to it.

Suppose that in all three studies an ***x-y*** plane has just been selected during the 90° pulse of a spin-echo sequence, with $\boldsymbol{G}_z$ on briefly, and is now sitting facing a uniform external field. The basic idea underlying what follows is that the contribution to the overall detected signal, $S(t)$, that comes from the ith voxel alone, namely $s(i)$, is proportional to $\boldsymbol{m}(i)$, which is itself linear in PD(i),

$$s(i) \propto \boldsymbol{m}(i) \propto \text{PD}(i). \qquad (12.2\text{a})$$

For convenience and to simplify the notation, we've gone beyond Equation (12.2a) and chosen a new set of units, for which $s_i = m_i$:

$$s(i) = \boldsymbol{m}(i) \propto \text{PD}(i). \qquad (12.2\text{b})$$

As before, lower case letters '*m*' and '*s*' will designate the magnetization and signal from one voxel,

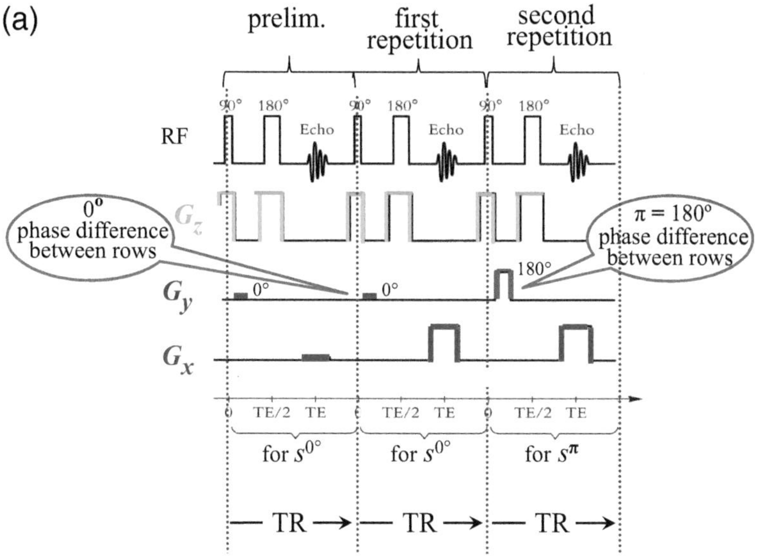

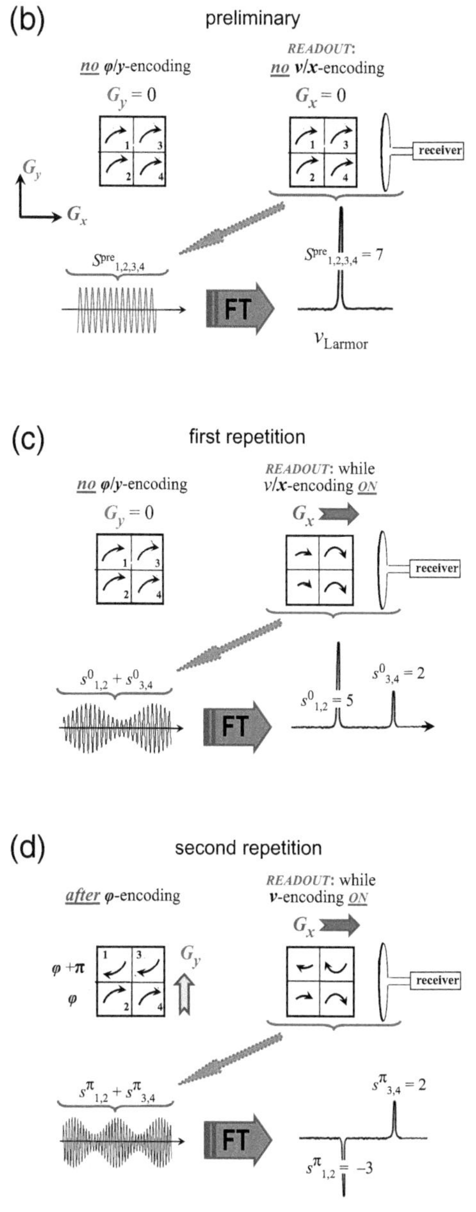

Figure 12.8 Spin-echo, spin-warp for our simplistic 2×2 matrix consists of a preliminary study plus two *repetition* sequences. (a) All three invoke the 90°–180° SE RF sequence, with the **z**-gradient on for both RF pulses, but the on-off timing of the gradients differs for each. The **x**-gradient is applied in both the first and second repetitions, but the **y**-gradient is turned on only for the second. Note the change in strength of the **y**-gradient between the S_0 and the S_π scans. The particular choice of values of TR and TE for the pulse-echo sequences determines, as we have seen, whether the resulting image is T1-weighted, T2-*w*, PD-*w*, or something else. (b) In the preliminary study, no gradients of either type are ever applied, and the four magnetizations precess at the *same frequency* and *in phase*. This gives rise to a simple, monochromatic sinusoidal signal, $S_{pre}(1,2,3,4)$, of frequency ν_{Larmor}. Its spectrum, which can be obtained by Fourier analysis, consists of a single peak which, for our example, is 7 units in amplitude. The subscript 'pre' on *S* indicates that here, there are no frequency or phase differences within and between the spins in the first and second rows. (c) The first pulse/gradient repetition sequence is straightforward spin-echo, in which $\mathbf{G_y}$, the gradient in the phase-encoding direction, remains *off*. The **x**-gradient is activated during readout, however, to separate signals from voxels in the first and second columns. The signal from this repetition is labeled S_0, again to indicate that G_y is not turned on and introduces no phase shift here. (d) In the second repetition, the **y**-gradient is turned on long enough to introduces a phase shift of π between the spins of the lower row and those of the upper, as indicated by the signal labeling, S_π. The simple beat pattern is like the one seen in Figure 7.7a.

respectively, while '*M*' and '*S*' are from all four at once.

pre The *preliminary* sequence consists of a straightforward 90–180° pulse pair, with both G_x and G_y remaining off throughout (Figure 12.8b). It will produce an MR the signal while the contents of all four voxels all precess at the same frequency and in phase. Nothing here to distinguish the voxels from one another.

i. For the *first iteration* (Figure 12.8c), the ***x***-gradient is turned on during readout, but the *y*-gradient is always left off. $G_y(t) = 0$, so the spins in the two rows of voxels will always remain in phase with one another. Application of G_x for readout, on the other hand, will cause the two columns to precess at somewhat different frequencies, resulting in a beat signal when they are mixed. This labels the ***x***-position of each voxel by way of *frequency-encoding*. Not much new here.

ii. But for the *second iteration* (Figure 12.8d), the ***y***-gradient is activated *before* the readout, for the precise time, Δt_{phase}, that is designed to be long enough for the spins in the upper row to gain π (180°) in phase relative to those in the lower row. This *phase-encoding* will provide the necessary information on the ***y***-position of each voxel. (More about Δt_{phase} in a minute.) Turning G_x on subsequently during readout will frequency-encode the echo signal and reveal ***x*** for each column.

In the current four-voxel example, since G_y is made to induce phase shifts of only two possible values—0° (first sequence) and π (second)—the respective signals will be labeled S_0 and S_π, to distinguish RF pulses (0° and 180°) from phase shifts (0 and π). The first and second iterations together will yield four linear equations in four unknowns, the m_i, which we would expect to be more than sufficient to solve the problem. (Another spoiler alert: it is not!) We then complete reconstruction by untangling all the signal data with a 2D FT to extract the part contributed by each voxel individually.

Working through our 4-voxel example in a bit more detail….

Preliminary study: G_y = G_y = 0

In the preliminary study, neither an *x*-gradient nor a ***y***-gradient is *ever* present, only a uniform external field and the four magnetization vectors precessing in the *x*-*y*-plane in synchrony: all at the same Larmor frequency and with no phase differences among them (Figure 12.8b). Around *t* = TE, in particular, they will together produce an echo signal of magnitude

$$\begin{aligned} S^{pre}{}_{1,2,3,4}(t \sim \mathrm{TE}) &= s^{pre}{}_1(t) + s^{pre}{}_2(t) + s^{pre}{}_3(t) + s^{pre}{}_4(t) \\ &= [m_1(t) + m_2(t) + m_3(t) + m_4(t)]. \end{aligned} \tag{12.3a}$$

Since the time parameter does nothing useful, let's drop it.

We can pass the signal through a Fourier analysis program and it will indicate, as expected, that the signal is a monochromatic temporal sine wave. Suppose that for our example, $S^{pre}{}_{1,2,3,4}$ is measured to be 7, say:

$$S^{pre}{}_{1,2,3,4}(\sim \mathrm{TE}) = 7. \tag{12.3b}$$

First S-E repetition: G_y = 0 throughout, but G_x on during readout

Things become interesting with the first repetition (Figure 12.8c). G_x is turned on for readout, an aspect of the process commonly called *frequency-encoding* of voxel position along the ***x***-axis in real-space. Here G_y remains off, however, and there is no phase encoding along the ***y***-axis. Or more correctly, there *is* phase encoding, but the phase shift between rows is by the amount 0°, signified by the superscript ^{0}.

Frequency-encoding along the *x*-axis works with G_x *on during readout*, as before. Part of the echo signal is generated by the pair of left-column voxels, 1 and 2, which precess completely in phase. Their combined contribution to the echo signal is measured to be the ***sum*** in their magnetizations, in this example found to be 5, say:

$$s^0{}_{1,2}(\sim \mathrm{TE}) = m_1 \oplus m_2 = 5. \tag{12.4a}$$

The two on the right resonate at a slightly higher frequency, as indicated by the longer curved arrows, and they combine to result in a value of 2:

$$s^0{}_{3,4}(\sim \mathrm{TE}) = m_3 + m_4 = 2. \tag{12.4b}$$

All four magnetizations remain in-phase up to and including the time that the echo arrives. The MRI coil and receiver will experience a composite RF signal arising from the superposition of those from the two columns, with the sum

$$S^0{}_{1,2,3,4}(\sim \mathrm{TE}) = s^0{}_{1,2} + s^0{}_{3,4}, \tag{12.4c}$$

which combine for the formation of a simple beat signal like that in the lower left-hand corner of Figure 12.8c.

After sampling and digitizing—and this is a critical point—a Fourier analysis breaks Equation (12.4c) back into Equations (12.4a) and (12.4b), yielding up separate values for $s^0{}_{1,2}$ and $s^0{}_{3,4}$. While this is a big step in the right direction, unfortunately it does not yet provide enough information to fully solve the problem. Equations (12.3a,b) comprise a system of two equations with four unknowns; to learn the values of the four unknowns m_1 through m_4, we need a total of four independent relationships among them. Spin-warp uses a clever trick to resolve the situation.

Second S-E repetition: G_y is on for the time t_{phase} before readout

After completion of the pulses and gradients that gave us the signal of Equations (12.4), we carry out a second repetition—another spin-echo measurement on the slice—but this time with a twist, literally.

G_y is activated for the *phase-encoding* period, t_{phase}, *before* readout, during which two magnetizations in the upper row of the matrix are made to precess exactly π (180°) farther than those in the lower row (hence the superscript on s^{π}) (Figure 12.11d). By our design, the upper row magnetizations both rotate through exactly one *extra* half-turn relative to those in the lower. After G_y is switched off, m_1 and m_3 of the top row are left pointing *opposite* to those of voxels 2 and 4. m_1 and m_2 now exactly counteract one another. This means that the net magnetization that they together produce as they precess (at the same frequency) is the *difference* of their separate magnetizations, rather than the sum. The strength $s^{\pi}{}_{1,2}$ of their combined contribution to the signal is

$$s^{\pi}{}_{1,2} = m_1 \ominus m_2 = -3. \qquad (12.5a)$$

Likewise, $[m_3 - m_4]$ is precessing at a slightly different frequency, with a contribution to the echo signal of

$$s^{\pi}{}_{3,4} = m_3 - m_4 = 2. \qquad (12.5b)$$

This time around, the combined signal reads

$$s^{\pi}{}_{1,2,3,4} = s^{\pi}{}_{1,2} + s^{\pi}{}_{3,4}. \qquad (12.5c)$$

$s^{\pi}{}_{1,2}$ and $s^{\pi}{}_{3,4}$ can also be determined separately by means of a Fourier transformation.

EXERCISE 12.4 What does the net signal from all voxels superimposed look like before readout? During readout?

EXERCISE 12.5 Is it meaningful that $S^{pre}{}_{1,2,3,4}$ and $S^0{}_{1,2,3,4}$ are both measured to be 7 near TE?

Note that the "π" labeling the SE *phase-encoding* y-gradient in this little 4-voxel example has nothing to do with the 180° SE *echo-generating* RF pulse that is about to be applied. Also, the phase shift happens to be π here only because this simplistic example involves a 2×2 tissue matrix. For a realistic 256×192 matrix, say, the phase shift between the bottom and the next highest voxel would be π/192, as explained below.

Solving the four equations for the four unknowns, m_i

In this four-pixel case, a total of two sets of measurements, with two different amounts of phase encoding, is adequate to generate the map of the voxel magnetizations existing at the time of echo formation. Equations (12.4a), (12.4b), (12.5a), and (12.5b) appear to be independent and, by inserting numerical values for s^0 and s^{π} obtained from the RF receiver, these equations should together be able to give values for m_1, m_2, m_3, and m_4 (Table 12.1). Combining the first and third of these yields $m_1 = 1$; similarly, from the other two, $m_3 = 2$. With a little more fiddling, we find all four, all in agreement with Figure 12.7b.

EXERCISE 12.6 Are the four equations independent? Is the preliminary study of any value here?

Table 12.1 Four independent equations in four unknowns, $m(j)$, should make it possible to solve for all of them

FT: $s^0{}_{1,2} + s^0{}_{3,4}$ → $s^0{}_{1,2} = m_1 + m_2 = 5$; $s^0{}_{3,4} = m_3 + m_4 = 2$

FT: $s^{\pi}{}_{1,2} + s^{\pi}{}_{3,4}$ → $s^{\pi}{}_{1,2} = m_1 - m_2 = -3$; $s^{\pi}{}_{3,4} = m_3 - m_4 = 2$

FT: $S^{pre}{}_{1,2,3,4}$ → $m_1 + m_2 + m_3 + m_4 = 7$

The above approach never really caught on in practice. Early attempts to formalize it as a *Fast Algebraic Reconstruction Technique* were seriously hampered, in particular by the inability to find a suitable acronym.

EXERCISE 12.7 What other problems might arise with the algebraic approach?

The phase-encoding process has been straightforward for the 2×2 case, but there is a need to present it in more rigorous terms for dealing with realistic voxel matrices. The shift in spin phase, $\Delta\varphi_y$, between adjacent rows (simply 180° here for 2 rows) in real-space will clearly be proportional their vertical separation, Δy, and the magnitude of the y-gradient. It is also linear in the *phase-encoding* period, t_{phase}, during which G_y is being activated. Indeed,

$$\Delta\varphi_y = (\gamma/2\pi)\, G_y \times \Delta y \times t_{phase}\,. \qquad (12.6)$$

EXERCISE 12.8 Demonstrate the validity of Equation (12.6).

EXERCISE 12.9 Suppose the matrix is 256×192, the y-gradient is that of Figure 3.5, the y-resolution is about 1 mm, and the minimum phase between lines is 180°/256. What is the value of t_{phase}?

EXERCISE 12.10 In Equation (12.6), one could change either G_y or t_{phase} between repetitions. Which is better, and why?

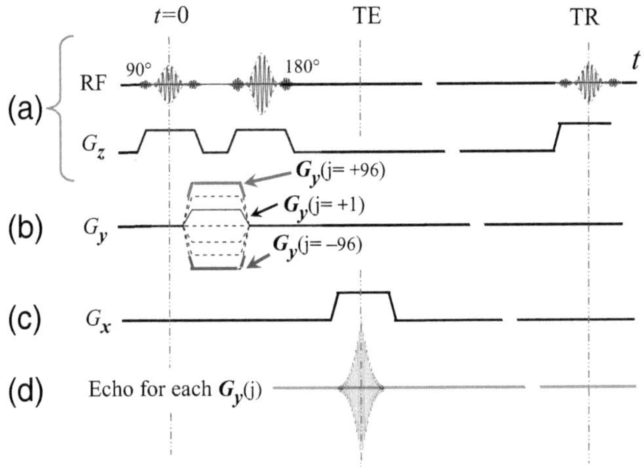

Figure 12.9 Essential steps in the spin-echo/spin-warp (SE/SW) 2D reconstruction of a transverse scan. A simplified version of one of the 192 repetitions of a standard gradient and RF pulse sequence for, say, a 256×192 phantom matrix. (a) The process begins with the standard 90° narrow-band RF pulse, perhaps 3 ms in duration, with the **z**-gradient activated. This leads to the excitation of protons only in a thin slice normal to the **z**-plane. (b) To begin work on the first (here, the lowest, j = −96) line in **k**-space (and before a 180° echo-generation pulse), the phase-encoding gradient G_y(j) is activated for the period Δt_{phase}. This first repetition is shown in red; the j = +1 scan appears in black, and the final one, j = +96 in blue. The increment in **y**-gradient between SE runs is $\Delta G_y = [G_y(j+1) - G_y(j)]$ for all j. (c) During readout, G_x is on for perhaps 8 ms, and each echo signal is sampled. (d) Data from this first set of MR echo signals are sampled at least 512 times, digitized, and entered into **k**-space as a horizontal line. After completion of this first *repetition*, at time TR, downloading of the next line in **k**-space is initiated, but now with a new magnitude for the gradient. **k**-space is filled in this fashion, echo by echo and line by line, up to the highest one, for which the phase has reached its greatest value, $\varphi_{y,max}$ = 180° or π.

12.5 The General Spin-Echo, Spin-Warp Echo Data Acquisition Process

Let's go over again the method of creating MRI images by means of spin-echo and spin-warp, but more generally, and for a 256×192-voxel matrix, which is quite typical. There is considerable variety in the fine details of the orchestration of RF pulse and gradient field sequences used in commercial MRI systems, but the one of Figure 12.9 is (apart from a few of its simplifications) fairly typical.

Imaging begins with the selection of the particular thin transverse slice of tissue at $z = z'$, as indicated in the top two rows (RF and $\boldsymbol{G_z}$) of Figure 12.9a. The z-gradient is turned at t = 0 and achieves its full magnitude with a switching time of several milliseconds. As soon as the z-gradient field has stabilized, the 90° RF excitation pulse is applied. Centered on the ν_{Larmor} carrier and with an envelope approximately of the Sinc(t) shape, it has been designed to optimize the balance among flat and parallel slice faces, good signal-to-noise, and other parameters that affect image quality. It will cause the net magnetization for all the protons in the chosen thin slice of tissue, and only them, to swing down into the x-y plane. The net transverse magnetization of $M(t) = \sum m_{xy}(x,t)$ continues to precess after cessation of the RF. Thus begins the spin-echo process.

When the z-gradient switches off, the first phase-encoding y-gradient ($\boldsymbol{G_y}$(j = −96) comes on for the first repetition, the red lowest $\boldsymbol{G_y}$-line in Figure 12.9b. Within the selected slice, the rate of proton precession and the phase angle accumulated over the phase-encoding time, t_{phase}, in any narrow strip of tissue lying parallel to the $\boldsymbol{x}$-axis will be determined by its y-coordinate, by the slope of $\boldsymbol{G}_y$, and by the exact duration of Δt_{phase}. Because the magnetization of a voxel located at a higher **y**-position in the selected plane is now in a slightly stronger local magnetic field than the ones lower down, it precesses a little faster, and it will pass through a greater total angle during the time that $\boldsymbol{G}_y$ is on. The phase-encoding time, Δt_{phase}, is the same for every repetition, and after each application, $\boldsymbol{G}_y$ is switched off. The phase angle of the magnetization vectors accumulated for any particular row of voxels is proportional to both the value of $\boldsymbol{G}_y$ and its **y**-position, as implied by Equation (12.6). After completion of this period of phase evolution, the **y**-gradient is removed.

The 180° echo-generating pulse is applied at time t = ½TE after the 90° pulse, and the system is left alone for nearly another ½TE (Figure 12.9c). Then, just before the echo is due to appear at t = TE, the **x**-gradient is turned on for signal readout. Because the x-gradient is present while the echo is being detected, the frequency (or k_x-value) of a voxel's contribution to the echo signal will be proportional to its x-coordinate. This, in effect, allows the sampling of the signal 2×256 = 512 times, for example, each occurring at a marginally different Larmor frequency, and corresponding to 256 different voxel values of x. (The factor of 2 is needed to satisfy the Shannon theorem.) The echo signal is detected by the pick-up coil, amplified, and digitized.

The next repetition begins the time TR after the previous 90° saturation pulse appeared, but now with a new value of the y-gradient, namely $\boldsymbol{G}_y$(j = –95) (Figure 12.9a), and so on. In 2D, there are many such horizontal $\boldsymbol{k}_x$-lines. In the case of a 256×192 matrix, there are 192 of them and, one at a time, sequentially they 'fill **k**-space.' Each $\boldsymbol{k}_x$-line employs exactly the same process as in 1D to deposit information along it. That is, to create a single complete 2D slice-scan in **k**-space, the entire pulse/gradient sequence is carried out for 192 repetitions, say, each with a different value of the **y**-gradient, and with each laying down a line in **k**-space. In this example, $\boldsymbol{G}_y$(j) increases for each repetition by an integer multiple of its smallest value, $\boldsymbol{G}_y$(j=1). So $\boldsymbol{G}_y$(j) = j×$\boldsymbol{G}_y$(j=1), for j = –96,…, +1,…,+95, where j=0 is included. Each $\boldsymbol{k}_x$-line employs exactly the same process as in 1D to deposit information along it.

This illustrates a simple prescription for a way to 'fill **k**-space,' one line at a time, one for every repetition and unique value of G_y. Once **k**-space is filled with parallel lines of MR raw data, by way of this or another path through it, the whole thing undergoes an inverse 2D Fourier transform that leads directly to the appearance of a clinically meaningful MR image in real-space.

The underlying algorithm is straightforward. Suppose you started at j = –96 and, having gone through 119 iterations, for example, you have reached j = 23:

1. For the current machine setting of $\boldsymbol{G}_y$, such as j = 23, carry out one (or most likely, several) complete SE sequences, gathering enough echoes to fill line 23 with 2×256 sample data.
2. Increment $\boldsymbol{G}_y$ by one $\Delta\boldsymbol{G}_y$.
3. Return to step 1 unless j = +95.
4. If j = +95, carry out a 2D inverse FT to transform from all of **k**-space to real space.

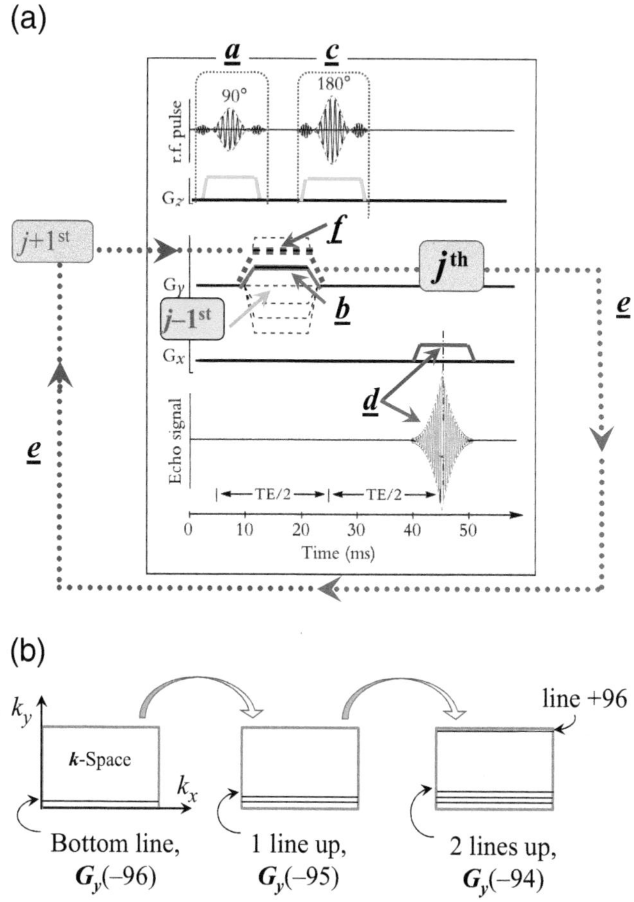

Figure 12.10 See Exercise 12.14 which asks you to write your own caption to this figure.

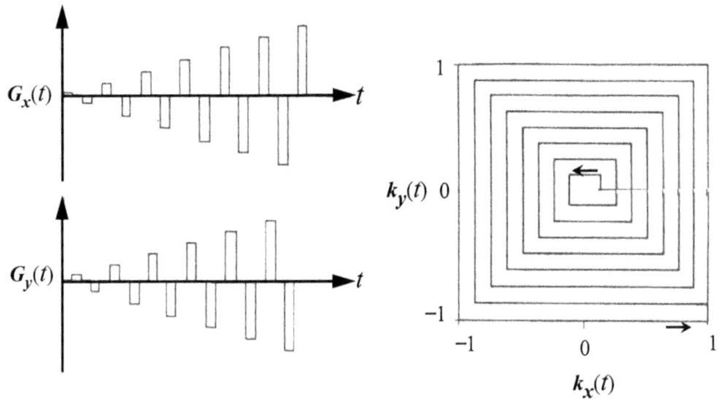

Figure 12.11 Another path in ***k***-space for data acquisition, to be contrasted with that of Figures 12.12. The machinations of $\mathbf{G_x}$ and $\mathbf{G_y}$ over time required to achieve this 'square spiral' will be discussed in the next chapter.

Figure 12.10 demonstrates a typical trajectory through 2D ***k***-space, a simple way to fill k-space, one line at a time. You will be asked in Exercise 12.14 to explain it. Bear in mind the definitions $k_x = \gamma/2\pi\ G_x\ t_x$ and $k_y = \gamma/2\pi\ G_y\ t_y$.

EXERCISE 12.11 What might be the impact of selecting G_y-values in the order of Figure 12.11?

*EXERCISE 12.12 Does a 256×192 matrix in **k**-space transform into a 256×192 matrix in 2D real-space?*

EXERCISE 12.13 Show that immediately after phase encoding, all spins for any value of y will have the same phase, $\varphi_{y,1}(y) = \varphi_y(0) + [(\gamma/2\pi)\ G_y(1)\ \times y \times \Delta t_{phase}]$.

In our example, values of φ_y run from 0° to 180° in every repetition. Other $\mathbf{G_y}$(j) sequences, or paths through ***k***-space, are possible, and may be preferable. One possible path is the 'square spiral' of Figure 12.11; this and several other variants on it will be addressed in the next chapter, which introduces techniques of 'fast' imaging. Each of these filling approaches has its benefits and disadvantages, and the choice depends on the clinical situation. But whatever the method, MR imaging consists of filling out the matrix of k_x- and k_y-values that comprise ***k***-space for the z-slice being imaged.

Finally, Figure 11.11 walks along a trajectory in 1D ***k***-space for a simple sequence. We now ask you to do the same for a standard SE-SW sequence in 2D in the following exercise. Figure 11.11 should allow you to carry it out, but we've included a few others to help along (Figure 12.12).

EXERCISE 12.14 Write a caption for Figure 12.12, explaining it verbally, and insert the RF pulses where they belong.

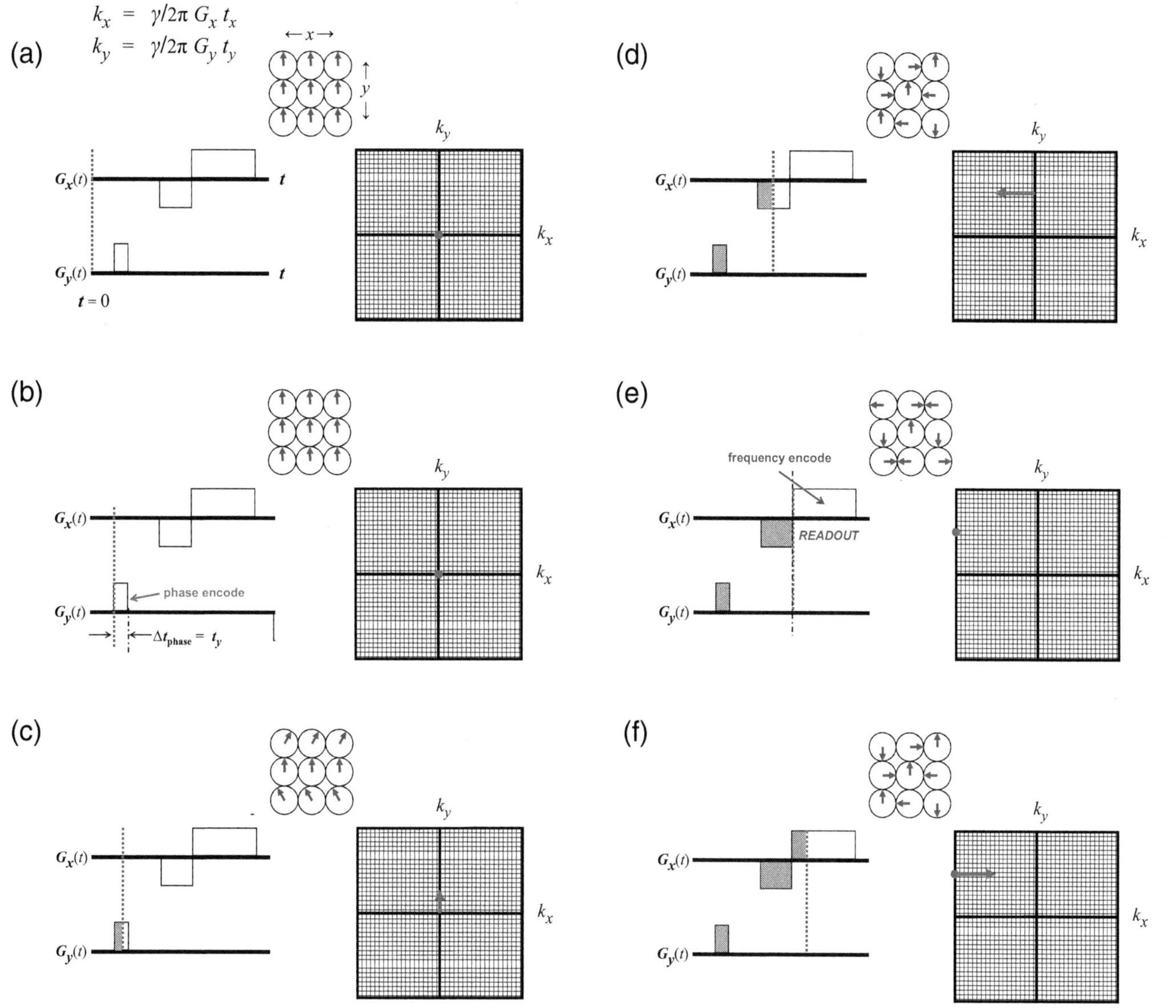

Figure 12.12 Some steps along one path through 2D **k**-space for SE-SW, to help with Exercise 12.14.

CHAPTER 13

Fast Imaging

A major objective of MR researchers and manufacturers has long been to reduce image acquisition time as much as possible, but with little or no loss of image quality. Developers continue to design specialty radiofrequency and gradient pulse sequences that by now are fast enough to capture the beating of a heart in cine mode, and with good enough images to satisfy a broad reach of clinical needs. Techniques devised to reduce scan speed come under the general heading of *fast MRI imaging.*

In the earliest days, pre-clinical *sensitive point reconstruction* obtained an image only one voxel at a time: by causing the three gradient fields to seesaw wildly everywhere except at one spot, the signal could be read out there, and only there, as if in a simple NMR study of something in a test tube. The sensitive point was then shifted over to an adjacent voxel, and so on and on. The obvious problem is the inordinate time required to measure signal strength at a hundred thousand or so positions within the body.

The first clinical MRI studies were not as slow as that, but still required tens of minutes for a single slice —which raised questions about the modality's potential for routine medical applications. But fast spin-warp and other forms of rapid imaging have come a very long way since then, and data acquisition times for some studies are now measured in fractions of seconds. Cardiac, abdominal, breast, and other forms of imaging large organs required moving beyond SE with fast gradient echo imaging. Many functional imaging techniques were fully realized only with the acquisition of all of ***k*****-**space in a single excitation. And breath-holding or higher-resolution acquisitions demanded images that could be reconstructed using under-sampled raw data, again to shorten scan time.

Fast imaging procedures can be rather complex, and this chapter may be a little more demanding than the others. Hopefully it will introduce the field in a manner that is not too taxing, however, and that it will impart a sense of the workings of the more important processes currently in wide use in the clinic. These are described more fully in the literature [Bernstein et al. 2004].

13.1 Time to Acquire a Slice-Image

What perhaps most distinguishes the operation of an MRI machine, from the physician's and radiographer's perspectives, is the number of possible combinations of parameter values to be selected from and the extent to which the nature of the information acquired is sensitive to the particular set settled upon.

The parameter values to be employed in a study is a complicated and evolving business, and it is driven by several factors—the clinical objective, the methods chosen to enhance the contrast-to-noise ratio or resolution, the need for gadolinium or another contrast agent, aspects of patient anatomy, etc. While some of these are fixed, such as the main field strength, many are under the immediate control of the operator, albeit subject to well-established general guidelines. These include

- the type of sequence (spin-echo, gradient-echo);
- ***k***-space trajectory (Cartesian, echo-planar, radial, spiral);
- the timing (TE, TR, TI) and, sometimes, amplitudes of RF pulses and applied gradient fields;
- the slice thickness and inter-slice separation;
- the dimensions of the voxel matrix and of the field of view (which together determine the in-plane voxel size, hence a limit on achievable resolution);
- the use of physiologic gating (cardiac, peripheral, breathing);
- surface coil element combinations to detect the RF echoes; and
- other factors.

The selection of a specific set of parameters will affect the fundamental image characteristics—such as the inherent signal and noise properties, contrast-to-noise ratio, resolution, the presence of a variety of artifacts, and the time it takes to acquire a slice image (Figure 13.1). Fast imaging generally consists of optimizing these parameter values in some fashion so as to decrease imaging time. Reduction of the repetition time TR, for example, will generally (but not always) allow for faster imaging, but it also modifies contrast, available signal, and noise.

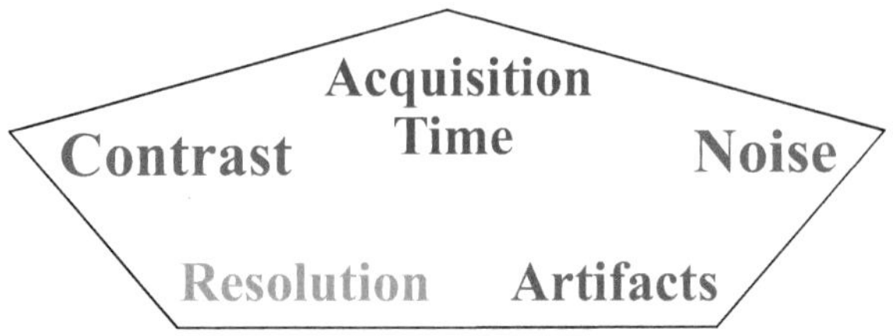

Figure 13.1 There are trade-offs among the important measures of image quality, including acquisition time. The balance achieved in a given study is determined largely by the specific operating parameters selected by the radiographer.

Let's start with a simple expression for the time needed to acquire an image, one that is determined ultimately by the nature of the pulse sequence. The imaging time can be viewed as the product of two factors that often seem independent, but that may interact in subtle ways. The first is the duration of a single *excitation/readout period*, which occurs over an excitation pulse repetition time, TR, discussed in connection with Figures 11.2, 12.11a, and 12.12a, and elsewhere. The other is the total *number of distinct excitation/readout periods*, $n_{e/r}$, needed to carry out the scan, suggested in Figures 11.4, 12.12, etc. Then the image *acquisition time* is the product of the readout period and the number of periods:

$$T_{acq} = n_{e/r} \times TR, \qquad (13.1a)$$

where the interesting details are rolled into $n_{e/r}$. Equation (13.1a) is generally adequate to cover a number of sequences and acquisition strategies (and we will refer back to it), but it is not, as it stands, flexible enough to yield the accumulation time for all possible kinds of scans.

With one standard rudimentary sequence for scanning a single 2D slice with Cartesian sampling, a single line of ***k***-space is read out from a single echo in the frequency encoding direction for each excitation of tissue; the $\boldsymbol{k}_y$ phase is then incremented to collect another line from the new position. This routine is repeated again and again until all of ***k***-space is filled. This is not the fastest way to capture an image, but it was a great improvement on sensitive point reconstruction, since a full line (i.e., multiple data points) of ***k***-space is acquired for each excitation. To fill all of ***k***-space based on the desired resolution and phase field of view, a total number of phase-encoding steps, N_{PE}, or, equivalently, N_y, is normally required, and

$$T_{acq} = N_y \times TR, \qquad (13.1b)$$

One way to reduce the incoherent stochastic noise is by acquiring multiple copies of the same image, with exactly the same machine settings, and then averaging them together (i.e., increased *sampling time*), Equation (5.27b). The number of copies, usually 2 to 8, is referred to as *NEX* (number of excitations), *NSA* (number of signal averages), or *ACQ* (acquisitions), depending on the scanner vendor. The total *acquisition time* for a single 2D-slice, multiple-*NEX* image is then

$$T_{acq} = N_y \times NEX \times TR. \qquad (13.1c)$$

To decrease the image acquisition time within this acquisition framework, one can reduce either the number of phase encodings (hence, vertical resolution), or the TR (signal and contrast), or perhaps the *NEX* (resulting in more noise). Understanding these trade-offs between image quality and acquisition time are a fundamental part of optimizing an MRI protocol. Choosing a data acquisition schema in which *NEX* = 1 may considerably shorten the imaging time, but it can lead to an unacceptable signal-to-noise ratio. Likewise, either decreasing the number of phase encodings (and vertical resolution) or altering how many are collected per TR are widely used methods for increasing image speed, but these choices can affect resolution and general appearance.

Figure 13.2 diagrams the excitation and readout process for one line in ***k***-space, where N_y such lines are

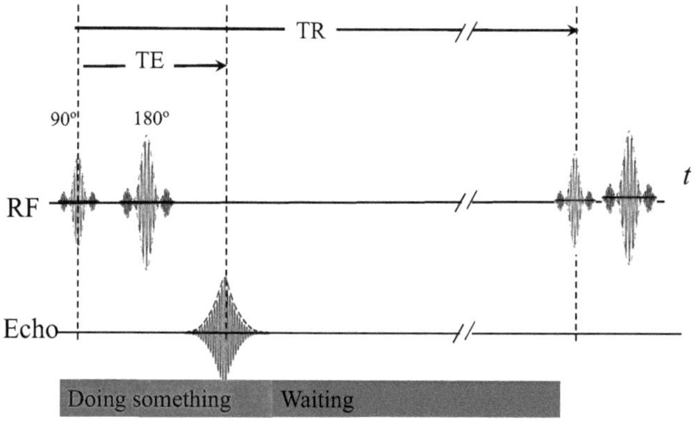

Figure 13.2 Duty cycle during a one-slice rudimentary spin-echo sequence. Between the 90°excitation pulse and the end of the echo readout, transverse magnetization exists that is used for imaging. After the readout, the residual transverse magnetization is *crushed*, and the scanner is quiescent (*waiting*) for the rest of the TR period required to generate the desired contrast. One can use this time to advantage for multi-slice imaging.

needed for a single SE 2D slice-image. Readout of the line into ***k***-space begins during the time interval TE, during which the scanner is exciting and manipulating the transverse magnetization. In general, TE is taken to begin at the center of the 90° excitation pulse and to terminate at the center of the echo (exceptions are discussed below in Section 13.10). For conventional imaging, the first half of the echo signal is read out by the time you reach t = TE. If you follow the horizontal path in ***k***-space along the k_x-encoding line that passes through the origin (i.e., the line with 0° phase encoding), the end of the TE period is the time when data are being acquired at the very center of ***k***-space, at ***k*** = 0.

The data in the central region of ***k***-space define the overall bulk contrast of the clinical image; it is the data in the periphery of ***k***-space that determines the fine details. It is the relative durations of TR and TE that determine the T1- vs. T2-weighting of the resulting image (Table 11.3).

For T2-weighted sequences, in particular, TR of a scan line can be quite long (Table 11.5 and Figure 11.6). This creates a problem, given Equation (13.1c) above: long TR leads to long T_{acq} and slow imaging. *Multi-slice imaging*, however, offers a significantly faster solution that maintains the TE- and TR-values for the correct kind of weighting contrast, by using the time far more economically. It just gets a little more complicated.

The echo readout window extends farther than the end of TE by an amount we shall call the *Residual Readout Time* (RTT). For optimal signal-to-noise ratio and to minimize artifacts, it is often desired to capture the whole echo and record it until time (TE + RRT).

13.2 Multiplexed Multiple Slice Imaging

If we were to acquire data from only one slice per each excitation pulse—that is that we sample from only one slice each TR period—then the time it would take to scan six slices would be $T_{6\ slices} = 6 \times N_y \times NEX \times TR$. Very slow! Multi-slice imaging attempts to excite and read out more data per excitation than single-slice techniques. Specifically, the acquisition of multiple slices is ordered in a particular way, or *multiplexed*, so as to gather multi-slice data within a single TR period! The system will not, in general, be able to examine *all* N_{slices} planes at the same time; instead, multi-slice imaging partitions a stack of slices into a number,

$N_{\mathbf{batches}}$, of smaller *batches*, in such a way that the slice measurements and calculations within a batch do not interfere with one another.

Each batch consists of $N_{\mathbf{slices/batch}}$ slices and, of course, there are

$$N_{\mathbf{batches}} = N_{\mathbf{slices}} / (N_{\mathbf{slices/batch}}) \quad (13.2)$$

batches altogether. After a batch is done, then its on to the next. The system operates on the batches one by one, so there is a need to determine how many batches there will be, $N_{\mathbf{batches}}$, and also $N_{\mathbf{slices/batch}}$. But how to do this in practice?

Suppose, as an example, we estimate in advance that it will take $N_{\mathbf{slices}} = 30$ slices of a certain thickness and distance apart (or *slice gap*), each at a different value of z within the patient, to cover the region of interest in the body. With a multiple-slice sequence, the operator selects values for TE and TR (which determines the T1- or T2-weighting) and enters the desired $N_{\mathbf{slices}}$. A straightforward computer program determines RRT based on receiver BW, gradient slew rate, and other parameters, and will also take care of $N_{\mathbf{batches}}$ and the rest.

In rare, fortuitous cases, it may *happen* that (TE + RRT) divides into TR an integral number of times. For example, if we want to scan $N_{\mathbf{slices}} = 30$ and $N_{\mathbf{slices/batch}}$ = TR/(TE + RRT) happens to be 6, as in Figure 13.3, then the system partitions the slices as

$$N_{\mathbf{batches}} = N_{\mathbf{slices}} / (N_{\mathbf{slices/batch}})$$

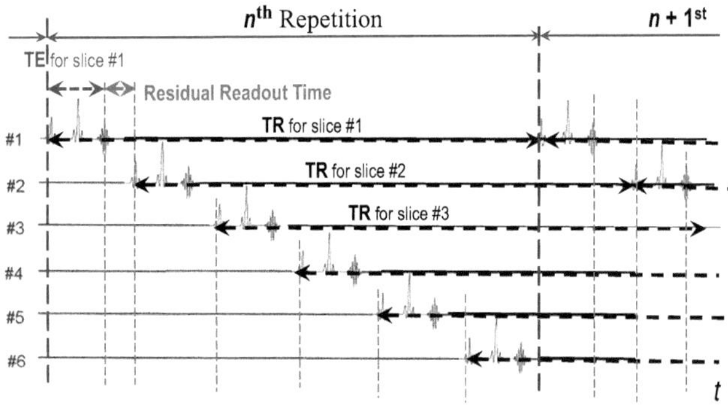

Figure 13.3 Temporally interleaved multi-slice acquisition for a simple spin-echo sequence. In this batch of six sequential scans, each one begins the time (TE + RRT) later than its predecessor, and also at a greater **z**-value (or depth in your 2D imaging slice stack) within the patient. In this particularly simple case, TR / (TE + RRT) happens naturally to be an integer, 6.

from Equation (13.2), or 30 / 6 = 5 batches. This is an idealized case, however, and in reality there is usually some slop in the timing. Suppose the system makes a preliminary estimate that the pulse sequence should put 5½ slices into each batch. That is meaningless, and the standard procedure is to truncate $N_{\mathbf{slices/batch}}$ down to the nearest integer value, bearing in mind that

$$\mathrm{TR} \geq N_{\mathbf{slices/batch}} \times (\mathrm{TE}+\mathrm{RRT}). \quad (13.3a)$$

There will most likely be a need to image more than one batch, so the total image acquisition time will be

$$\mathrm{T}_{\mathbf{acq}} = N_{\mathbf{batches}} \times \mathrm{T}_{\mathbf{batch}} = N_{\mathbf{batches}} \times N_y \times NEX \times \mathrm{TR}. \quad (13.3b)$$

How this is done is sketched below. But first, staying momentarily with a single batch....

The top slice-line in Figure 13.3, indicated as 'slice #1,' is largely the same as Figure 13.2. Here, however, it appears as the first scan-line in a batch of six (in this example) planes, each at its own specific value of z. All six are designed to have the same settings for TR and TE, and they are to be imaged more or less as a group. After the system has finished taking care of these six, it will turn to the next batch. Figure 13.4 shows the same kind of thing in a slightly different fashion.

The trick is to distinguish and separate out one particular slice within a batch at a given time, and excite and receive data from it alone. Two things are going on here that together bring about that goal.

First, just after a slice is excited and the echo read out, the z-component of the local magnetic field, $\boldsymbol{B}_z(t)$, is incremented immediately before addressing the next slice. That is, the excitation frequency is changed by a small amount before excitation of the following slice. Since the slices reside at the six distinct z-levels, and experience different values of $\boldsymbol{B}_z(z)$, each has its own Larmor frequency for its 90° and 180° RF pulses and echo. And Fourier methods can tell them apart. This is the same sort of thing used earlier to select out a single slice for imaging, Figure 12.6 and Equations (12.1).

Second, immediately after the readout is complete for slice #1, the excitation of slice #2 and its readout can begin within the same TR period. And so on. That is, there is a temporal shift between the start times of work on the slices in a batch: work on #2 slice is not initiated until time t = TE + RRT, which is after the slice #1 echo has been read completely. As evident from Figure 13.3, the 'TR' period commences later and

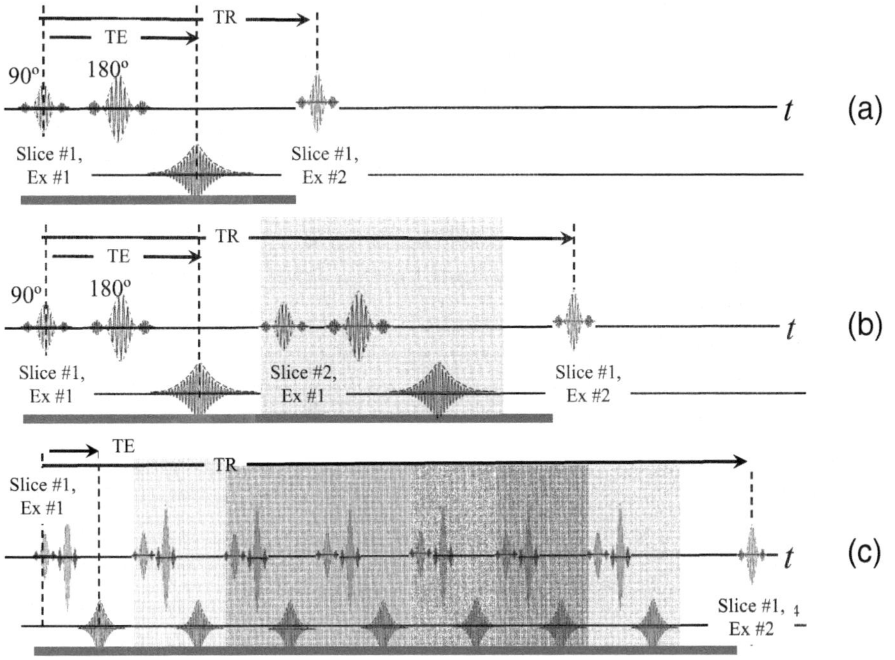

Figure 13.4 Another way of displaying multi-slice acquisition timing. TR is fixed and TE is adjusted as necessary for a desired T1- or T2-weighting. In the previous diagram, the several slices in the batch are represented as separate parallel lines, each with a unique value of **z** within the patient. Here the slices in a batch appear, instead, in different color blocks. (a) The horizontal orange line shows time when coherent transverse magnetization exists on slice #1. TE > ½TR here, so there is time for only one line of ***k***-space to be excited and read out in a TR period. In this case, tissue from the same slice will be excited again to preserve any T1 contrast for a given TR value. (b) TE < ½TR, so during one TR period, the excitation and readout of data for a batch of two difference slices can be interleaved. (c) In this diagram, TE is taken to be significantly shorter than TR, allowing for the interleaving of a batch of seven slice readouts.

later as the value of z increases, but it is always of the same duration; so while we could indicate a TR(z) for each, it suffices just to call them all TR. The lengths of the TR- and TE-periods (i.e., the timing of the RF pulses) and of RRT do not change, nor do the $G_x(t)$ and $G_y(t)$ settings. So exactly the same kind of information is produced in each case, except that the slices are located on six separate z-planes. The end of TR period for slice #3 occurs at t = TR + 2 × (TE + RRT), and so on. One can begin this process for all six slices in the batch before the end of the slice #1 TR, and it seems the TR period for slice #6 ends a time almost 2 × TR (= 2 × TR – (TE + RRT)$_{\text{slice \#6}}$) after the beginning of #1. It would thus appear, then, that each six-slice batch can be imaged in the total time span of $T_{\text{batch}} = 2 \times N_y \times NEX \times \text{TR}$. While eminently reasonable, this is nevertheless incorrect: after the last necessary TE+RRT period for slice #6—the last phase encoding, for the last line in k-space—the scanner just stops. Since there is no more data to collect, we're done. On Figure 13.3, imagine that the ***n***th repetition is the last. The readout for slice #6 occurs during the TR for slice #1, and we need go no further. So, $T_{\text{batch}} = N_y \times NEX \times \text{TR}$.

As just shown, the transverse magnetization at any given time arises from only a single slice. In spin-echo, we excite and refocus (with a 180° pulse) the transverse magnetization in the one slice and then read out its echo, including over the residual readout time (RRT). After that, we can excite magnetization from other slices. We might call this process something like "temporally interleaved multi-slice acquisition." Or, trying to avoid such a mouthful, we could just consider this to be a natural part of multi-slice acquisition, to be invoked whenever possible.

You may be wondering how to prevent transverse magnetization that might be remaining in slice #1 from affecting other slices, considering that we're interleaving slices within one TR? Any time multiple RF pulses are invoked during a TR, they can create unwanted echoes that lead to ghosting (even for single-slice imaging)! For example, if a 180° pulse is not perfect (more like 179°, for example), this can inadvertently excite residual longitudinal magnetization at the time of that pulse into a small bit of coherent transverse magnetization. This bit of transverse magnetization behaves like a tiny free-induction decay signal, just like what was created after the 90° pulse. One can mitigate these spurious echoes by inserting *crusher* gradients around each 180°. The two crusher gradients around the 180° pulse are mirror images of each other, and whatever dephasing they create for transverse magnetization before the 180° is re-phased afterward. However, the crusher gradient after the 180° pulse destroys any *newly generated bit* of transverse magnetization from the imperfect 180°. For multi-slice imaging, any gradient pulses that play out for slice #2 act as crushers of the residual transverse magnetization that may remain for slice #1, and so on. This is a fairly complicated process, the details of which are beyond the scope of this discussion.

Ideally, no residual transverse magnetization generally carries over from slice to slice, and we could excite magnetization from other slices. However, the order of slice excitation is critical. Try as we might, the slice profile cannot be made perfectly rectangular, which would require a perfect RF Sinc pulse of infinite duration to play out. The time savings arises from a truncated Sinc pulse, which gives "shoulders" to the slice profile. If slices are placed next to each other, these shoulders will overlap, and overlapping regions could experience excitation for adjacent slices. To prevent this, the slice gap puts some distance between adjacent shoulders (anatomy in this gap is not imaged). Another solution is to excite slices in an interleaved fashion so that adjacent slices are excited farther in time from each other. For example, if a stack of 10 slices is excited during one TR period, the time difference between excitations for sequentially acquired slices is TR/10. With interleaved imaging, every other slice is excited before returning to the one in between. In this scheme, the time difference between excitation is TR/2, which might be enough time for significant T1 relaxation to occur in the overlapping region.

Suffice it to say, no residual transverse magnetization generally carries over from slice to slice.

On Calculating the Numbers of Batches and Slices per Batch

This subsection covers more details on how to calculate the numbers of batches and of the slices-per-batch. You may wish to skip over it on a first reading.

The scanner is focused on *do*ing something with the transverse magnetization in one slice between excitation and the end of the readout window. For shorthand, we will refer to the period TE + RRT (which is rather clunky to repeat over and over) as T_{do}.

Equation (13.1c) suggests that it might be possible to go faster by shortening TR. This is the case when scanning a single slice. But multi-slice imaging is one of those few applications where it is ***not*** necessarily true and, indeed, shortening TR here may actually elongate the acquisition time! Equation (13.3b) is valid only for integer values for $N_{slices/batch} = TR/T_{do}$, so it is necessary to deal with the case of non-integer TR/T_{do}.

With multi-slice imaging, the number of slices in a batch is controlled by three factors: the number of slices that the user specifies, N_{slices}, the TR duration, and T_{do} (which is intrinsic to the pulse sequence timing and which is always less than TR by definition). A formal expression is given as the following:

$$N_{\text{batches}} \equiv \text{ceiling}\left\{N_{slices} / \text{floor}\left[\min(TR/T_{do}, N_{slices})\right]\right\}. \tag{13.4}$$

The function min() chooses the lesser of the two quantities within its parentheses; the floor[ ] function rounds the expression *down* to the next lowest integer; and ceiling{ } rounds the expression up to the next highest integer. The formulation of N_{batches} is somewhat complicated, but here is the gist of it: to image 12 slices say, one selects a particular TR to give the appropriate weighting. But suppose we have time to excite only 4 slices during that particular TR. In that case, we'd need to scan $N_{\text{batches}} = 12/4 = 3$ batches of 4 images.

Difficulties arise when the number of slices does not fit evenly into batches of slices. Consider an example in which we want to scan 11 slices using a TR that we chose for a particular weighting. Two extreme cases will illustrate how Equation (13.4) works. If all 11 slices can be excited and read out within the span of one TR, then TR/T_{do} is $\geq N_{slices}$, and the min() term = N_{slices}. Likewise, the floor[ ] term will then be N_{slices}, and the entire term will be evaluated as ceiling$[N_{slices} / N_{slices}] = 1$. Our expression Equation (13.3b) is equal to Equation (13.1c). Here, the number of slices does not affect scan time at all! In practice, there is often a range for TR that is acceptable for the protocol, allowing the technologist to adjust the TR slightly to increase/decrease the number of slices that will fit into the TR period and, thus, optimize the acquisition for efficiency.

If, however, only one slice can be excited and read out during one TR, then T_{do} is very close to TR, and the min() term = $TR/T_{do} = 1$. Finally, if the denominator goes to 1, and the ceiling term becomes N_{slices}, and gives back Equation (13.3b) with N_{batches} set to N_{slices}. This is common with gradient echo imaging with short TR, where the scan time will scale directly with the number of slices.

For all cases in between, where some slices can be read out in a batch during one TR, the actual scan time scales in a strange, stepwise manner with increasing numbers of slices—it scales with the number of batches. In other words, adding a small number of slices may not increase the scan time at all, but adding more beyond that will increase N_{batches}, causing the scan time to jump to a longer duration.

In practice, Equation (13.4) is not widely used because T_{do} is difficult to determine precisely unless you have a diagram of the pulse sequence. But we can intuit a few concepts from the expression. For one-slice imaging or for imaging where $T_{do} \ll$ TR, Equation (13.3b) reduces to Equation (13.1c), and decreases of TR can lead directly to speed improvements as expected. But, if $1 < \text{TR}/\text{T}_{do} < N_{slices}$, the whole stack of slices cannot be excited within one TR, and scanning must occur in batches of a few slices at a time. Furthermore, decreases of TR can affect the number of slices per batch, as in Equation (13.4), leading to the non-intuitive result that overall imaging time can *increase* with decreasing TR!

13.3 Stack of 2D Slices vs. 3D Block

With volume-imaging modalities such as multi-slice CT, MRI, SPECT, and PET, regions of interest can be shown either in the form of a set of single-slices or in three dimensions. A three-dimensional rendering of organs and bones can assist in preoperative planning for surgical, radiotherapy, and other treatments. 3D MR images are commonly created these days with an inherently 3D pulse sequence, developed to capture an entire 3D block of tissue in a single scan. These are typically high in spatial resolution and in SNR, and they have proven helpful in examining cardiovascular problems, hepatobiliary physiology, orthopedic damage, the brain and peripheral nervous system, and other systems.

An early method of obtaining 3D images is to combine multiple adjacent 2D slices. The surface-rendered image of Figure 13.5a, for example, was built out of a number of separate CT cranial slices of the skull of a 2,200-year-old individual, until recently a resident of Luxor, Egypt, and currently at home in the Smithsonian Institution in Washington, D.C. On every slice, a computer program automatically located each interface between bone and what had been soft tissue (or, in this case, air), where the x-ray attenuation coefficient changes abruptly, and drew a contour line there (Figure 13.5b). It stacked the curves in three dimensions (Figure 13.5c) and then joined them optically so as to create a cover representing bone surface (Figure 13.5d). Finally, it smoothed off the edges and modulated the shading to give the impression of smooth overlap and

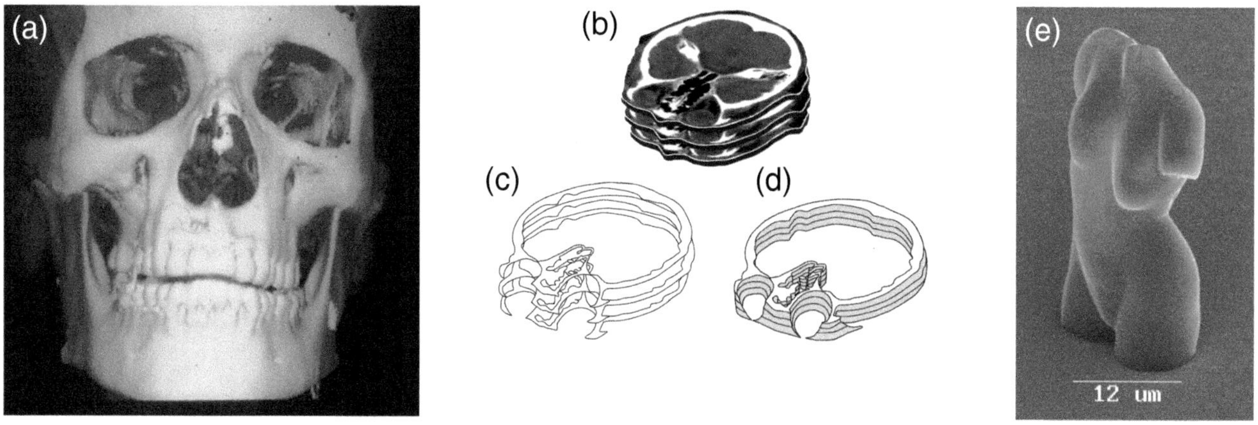

Figure 13.5 Three-dimensional display of the skull of a 45-year-old former resident of Luxor, Egypt, who died 2200 years ago, apparently of natural causes. He is now part of the permanent exhibit of mummies at the Smithsonian Institution in Washington, D.C. (a) Surface rendering of the skull. [Photograph courtesy of Wayne Olan, M.D., George Washington Medical Center, Washington, DC.] (b) The process began with a set of separate, adjacent transverse CT slices. CT was a good choice since no soft tissue remains, only bone. (c) On each, the computer automatically drew lines at the bone surfaces, stacked the curves in three dimensions, and from this produced a smooth chicken-wire-frame image of the skull. (d) It then paved over the areas between neighboring contours and added shading. (e) Three-dimensional Venus only micrometers (μm!) tall, in this case from an optical surface determination and then a laser rendering with UV-sensitive resins. Short laser pulses are focused into the volume of the liquid resin and, by moving the laser focus along her, the outer shell is polymerized; the non-irradiated resin is then washed away with solvents. This method, one of numerous now commercially available with 3D printers, can achieve a resolution down to 0.1 μm or better. [Photo kindly provided by Laser Zentrum Hannover e.V.]

depth. It's rather like creating a papier-mâché object by papering over a chicken-wire framework, but here it's all done by the computer. You may have created animals in a similar fashion as a child, gluing together shaped cardboard cutouts.

A step beyond a three-dimensional image display of a skull is a 3D plastic replication of it (Figure 13.5). From multi-slice CT or MRI information, a photolithography *3D printer* can create a life-size (or any other size) model skull or organ that a surgeon or radiation oncologist might want to be able to play around with—a nearly exact copy of the real thing! Because everything starts off in digital form, replications can be produced rapidly anywhere there is a compatible fabricator.

For MRI, in particular, one can create a 3D representation either out of a stack of 2D slices or from a single 3D pulse sequence. A principal difference between the two is how the tissue volume is visualized. In the first case, each 2D sequence excites a thin slice using a z-axis slice-selection process, and then performs 2D spatial encoding within the ***x-y*** plane. A 3D sequence, on the other hand, excites a thick slab, then performs something like the typical 2D spatial encoding across the FOV; if nothing else were done, a 3D sequence would just produce one very thick, low-resolution slice as per a 2D sequence. But an additional phase-encoding step is performed in the slice-select direction to localize the signal more precisely along the z-direction.

Because of this extra phase-encoding step, the imaging time is slightly different from that of Equation (13.1c):

$$T_{acq} = N_y \times N_z \times NEX \times TR. \qquad (13.5)$$

N_z is the phase-encoding matrix in the slice-select z-direction, whereas N_y is the phase-encoding matrix within the image plane. The extra multiplier may appear to increase the imaging time when compared with Equation (13.1c); however, TR is generally shorter for 3D sequences in practice. All effects considered, 3D imaging times are somewhat longer than those for 2D imaging.

13.4 Fast Spin Echo vs. Conventional Spin Echo—More Readouts per Excitation

Fast spin-echo (FSE) imaging techniques, including RARE (Rapid Acquisition with Relaxation Enhancement) and its derivatives such as TSE (Turbo Spin-Echo), produce images in much less time than standard spin-echo. One might employ a single 90° excitation followed by a train of a number of 180° pulses, every one created with a new and separate phase encode, $G_y(t)$. This generates a corresponding sequence of multiple echoes, each to be read out with the frequency-encoding gradients on (not shown, Figure 13.6). The overall scan time for a slice is reduced roughly by the number of echoes, or *echo train length* (ETL), for each

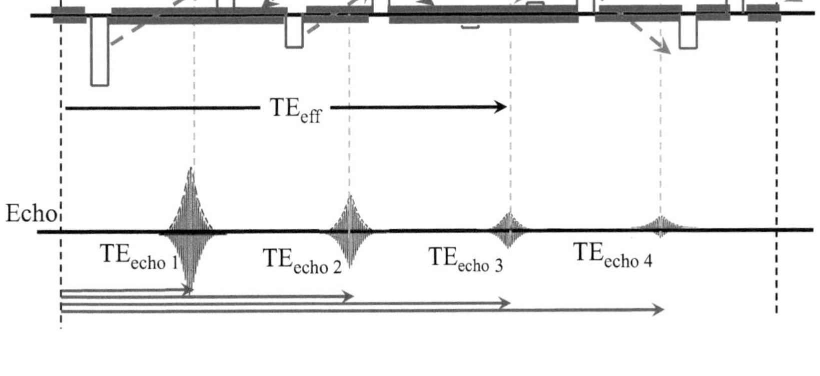

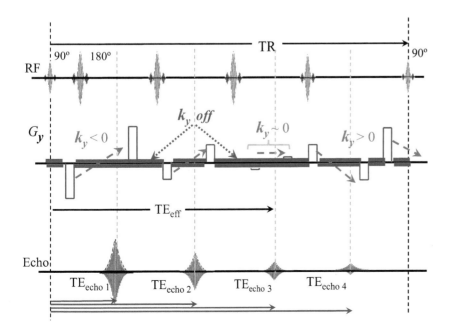

Figure 13.6 Pulse-sequence diagram for an FSE sequence. (Slice select gradients, G_z, are not shown, nor are frequency-encoding gradients, G_x.) The phase encoding from the previous step is reversed after each readout (red dashed arrows), to permit a new value of phase encoding in the next echo. TE shows the time between consecutive echo signals, each one occurring at a different time after the excitation (labeled here as $TE_{echo\ n}$, where n is the number of the echo). TE_{eff} is the time to the echo occurring while traversing the middle of k-space (i.e., where there is no phase-encode gradient).

sequence. Commonly the ETL = 4 to 24 for 2D FSE sequences—although by the time you read this, that number may be greater. The data acquisition time for the FSE can thereby be reduced from something like 12 minutes to just 2 or 3.

Another flavor of fast SE imaging is a sequence called *Single-Shot Fast Spin Echo* (SSFSE), where an image is acquired after only one excitation. In this case, the ETL is quite long (~60 or more, with help from other techniques), and the quality is correspondingly worse than conventional FSE.

The ETL obviously affects the scan time. The factor of N_y appears in Equation (13.1c) because one phase encoding is performed per slice for every TR period. Fast spin-echo imaging uses an echo train created by multiple 180° pulses to generate a sequence of multiple echoes free from T2* relaxation effects. This modifies the single-slice imaging time equation to

$$T_{acq} = (N_y \times NEX \times \mathrm{TR}) / \mathrm{ETL}. \qquad (13.6)$$

SE is the clinical workhorse sequence for acquiring T2-weighted images because of the excellent speed, in spite of the necessarily long TR because of the echo train.

3D FSE sequences can use much higher ETL values, on the order of ~100, but we will not discuss these in detail here.

Nothing is easy or free in this world, and one of the complications of FSE sequences arises if the 180° refocusing flip angles aren't exactly as strong as advertised. Without special care, this can lead to ghosting created by other spurious signals known as *stimulated echoes*. To prevent these, a set of requirements called the CPMG (Carr-Purcell-Meiboom-Gill) conditions must be met. These include the stipulation that the refocusing pulses are 90° out of phase with the excitation pulse, and the refocusing pulses must occur regularly at two times the interval between the excitation and refocusing pulse. The CPMG condition constrains any stimulated echoes to occur simultaneously with the spin echo.

Even though a series of echoes is acquired, there is often still time left over within a TR period for several more in the ETL. Therefore, FSE commonly excites multiple slices during one TR. This leads to improvements in speed analogous to the predictions of Equation (13.3b), but with ETL taken into consideration in determining the number of slices in a batch that can be scanned in TR. Considering the extremely long TR (often 8 sec or more) for FSE-based FLAIR sequences (Figure 11.12), multi-slice imaging is also desirable and possible for them.

The timing of interleaving slice excitations in one TR is more complicated than our earlier six-slice discussion would suggest; the inversion pulse does not create transverse magnetization, which allows for excitation and relaxation of multiple slices during one TI interval of the sequence. Equation (13.3b) is, therefore, appropriate for FLAIR imaging. Without such speed improvements, FLAIR would take way too long in the clinic.

The definition of image contrast for FSE sequences tends to be more abstract than for SE. The echoes shown in Figure 13.6 are decreasing monotonically due to T2 relaxation, which means that different lines in ***k***-space experience different amounts of relaxation. Since multiple echoes are acquired during one excitation, each one echo will form at a slightly different time as the sequence plays out. Technically, the sequence behaves as if data on different lines of ***k***-space are collected using different TE values. Since data from different lines in ***k***-space contribute to different wavelength scales in the image, structures of different sizes (in particular, fine detail in an image) will show differing amounts of T2 relaxation. It's a little confusing. Some degree of blurring may occur, known appropriately as *T2 blurring*. Again, however, the predominant overall *spatial contrast* of an image (as opposed to the *details*) is determined largely by the time associated with *echoes that correspond to points near the middle of **k**-space* (where large-scale contrast information resides). The time at which this particular echo forms after excitation goes by the name of *effective* TE (TE_{eff}), because the image contrast looks as though an echo time of TE_{eff} was used.

Here's an example of how an FSE sequence works. For a given resolution and bandwidth, an FSE will have a particular spacing between echoes, such as 10 msec. Once the ETL is chosen (e.g., 8 echoes), the scanner calculates the times at which the echoes are formed. Here, the echoes would occur at t = 10, 20, 30, 40, 50, 60, 70, and 80 msec. Now, we specify that we want a TE_{eff} of 62 milliseconds to give us the desired contrast. The scanner picks the echo that occurs closest to this time—here, it would be 60 msec—and it sets up phase encoding such that the 60 msec echo is read out when the device is scanning through the middle of ***k***-space.

It's a little frustrating because we don't get our 62 msec echo time exactly, but it's close enough. If we wanted a different TE_{eff}, we would simply read out one of the other echoes in the middle of ***k***-space. Because the order of acquiring ***k***-space lines does not follow the convention for plain spin-echo imaging (where ***k***-space tends to be filled in order from top to bottom), this FSE filling scheme is referred to as ***k****-space reordering*.

Large ETLs lead to substantial savings of time, but this efficiency comes at a cost. At some time after excitation, tissue will have relaxed due to T2 to the point that an echo will provide virtually no signal. Unless an extremely long TE_{eff} is desired, phase encoding will associate these long TE echoes with the high-***k*** data that are to be found in the peripheries of ***k***-space. Images with no data from the edges of ***k***-space suffer from a lack of detail and feature resolution. This effect is known as *T2-blurring*. One particular implementation of FSE, known as Single-Shot Fast Spin-Echo (SSFSE), attempts to acquire data using very large ETL, enough to construct a single image in one excitation. This sequence is generally resistant to the patient motion of trauma victims or non-sedated infants. The downside is that the image sharpness is severely compromised by T2-blurring. While FSE and SSFSE imaging have enjoyed many improvements made to the pulse sequence technique over the years, there is a very real connection between the performance of these sequences and the strength of the gradients (ETL played out in a shorter time) as well as stability of the gradients and RF over multiple echoes (ghosting). Another consideration with multi-slice FSE sequences, particularly those with short TR periods, is that, when optimized, a substantially higher amount of RF energy is being deposited in the patient per unit time than standard spin-echo acquisitions. This can be a patient safety and comfort issue, particularly at higher field strengths, that creates the need to put limits on the acquisitions or come up with methods to reduce RF energy delivered to the patient (i.e., purposely using reduced refocusing pulse angles).

13.5 Gradient Echo (GE) / Gradient Recall Echo (GRE)—Reducing the Repetition Time

In the never-ending quest for more rapid imaging, a variety of methods have been introduced to shorten examination times. The first 'fast' MRI method to become widely accepted was actually a form of *gradient-echo* (GE), or *gradient-recall echo* (GRE) imaging, rather than spin-echo. Gradient-echo is unlike SE in several important regards, the first being the design of the pulse sequence itself.

Shortly following the excitation pulse, GE intentionally applies an *x*-gradient of negative polarity, $-G_x$, for a short while (Figure 13.7a,b). This causes a con-

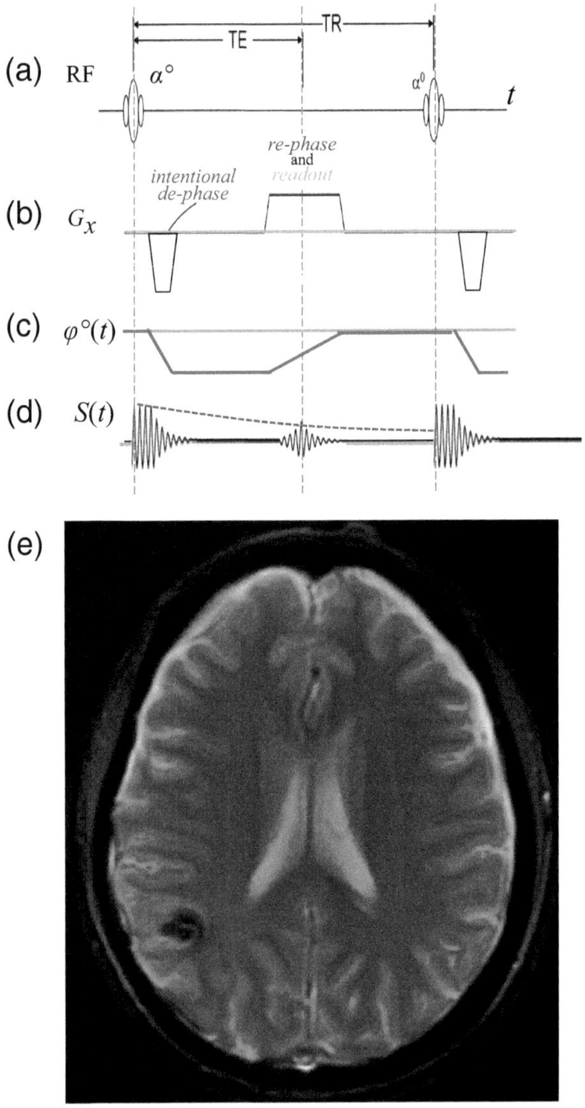

Figure 13.7 Gradient-echo. (a) The gradient-echo (GE) pulse sequence, with *flip angle* $\alpha°$, commonly less than 90°, and repetition time TR. (b) Shortly thereafter, the ***x***-gradient is activated in the 'wrong' (*negative* ***x***) direction, as $-G_x$, thereby intentionally *de*-phasing the transverse magnetization between voxels. Subsequently turning on $+G_x$ in the positive (opposite) direction before TE causes the voxel phases to start coming back together. (c) The phase of the spin packets over time for one set of $G_x(t)$ values. (d) The center of the echo signal occurs at TE. (e) T2*-*w* GRE image in which a susceptibility effect from hemoglobin reveals the location of a hemorrhage.

trolled amount of phase-shifting across all voxels in the FOV. This is not the time-independent T2-type of de-phasing, but rather like that of SFd-P, Equation (9.4)—in which the mean phase of the transverse magnetization in each voxel is slightly different from its neighbors, but here in a reversible fashion. The rate of spreading at a voxel, hence the phase change induced by the gradient while it is on, is proportional to $(x \cdot G_x)$. The farther a voxel is from the center of the row, where $\boldsymbol{x} = 0$, the faster or slower its magnetization will precess relative to that at the origin, and the more rapidly the phase of its spins will speed away from those at $\boldsymbol{x} = 0$ and from one another.

Now comes the need to somehow reverse the process, so that the spins come back together and generate an echo. There is no 180° re-focusing RF pulse with GE, however, as there is in SE. Instead, a $\boldsymbol{G_x}$ gradient of opposite (now positive) polarity is applied during readout, one whose strength and duration are what is required to exactly counteract the earlier de-phasing (assuming no other relaxations), at the time that the center of the echo arrives.

Also, with spin echo, the RF excitation pulse is normally of an intensity or duration to flip the near-equilibrium magnetization $\boldsymbol{m_z}(0)$ down through 90°, into $\boldsymbol{m_{xy}}(0^+)$ in the $\boldsymbol{x}$-$\boldsymbol{y}$ plane (Figure 13.8, top row). GE differs, first of all, in that the initiating RF pulse creates a *flip angle*, α, well *below* 90°, typically only a few tens of degrees or less (bottom row). This smaller angle results in a much shorter projection of $\boldsymbol{m}(0^+)$, onto the $\boldsymbol{x}$-$\boldsymbol{y}$ plane. It produces a somewhat weaker but still usable signal, but the total spin system will need much less time to recover toward thermal equilibrium. That way the TR that produces optimal signal in the steady state can be shorter, and the whole process can be a good deal faster. The signal-to-noise ratio is at a maximum for a flip near the *Ernst angle,* Equation (9.11).

The smaller flip angle also allows for an *apparent* faster recovery of longitudinal magnetization. This advantage allows for T1-w image contrast using much shorter TRs, comparable in appearance to images with a longer TR and a 90° RF-pulse…but much faster!

GE produces a valuable echo signal, like SE, but it does *not* eliminate susceptibility and field inhomogeneity effects the way SE does. The images it produces can, therefore, not be pure T2-*w*. Rather, GE images can be T1-*w* or T2*-*w*, where T2* was defined in Equation (10.4). GE image weighting can be selected

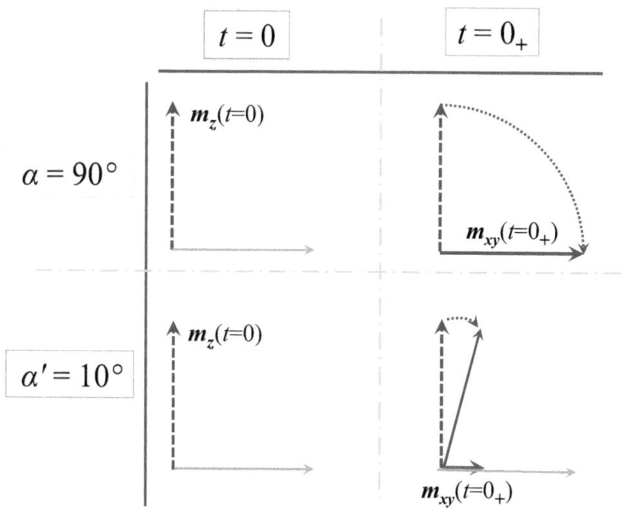

Figure 13.8 Spin-Echo traditionally begins with a 90° pulse at time $t = 0$, top row. Following a previous RF pulse, here the longitudinal magnetization has had time to regrow to $\boldsymbol{m_z}(0)$. The excitation pulse swings it down into the transverse plane, where it morphs into $\boldsymbol{m_{xy}}(0_+)$, where $|\boldsymbol{m_{xy}}(0_+)| = |\boldsymbol{m_z}(0)|$. Spin excitation in a GRE sequence, however, can start off with a much smaller *flip angle,* lower row. An α' of 10°, for example, gives rise to a transverse magnetization, $\boldsymbol{m_{xy}}(0_+)$, that is large enough to produce a detectable signal, but requires only a relatively short time for the system to relax back to near-equilibrium.

by manipulating TR and TE, as with SE, but here the flip angle is also an important adjustable parameter.

There are a fair number of variations on the GE theme available on commercial machines going by acronyms like

- GRASS (gradient-recalled acquisition in the steady state),
- SPGR (spoiled gradient recalled echo),
- FISP (fast imaging with steady state precession),
- field echo,
- FLASH (fast low-angle shot), and
- T1 CE-FFE (T1-*w* contrast enhanced fast field echo).

Example GE applications include the use of 2D T2* imaging for characterizing hemorrhage, 3D FSPGR scanning for dynamic T1 contrast changes, and steady-state imaging for cardiac function. Essentially, the differences arise from what the machine does after readout of data: does it destroy the residual transverse magnetization (called *spoiling*), or does it try to preserve it for higher signal in subsequent readouts? Many of these overlap, with various vendors attaching different names to what is inherently the same thing.

The key to the GE savings in time is that this process can start anew in a fairly short TR, on the order of 10 to 100 msec, without the need for the 180° RF-pulse and the re-focusing period after it. For 2D GE imaging and a short TR, Equation (13.3b) dictates the scan time.

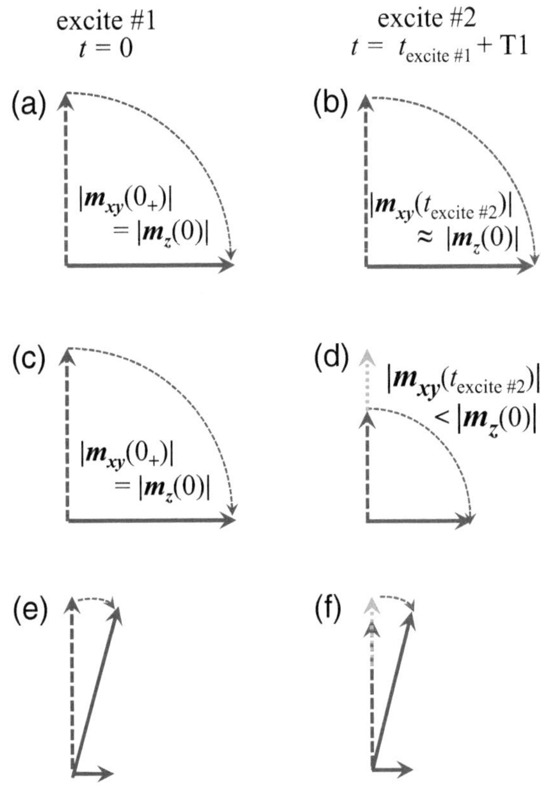

Figure 13.9 Evolution of longitudinal magnetization for three different multiple RF pulse and spin relaxation situations. In each case, the extant magnetization is initially fully longitudinal. (a) In the first scenario, the separation of RF pulses is significantly greater than T1. Spin-echo begins with a flip angle of α = 90°. With this first excitation, all of $\boldsymbol{m}_z(0)$ is nearly instantaneously converted from longitudinal magnetization (dashed red arrow) to transverse magnetization (solid red arrow), and $|\boldsymbol{m}_{xy}(0_+)| = |\boldsymbol{m}_z(0)|$. (b) Because T1 is relatively short, the system has time to relaxes nearly back to $\boldsymbol{m}_z(0)$ before the next excitation occurs, so $|\boldsymbol{m}_{xy}(t_{\text{excitation \#2}})| \approx |\boldsymbol{m}_z(0)|$. (c) Again α = 90°, but now the time between excitations is *less* than T1. The green dotted arrow indicates the amount of longitudinal magnetization that has *not* recovered before the next excitation. If we waited t = T1 seconds after excitation, (d) the shorter dashed arrow would represent 63% (1 – e) of the longitudinal magnetization at reset. (e) Virtually all of the longitudinal magnetization for small flip angles is recovered because so little of it was affected in the first excitation. (f) After the second pulse, the transverse magnetization will be less than after the first magnetization in the case of the 90° flip because of partial relaxation, whereas the transverse magnetization for small flip angles will be almost the same as before.

But now there arises a new problem. If TR is very short, won't this cause all images to have a strong T1 weighting associated with them?

We can't change the T1 of tissue, of course, so how does longitudinal magnetization recover more quickly? The explanation requires use of the quasi-quantum spin model. Let's revisit how TR affects T1 weighting in a single-voxel SE (*not* GE) study. First, a 90° RF pulse induces a transverse magnetization, and immediately the voxel signal is measured to be $s_1(t)$, proportional to the transverse magnetization, $\boldsymbol{m}_{xy}(0+)$ (Figure 13.9a). Then, over a *long* time, comparable to T1, individual spins flip over until the longitudinal magnetization is back near equilibrium, $\boldsymbol{m}_z(0)$. Now, repeat the process. The signal we measure after the second excitation, $s_2(t)$, is still about the same value as after the first, $s_1(t)$. The reason is that the transverse magnetization after the second excitation is about the same as after the first excitation (Figure 13.9a,b).

If we had waited only a short while after the first 90° RF excitation, on the other hand, then the longitudinal magnetization would be much smaller when the second excitation occurs, and the signal would be correspondingly weaker (Figure 13.9c,d). In fact, if we waited a time of T1 between excitations, the longitudinal magnetization before the second excitation will be only about 63% of what it was before the first excitation. The transverse magnetization will be smaller by the same proportion, as will be the MR signal, $s_2(t)$.

If, instead, a much smaller flip angle is tried (Figure 13.9e,f), such as, say, α = 5°, the transverse magnetization ends up as $\boldsymbol{m}_{xy}(t_{\text{excite \#1}}) = m_z(0)\sin(5°)$, or 9% of the longitudinal magnetization before excitation. If we wait a period T1, then 63% of what was briefly transverse magnetization will relax back to longitudinal. Now we prepare to excite the tissue again at $t_{excite\#2} = t_{excite\#1}$ + T1. The small time, δt, just before the second excitation, the longitudinal magnetization is

$$\begin{aligned} m_z(t_{\text{excite \#2}} - \delta t) &= m_z(t_{\text{excite \#1}}) + m_{z,\text{recovered from } xy}(t_{\text{excite \#2}} - \delta t) \\ &= \left[m_z(0) - m_{xy}(t_{\text{excite \#1}})\right] + 0.63\, m_{xy}(t_{\text{excite \#1}}) \\ &= 0.97\, m_z(0). \end{aligned} \tag{13.7}$$

Since the longitudinal magnetization is essentially the same as before the first excitation, the transverse magnetization after the second excitation will also essentially be the same as after the first. Finally, if we wait for t = T1 after the second excitation, 63% of the transverse magnetization that was newly excited would relax back to longitudinal *in addition* to another, about 63% of the residual transverse magnetization from the first excitation. After many excitations, we will achieve a *steady-state* value for the longitudinal magnetization before any future excitation that will be close to full longitudinal magnetization $m_z(0)$. For this description to hold true, it is important to recognize that residual transverse magnetization from previous excitations still exists.

EXERCISE 13.1 What is the ultimate reason that if we waited t = T1 seconds after excitation in Figure 13.9b, the red arrow in the center panel would represent 63% of the original longitudinal magnetization at reset?

Gradient-echo sequences can generate a wide variety of image weighting-contrast. Unlike SE, most GRE sequences do *not* eliminate T_{SFd-P} effects, which yields T2*-weighted rather than T2-*w* images for suitable imaging parameters. But T2*-*w* studies are still of some clinical usefulness. Indeed, T2*-*w* sequences are sensitive to pathology that causes local magnetic susceptibility variance, such as micro-bleeds, vascular malformation, calcifications, and iron deposition in the liver and other organs (Figure 13.7e). Gradient-echo sequences can also generate T1 weighting on images. Although smaller flip angles lead to less T1 weighting in general, this allows the use of much shorter TR values than for a T1-*w* SE sequence. Indeed, a fast form of GRE sequences known as FSPGR (Fast SPoiled Gradient Recall) or TFE (Turbo Field Echo) or Turbo-FLASH (Turbo Fast Low-Angle SHot) is the principal clinical tool for producing 3D T1-*w* images.

Gradient-echo sequence trains are of two general categories that greatly impact the image characteristics and contrast, known as *steady-state* and *spoiled.* For the first of these, after the initial few GRE sequences, during which the process stabilizes, $\boldsymbol{m}_z(t)$ is tipped consistently through the same flip angle by subsequent excitation pulses. Spoiled GE applies special RF and gradient pulses, such as a *spoiler gradient* after readout to remove any residual transverse magnetization at the end of the TR period, so that only the *z*-component remains at the time of the next excitation. The TR is short, and so the residual transverse magnetization can be quite large. This way, residual coherent T2*-weighted magnetization doesn't get flipped back to the *z*-axis, allowing better T1 contrast. Non-spoiled techniques tend to utilize this additional weighting from T2. These topics get pretty involved and will not be pursued much here.

13.6 Echo Planar Imaging (EPI)—Many More Readouts per Excitation

The fastest acquisition methods currently in regular clinical use belong to the *echo planar imaging* family, although perhaps they should more appropriately be described as $\boldsymbol{k}$-space readout methods, rather than a sequence. An EPI technique uses gradients to refocus the signal while sampling data to fill one line of $\boldsymbol{k}$-space, as in a GRE acquisition. The phase encoding begins with one large *positive* G_y pulse, allowing the first readout of data into the top $\boldsymbol{k}_x$-line in $\boldsymbol{k}$-space. After a readout, the phase difference in tissues between rows of voxels along the phase-encoding axis is shifted down slightly from its previous value, with a small *negative* phase-encoding G_y nudge, and the next line of $\boldsymbol{k}$-space is laid down (Figure 13.10a). Like an FSE sequence, one excitation accompanies a long echo train of data, filling much of $\boldsymbol{k}$-space in a short period of time. An entire slice can be obtained in as little as 50 milliseconds, but it will have more noise and lower spatial resolution than typical anatomical images. Also, there may be artifacts associated with the $\boldsymbol{k}$-space trajectory adopted. EPI is used commonly in functional imaging, such as for blood-oxygen level dependent (BOLD) contrast, contrast-enhanced perfusion imaging, and for diffusion imaging.

Can we call an EPI part of the GRE family? Yes, and maybe. Each line uses a gradient to re-phase the echo, so in this sense, an EPI is a GRE sequence. A variant of EPI known as SE-EPI, however, inserts a 180° RF pulse after the excitation to create a spin echo at a time TE during readout (Figure 13.10b). In this type of sequence, the middle of $\boldsymbol{k}$-space is read exactly at TE to give a higher weighting to the desired T2-dependent contrast. T2* effects are nulled at the center of the readout as a result, although other echo readouts occurring at times other than TE experience varying amounts of T2* relaxation (unlike FSE). So, contrast in this SE-EPI behaves more like that of a spin-echo sequence with some susceptibility effects, than of a gradient-echo. Diffusion-weighted EPI sequences, which require reasonably long TE to play out the additional gradient pulses for diffusion, belong to the class

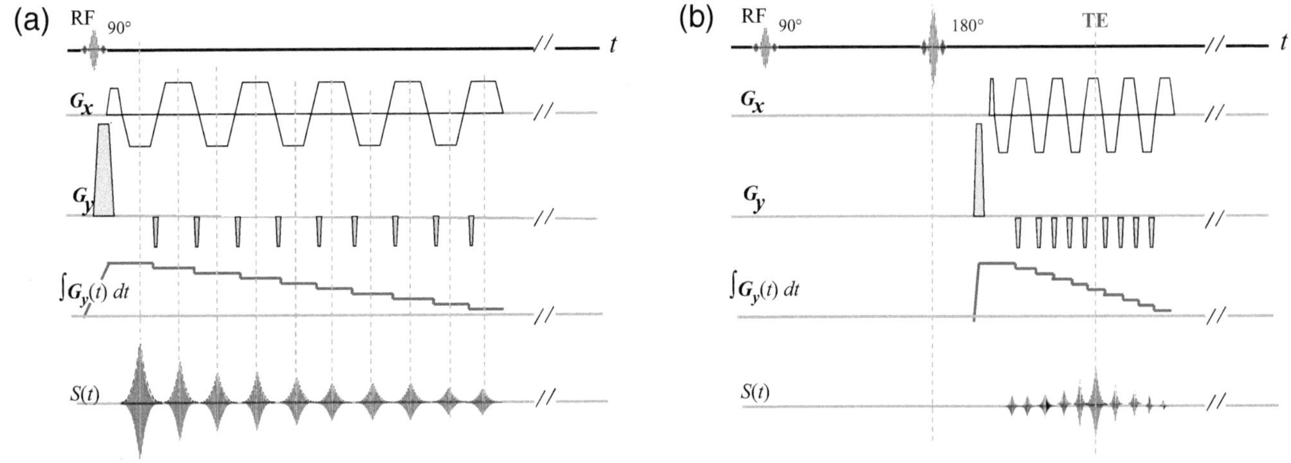

Figure 13.10 Two distinct forms of echo planer imaging (EPI). (a) Gradient echo EPI (GE-EPI), which starts with a single RF excitation pulse. The phase-encoding gradient, $\boldsymbol{G}_y$, is turned on to a very large value for a short duration, and the frequency-encoding gradient $\boldsymbol{G}_x$ is activated (and then reversed) to create the first echo, $S(t)$. If the pulse diagram were cut off after the first echo, the gradient patterns would form a simple gradient echo diagram. Whereas the first phase-encoding was large, subsequent blips of $\boldsymbol{G}_y$ cumulatively integrate over time, as $\int \boldsymbol{G}_y(t)\, dt$, to create a decreasing phase encoding after each blip. Because $\boldsymbol{k}_y$ (i.e., k along the phase-encoding axis) is proportional to the integral of $\boldsymbol{G}_y$ over time, a new and different line of **k**-space is selected with each blip. The frequency-encoding gradient is also switched after every phase-encoding blip to read out a new echo. Echoes decrease with time from T2* relaxation, identical to GRE sequences, but the extended time sampling of the echo train length (ETL) leads to substantial T2* blurring effects. (b) Spin echo EPI (SE-EPI) is similar to the GE-EPI except for the addition of a 180° pulse after the initial 90° pulse. For the SE-EPI, the EPI readout is placed around TE (or 2x the time between the 90° and 180° pulses). Here, exactly at TE, the 180° pulse mitigates T2* relaxation (but not T2), whereas some T2* relaxation occurs in the echoes surrounding the TE time, making this form of EPI a hybrid sequence—the echo through the center of **k**-space at TE is a spin echo, while other echoes are closer to gradient echoes in function. Because phase is not refocused after each line of **k**-space, phase accumulates during readout in the phase-encode direction due to off-resonance, leading to substantial chemical shift and susceptibility distortion in this direction (unlike standard SE/GE sequences). Parallel and half-Fourier techniques discussed below can aid in reducing the time required to collect all of **k**-space in the phase-encode direction and help mitigate these issues. However, similar to the FSE/SSFSE family of sequences, increased gradient power also leads to a faster collection of the ETL and helps mitigate artifacts. Similarly, RF, gradient stability, and eddy current management are key in reducing ghosting artifacts in the sequences.

of SE-EPI sequences. Without the 180° RF pulse (used for SNR and to mitigate susceptibility contrast effects), the sequence is purely gradient-echo.

GE-EPI is a curious sequence in that all of the readout data may be acquired in one stream, after one excitation. As the readout occurs, spins are going to gain phase because the field is not perfectly homogeneous (i.e., off-resonance). In a conventional scan, the amount of phase accrual isn't dramatic along the frequency-encoding axis. For EPI, however, phase continues to accumulate during the acquisition of each phase-encoding line—think of the acquisition as one long frequency encode process broken up into phase-encoding segments. In as much as spatial location depends on phase, this will lead to distortion along the phase-encoding direction of the image that can be dramatic. To alleviate this problem, one can break up the EPI scan into multiple excitations followed by shorter sets of phase-encoding line acquisitions, known as multi-shot EPI. It takes longer and might create other problems if the patient moves, but this approach can lessen the distortion.

13.7 *k*-Space Has (Approximate) Conjugate Symmetry, and its Data Are Redundant

The fast imaging techniques discussed so far either shorten the period of excitation, TR, or increase the amount of data (number of 'views') read per excitation, i.e., multiple-N_y per TR. Equation (13.1b) indicates that including fewer phase encodings (under-sampling ***k***-space) also results in a time saving. As we shall soon see, some MR techniques deliberately under-sample to save time. This choice, however, has consequences on image quality and may lead to artifacts.

MR images can be reconstructed with data from ***k***-space that is only half-filled. This can occur in three (of many possible) different ways (Figure 13.11). In this figure, white and black lines or blocks represent

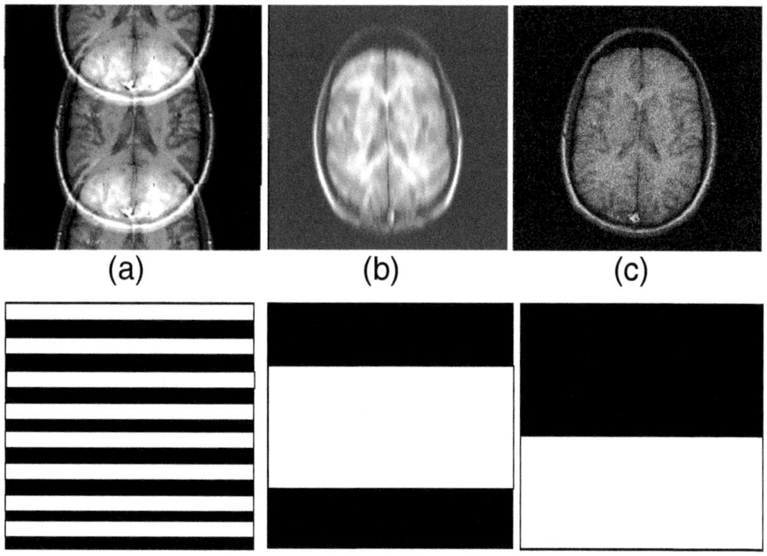

Figure 13.11 MR images associated with three different ways to fill 50% of ***k***-space. Images (top row) and corresponding ***k***-space patterns are shown. White and black bands represent regions where data are, and are not, acquired, respectively. (a) Skipping alternating lines in ***k***-space (or multiple alternating lines) results in an aliased image. (b) Collecting data only in the center of ***k***-space provides spatial contrast, but reduces the resolution along the under-sampled axis. (c) Filling only one side of ***k***-space decreases the SNR for the image, but otherwise, it retains its resolution and FOV.

regions where data are, and are not, acquired, respectively. Skipping alternate lines (or multiple alternate lines) in ***k***-space results in an aliased image because the field of view is halved (Figure 13.11a). Collecting data only in the *center lines* in ***k***-space reduces the resolution along the under-sampled axis (Figure 13.11b). Filling only one side of ***k***-space decreases the SNR for the image, but otherwise it retains its resolution and FOV (Figure 13.11c). With a conventional phase- and frequency-encoding scheme, all three of these methods shorten the imaging time by half, because skipping lines of ***k***-space is equivalent to skipping phase encodings.

The image characteristics of all three under-sampling patterns lead to dramatic/destructive change in image quality. Acquiring every other row or every few rows creates aliasing on the image by shrinking the FOV along the phase-encode axis; this confuses visualization for much of the FOV where anatomy overlaps. If the phase-encoding axis is along the narrow direction of anatomy (L-R instead of A-P), however, aliasing may not be observed, and staying within a "rectangular FOV" (or 'phase FOV') may be a useful time-saving tool. This technique is commonly employed in breath-hold abdominal imaging, where the patient width in the left/right direction exceeds the width in the anterior-posterior direction.

Skipping the peripheral phase-encoding lines of ***k***-space, as in Figure 13.11b, leads to decreased resolution along the phase-encoding axis. When attempting to minimize scan time, it is common to decrease the phase-encoding matrix and leave the frequency-encode matrix the same for a slight time savings, although the effect in Figure 13.11b is so dramatic as to render the image clinically un-useful. Similarly, one may start with matrices of a certain size and boost frequency-encoding matrix to get slightly better resolution for the same scan time.

The under-sampling pattern of Figure 13.11c yields the most anatomically uncorrupted image in most circumstances. Scanning the top or bottom half of ***k***-space leads to increased noise and potential ghosting. The details in the image are largely intact because ***k***-space has *conjugate symmetry*, meaning that

$$f(-k) = f(k), \tag{13.8}$$

where this $f\,(\boldsymbol{k})$ is also said to be an 'even' function, Equation (5.2h). In other words, for every datum in one location of ***k***-space, there is a corresponding partner datum that exists in another, 'opposite' location. If you draw a line from the first datum through the center of ***k***-space and continue an equivalent distance on the other side in the same direction, you will connect the partnered data points. Likewise, Fourier components of a particular wavelength and direction are naturally mirrored across the center of ***k***-space.

The technique of skipping a portion of ***k***-space on one side, as in Figure 13.11c, is known as *partial Fourier*' or *partial*-NEX, because NEX<1 when this is done in the phase-encoding direction. For high-contrast imaging (e.g., as needed for MR angiography) or for images with a large FOV, the small loss of signal-to-noise associated with the reduced views may be an acceptable trade-off for shortened TE and TR (partial Fourier in frequency encoding direction), or reduced acquisition time (partial Fourier in phase-encoding direction). One of the uses of partial Fourier imaging is to allow for collection of images that normally would not otherwise be available. For example, sequences such as HASTE (Half-Fourier Acquisition Single-shot Turbo spin Echo) combine a SSFSE approach with partial Fourier imaging for abdominal studies, in order to shorten the period of patient breath-hold needed. These sequences are useful also for patients who are non-compliant or who move every few seconds, such as some children.

A variant of the partial Fourier technique, known as *keyhole imaging,* applies the under-sampling of ***k***-space in a dynamic series of scans. First, a full image is acquired. Next, for all or many subsequent images, only a number of truncated lines in the center of ***k***-space are acquired in the interest of saving time (Figure 13.12). This approach would ordinarily produce low-resolution images, but they would capture the general contrast. If the remainder of ***k***-space for the subsequent images is filled with the peripheral data from the *initial* ***k***-space, however, high-resolution details are recovered, albeit only those reproduced from the first, full image. This method theoretically allows for high-speed imaging of dynamic contrast evolution, such as breast imaging as a bolus of contrast agent passes quickly through blood vessels. However, while the fast acquisition methods employed today look similar to keyhole imaging, all these methods have varying update strategies for the edges of ***k***-space.

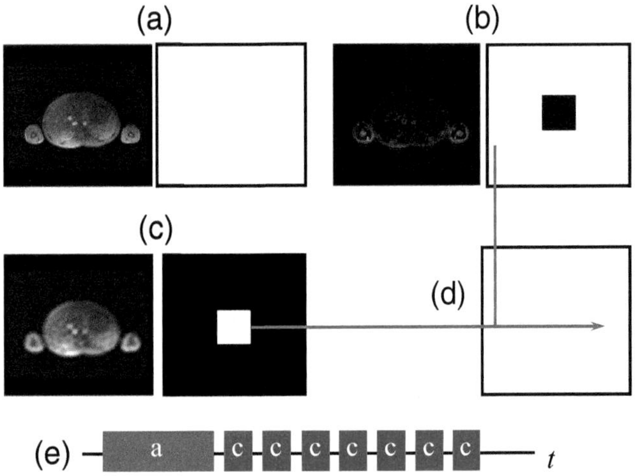

Figure 13.12 Keyhole imaging. (a) A fully-sampled ***k***-space, where the amount of *space sampled* is shown in white, leads to a complete image. (b) If the center is cut out and replaced with zeros, the corresponding image shows only the general structure. (c) Sampling the center of ***k***-space takes much less time than a full image because of the limited number of phase encodings. Although the image will be of low resolution, the basic image contrast is preserved. (d) A new image can be created by taking the center of ***k***-space, (c), and placing the data in ***k***-space with the center cut out, (b). (e) In *keyhole imaging*, a full image is first acquired (shown by "a"), then only the center of ***k***-space is sampled afterwards ("c"). The centers are scanned quickly, so dynamics of image contrast are captured (e.g., contrast agent passing through a blood vessel). By combining the centers of ***k***-space with the periphery from the first image, full-resolution images can be reconstructed that show the quick changes in contrast.

We have assumed that lines in ***k***-space that are missing lie parallel to the $\boldsymbol{k}_x$-axis (i.e., the phase-encoding axis), because such choices lead to reduced scan time, Equation (13.1c). However, the same image results arise if one under-samples along the frequency-encoding direction, rather than the phase-encoding direction. This choice does not shorten the scan time, but it reduces the duration of the readout window, allowing for shorter TE periods. Applications include imaging of cartilage with ultra-short TE sequences.

13.8 Under-sampling of *k*-space in Combination with Phased Array Coils

Advances in hardware have also reduced imaging times for all imaging sequences. Perhaps most significant has been the advent of *phased arrays* (PA) of receiver coils and of *parallel imaging* (pMRI) methods [Larkman et al. 2001; Yanasak et al. 2014]. These include ASSET (Array Spatial Sensitivity Encoding Technique), SENSE (SENSitivity Encoding), and GRAPPA (GeneRalized Auto-calibrating Partial Parallel Acquisition). They all share some aspects of implementation, but they are given different names by the vendors.

As discussed in chapter 8, multiple small RF pickup coil elements—the 'array' in *phased array*—are spread out over the full region being examined (Figure 13.13), and the signal from each is recorded separately and sent to its own independent receiver. With each element acting as an independent channel, the signals obtained from all can be reconstructed independently at the same time; the regions reconstructed from each coil could be fitted together into one large image. With no other considerations, the scan time

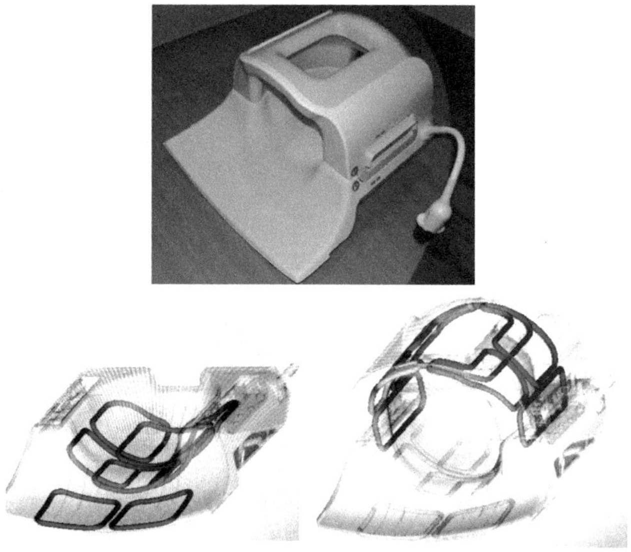

Figure 13.13 A 16-channel phased-array coil. Individual coil elements appear as distinct blue loops.

would be the same as for any technique described above. In practice, however, coil elements receive signals from overlapping regions, so some of the data between elements is redundant.

Phase-encoding steps can be ignored in the schemes shown in Figure 13.11a to save time, and a full image can be assembled using the under-sampled ***k***-space data from all of the channels.

In all likelihood, it is not intuitively obvious how pMRI works: the data sampled from each of the eight coil elements in Figure 13.14a is able to construct a full-sized image by itself. However, the images from each element will show variation in signal intensity, as determined by its location within the whole assembly (i.e., the signal is spatially localized to an extent due to the receiver sensitivity profile, and this can be used as encoding information). Some regions appear quite dark for particular elements, so those elements are not adequately sensitive in sampling particular regions. Using any number of image-processing methods, the information from all elements is equilibrated and combined, and an optimally balanced and informative image appears (Figure 13.14b).

Noise is sometimes a problem in pMRI. The reason for this is obvious—the principal objective is to achieve high speed by eliminating some of the data that could have been incorporated for higher quality, which results in worse SNR.

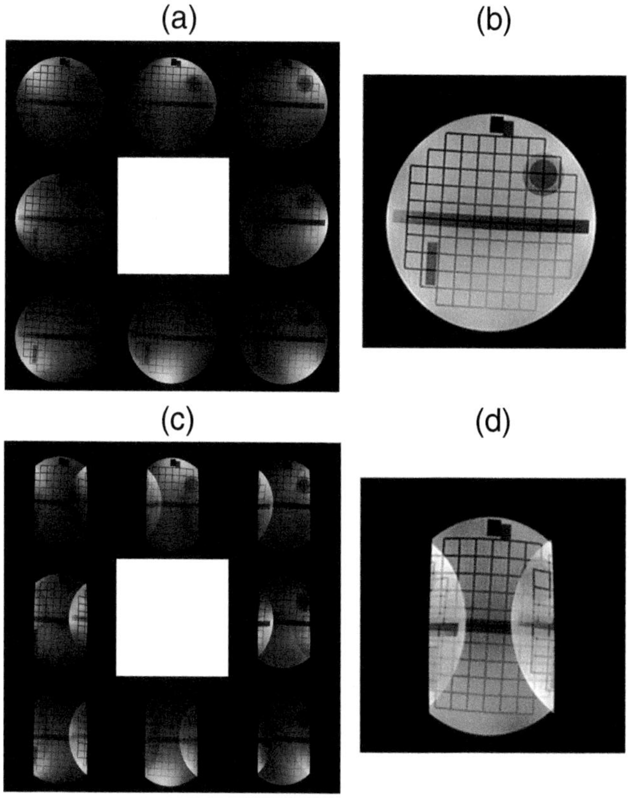

Figure 13.14 ACR phantom images from individual coil elements and their balanced composites. (a) With fully-sampled data, the images from each coil element are most bright in the area where the element is located around the phantom. (b) The composite obtained by balancing and combining them looks fairly uniform and leads to a non-aliased image covering the full field of view. (c) Under-sampled data yield an aliased image from each of the coil elements, which shows variation depending on coil location. (d) The composite image displays the aliasing common to them all. From [Yanasak et al. 2014].

In Figures 13.14c and d, to implement faster image acquisition, data are under-sampled by periodically skipping phase encodings. What are the consequences? Back in chapter 7, we related how the spacing of data in ***k***-space relates to the FOV in image space, Equation (7.18c). If we skip phase-encoding lines, then Δk is getting larger along the phase-encoding axis, and this leads to a smaller FOV along the PE direction via Equation (7.18c). If we have tissue, originally within the FOV, that is now excluded from the FOV after skipping phase encodings, this process leads to aliased images in the phase-encoding direction (Figure 5.22). The image from each element is fully sampled, but the signal intensity varies across each of them because its sensitivity is not spatially uniform.

The SENSE implementation of pMRI (known as an "image-based" method) pulls apart the aliased images and reconstructs a full image using the coil sensitivity profile in image space. For other pMRI methods, similar tricks at different stages of reconstruction allow for the same result, but some, such as GRAPPA (referred to as a "***k***-space-based" method) may perform this process in ***k***-space, with interpolation of the missing lines using the center of ***k***-space sampled by each coil. To bring this all about, however, the *spatial sensitivity* of all coil elements must be obtained, either by design or measured at some point in the exam. Array coil sensitivity maps are mathematically essential for the untangling of aliasing from multiple images.

The signal-to-noise in parallel imaging, SNR_{pMRI}, may appear quite a bit lower than that of conventional imaging, SNR. It is affected by all the factors listed in and below Equation (8.8b) for non-pMRI, but it incorporates two other entities: the *reduction* or *acceleration factor*, *R*, and the spatially-dependent *g*-factor:

$$SNR_{pMRI} = SNR / g R^{1/2}, \qquad (13.9)$$

[Aja-Fernandez et al. 2014; Breuer 2009]. In parallel imaging, just as in partial Fourier or keyhole imaging, scan time is cut down by omitting some phase-encoding lines. The *R* factor is the ratio of the number of lines in full ***k***-space to the number acquired for any element, hence the reduction in scan time. If only 1/R of the ***k***-space lines are sampled, then the number of phase steps is reduced by this amount, in effect, with the consequent decline in image quality.

The magnitude and the spatial pattern of *sensitivity overlap* between coil elements is parameterized by the *geometry factor*, or *g-factor*, which is a nonconstant scalar that depends on voxel position in an image. It is determined by the number, sizes, and locations of the surface coil elements in the phased array, and on factors related to coil design and construction, such as

- how much overlap there is in the sensitivity profiles of the coil elements,
- noise correlations between coils,
- the tissue geometry in the FOV,
- the phase-encoding direction,
- the amount of aliasing in tissue in under-sampled images and the pattern of under-sampling, and
- the particular pMRI reconstruction method employed.

Image uniformity will be compromised for tissue in close proximity to the individual elements. The combined sensitivity of other elements to tissue surrounded by such an array will counteract some of the image nonuniformity, particularly at the FOV's center. The g-factors vary typically from 1.0 to 2.0 across tissue volumes with standard scan procedures. However, faster acceleration will increase *g* to larger values, which may be acceptable in the clinic if the coil has many elements.

Parallel imaging artifacts may arise if too large an acceleration factor is selected for the available coil configuration, or too small an FOV is chosen, so the reduction in imaging times is generally limited to a factor of 2 to 3 when eight small coil elements are operating, for example. The noise increase associated with the reconstruction is offset to some degree by the stronger signal produced by the smaller coils in the array, although greater speeds generate noisier images. In any case, the increased noise, or residual aliasing, shows up near the center of the FOV where the aliased signal from outside the FOV was attempted to be unaliased by the parallel imaging unwrapping algorithms.

We've already reviewed one technique for improving speed in dynamic series—keyhole imaging—where only the center of ***k***-space is scanned on many of the subsequent images. *Partial-Fourier* is similar to this technique in that the center of ***k***-space must be collected, but the technique works because the top half of ***k***-space is correlated with the bottom half, allowing only half the ***k***-space periphery to be acquired. Because of this near redundancy, we can get rid of approximately half of ***k***-space and still reconstruct an acceptable image. Parallel imaging techniques that work in ***k***-space also tend to acquire the central part of ***k***-space (i.e., the keyhole) for self-calibration with the edges of ***k***-space being under-sampled ('aliased'). Of course, this technique applies to the slice encoding direction for 3D sampling as well. Additionally, parallel imaging can be applied to *dynamic* series. In the traditional sense of pMRI, we could skip the same phase-encoding lines for each image in the temporal series.

The images of a dynamic series display temporal correlations in the images as well. A gated cardiac cine series shows movement of the heart, for example, with much of the abdominal anatomy stationary from frame to frame. As with partial Fourier techniques, one can under-sample the image series in time and still produce adequate images.

We can combine both of these advantages and under-sample both in ***k***-space and temporally. Operationally, this technique skips certain phase-encoding lines in k-space, and the pattern of skipped lines changes for each image in the series to achieve under-sampling in the temporal domain. This allows for higher imaging speeds, as with ***k-t*** SENSE and ***k-t*** GRAPPA. In addition to cine imaging, the approach is useful for many dynamic series with small differences between dynamic images, such as functional MRI (fMRI) or phase-contrast MR angiography. However, in general, the center of k-space must be collected regularly while the update rate for the edges of k-space may vary to suit the spatiotemporal needs of the application.

What about 2D slice encoding? An aliasing artifact appears from skipping every other phase encoding line, such as in Figure 13.11a, where in many cases the image resembles a superposition of two pictures—one in the center of the FOV and the other in two halves, each with its center pushed to a boundary of the FOV. We can recognize the aliasing artifact because those two superimposed "pictures" are of the same tissue volume. With pMRI, one can un-alias the full image since the sensitivities of the individual coil elements in overlapping voxels are known. What if, rather than duplicating an image and pasting the duplicate on top of the other image after translating it, we pasted images from two different slices together? Could the same approach of conventional pMRI reconstruction be used to un-paste the two different pictures? Yes, if we knew what the sensitivities of coil elements were for the two different slices.

This example leads to the imaging of multiple excited slices simultaneously. Rather than placing voxels on top of one another by skipping phase-encoding lines and creating aliasing, multi-slice excitation methods excite multiple slices and receive signals from all slices at the same time. In a common implementation known as CAIPIRINHA (Controlled Aliasing In Parallel Imaging Results In Higher Acceleration) [Breuer et al. 2005], the different slices are excited *at the same time* with different precessional phases (i.e., the mean phase of transverse magnetization immediately after excitation is made slightly different from slice to slice on purpose). According to the *shift theorem* (discussion near Figure 5.8), the apparent positions of all slices will shift across the FOV as per an aliasing artifact. The benefit is a decrease in scan time that is equal to the total slices in the stack divided by the number of slices excited simultaneously. Like the temporal under-sampling methods, CAIPIRINHA can also be combined with skipping of phase-encoding lines, as well for additional pMRI scan accelerations. In this case, an image from any element reconstructed before processing would be a combination of multiple overlaid slices and aliasing.

In practice, how does an MRI excite multiple slices simultaneously with CAIPIRINHA? Section 11.2 describes the process of exciting a single slice, where a slice-select gradient is turned on and a band of frequencies transmitted to excite tissue only within a slice at a particular location, z_0. To excite multiple slices, multiple *bands* of frequencies must be transmitted while a gradient is turned on. Whereas a traditional MRI scan uses a Sinc function for all RF transmit pulses, CAIPIRINHA uses a completely different transmit pulse shape that depends on the number and thickness of simultaneous slices desired.

13.9 Frequency- and Phase-Encoding Similarities

Although considered routine by current standards, spin-echo/spin-warp described in chapters 11 and 12 may be thought of as the first reasonably rapid imaging technique. Conventional, 'non-fast' spin-echo consists of acquiring a single k_x-oriented line in k-space from one echo, with one value of phase encoding. Then, using an incremented amount of phase encoding, a second, complete k_x-line parallel to the first is read out and laid down, and so on. This ordinary spin-echo/spin-warp, with a set of lines acquired in sequence one at a time, remains in widespread use, but other more sophisticated approaches utilizing different trajectories through k-space have been developed for special applications. Many key image properties track with the trajectory through k-space and rate at which k-space is acquired. With square spiral imaging, for example, point-by-point data acquisition occurs following a spiral path in k-space, starting at the origin, where the pattern is made up of increasing long segments (Figure 12.11) and full line-by-line scanning plays no role. The initial time spent at the center of k-space and the oscillating nature of the gradients leads to excellent motion properties of the sequence, at the expense of T2* filtering of the signal and sensitivity to off-resonance at the edges of k-space (like EPI).

For any of the sequences that use non-standard sorts of readout paths in k-space, (i.e., not traditional, parallel $\boldsymbol{k}_x$-lines associated with Cartesian sampling), the traditional concepts of phase and frequency encoding begin to lose their intuitive value in describing image properties. For conventional spin-echo, the voxel magnetization precesses at different frequencies depending on position along the $\boldsymbol{x}$-axis (i.e., frequency encoding); analogously, the initial phase offset of those magnetizations varies along the perpendicular spatial axis (i.e., phase encoding). In image space, the frequency-encoding axis is the one along which we read out data into $\boldsymbol{k}$-space. In other words, if we are filling up $\boldsymbol{k}$-space along some axis, then after an inverse Fourier transform, that same axis in the image is the frequency-encoding axis. Spatially, it is the direction that the magnetic gradient points along during readout. When generalizing to arbitrary readout trajectories in $\boldsymbol{k}$-space, this fact is still true—the direction of the readout gradient at any time is in the frequency-encoding direction. But this *direction* is changing *during* the readout period itself. So, spatial position may be localized by frequency and phase along two axes. In effect, this changes the effective point spread function for sampling. For instance, in radial imaging a line of $\boldsymbol{k}$-space is collected and then rotated with respect to its center point to acquire the edges of $\boldsymbol{k}$-space. The traditional Cartesian ghosting artifacts in the phase-encoding direction and off-resonance distortion artifacts in the frequency encode direction turn into subtle blurring of the structures or radial streaking in the image.

The difference of phase and frequency encoding in practice is that the gradient along the phase-encoding axis, $\boldsymbol{k}_y$, is turned on *only for a limited period* after an excitation, while the frequency-encoding gradient, $\boldsymbol{G}_x$, is left on throughout a readout period. There is a deeper concept here, one that relates the two forms of encoding. One can think of frequency encoding as a series of individual, discrete phase-altering encoding steps along $\boldsymbol{k}_x$ following one another, and added together. In other words, start with a series of $\boldsymbol{G}_x$-gradient pulses, switching on and off. Now, place these pulses immediately next to one another in time such that the next pulse begins as the previous one ended. The result, in effect, is that there is only one long $\boldsymbol{x}$-gradient pulse in play. Each one of the discrete gradient pulses is a form of phase encoding; if you fit the multiple pulses together in sequence, the result is a continuous $\boldsymbol{x}$-gradient pulse, equivalent to what goes on during frequency encoding. So, the frequency-encoding process can be thought of as a string of multiple phase-encoding steps along the frequency-encoding axis. This insight isn't necessarily useful for spin-warp imaging, but it comes in handy for other data-readout schemes.

Spin warp imaging is a prototypical 'fast' technique because the use of frequency encoding fills up one row of $\boldsymbol{k}$-space at a time with multiple data points. We sample the MR signal with the frequency-encoding gradient left on, producing a series of discrete observations. Each datum in the row could be considered to be a single discrete 'phase-encoding' step along the traditional frequency-encoding axis. By leaving the frequency-encoding gradient on, we are actually changing the value of phase encoding over time along this axis, during *one* continuous readout. We could actually image even more slowly than with spin warp (albeit with no added clinical benefit) by acquiring one datum in $\boldsymbol{k}$-space per excitation (instead of a full row). This would involve activating the phase-encoding gradient, then *turning on and off* the frequency-encoding gradient, and finally performing the readout. This would be extraordinarily slow...but it forms the basis for how to think of what magnetization precession is doing during readout for a sequence that uses a non-Cartesian $\boldsymbol{k}$-space trajectory, as in Figure 12.11. Let's explore the mathematics behind this new acquisition method: the 'one datum' collection scheme.

As discussed in chapter 12, a narrow-band 90° pulse with the z-gradient will briefly activate and excite tissue in a thin 2D slice normal to the z-axis. The tissue magnetization, $\boldsymbol{m}_{xy}(\boldsymbol{r}, t)$ for the voxel at position $\boldsymbol{r}$ in the transverse $\boldsymbol{x}$-$\boldsymbol{y}$ plane ends precessing in accord with a 2D generalization of Equations (7.5):

$$\begin{aligned} \boldsymbol{m}_{\boldsymbol{xy}\,\text{fixed}}(t)\cdot\boldsymbol{y} &= m_{xy}(0)\mathrm{e}^{-2\pi \mathrm{i}\,\nu_{\text{fixed}} t}\cdot\boldsymbol{y}, \\ \boldsymbol{m}_{\boldsymbol{xy}}(\boldsymbol{r},t) &= \boldsymbol{m}_{\boldsymbol{xy}}(\boldsymbol{r},0)\mathrm{e}^{-2\pi \mathrm{i}\,\nu(\boldsymbol{r},t)t}. \end{aligned} \tag{13.10a}$$

where $\boldsymbol{m}_{xy}(\boldsymbol{r},0)$ is the real (non-imaginary) number representing the magnitude of the magnetization at $t = 0$ and precessing in the transverse plane, for the voxel at position $\boldsymbol{r}$ in the patient. The precessional frequency is given by a 3D generalization of Equation (12.1a):

$$2\pi\,\nu(\boldsymbol{r},t) = \gamma\left[B_0 + \boldsymbol{G}(t)\cdot\boldsymbol{r}\right], \tag{13.10b}$$

where '•' is the scalar or 'dot' product between two vectors. Plugging this into Equation (13.8) and expanding (and assuming only $\boldsymbol{x}$- and $\boldsymbol{y}$-gradients):

$$m_{xy}(\boldsymbol{r},t) = m_{xy}(\boldsymbol{r},0)\left[e^{-i\gamma B_0 t}\right]\left[e^{-i\gamma[\boldsymbol{G}(t)\cdot \boldsymbol{r}]t}\right] \quad (13.11a)$$
$$= m_{xy}(\boldsymbol{r},0)\left[e^{-i\gamma B_0 t}\right]\left[e^{-i\gamma[G_x(t)x]t}\right]\left[e^{-i\gamma[G_y(t)y]t}\right].$$

In the rotating frame, $\boldsymbol{B_0}$ vanishes, leaving

$$m_{xy}(\boldsymbol{r},t) = m_{xy}(\boldsymbol{r},0)\left[e^{-i\gamma[G_x(t)x]t}\right]\left[e^{-i\gamma[G_y(t)y]t}\right]. \quad (13.11b)$$

where the real and imaginary components of $m_{xy}(\boldsymbol{r},t)$ represent the $\boldsymbol{x}$- and $\boldsymbol{y}$-components of the magnetization, respectively.

As before, the Phase-Encode (PE)-direction is aligned along the $\boldsymbol{y}$-axis. Shortly after $t = 0$, at time $t = t_{PE}$, the $\boldsymbol{y}$-gradient is turned on and held at the constant value $\boldsymbol{G_y}$ for a Phase-Encoding period of duration Δt_{PE}. After that, the transverse magnetization will have evolved into

$$m_{xy}(\boldsymbol{r},t_{PE}+\Delta t_{PE}) = m_{xy}(\boldsymbol{r},t_{PE})\left[e^{-i\gamma[G_y y]\Delta t_{PE}}\right], \quad (13.12)$$

where the $e^{-i\gamma[G_x(t)x]t}$ of Equation (13.11) will have no impact, since the $\boldsymbol{x}$-gradient has not been activated. The exponential function in Equation (13.12) says that the phase of the magnetization varies as a function of position along the $\boldsymbol{y}$-axis. The exponent can be expressed as a phase:

$$\varphi_{PE}(y) \equiv \gamma[G_y y]\Delta t_{PE}, \quad (13.13)$$

which is a function only of the $\boldsymbol{y}$-variable, and definitely *not* of $\boldsymbol{x}$. Immediately after phase encoding, the phase of the transverse magnetization would appear as in Figure 13.15b. The orange bracket indicates two rows for which the phases of the magnetization differ by exactly π.

Consider frequency encoding (FE) again along the x-axis, but this time from a new perspective. As usual, the $\boldsymbol{x}$-gradient is turned on at $t = t_{FE}$. The system evolves over the short span of duration Δt_{FE}, after which the transverse magnetization will have become

$$m_{xy}(\boldsymbol{r},t_{FE}+\Delta t_{FE}) = m_{xy}(\boldsymbol{r},0)\left[e^{-i\varphi_{PE}(y)}\right]\left[e^{-i\gamma G_x \Delta t_{FE}}\right]. \quad (13.14)$$

At this instant, the exponent represents a new and different sort of phase, one that depends on the $\boldsymbol{x}$- (rather than the $\boldsymbol{y}$-) position of the voxel:

$$\varphi_{FE}(x) = \gamma G_x x\, \Delta t_{FE}. \quad (13.15)$$

At $t = t_{FE} + \Delta t_{FE}$, the phase across the FOV would appear as in Figure 13.15c.

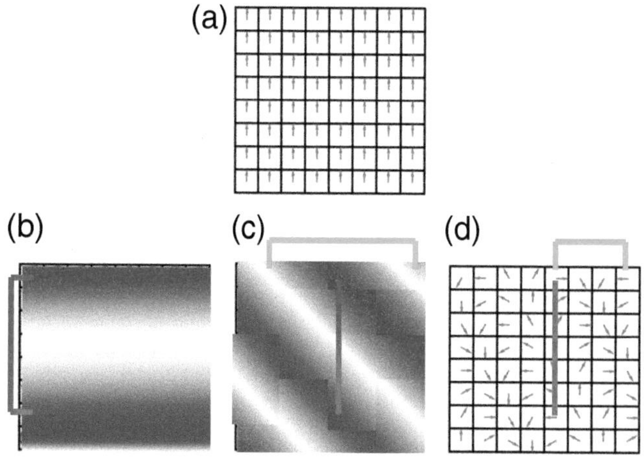

Figure 13.15 Evolution of in-plane magnetization phase differences during phase and frequency encoding. Boxes represent voxels containing magnetization at a certain phase (the phase is indicated using a small green arrow in each box) in a particular rotating reference frame. (a) Immediately after excitation and before spatial encoding, magnetization phases are the same in all voxels across the FOV. (b) Phase encoding occurs along the **y**-axis, when a **y**-gradient is turned on and off. Colors indicate whether the phase is leading (red) or lagging (blue). A difference of π radians of phase is shown as an orange bracket. (c) Frequency encoding is now turned on for a short period. Phase within the voxels now changes along the **x**-axis, with a difference in **x**-phase of π radians shown with a green bracket. (d) While the frequency-encoding gradient is turned on, phase differences along ***x*** continue to grow.

At this moment, we choose to *sample* our total MR signal *once*—this is a new 'one-datum' approach to readout. The amplitude of the signal is proportional to the sum of the transverse magnetizations in all voxels together:

$$S(t_{FE}+\Delta t_{FE}) = \sum \boldsymbol{m_{xy}}(\boldsymbol{r},0)\left[e^{-i\varphi_{PE}(y)}\right]\left[e^{-i\varphi_{FE}(x)}\right]. \quad (13.16)$$

The single sample corresponds to *one* measurement in $\boldsymbol{k}$-space. From this point of view, the frequency-encoding process for our one sample accomplishes the same thing as phase encoding: it creates a variation in the phase of the magnetization across the FOV, but along the $\boldsymbol{x}$-axis instead of $\boldsymbol{y}$-axis.

Turn on the frequency-encoding gradient again—another blip of duration Δt_{FE}—and perform another readout of one datum. The next signal that is observed would be:

$$S(t_{FE}+2\Delta t_{FE}) = \sum m_{xy}(\boldsymbol{r},0)[e^{-i\varphi_{PE}(y)}][e^{-i\varphi_{FE}(x)}], \quad (13.17)$$

and the new phase along the $\boldsymbol{x}$-axis (formerly known as the 'frequency-encoding axis') would be

$$\varphi_{FE}(x) = \gamma(G_x x)2\Delta t_{FE} . \qquad (13.18)$$

If we continue to carry out this process—turn on and off the frequency-encoding gradient, read out one datum, then repeat—we could read in a full row of data just like spin-warp imaging. In other words, in spin-warp imaging, what we're really doing is akin to performing a series of discrete phase-encoding steps along the frequency-encoding axis, with each step resulting in one datum in ***k***-space. If this series of steps is performed back-to-back (i.e., we leave the gradient on and sample continuously), this is just the 'frequency-encoding' process.

13.10 Cartesian vs. Non-Cartesian Readouts

There are two important points to consider from the previous section. First, spin-warp imaging—reading in a *row* of data within ***k***-space—is faster than reading in a single *datum* per excitation. Some early forms of MRI imaging experimented with building up an image voxel-by-voxel, which acquired single samples in real space as opposed to ***k***-space. Nevertheless, the speed of these "sensitive-point" methods would be akin to the single datum readout scheme described above. If it were not for the invention of phase and frequency encoding, MRI would never have been practical because of long scan times.

The second important point is that the use of phase- and frequency-encoding is a very particular approach to filling ***k***-space. There are others that work more quickly—and in some cases, it may be desirable *not* to fill it completely, anyway! For these techniques, the phase- and frequency-encoding interpretation does not make any intuitive sense. For non-Cartesian readouts, Equation (13.16) captures the essence of what is going on. Gradients can be turned on or off (or reversed in polarity!) for various amounts of time, and what happens then is that the magnetization phase is changing along both ***x***- and ***y***-axes in some complicated way. In other words, the signal in the rotating frame can be described as:

$$S(t) = \sum m_{\boldsymbol{xy}}(\boldsymbol{r},0)[e^{-i\varphi_x(x,t)}][e^{-i\varphi_y(y,t)}], \qquad (13.19)$$

where the instantaneous phase difference at any time t along the ***x***-axis is given rigorously by an integral over the time from excitation to t:

$$\varphi_x(x,t) = -x\gamma\int_0^t G_x(t')dt'. \qquad (13.20)$$

A similar equation holds for the ***y***-axis. Varying the time-dependent gradient waveforms will alter the spatial pattern of phase across the FOV and its temporal evolution, equivalent to sampling data along a specific trajectory in ***k***-space. The relationship between the position along the trajectory at a given time, $\boldsymbol{k}(t)$, and $\boldsymbol{G}(t)$ can be generalized as follows:

$$\boldsymbol{k}(t) = (\gamma/2\pi)\int_0^t \boldsymbol{G}(t')dt'. \qquad (13.21)$$

All three under-sampling patterns in ***k***-space (Figure 13.11) can lead to a reduction in scan time from that of a traditional spin-warp approach because phase-encoded rows are thrown out. Another way to under-sample ***k***-space by half would be to skip over every other data point. Nevertheless, one would still need to perform a full set of phase encodings, so this scheme does not save time. Reading out only half of every phase-encoding row does not speed things up, either. Although spin-warp was an important time-efficient development that is still widely used today, the order of reading data into ***k***-space is merely a technical choice albeit with implications on image contrast and artifact appearance (e.g., motion).

At the end of the day, if ***k***-space is filled with data, then a full and optimal (for the given data) image can be reconstructed. Furthermore, it is possible to combine other fast techniques, such as partial-Fourier encoding or dynamic keyhole, with readouts that order the examination of ***k***-space differently. Particular readout trajectories bestow certain benefits and may also impose technical challenges. Because virtually any trajectory is possible, we also can have trajectories that do not strictly read out data along a rectilinear grid—because of which we refer to these as *non-Cartesian* (Figure 13.16).

We have already spent time discussing Cartesian imaging—one excitation, one line of ***k***-space read out, using phase- and frequency-encoding (Figure 13.16a). Likewise with multi-line readout per excitation, such as FSE (Figure 13.16b). Things get more interesting when you progress beyond that.

For any readout technique, the effective frequency-encoding direction is the path in ***k***-space along which the data is read out, whereas the phase-encoding direc-

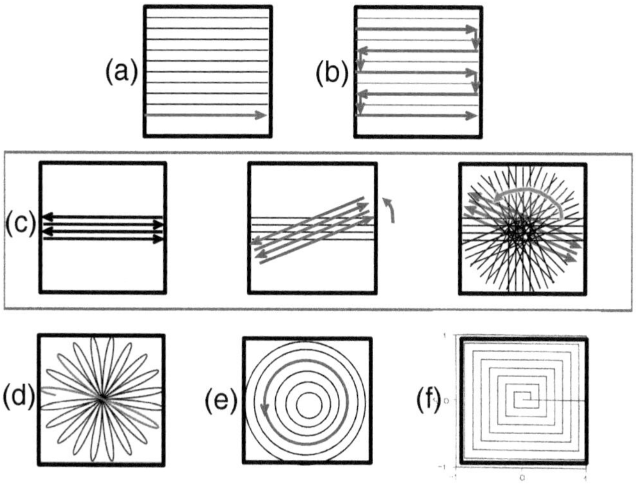

Figure 13.16 Various **k**-space trajectories. Data associated with a single excitation are shown in red for all drawings except (d) (where a portion of the readout is shown). (a) Simple Cartesian imaging (one excitation, one line of **k**-space read out, using phase- and frequency-encoding). (b) Conventional multi-line readout per excitation, such as FSE. (c) PROPELLER imaging, at various stages during data collection. Data are read into "blades" of **k**-space for each excitation. Subsequent excitations rotate the blade around until enough data are acquired for reconstruction. (d) A rosette trajectory [Noll et al. 1998]. The red line indicates a portion of the trajectory. (e) A circular trajectory. (f) A square spiral, Figure 12.14 again.

tion is perpendicular to that. For example, the particular **k**-space trajectory for a sequence known as PROPELLER (Periodically Rotated Overlapping ParallEL Lines with Enhanced Reconstruction) appears in Figure 13.16c [Pipe 1999; Hirokawa et al. 2008]. This pathway is similar to a keyhole Cartesian readout pattern for an FSE sequence with a small echo train centered in **k**-space, except that the lines being read out for each excitation rotate around the center between excitations until the edges of **k**-space are eventually filled. This choice has two effects. First, the center of **k**-space is highly over-sampled, being collected once for each excitation. The over-sampling leads to PROPELLER being slower than a typical FSE sequence (so, no improvements in speed!). But on the other hand, each readout can be executed quickly, so even a patient in motion has little time to move during the excitation and readout period for each TR. The feature of this sequence is that all of the samples can be combined such that in-plane motion *between* readouts is minimized. The key to PROPELLER is that motion results in displacements of the patient, which causes phase shifts in **k**-space data (shift theorem). Because the center of **k**-space is sampled redundantly, phase shifts between all readouts can be corrected, resulting in a composite image as if the patient had not moved at all.

One of the challenges to a non-Cartesian **k**-space readout is the performance of an inverse Fourier transform. As discussed in section 5.4, multi-dimensional transforms are based on **k**-space data that are laid out in a discrete grid. The subject of reconstructing images from non-Cartesian data has a rich history, but current methods generally involve reforming the problem as if data were acquired on a Cartesian grid. Solutions include resampling data onto a rectilinear grid *before* reconstruction, or re-gridding the data *during* it. Another common technical problem is that sampling density of **k**-space may be higher in some areas than in others for a given trajectory, as with PROPELLER.

PROPELLER is subject to conventional artifacts, including the *zipper*; its appearance, however, is different from the one more commonly encountered. The frequency-encoding direction runs through the center of **k**-space in a particular direction that is rotated with each excitation. So instead of one or more zippers running parallel to one axis in the image, the image has a collection of zipper-like structures at various rotation angles. The aspect of a zipper structure at a particular orientation corresponds to contamination of the data during a particular excitation, where the readout occurred along a particular orientation, as in Figure 13.17b.

Equation (13.21) indicates what the **x**- and **y**-gradients must do for the system to move along a *circular* trajectory in **k**-space at a radius, $\boldsymbol{k_r}$, and at constant angular frequency, ω:

$$k_x(t) = k_r \sin(\omega t), \quad k_y(t) = k_r \cos(\omega t) \qquad (13.22)$$

$$G_x(t) = -k_r(\gamma/2\pi\omega)\cos(\omega t), \quad G_y(t) = k_r(\gamma/2\pi\omega)\sin(\omega t), \qquad (13.23)$$

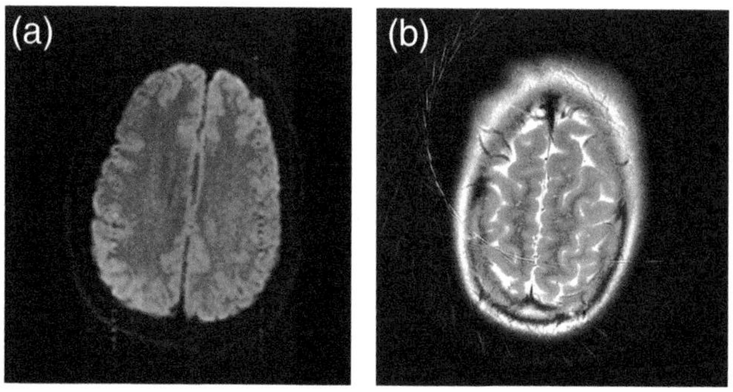

Figure 13.17 Comparison between (a) zippers on a Cartesian-readout EPI image and (b) a PROPELLER image. From the "MRI Artifacts" AAPM/RSNA web modules.

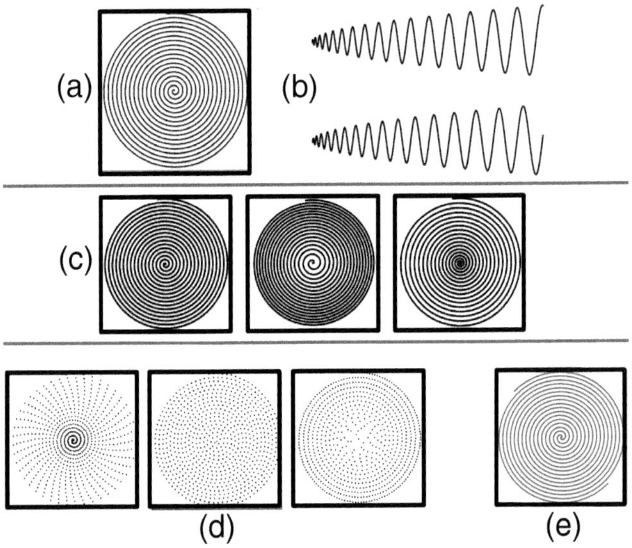

Figure 13.18 Spiral **k**-space trajectory and its properties. (a) A simple spiral trajectory with monotonically increasing radius. (b) Gradient waveforms for G_x (top) and G_y (bottom), constrained so as to yield a uniform distribution of data over **k**-space. (c) Adjustments of the radius of the spiral, with constant angular velocity, can change the sampling density over **k**-space via the trajectory position. (d) Varying the angular velocity can also lead to variable sampling density. All three of these samples use the trajectory in (a). However, the left-most image shows the data sampling pattern if the sampling rate is uniform over time, while the other images increase the sampling rate during the readout. (e) Two-excitation spiral trajectory, with each readout half of the duration of (a).

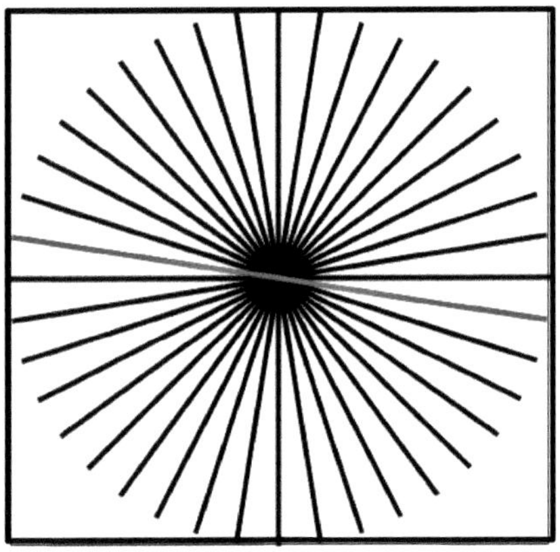

Figure 13.19 A pure radial **k**-space trajectory, with one readout line in red. Readouts could also start in the middle of **k**-space for short TE scanning.

By changing the radius over time, i.e., $k_r = k_r(t)$, one could cover all of ***k***-space either in a stepwise fashion (i.e., a series of circles, as in Figure 13.16e) or continuously. Allowing the radius to grow continuously, either monotonically or at some other rate, results in a *spiral* trajectory (Figure 13.18a,b). The spiral readout is a time-efficient sampling method, and it does not put a heavy performance requirement on the gradients, as with EPI. As a result, single-excitation readouts with a spiral can be performed at very high speeds characteristic of EPI. The sampling density can be controlled by changing $k_r(t)$ (Figure 13.18c) or by varying the angular velocity, $\omega(t)$, over time, employing the gradient slew rate (Figure 13.18d). Spiral sequences have the benefit of being less sensitive to motion. Spiral sequences are used in various commercial offerings, such as the perfusion technique called Arterial Spin Labeling (ASL) with General Electric scanners.

Another scheme has a *radial* readout (Figure 13.19). Like PROPELLER, radial readout sequences have a much higher readout density in the center than on the edge. This makes radial imaging potentially slow. Akin to x-ray tomography techniques, undersampling of data for speed leads to *streaking artifacts* with radial readouts, rather than the aliasing seen with Cartesian readouts. Often, the streaking is tolerable and can be filtered. Oversampling in the center also leads to insensitivity to motion, for reasons similar in both spiral and to PROPELLER. Because each radial line goes through the center of ***k***-space, contrast information is contained in every image, making radial readout sequences very suitable for keyhole imaging. Finally, if we consider the echo time as a factor that we may wish to accelerate, a radial readout trajectory can be specified that begins at the center of ***k***-space instead of tracing fully across ***k***-space, as mentioned earlier. In combination with short RF excitation pulses, sub-millisecond TE values can be achieved, allowing for the visualization of tissues with very quick T2* relaxation, such as bone or cartilage.

With Cartesian ***k***-space readouts, many artifacts appear preferentially oriented along either the phase- or frequency-encoding direction. Examples include aliasing, motion and flow ghosting, chemical shift, and zippers. What happens to the appearance of these artifacts with non-Cartesian ***k***-space trajectories? The simple answer is that many of the artifacts take on different looks, some of them unrecognizable. For example, the *zipper* artifact appears as a dashed line or set of lines occurring at a particular position along the frequency-encoding axis (and running along the full phase-encoding axis). The reason for this is that zippers occur when a spurious RF frequency signal mixes with the MRI signal. When the data are frequency encoded, the data become corrupted at a spatial location corresponding to the frequency of the spurious signal. For non-Cartesian trajectories, a zipper will not be confined to a particular axis, with a variety of different appearances.

CHAPTER 14

Fluid Motion

Magnetic Resonance Imaging (MRI) has an well-earned reputation for evolving in fresh directions, allowing for the development of additional MRI-based imaging methodologies. The detection and display of the rate at which something visible is rapidly *changing* gives rise to forms of *contrast* among different tissues. Factors that can radically alter the apparent relaxation times for a voxel of tissue, say, are the flow or perfusion of blood through it, or the diffusion of water or other materials. Specialized pulse sequences can reveal such movements and make it possible to devise various kinds of contrast that are clinically relevant. Examples of methods with success or promise for improved clinical diagnostics include:

- *magnetic resonance angiography* (MRA), the structural and quantitative imaging of blood flow (a form of change) through the arteries and large veins;
- the characterization of blood *perfusion* throughout an organ or tissue or the lymphatic system, with or without the use of exogenous contrast;
- *diffusion tensor imaging* (DTI) of the movement of intra-cellular water along neural tracts, leading to the visualization of myelinated white matter in the brain; and
- *functional MRI* (fMRI) makes possible the assessment of detectable changes in blood-oxygenation level in localized regions of brain tissue undergoing vasodilation caused by certain internal or external stimuli.

While all of these are 4D in nature, the essence of each can be illustrated here in terms of the motion of protons in 1D, along the x-axis. But first....

14.1 Water and Blood Flow

Hemodynamics is a field of physiology that plays important roles in revealing the workings of the circulatory system. Most of it just involves small variations on the theory of standard water hydrodynamics.

The average flow at the point x in a conduit can be defined as the product

$$\text{Flow} \equiv \mathsf{A}(x) \times \mathsf{v}(x) \qquad (14.1a)$$

of the cross-sectional area there and the median fluid velocity (Figure 14.1). Equivalently, it is $\Delta\mathsf{V}/\Delta t$, the

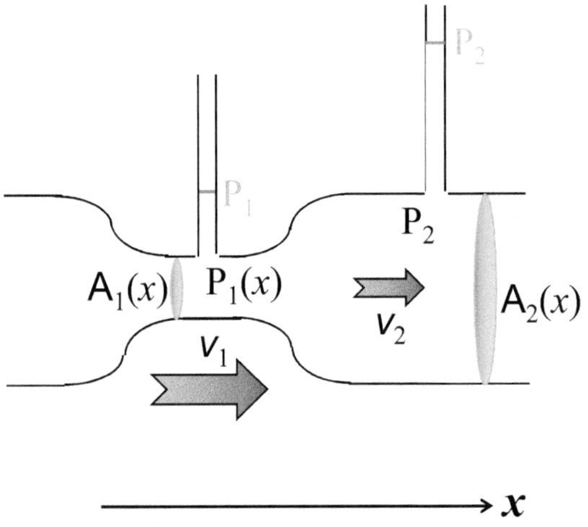

Figure 14.1 The *equation of continuity* reveals that at a constriction in a conduit, the flow of fluid increases. In addition, by the *Bernoulli equation*, the pressure P there decreases. Is it misleading here to locate the pale P_1 and P_2 at the *tops* of the water columns?

volume of fluid that passes by x in a short amount of time. An *equation of continuity* states that for an ideal fluid (incompressible, no viscosity) passing along a conduit of varying cross-sectional area,

$$\Delta V/\Delta t = A(x_1)\times v(x_1) = A(x_2)\times v(x_2). \quad (14.1b)$$

This is useful, as is, if the fluid proceeds along its channel in a *laminar* fashion, like smoke rising in nearly parallel lines just above a candle wick, or with water that is moving lazily along the middle of a stream. But much of the time blood moves faster, or closer to a boundary surface, and it is thereby driven into a state of chaotic or *turbulent* flow, like what goes on higher up above the candle or at the edges of the stream. In such cases, the notion of 'average' flow may not be too revealing a construct.

EXERCISE 14.1 Arteriosclerosis is an often fatal disease in which fatty deposits, low-density lipoprotein (LDL or 'bad' cholesterol), calcium, and other materials build up on the interior surfaces of arteries. Such plaque gives rise to a narrowing, clogging, and hardening of the vessels, and can lead to heart attacks and strokes. Suppose the velocity of blood is in a healthy portion of artery is 100 cm/s. In another part, there is a 75% blockage. What is speed of fluid traversing this obstacle?

A parameter that can predict the velocity at which fluid transforms from laminar to turbulent flow is the *Reynolds number*,

$$R \equiv \rho d(x)\, v(x)/\eta. \quad (14.1c)$$

Eta (η) is the fluid *viscosity,* and $d(x)$ the diameter of the conduit at x. Flow is laminar for $R < 2000$, typically, and turbulent for $R > 2500$. Between these two, a fluid may undergo vortex flow, displaying the kind of eddies that occur downstream of the rocks in the stream.

The local laminar velocity of blood tends to vary across its vessel as a parabola—very slow near its walls, and peaking to perhaps 1.5 times its median value near the center. It is greatest during *systole*, and ranges downward over several orders of magnitude in the various vessels, from 500 cm/sec in the ascending aorta, where flow is turbulent. An audible *bruit* caused by turbulence in a carotid artery can be an early diagnostic sign of a stenosis within. In the major arteries, velocity averages perhaps 150 cm/sec or so, and it is about 50 cm/sec within the cranium. Return flow in normal veins drops to around 20 cm/sec and, despite their small diameters, it is commonly laminar.

Bernoulli's equation,

$$P(x)+\rho g y(x)+\tfrac{1}{2}\rho v(x)^2 = \text{constant}, \quad (14.1d)$$

can relate the blood pressure $P(x)$ at position x within a vessel to its velocity, $v(x)$, and its vertical position $y(x)$ above a reference level, and the strength of gravity, g [Young et al. 2015]. For a patient supine on the MRI table, there is no need to worry about y.

The Bernoulli and continuity equations together tell us that in regions where there is a narrowing or partial blockage, the reduced flow (at increased speed) is accompanied by lowered pressure. In the extreme case, internal pressure may not be enough to prevent the artery from collapsing inward rapidly and abruptly shutting off, even though the blockage is not complete.

EXERCISE 14.2 In the situation of Exercise 14.1, how great is the drop in pressure at the stenosis? Assume that the density of blood is about the same as it is for water.

14.2 Magnetic Resonance Angiography

Ever since 1923, when researchers at the Mayo Clinic found iodine to be a safe x-ray contrast agent, it has been a straightforward matter to bring arteries and veins into view. Modern *digital subtraction angiography* (DSA), CT angiography, and other techniques can examine blood vessel anatomy in exquisite detail, making possible a broad range of interventional procedures. By injecting a bolus of contrast agent into a vessel and following it with multiple images over time, moreover, x-ray DSA, CT, etc. can indicate, in addition, the *rate of flow* along its paths, as can Doppler ultrasound, albeit with less precision.

MR angiography (MRA) can do the same [Markl et al. 2004; Carr et al. 2012]. It is widely employed to search for a *stenosis* (abnormal constriction of an artery), an *occlusion* (artery obstruction, such as with a blood clot), an *aneurysm* (ballooning out of a small area of the artery wall, putting it at risk of a rupture), or other causes of flow (mostly arterial) irregularities (Figure 14.2a). MRA flow measurement methods may require gadolinium contrast agent, with the associated small risks of vascular injury, reactions to the contrast agent, or stroke, or the need to await contrast wash-out. Other MRA approaches are contrast-free, however, with the flowing blood itself, in effect, acting the part. The particular method chosen depends, of course, on which is best for diagnosis in a particular clinical situation, and on what software is available on the MR device.

So MRA does *not* simply map out T1-*w* and T2-*w* regions of blood and soft tissues, like standard MRI. The *apparent* relaxation times of protons in fluids will be affected primarily by the bulk flow of their water molecules within spatially or temporally varying magnetic fields, rather than by their T1 or T2 values. Instead, flow can alter the measured magnitude and phase of the MR signal from a small volume of blood. The *time-of-flight* (TOF) angiography techniques capitalizes on the changes in signal magnitude produced by flow (Figure 14.2a). *Phase contrast* (PC) angiography makes use also of variations in the phase. PC MRA has the advantage of indicating not only the speed of flow, but also its direction (Figure 14.2b); the downside of PC methods is that they take up twice as much time as those of TOF. We shall illustrate the TOF approach and let the reader imagine others [Tan 2010].

Considering TOF MRA, imagine that we want to create a flow image with strong T1 weighting. We excite a slice of the desired tissue (regions **a** and **b** in Figure 14.3). After reading out the signal and waiting a short TR period, we then excite the tissue again. The brief time following the second (and all subsequent) excitations does not allow the longitudinal magnetization, $M_z(t)$, to recover fully via T1 relaxation, resulting in a less-than-maximal transverse magnetization. Ordinarily, with an appropriate TR, some tissue types will have experienced more recovery of $M_z(t)$—and consequently more signal from transverse magnetization—than other tissues. We now have an T1-weighted image, with contrast between different tissues governed by T1 relaxation.

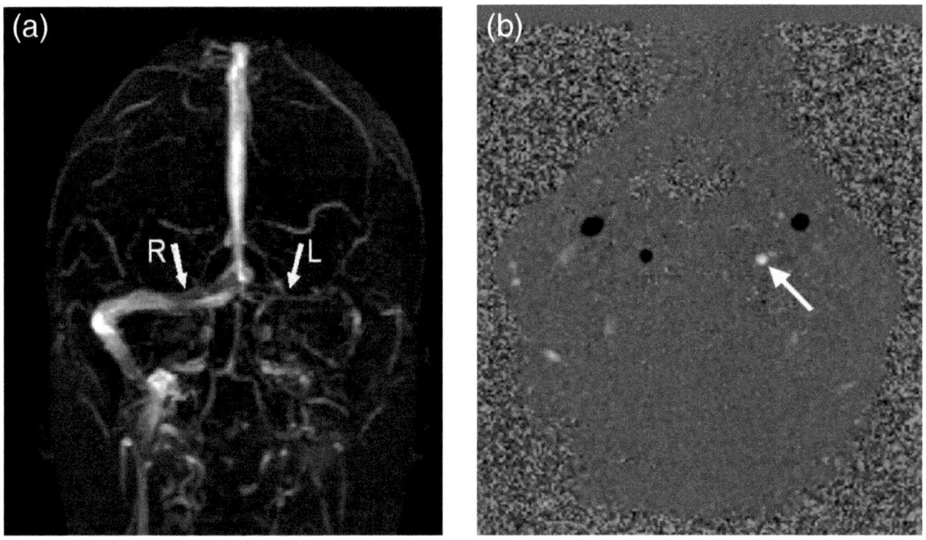

Figure 14.2 Magnetic resonance angiography (MRA). (a) Gadolinium-enhanced *time-of-flight* (TOF) MRA image demonstrating a stenosis of the left interior carotid artery. (b) *Phase-contrast* (PC) MRA indicates not only the magnitude of the rate of flow, but also its direction. Flow in the right vertebral artery is coming out of the page, and it shows up dark. The left artery (arrow) is bright, however, indicating retrograde flow instead, in the cranial-to-caudal direction. This results from a vascular error known as 'subclavian steal.'

This process continues for each subsequent TR period, leading to virtually no signal from the non-mobile tissue in the slice and good signal from the blood refreshing the vessel.

Let's now take this process in a crazy direction. We set TR to a dramatically short value—on the order of a few milliseconds, with a reasonably large flip angle. In this case, the longitudinal magnetization will have little time to recover in any tissues, leading to very small signal after a few TR periods. But this procedure offers an advantage. Unlike the case of stationary tissue, blood in a vessel passes through an imaged slice, and the blood in the slice at any moment will be replaced soon thereafter. Saturated blood is rapidly replaced with fresh, unsaturated blood from upstream. The rate parameter characterizing regrowth of the $M_z(t)$ curve there will be determined largely by the blood's velocity, not by its T1 at rest. The shorter the *apparent* time constant is, the brighter the blood will appear on what is essentially a T1-weighted image. Meanwhile, stationary tissue is very saturated and suppressed, so that what the image displays is bright blood vessels superimposed on faint structure from the non-moving parenchyma. With an MRA *bright blood* technique, flowing blood appears light against a dark background. Conversely with *dark blood* approaches, it materializes against a bright background. The difference is not one of inverting the grayscale, but rather indicates separate imaging approaches.

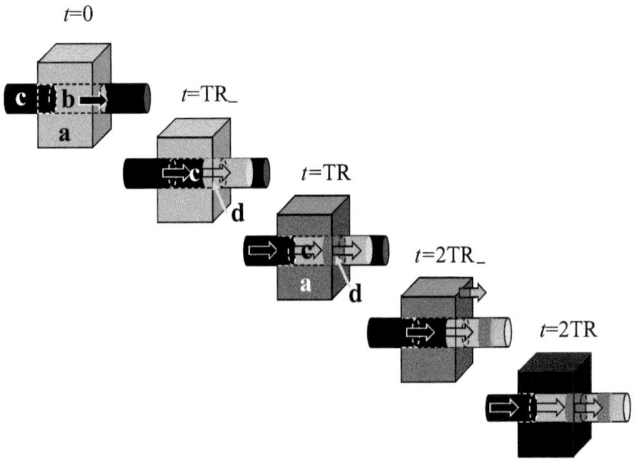

Figure 14.3 Temporal evolution of signal in Time-of-Flight Magnetic resonance angiography (TOF-MRA). Shades of gray reflect the magnitude of transverse magnetization (bright = much transverse). For this sequence, TR is short compared to T1 of the blood or the tissue, and the longitudinal magnetization of tissue that is excited in one TR period will not recover before the next TR period. At $t=0$, a slice of tissue is excited, consisting of non-mobile tissue (a) and blood currently flowing along a vessel and through the slice, (b). At this moment, all of the blood marked (c) is fully to the left of the rectangular cuboid of tissue, and has not yet entered it. After reading out the signal and shortly before the next excitation ($t=TR_-$), most of the blood that was excited flows out of the slice, while (c) non-excited blood upstream in the vessel flows into it. The amount of excited blood that flows out of the slice depends on the rate of flow and the value of TR. If the blood flow is very fast, all of the excited blood may be replaced by non-excited blood. At the next excitation ($t=TR$), the same slice is excited again. Because of the short TR, the longitudinal magnetization in non-mobile tissue, (a), and some fraction of excited blood remaining in the slice (d) does not have much time to recover, leading to less transverse magnetization in these regions (shown as a darker shade of gray compared with $t=TR_-$). However, the non-excited blood, (c), that flowed into the slice has much more transverse magnetization after the excitation because it was previously unexcited (shown as lighter gray compared with $t=TR$.

Phase contrast MRA is somewhat more complicated. PC MRA differs from TOF in that it indicates the *direction* of flow, as well as the rate. Flow in the right vertebral artery of Figure 14.2b, for example, is in the caudal-to-cranial direction, as is normal, and shows up dark. (Bear in mind that we are viewing the patient from the feet upward.) On the left side (arrow), however, it is bright, indicating retrograde flow in the wrong direction, resulting from a vascular pathology known as 'subclavian steal.'

The mean phase of the transverse magnetization is used to encode velocity, both magnitude and direction. Because the natural range of non-degenerate phase angles ranges from –180 to 180 degrees, the concept of encoding velocity with phase requires that a particular maximum velocity magnitude be specified (to label tissue with the maximum phase angle). The *Velocity ENCoding* (VENC) parameter that controls this is chosen to be slightly higher than the maximum velocity expected for vessels of interest. This PC-MRA sequence is a relative of the diffusion weighted sequence described below, although the flow is usually unidirectional within a vessel, whereas it is directionally random during diffusion. These differences lead to coherent transverse magnetization that incurs a velocity-dependent phase shift for PC-MRA, and incoherent transverse magnetization due to phase dispersion for diffusion imaging.

As with nuclear medicine and computerized tomography cardiac studies, an electrocardiogram (ECG) signal is commonly harnessed to gate highly specialized, fast RF pulse and gradient sequences. This can yield images (for any planar section) of the heart, in particular, in different phases of the cardiac cycle. A recent National Institutes of Health (NIH) study indicated, for example, that cardiac MRI can distinguish between patients with severe coronary artery disease from those presenting with other kinds of chest pain—and it can do so more effectively than can ECG, blood enzyme levels, and other measures.

14.3 Perfusion Imaging

MRI of perfusion (blood flow through capillaries) refers to a cluster of MR methods that find widespread use in neurological maps of cerebral blood flow, of cerebral blood volume, of mean transit time, and elsewhere [Parkes et al. 2002; Baker 2013].

Common methods of perfusion imaging follow the progress of a bolus of a paramagnetic contrast agent through blood vessels and capillary beds. This dramatically increases T2* relaxation there for a short period of time, but it also affects the T1 relaxation value as well. With the *echo-planar imaging* (EPI) sequence, one can acquire several images within a second in order to visualize the passing bolus as tissue darkening and brightening via T2* relaxation. A similar non-EPI technique acquires a 2D stack or a 3D slab of T1-w images to visualize blood passing through tissue volume, although each image volume (stack or slab) requires multiple seconds.

With a form of MRI perfusion imaging known as *dynamic susceptibility contrast* (DSC) *perfusion*, one examines T2* images of the same slice rapidly many times in sequence. Regions in which there are temporary decreases in MRI signal (due to T2* de-phasing) indicate tissues into which the contrast agent is infiltrating. *Dynamic Contrast Enhanced* (DCE) perfusion is an MRI sequence that acquires multiple T1-*w* MR images during and after injection of Gd to follow the flow and migration of contrast agents into certain tissues, and the subsequent washout. DCE requires the analysis of the shape of the enhancement curve, and the subsequent washout, and of the kinetics it describes. Signal intensity increases as the agent migrates away from vessels into extracellular spaces, such as in breast tissue. The extracted concentration curves can provide quantitative and clinically valuable measures of blood flow, which are useful for differentiation of soft tissue pathology and tumor viability.

DSC is widely used for examining acute stroke or differentiating brain lesions [Collins et al. 2004]. In some cases of *ischemic stroke*, in which a vessel is blocked by a clot, MRI is able to reveal a region of reduced flow long *before* the fine vessels lose their structural integrity, which they must do for CT contrast agent to cross the blood-brain barrier and be seen. For these and other reasons, it is expected that increasing numbers of architectural drawings for new emergency rooms will include rapid access to MRI machines.

DSC is effective also in the detection, classification, and prognosis of some cancers, in the evaluation of synovial activity in patients with rheumatoid arthritis, and elsewhere.

14.4 Diffusion Imaging

The seductive aura of garlic enshrouding a marinara pizza falls off gradually as you stroll by, as does the memory of it. Likewise, a small drop of dye in a glass of water slowly fades away with both time and distance (Figure 14.4a). It does so because whenever a dye mol-

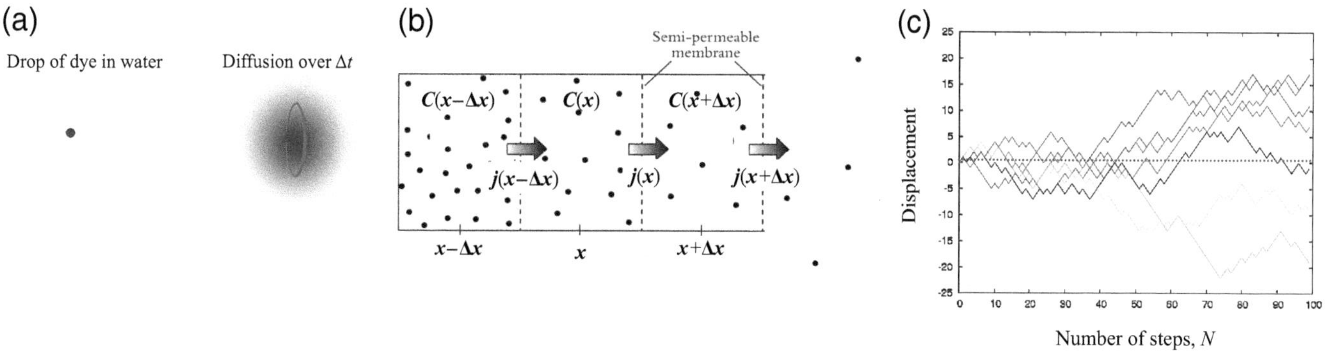

Figure 14.4 Diffusion. (a) A drop of dye in water slowly disperses outward from its initial high-concentration point, as the entropy (degree of disorder) increases. The red ring on the right indicates the average distance that dye molecules have traveled by a certain time. (b) A substance diffuses from a region where its concentration (in air, water, etc.) is high to another where it is lower. This is purely a matter of the random motion of its molecules, the evolution of the system to more probable configurations of its particles, and a state of higher entropy. The steeper the concentration gradient, the faster the rate of diffusion flow. (c) A complementary explanation of diffusion employs a branch of probability theory that deals with 'random walks,' seven of which are shown here.

ecule collides with a tumbling water molecule, it is equally likely to end up moving away from the center of the spot as back toward it. Over time, the spherically symmetric and radially Gaussian cloud broadens and thins. Likewise within tissue, free intercellular water itself drifts continuously from place to place. The principal driver in this is the process of *diffusion.*

Diffusion is a molecular-scale, nonequilibrium process that takes place when there is a concentration gradient of a substance (e.g., dye) between regions (the center of the original drop and the surrounding clear water), and one or more channels along which the substance can slowly wend its way.

Suppose that all the compartments in Figure 14.4b originally contain pure water, but a drop of dye is introduced into the one far to the left. The membranes separating them are slightly permeable to the dye. Because of thermal jostling, all the molecules are in constant random motion, going any which way, and any dye molecule that happens to be near a pore in the boundary has an equal probability of bouncing up or down, to the left or the right, of crossing over or not. But because the chamber to the left contains a higher concentration of dye, more of it will approach the boundary from that direction—and the net flow of it will be to the right, in the **+x** direction.

$C(x,t)$ describes the concentration of dye in the compartments, within an impermeable narrow tube, that are separated by membranes a short distance, Δx, apart. These allow the free movement of water, but only somewhat so for the dye molecules. The difference, $\Delta C(x) \equiv [C(x+\Delta x) - C(x)]$, between compartments will be small. As we might expect, the rate of diffusion flow of dye along the system, $j_x(x)$, increases with $\Delta C(x)$ between segments, and diminishes as each grows longer:

$$j_x(x) = -D_x \Delta C \times 1/\Delta_x \rightarrow -D_x \partial C/\partial x. \quad (14.2a)$$

where $j(x)$ is the current density, $\partial C/\partial x$ is the concentration gradient in the **x**-direction, and the parameter D_x is the *diffusion coefficient*. Here it is assumed not to be isotropic, a scalar, but rather to refer to transport along the **x**-axis. The minus sign shows that the net flow of dye is in the direction in which the concentration decreases. This expression is known as *Fick's law*, which is the starting point for several kinds of quantitative analysis of diffusion. Fick's law combines with the law of *conservation of mass* (another expression of the continuity equation), as

$$\partial C/\partial t = -\partial j/\partial x, \quad (14.2b)$$

to give the *diffusion equation* for diffusion along the **x**-axis

$$\partial C(x,t)/\partial t = D_x \, \partial^2 C(x,t)/\partial x^2. \quad (14.2c)$$

Diffusion can be included in the Bloch equation for a voxel by adding of a term that includes the three-dimensional magnetization as $\nabla(D\nabla \boldsymbol{m})$:

$$\begin{aligned} d\boldsymbol{m}/dt = {} & \gamma(m \times B) - (m_x\mathbf{i} + m_y\mathbf{j})/T2 \\ & - (m_z - m_0)\mathbf{k}/T1 + \partial(D_x \nabla \boldsymbol{m})/\partial x, \end{aligned} \quad (14.3a)$$

where **×** is the vector cross product and **i**, **j**, and **k** are Cartesian unit vectors.

EXERCISE 14.3 For the steady state condition, where $\partial C(x,t) / \partial t = 0$, find an expression for $C(x,t)$.

There is an alternative way to look at this. Recall the legendary late-night reveler trying to get home after an evening at a new tavern. He meanders up the street to the nearest lamppost and rests there awhile. But when beginning the next leg of his journey, he forgets which way he was headed, flips a coin to decide how to continue this time, and moves on. And so on, again and again. Figure 14.4c follows his routes for this one and for the six other fresh pubs of the week. It happens in this probabilistic 1D *random walk* exercise that, on *average*, he will end up a distance of about 0.8 $N^{1/2}$ from the starting point after N steps—but there's no telling which side of the starting point he will be on! The same happens when a water proton (or dye molecule) undergoes collisions with others, ends up spinning at a different rate about a new axis of rotation and drifting in a new direction, accumulating a bit more dephasing—but whether it acquires a positive or negative increment of phase angle is anybody's guess. This is a kind of *Brownian motion*, observed first by the Rev. Robert Brown in 1827 for floating pollen, and explained by Albert Einstein in one of his great papers of 1905, the *anno mirabilis.*

The propensity of a substance to diffuse (manifest in the magnitude of D) depends, of course, on its environment. The diffusion of intracellular water in tissue, for example, is more restricted than that of extracellular, and diseases and other pathologies can alter the behavior of both [Douek et al. 1991; Le Bihan et al. 2002]. While the beginning stages of an acute stroke may or may not be noticeable on a conventional T2 image (Figure 14.5a), it is quite unmistakable with *diffusion-weighted imaging* (Figure 14.5b). DWI can detect early on that the sodium-potassium pump has failed and that water is not moving properly in and out of the afflicted cells (Figure 14.5).

The movement of water molecules in biological tissues is often restricted by physiological boundaries, such as axonal or other cell membranes. The mathematical description is similar to that of pure diffusion, but it is parameterized using the scalar *apparent diffusion coefficient* (ADC), where a high ADC is indicative of little restriction to diffusion. Later on, the cells are likely to burst, and the stroke area becomes very dark for DWI.

Water itself can diffuse easily in an extracellular aqueous solution. The diffusion of intracellular water is generally more restricted—an important special case of that being along the length of an axon. Structures such as cell membranes and myelin layers form powerful barriers that restrict leakage out of the cells, so diffusion-like motion of water along the trunks of long nerve axons in the brain is generally much quicker than radially across the cell membrane and myelin sheath.

Pulse sequences, as with the 1D example of Figure 14.6, have been designed to generate images sensitive to molecular diffusion. One formulation of the sequence involves positioning *diffusion gradient* pulses, G_{diff}, symmetrically on the two sides of a 180° refocusing pulse; only the molecular spins that do not move will be completely refocused, and able to make fully in-phase contributions to the detected signal. Water molecules that do diffuse far will contribute less signal than those that go short distances.

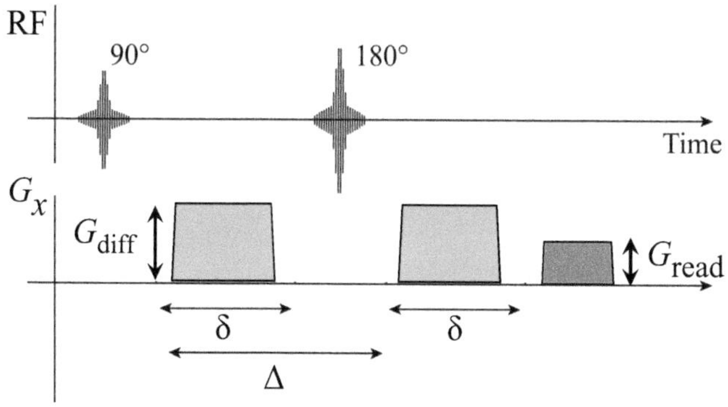

Figure 14.6 Pulse sequence for imaging diffusion, with a relatively strong G_{diff}. The essential parameter *b* is itself a function of G_{diff} and the time intervals δ and Δ.

To improve the sensitivity of a pulse sequence to diffusion, one could apply gradient pulses of greater amplitude, or of longer duration, δ, or more separated in time, Δ. The first of these is what is normally implemented on modern clinical scanners. The changes are captured in a parameter called the *b-value*, where the signal strength at time t = TE depends on the diffusion coefficient as:

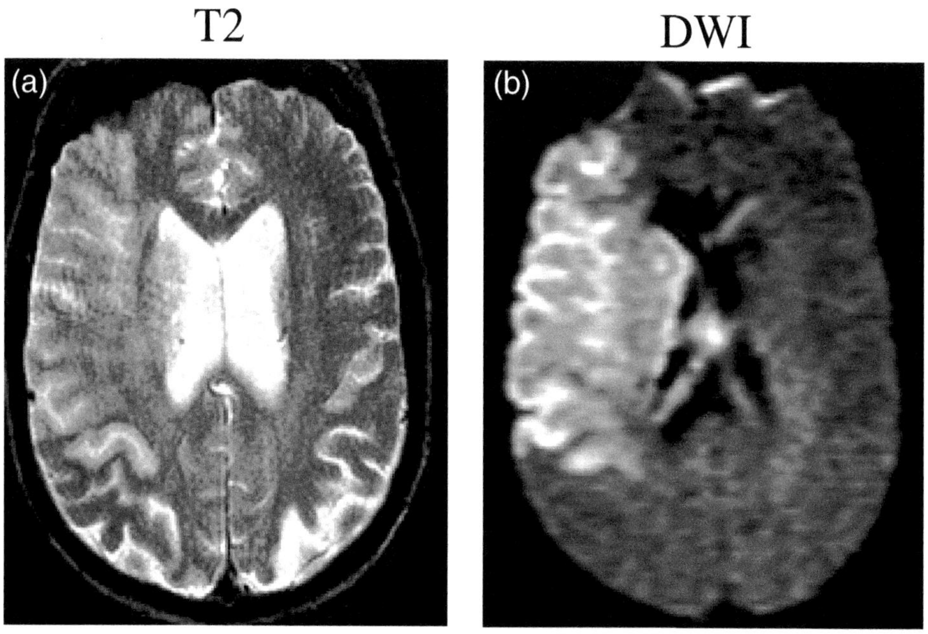

Figure 14.5 For a patient with an acute stroke, comparison of (a) conventional T2 and (b) diffusion-weighted images. With DWI, contrast is created by identifying voxels through which water is self-diffusing rapidly, such as through failing cells or along nerve tracts. [*Courtesy of David K. Powell, University of Kentucky MRISC.*]

$$S(\text{TE}) \propto (1-e^{-\text{TR/T1}})(e^{-\text{TE/T2}})(e^{-bD}). \quad (14.4a)$$

where TR, TE, and D are characteristics of the tissue through which a fluid is diffusing. T1, T2, and b, on the other hand, are parameters of the sequence chosen by the operator. The b-value in the diffusion encoding term is determined by G_{diff}, δ, and Δ as

$$b \propto \gamma^2 G_{diff}{}^2\, \delta^2 (\Delta - \delta/3). \quad (14.4b)$$

There is no universally applicable value for b, and newer scanners allow for selection of b-factors ranging typically from 0 to 4000 s/mm^2 or so [Kingsley et al. 2004]. A sequence with $b = 0$ would produce an image that is essentially T2-*w*, with no diffusion weighting. As b increases, the diffusion weighting increases, but the effects from the T2-weighting, referred to as T2-*shine-through*, may still be observed and be misleading. By acquiring images with multiple b-values, one can calculate an apparent diffusion coefficient for each pixel, rendering an image in which brighter areas correspond to tissues with greater diffusion. These ADC-maps eliminate T2-shine-through.

How do protons behave as they contemplate undergoing diffusion? Let's assume that we are viewing from the rotating frame, that there are no relaxation processes at work here, and that the main magnetic field and the z-axis are collinear with the time axis in Figure 14.7. Although it could be pointing in any direction, consider here the diffusion gradient G_{diff} to be applied along the $\boldsymbol{x}$-axis. We shall follow behavior of a single cohort of protons over time, as it moves upward and to the right.

If $G_{diff} = 0$ during execution of a pulse sequence, as in Figure 14.7a, then immediately after a 90° RF pulse, the immobile spin packet, fixed at some position, x', remains unaffected, and does not change as the time increments along the diagonal time axis. (The ′ on the x' axis is a reminder that we are in the rotating frame.)

If, however, G_{diff} is turned on just after the 90° pulse (Figure 14.7b), the gradient will cause the protons sitting *immobile* at x' to precess slightly faster (or slower) than they would ordinarily. But a subsequent 180° pulse and a reversal of the gradient direction will turn the process around. At the end of application of the second gradient, the packet will be back where it began, as with spin-echo, and with no accumulated change of phase.

The situation is radically different if the proton packet is free to move, and happens to be *drifting*, say, to the right in Figure 14.7c. In that case, the strength of field that the protons experience varies continuously and randomly, and the accrued de-phasing will not be undone by the second gradient. If we assume that protons may be drifting either to the right or to the left, the accrued phase may be positive or negative. And, the more motion during the time when the diffusion gradient is on, the more phase shift is accrued. Averaged over a voxel, the behavior may lead to complete phase incoherence and no signal if a particular diffusivity exists. It is this permanent de-phasing of moving protons that diffusion imaging is sensitive to.

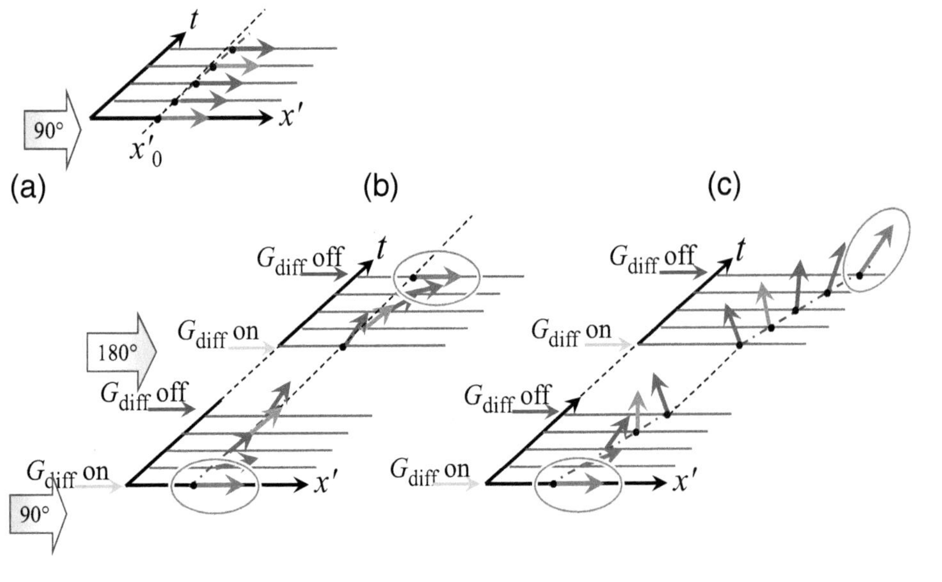

Figure 14.7 A heuristic explanation of how the phase relationships over time within a packet of protons indicate diffusion. G_{diff} is a specific type of $\boldsymbol{x}$-gradient. (a) Immobile spins, $G_{diff} = 0$, seen at five times after the application of a 90° excitation pulse. (b) Tied-down, immobile spins, where a finite G_{diff} is applied for the time δ after a new 90° pulse. After the protons have sat there for Δ, during which a 180° pulse was applied, a second δ-long gradient is delivered. By the end of the sequence, the phase of the spin-packet returns to its original value. (c) If the protons are able to drift in the gradient during the sequence, they will not come back to their initial phase, and that can lead to an estimate of their rate of diffusion.

EXERCISE 14.4 The same G_{diff} is used twice in Figure 14.5. Does one get a similar result by inverting the second gradient and deleting the 180° RF pulse?

The scalar coefficient D_x can delineate preferential diffusional motion of water molecules along a narrow tube (Figure 14.4b). The sagittal, thin-slice DTI of Figure 1.3b, for example, indicates that a glioma lies directly adjacent to, and possibly infiltrating, superior portions of the optic radiation. Alternatively, diffusion may show up against a background of other molecules that are diffusing isotropically (Figure 14.5b). These are special cases with the simplest of geometries, but local anatomy has a way of being more interesting.

While sometimes either isotropic (and parameterize by a scalar D) or uni-direction (D_x), the direction and magnitude of diffusion can vary throughout a region of tissue. In that case, *diffusion tensor imaging* (DTI), is built around a 3D *diffusion tensor*, with elements D_{xx}, D_{xy}, D_{xz},...D_{zz}, as is suggested by Figure 14.8a. The tensor can characterize directional differences in diffusion that arise from characteristics like organized cellular structure [Powell et al. 2012]. While originally used for brain and neuron studies, such as those in Figures 1.8b and 14.8b, DTI has extended its range and now finds some application in the study of carpal tunnel syndrome, compressed lumbar nerves, cervical spondylotic myelopathy, neural tumors, and other maladies affecting the nervous and musculoskeletal systems.

14.5 Functional MRI (fMRI) with the Blood Oxygenation Level Dependent (BOLD) Effect

Positron emission tomography and functional MRI (fMRI) are medical tools that have been bringing about a revolution in the neurosciences. Research and clinical psychologists and psychiatrists are employing them to correlate neural metabolism and the activities of local cranial microvasculature with normal and abnormal mental functions, including sensations, emotional responses, cognitive tasks, and a broad range of neurological diseases. It can be extremely useful, moreover, in sparing functional areas of brain tissue while planning for or undertaking a biopsy, surgery, radiotherapy, etc.

fMRI is similar to positron emission tomography in some regards, but differs in its fundamental mechanism: while PET commonly reveals changes in cellular *glucose uptake* and *metabolism* in various regions of the brain in response to stimuli, fMRI monitors the stimulus-induced variations in the *ratio of oxygenated-to-deoxygenated blood* needed there for that metabolism to occur during a mental process. This provides another, quite different type of MRI contrast. So fMRI and PET are complementary and often correlated, but they are not the same.

fMRI, and in particular *blood oxygen level-dependent (BOLD)* imaging, commonly entails either the periodic application of some stimulus that gives rise to neural activation, or the patient herself initiating such a phenomenon by counting, tapping a finger, experiencing mild pain, listening to Bach, etc. [Jezzard et al.

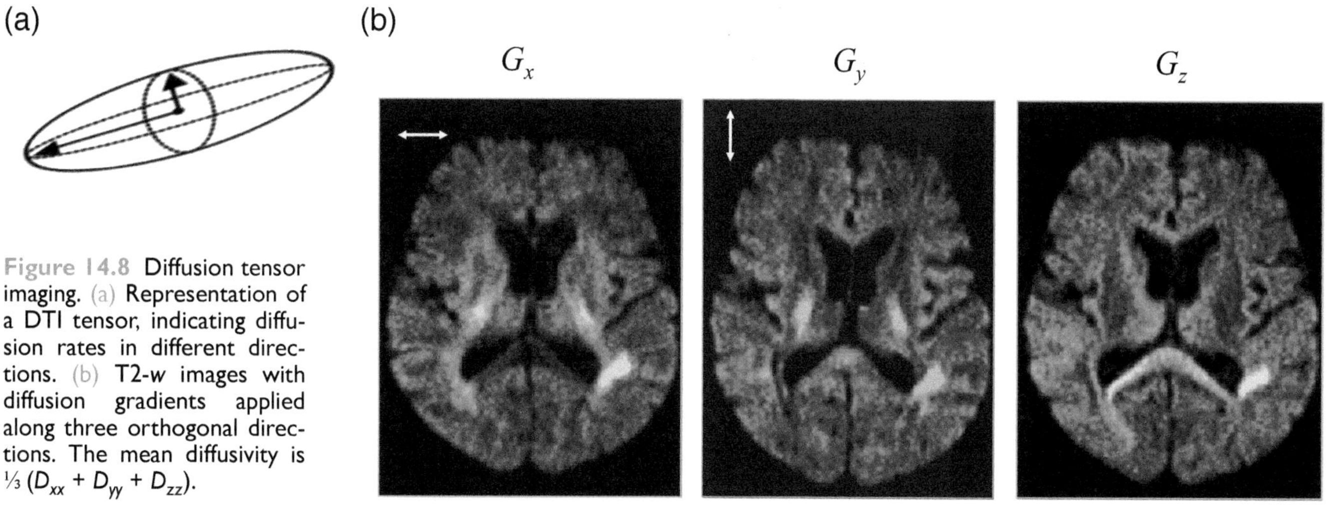

Figure 14.8 Diffusion tensor imaging. (a) Representation of a DTI tensor, indicating diffusion rates in different directions. (b) T2-w images with diffusion gradients applied along three orthogonal directions. The mean diffusivity is ⅓ (D_{xx} + D_{yy} + D_{zz}).

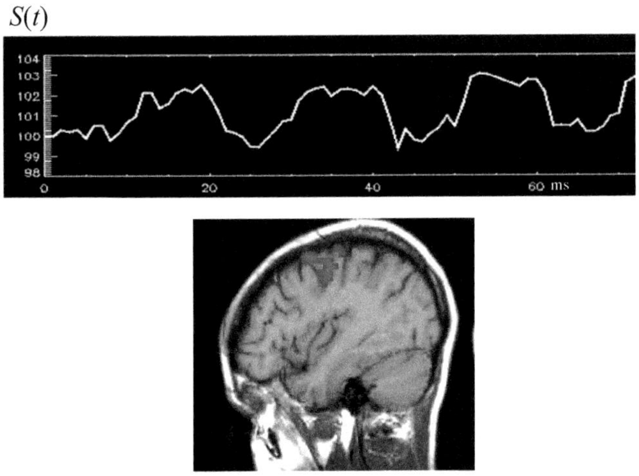

Figure 14.9 A periodic stimulus evokes a corresponding response in the relevant small region of the brain. The detected signal is very weak, but accumulating and averaging a number of repetitions—like adopting a high NEX in Equation (5.27b)—can improve the SNR enough to allow a pinpointing of where the activity is located.

2003; Buxton 2009] (Figure 14.9). Either way, this leads to an alteration in the hemodynamic balance in a relatively small region of the brain.

fMRI exploits the difference in magnetic properties between oxygenated and non-oxygenated blood. Under 'resting' conditions, each voxel of normal brain tissue maintains a balance of oxyhemoglobin and deoxyhemoglobin within erythrocytes. Deoxyhemoglobin is paramagnetic, containing an unpaired electron on the iron atom which, somewhat like gadolinium contrast agent, produces a magnetic field orders of magnitude greater than that of a proton. Oxyhemoglobin, on the other hand, is diamagnetic, and the magnetic field produced by its cloud of spin-paired electrons is extremely weak. When brain tissues are active, they need to consume extra oxygen, thereby transforming oxyhemoglobin that brings it into deoxyhemoglobin. This disturbs the normal equilibrium between the two, which can lead to noticeable changes in MRI signals.

You might expect that neurons burn more oxygen where thoughts or emotions have been triggered (Figure 14.10a). Since deoxyhemoglobin is paramagnetic, you might expect relaxation processes to speed up, yielding a faster rate 1/T2* and a shorter T2* or T2 coming from the faster de-phasing of $M_{xy}(t)$. A shorter T2* gives rise to a weaker MR signal, and to a darker region in an image. Makes perfect sense, but it's pretty much the opposite of what actually happens.

EXERCISE 14.5 Why does a shorter T2 lead to a darker region in the image?*

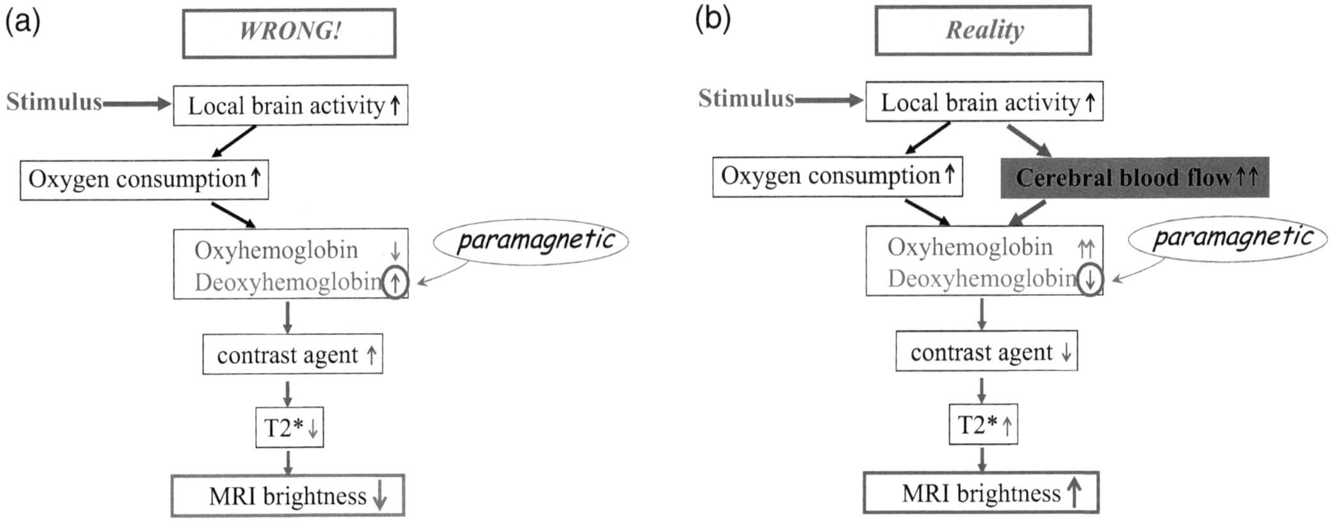

Figure 14.10 Functional MRI (fMRI). Oxyhemoglobin, which is diamagnetic, and deoxyhemoglobin (paramagnetic) have magnetic properties sufficiently dissimilar that they have different impacts on the relaxation times of nearby protons. When a patient receives a stimulus, certain parts of the brain respond, consuming oxygen, and this indirectly modifies the local balance of oxygenated to oxygen-depleted blood. That is sensed by the system and, through vasodilation and vasoconstriction, automatically altering the local blood flow. It also affects the proton relaxation times, hence the MRI signal. (a) The obvious, but incorrect, mechanism of BOLD. (b) What really happens, with an alteration of the hemodynamics triggered by a paucity of oxygen.

The brain's auto-regulated vascular system is actually much more clever than that, and the process is more subtle. The local consumption of oxyhemoglobin sends out biochemical signals that manipulate the muscles controlling the local vasculature so as to bring more oxygenated blood to the region. This response overcompensates, however, and removes deoxyhemoglobin from the region so fast that its concentration falls below what it is in the resting condition. The local oxygen concentration actually *increases*, and this is the dominant effect (Figure 14.10b). This is a good thing; otherwise, when you required more oxygen to remember something, you'd be oxygen-starved without the extra reserve. Indeed, that is the case with people who have compromised vascular systems, such as with atherosclerosis or Alzheimer's disease. Since the presence of paramagnetic deoxyhemoglobin tends to increase the rate of T2*-de-phasing, in effect shortening T2*, its removal from a region would lead to a slight enhancement in the MRI signal from there. fMRI can detect where such changes are occurring. In other words, your brain is smart enough not only to think deep thoughts, but also to keep track of what part of it needs special attention to keep the deep thoughts coming.

This makes possible an altogether new and different type of MRI contrast. Suppose a subject is receiving regularly spaced stimuli, such as periodic touches to the right middle toe, or regularly flashing images of a man pointing a shotgun at your face. Alternatively, the subject himself might be intermittently bending and straightening the left thumb. Either way, there will be an associated activity in some part of the sensory or motor regions of the brain, which will, therefore, consume a little more oxygen than normal, thereby increasing the level of deoxyhemoglobin. If there is a cyclical pattern to the performance of the task (e.g., 10 seconds of finger tapping, 10 seconds of rest, 10 seconds tapping, etc.) then there will be a cyclical pattern to the signal over time, of the same periodicity, and that will make detection much easier. BOLD images are usually T2*-*w*, and areas of the brain with signal variations that correlate well with the cyclical task pattern can be displayed as an overlay on top of an anatomical MRI image.

The response of a patient to an intermittent visual stimulus appears in a 1 mm thin sagittal fMRI slice through a lesion in Figure 1.6a, and indicates that an 'optical radiation' of the brain lies nearly adjacent to it. The temporal *variations* (which is what is of interest here) in the MRI signal are only a few percent of the average value of the signal itself, so effective noise-rejection and statistical information-processing programs must be invoked in data analysis. Spatial resolution is typically a few millimeters, and single-shot EPI provides 10 or 15 slices per second. The sensitivity (contrast-to-noise ratio) is notably better at 3 T.

Stimulus-induced changes in signal strength are only a few percent of the average signal. To monitor rapid changes in the brain that are associated with these stimuli, one must collect multiple images. To look at regional changes in brain metabolism, the experimental method called 'block-design' yields reasonable signal-to-noise for many tasks. In such a design, periodic tasks are repeated in a sequential fashion, such as *s* seconds of stimulus, *s* of rest, *s* of stimulus, and so on, easily carried out with a 'lock-in amplifier.' Achieving adequate SNR also requires the use of powerful statistical methods to separate information from noise. Typically, one would obtain a stack of 20 to 30 2D images, covering the whole brain, every few seconds. For an fMRI experiment lasting four minutes, say, that could correspond to more than a thousand images.

BOLD is a powerful research tool that makes it possible to track precisely which parts of the brain are responding to various specific external or self-initiated stimuli. But fMRI has a growing number of other clinical applications as well. Enhancing standard MRI with fMRI can assist in the planning of delicate surgery; not only can a neurosurgeon see exactly where a spurious lesion responsible for epileptic seizures may reside, but she can also take more fully into account potential losses of cognitive, sensory, or motor abilities that might be caused by damage to nearby healthy tissues from various surgical strategies. She might even be able to see if the nearby eloquent cortex is unexpectedly active, indicating some compensation that the brain is already performing to circumvent loss of function.

Other modalities, such as electroencephalography (EEG) and magnetoencephalography (MEG), are being utilized for mapping brain function as well. Establishing connections between their results and those of higher-resolution fMRI and PET will be essential for their progress. fMRI, incidentally, is generally much easier to perform that PET, as it requires neither the use of a scarce PET scanner, nor costly, short-lived positron-emitting radionuclides, nor the hassles of a hot lab to manage the isotope samples.

CHAPTER 15

MRI Device Quality Assurance and Safety

An MRI machine is arguably the most complex and sophisticated of clinical medical imaging instruments, and it poses a range of challenges even to the most skilled physicists. Starting up a new system requires specialized training and hands-on experience both to develop and carry out *quality assurance* (QA) and safety programs.

When a medical center decides to acquire a new or replacement MR device, the physicist may be asked to participate in a number of activities [Hand 2012]. She will examine and compare the specifications of a number of models during the purchasing process. She may consult with a hospital planning committee and engineers to confirm that the proposed site is suitable (electricity, water, size, load-bearing floor, door access, magnetic shielding, etc.). She is likely to be involved with preparation of the site and with the physical installation process. Once the machine is up and running, she will design and carry out an extensive acceptance testing and commissioning program to verify its proper operation. A comprehensive safety program is also essential for accreditation and to protect patients, visitors, and staff.

For both QA and safety, she must establish and oversee a program for the routine (daily, weekly, monthly, semi-annually, annually, or some combination thereof) evaluation of image quality and to establish and continually review the safety practices of the institution. Nearly all the daily and weekly checks can be performed by the technologists who regularly operate the machine, with considerable help from the machine's own computer automation, but their reports will be examined by the medical physicist. It is part of her job, moreover, to carry out several more elaborate but less frequent tests. Finally, there will be periodic *preventative maintenance* (PM) visits by the manufacturer's own engineers, to the satisfaction of the physicist.

Executing the necessary studies requires knowledge of the standard test protocols, which may vary among MRI centers, and that can be rather confusing. Fortunately, a few major organizations—in particular the American Association of Physicists in Medicine [AAPM 2010] and the American College of Radiology [ACR 2018]—have prepared detailed directions for comprehensive QA and safety programs, and their documentation provides nearly everything a physicist needs to know. The ACR is critical in that any imaging center is required to pass its accreditation program every three years to continue receiving reimbursement for MRI patient examinations. Also of interest is the undated sketch of FDA Guidelines for Magnetic Resonance Equipment Safety [Zaremba].

The following sketch of MRI QA and safety is far from comprehensive, and for more detail, one should refer to other QA documents as well, e.g., [Kanal et al. 2013, Steckner 2012]. The 2013 white paper by Kanal et al. is the current reference standard representing the ACR's position on MR safety. It was put together by a blue-ribbon panel of medical professional experts covering a comprehensive spectrum of expertise (e.g., physicists, radiologists, nurses, legal counsel, safety experts, etc.). That being said, it is only guidance, and to quote the document:

> *"These guidelines were developed to help guide MR practitioners regarding these issues and to provide a basis for them to develop and implement their own MR policies and practices."*

Translation: a lot of thought was put into this document, but no two MR sites are the same, not every scenario can be accounted for, and, importantly, practice standards change over time as we learn and experience more. It is one of the critical jobs of the medical physicist to help a facility keep abreast of what safety practices are important, applicable, comprehensive, clearly stated, and feasible.

Finally, the Atomic Energy Act of 1954 provided the U.S. Nuclear Regulatory Commission (NRC) with the authority and the requirement to promulgate regulations on the safe and peaceful use of radioactive materials. The NRC thereupon created the rules now found

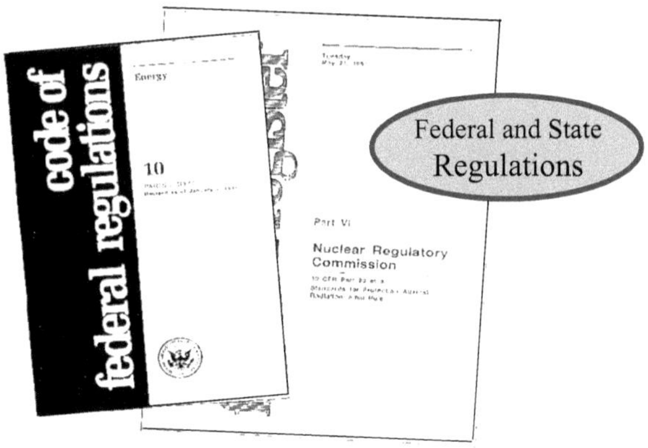

Figure 15.1 Changes in the *regulations* of the FDA and other federal agencies appear first in the *Federal Register* (published daily), after which they are soon compiled in the U.S. *Code of Federal Regulations* (*CFR*), one volume of which appears here. The states have similar systems, sometimes with more restrictive rules. These are commonly based on the guidance and recommendations of expert advisory groups such as the American College of Radiology (ACR).

in the Code of Federal Regulations (Figure 15.1) [10 CFR 20]. Congress, however, did not do the same for the FDA, and there are no central standards or regulations that do the same for MRI devices [ISO TS 10974 2018; Delfino et al. 2016; Woods 2007]. As result, guidance for the use or manufacture of scanners consists of a set of generally agreed-upon upper limits on a number of variables, such as clinical field strength (3.0 T), changing magnetic fields (20 T/s), RF power deposition (SAR < 3 W/kg, head) and tissue temperatures (38 °C, head), and average acoustic noise level (105 dBA). These will be discussed further below.

15.1 Daily/Weekly/Monthly Quality Assurance Checks by Technologists, and the Semi-Annual QA by the Medical Physicist

Standard quality assurance (QA) procedures address the same sorts of issues (contrast, resolution, noise, geometric distortions, artifacts, etc.) as do those for CT and the other modalities. An MRI machine tends to be quite stable, fortunately, and its routine QA under normal conditions is not very onerous. Daily checks are largely automated and carried out under the control of software provided by the vendor. Typical tests that are performed daily or weekly by the MRI technologists are indicated in the left-hand column of Table 15.1.

These are all self-explanatory, but details are published as standard QC protocols in the references.

Table 15.1 Some standard QA tests to be undertaken daily and weekly by technologists, and others semi-annually or annually by specialists. There is much published MRI QA documentation that provides detailed instruction on how to carry these out. The ACR triennial MR accreditation program is more demanding.

Daily/Weekly	Semi-Annual/Annual
visual checklist	The weekly tests ***plus***...
safety program	magnetic field homogeneity
center frequency and field strength	RF bandwidth, tuning
transmitter gain	gradient linearity
high-contrast resolution	RF coil efficiency
artifact evaluation	image intensity uniformity
setup and table position accuracy	low-contrast detectability
geometric accuracy	slice thickness/position accuracy
display/film printer QA	inter-slice RF interference
	monitors QA
	re-evaluation of weekly QA

Many of the more advanced measurements and maintenance procedures are undertaken semi-annually or annually, and these require the services of a physicist or engineer with specialized training and experience (also indicated by Tables 15.1 and 15.2). These tests include, but are not limited to, all the routine tests plus

- the measurement of homogeneity of the main magnetic field;
- slice thickness;
- spatial resolution in three dimensions;
- image uniformity and linearity;
- image signal-to-noise ratio;
- RF pulse parameters, including their strength, shape, duration and frequency makeup;
- performance characteristics of the detection and RF coils and electronics;
- contrast and SNR from test phantoms that contain regions of materials with standardized relaxation characteristics;
- video display (Figure 15.2) and computer performance; and
- all emergency and patient safety equipment.

Table 15.2 Commonly accepted guidelines on the manufacture and operation of MRI scanners, some traceable (not necessarily easily!) to the FDA and other bodies [ISO TS 10974 2018; Delfino et al. 2016; ISMRM 2015; FDA 2014; IEC 2010; Woods 2007; Bushberg et al. 2012]. Numbers cited here are from the Medicines and Healthcare Products Regulatory Agency (MHRA) report *Safety Guidelines for Magnetic Resonance Imaging Equipment in Clinical Use* (2015), based on the 3rd edition of the International Electrotechnical Commission (IEC) standard for manufacturers of MRI equipment (IEC 60601-2-33, 2010). The IEC standard defines three modes of operation: normal operating mode, 1st level controlled operating mode, and 2nd level controlled operating mode. Under normal mode, the operation of MRI equipment should cause no physiological stress to patients. The 1st level mode may cause some physiological stress that requires medical supervision. At the 2nd level mode, the potential for physiological stress may result in significant patient risk and requires approval from an ethics committee or institutional review board (IRB) for human research.

		Normal Level	1st Level	2nd Level
Static magnetic field	(in Tesla)	<3	3–4	>4
Pulsed gradient field	dB/dt (in T/s)	<80	80–100	>100
Temperature	(degree C)			
	Rise in whole body	0.5	1.0	>1.0
	Absolute, head/fetus	38	38	>38
	Absolute, trunk	39	39	>39
	Absolute, limbs	40	40	>40
RF field (SAR)	(in W/kg)			
	Whole body	2	4	>4
	Partial body (head)	3.2	3.2	>3.2
	Head only/local trunk	10	20	>20
	Extremities	20	40	>40
Acoustic noise	(in dB)			
	Protection required	>99	–	–

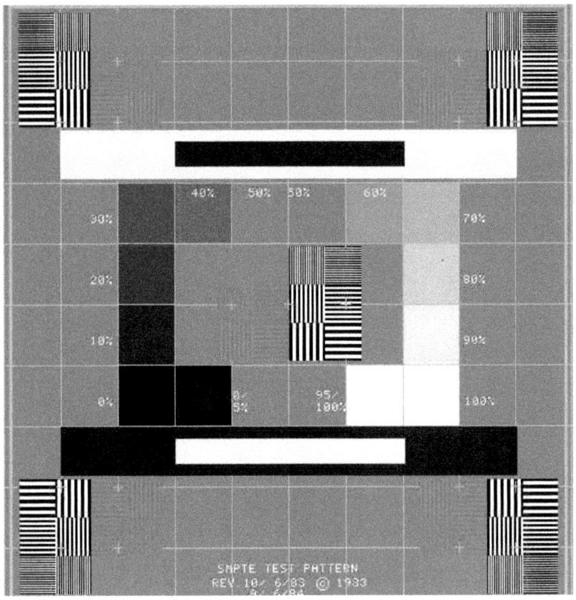

Figure 15.2 The ubiquitous electronic test pattern designed by the Society of Motion Picture and Television Engineers (SMPTE) to assess quantitatively the behavior of electronic display devices such as liquid crystal (LCD), light emitting diode (LED), plasma displays, et al. The pattern, produced by an electronic signal generator with no medical image input, can be used to evaluate a video system's ability to handle contrast, resolution, uniformity, distortion, noise, etc.

Nearly all of this is described in detail in publicly accessible documentation, such as the AAPM's *Acceptance Testing and Quality Assurance Procedures for Magnetic Resonance Imaging Facilities*, Report No. 100 [2010].

15.2 Accreditation by the American College of Radiology Every Three Years

The American College of Radiology is a nonprofit association of medical professionals who utilize ionizing and nonionizing radiation to create diagnostic images and to treat patients. Founded in 1923, the ACR now lists a membership of 30,000 diagnostic and interventional radiologists, nuclear medicine practitioners, radiation oncologists, and medical physicists.

The organization provides important services for its members and for the general practice of medicine. Their published *Appropriateness Criteria* (ACR-AC) continuously updates comprehensive evidence-based guidelines to assist physicians and others in making optimal selection and use of diagnostic imaging, radiotherapy protocols, and image-guided interventional procedures for a large number of clinical conditions. The *Accreditation Council for Continuing Medical Education* (ACCME) has commended ACR for excellence in the CME opportunities it provides. In addition, ACR advocates with Congress, the federal agencies and regulatory bodies, and their counterparts in the states on matters of interest to its membership and the profession.

To the clinical medical physicist, however, the issue of greatest importance is *ACR accreditation*, generally considered to be a gold standard of medical practice. Accreditation is offered in CT, MRI, Nuclear Medicine and PET, Ultrasound, Mammography, Stereotactic Breast Biopsy, Breast MRI, Breast Ultrasound, and Radiation Oncology Practice. This certification is not merely a badge of honor and a recognition of competency, but, in fact, it is required for provider centers that bill under part B of Medicare, such as for MRI services, to receive technical component reimbursement.

ACR's documents are not clear on whether it is a scanner that gets accredited or a facility, but it probably never really matters [ACR 2018]. ACR offers MRI accreditation for six anatomic applications (not including breast), with a separate testing module for each: head, body, spine, magnetic resonance angiography (MRA), musculoskeletal (MSK), and cardiac. Each scanner (or, "unit" in ACR terms) at an accredited site may have different modules associated with it. A scanner should have the appropriate associated module to qualify for site reimbursement for performing the corresponding clinical study.

Accreditation of MRI devices is a two-step process, the first of which is paperwork. The center's administrator must fill out and submit the ACR application form that covers general information regarding the facility, the qualifications and experience of personnel (supervising and other physicians, technologists, medical physicists/MR scientists), data on the individual pieces of equipment, the center's scanning protocols, etc. In addition, the comprehensive program for weekly and annual QA, and a description of MR safety policies, must be sent in, along with copies of representative reports.

The second step obliges the center to provide samples both of representative clinical studies and of physics-phantom test results. The first of these consists of typical examinations prepared by physicians and

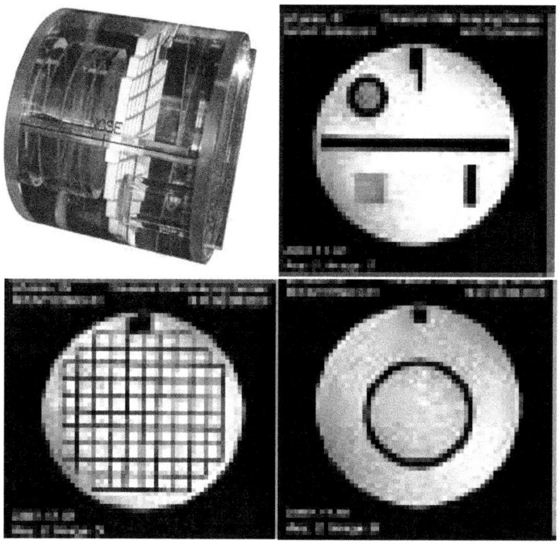

Figure 15.3 The American College of Radiology has devised a single MRI test phantom that makes seven important checks relatively straightforward. These include: low-contrast object detectability, image intensity uniformity, high-contrast spatial resolution, slice thickness accuracy, slice position accuracy, geometric accuracy, and percent signal ghosting.

obtained from patients with the center's standard MR imaging protocols. The others, of particular interest to physicists and engineers, comprise a mandatory set of measurements designed to assess machine performance. These are similar to those of an annual assessment, but include imaging tests completed on the ACR's own specially designed ACR MRI test phantom (Figure 15.3). This test device makes seven important checks relatively straightforward, namely: low-contrast object detectability, image intensity uniformity, high-contrast spatial resolution, slice thickness accuracy, slice position accuracy, geometric accuracy, and percent signal ghosting.

ACR has the right to undertake on-site surveys both before and after accreditation, moreover, so it is essential to have all the relevant QA documentation always current and available for immediate consultation.

15.3 MR Safety Requires Eternal Vigilance (Regulations, Restricted Access, Records...)

The absence of ionizing radiation is frequently touted as one of MRI's many virtues, but the machine can nonetheless kill people, and much more quickly and directly than any CT ever did.

The FDA has reported that there are about 40 MRI-related incidents per year in the United States, of which about 10% involve damage from ferromagnetic 'missiles.' Another 70% result from RF-induced burns. The effects on tissues of intense static magnetic fields, of rapidly switched gradient fields, and of RF power have been studied extensively. The FDA, the ACR, the NIH, state Departments of Health, and other bodies publish recommendations and requirements regarding limits on the main magnetic field strength, the rate of switching of gradient fields, RF power levels, and the like [Shellock et al. 2004; ICNIRP 2009; IEC 2010; MHRA 2015; Shellock 2016]. Studies of their long-term biologic effects will continue, but meanwhile it makes sense to exercise great caution, especially in scanning pregnant women, small infants, patients on life support systems, and patients, staff, and visitors within whom there may possibly reside metallic objects.

To preclude accidents from happening, many highly visible or audible warnings on the potential hazards should be in place. Three categories of labels have been designed to indicate that an object is *MR Safe* around a scanner, *MR Unsafe*, or *MR Conditional*, meaning that it is safe under certain circumstances. MRI staff should also institute and enforce appropriate physical restrictions on site access, and they should be drilled on preventing unauthorized persons or equipment from entering high-field areas. They should provide training for those who may be coming in and out regularly, including on the need for special watchfulness regarding people or objects who are *not* regulars. And even during an emergency—rather, *especially* during an emergency—it is important to scrutinize all individuals seeking to gain entry. It really can be very, *very* cost-effective to install a ferrous metal detector system at the entrance to the imaging suite to augment other safety procedures. Check staff and patient histories carefully. Always, always assume that the main magnet and its fringe fields are on and potentially very hazardous.

Serious problems seem to arise when untrained personnel—housekeeping, maintenance, security, police, firemen, parents, non-MRI physicians, and other medical staff—gain unsupervised access to areas where entry should be strictly restricted. Unless they are carefully screened, such as with a double-check of records, a few critical questions, and a metal-detector archway, they may inadvertently bring clipboards, intravenous poles, hand cuffs, guns, vacuum cleaners, oxygen bottles, whatever, into the vicinity of the magnet (Figure 1.10). And if those objects happen to have parts that are ferromagnetic, in an instant they can turn into lethal projectiles.

The FDA, state Departments of Health, and other bodies publish both guidance and recommendations, on the one hand, and legally binding *regulations*, also known as *rules*, on the other. These cover limits on the main magnetic field strength and the rate of switching gradient fields, RF power levels, and so on, which are thought to be protective of patients and others. Federal regulations are published regularly in the *US Code of Federal Regulations* (CFR), while recommendations are circulated by other means.

Table 15.3 CAVEAT! There are many items, possibly within patients or staff or others, that may be inadvertently brought into the imaging suite and that may be magnetizable and can, therefore, cause lethal harm. The best way to prevent such accidents from occurring is to keep such people or objects from entering the suite in the first place, especially through thorough record checking and the use of a sensitive portal metal detector.

Within/with patient, others	In/into imaging suite
aneurysm clip, shrapnel, cochlear implant, prostheses, artificial heart valve, stent, permanent denture, pacemaker, defibrillator, electrodes, neurostimulator, medical infusion pump, drug-delivery patch, tattoo	O_2 tank, IV pole, gurney, stethoscope, clipboard, hemostat, scalpel, syringe, scissors, pen, tools, bucket, phone, laptop, electronics, fire extinguisher, ax, keys, gun, handcuffs, flashlight, watch, credit card

15.4 Danger from the Main Magnet and Its Fringe Fields: Beware of Aneurysm Clips and Pacemakers, and of Flying Screwdrivers, Wheelchairs, and Oxygen Bottles

The most immediate and obvious source of potential danger is the main magnetic field. It is extremely strong, fills a large volume, and may extend some distance beyond the bore. But it is not the field strength at the center that is of main concern—a paper clip will come to rest at the middle of a magnet's donut-hole, where all the forces acting on it balance one another, and the net force is zero. Rather, it is the fringe gradient fields at either end of the bore, where unwanted non-uniform fields can exert very strong forces on magnetizable objects. The pull along the axis of a cylindrical magnet exerted on a magnetized object of classical dipole moment, μ, for example, is

$$F_z = \mu(\partial B/\partial z), \tag{15.1}$$

similar to the expression with which Stern and Gerlach explained the deflection of a beam of silver atoms in a strong gradient. The force on a wheelchair is a bit more complicated, moreover, and most likely a great deal stronger. At the center of the bore, by the symmetry of the situation, $(\partial B/\partial z) = 0$.

Other things being equal, the fringe fields naturally increase with the magnet's nominal strength. The fringe field of a 1.5 tesla magnet can exert forces powerful enough to fling scissors or an IV pole into the bore. Similarly, fringe fields can yank a ferromagnetic aneurysm clip (they still exist) or piece of shrapnel out of the body (and in the case of an aneurysm clip, this can lead to hemorrhage and death) (Table 15.3). Some medical implants require a certain number of weeks post-surgery for the patient before an MRI is undertaken—this may allow for scar tissue to form that immobilizes the object. Patients and others (medical staff, cleaners, police, etc.) within whom there might be pacemakers, surgical pins, metallic intrauterine devices, and heart valves or other possibly magnetizable items should come nowhere near an MRI device until these objects are shown to be safe. Far less critical, of course, but still important is the disruption that MRI fields can cause to hearing aids, magnetic credit cards, watches, old computer disks, and medical electronic gear such as x-ray and ultrasound units, CT scanners, gamma cameras, etc.

The main magnet can fling a fireman's ax across the room and into the bore hole, and flying car keys, scissors, and screwdrivers can also cause lethal damage. A child was killed in New York when an oxygen cylinder brought in for respiratory support was magnetically yanked into the bore and smashed into his head. A German firefighter was nearly dispatched when the air bottle strapped to his back was dragged into the bore, and he with it. And a nurse's assistant placed an oxygen cylinder under a patient's gurney before they went into the scanning room; it was yanked into the bore at high velocity, luckily before the patient had been positioned there.

MRI magnetic fields can be confined somewhat (and other static fields excluded) with passive and active *self-shielding*, designed to bring about rapid fall-off of the *spatial field gradient* (SFG) outside the main magnet (Figure 8.3). This can be fine-tuned with field-producing electric coils (active).

The FDA has recommended that areas where the main magnetic field is still above 0.0005 T = 0.5 mT = 5 gauss, about 10 times Earth's own field, should not be considered safe for the general public. Access to regions above 0.5 mT near an MRI machine should be denied to *all* (nurses, transport personnel) except those few people known to be safe there (i.e., patients and certain family members who have been screened, and MRI-trained technologists). It is important that the displacement forces near the main magnet be tested [ASTM 2006].

In some situations, an MR study would be helpful for certain patients, despite the counter-indication of a pacemaker or other electrical implant. Recent developments in pacemaker design, however, have made it possible to perform conditional scans if the new safety standards (which still emphasize the potential risks) are closely adhered to [DOTmed.com 2011; ISO 2012]. In any case, it would be wise to consult the vendor's safety officer before proceeding.

Rarely, a superconducting magnet is *quenched.* Some portion of the wire warms to above its critical superconducting temperature and becomes resistive. Much of the energy stored in the magnetic field is expended in heating and then violently blasting the extremely cold helium gas out through vent pipes. At system installation, methods must be put in place to let the gas escape and to evacuate the patient immediately, to avoid freezing and asphyxiation. It should be possible to extract the patient from the bore and, if absolutely necessary, quench the main magnetic field in a matter of seconds by hitting the *fast magnet rundown switch.*

The bottom line is that no processes occurring in MRI appear to pose serious risks to patients or staff under current normal clinical conditions when standard operational guidelines are followed and due caution is exercised. But eternal vigilance is essential, since accidents do occur.

15.5 Biological Hazards from Switching Gradients and RF Fields

MRI involves the detection of weak RF signals. It is important to keep out the magnetic fields and RF noise produced by passing trucks, elevators, linear accelerators, and even flickering fluorescent lights. At the same time, magnetic credit cards and medical electronic gear (which may be very responsive to the presence of magnetic fields) in the area must be protected from the magnetic and RF fields produced by the MRI system. The necessary isolation can be achieved by means of suitable active and passive magnetic and *electromagnetic* shielding, which squeezes the magnetic fields outside of the bore into a much smaller external volume to avoid interactions. The RF problem can be solved with a *Faraday cage*, a wire-mesh or solid metal panel enclosure around the patient (and RF coil) area.

Under some conditions, RF fields and rapidly switching field gradients can induce eddy currents that heat tissue and electrode wires or other metal objects on or within the patient; this may lead to serious (albeit localized) burns. In addition, they may strongly disrupt the operation of cardiac pacemakers and other electronic devices which, therefore, should not be allowed to come near the MRI. Methods have been developed to envelop some of these items with special materials that safeguard them from the effects of RF fields.

Above and beyond dramatic heating of wires (leading to burns), the effect of an MRI-based RF system on a patient with no implants is to heat him up gradually during scanning. In ordinary circumstances, this is not a problem. However, in the case of patients with poor thermal regulation or with infants, generalized heating of a patient is a risk in its own right. The FDA couches its RF recommendations in two general ways: limits on maximum *temperatures* and *temperature changes* in various parts of the body, and on the *specific absorption rate* (SAR).

The temperature of the head, trunk, and extremities should remain $T_{head} < 38$ °C, $T_{trunk} < 39$ °C, and $T_{extremities} < 40$ °C, respectively (Table 15.2). The thought process behind the differences in temperature relate to tissue thermal sensitivity and to the ability for a body part to dissipate heat readily. They also have found that the rapid switching on or off of gradient fields can cause peripheral nerve stimulation, which must be minimized to prevent patient discomfort.

In addition, as mentioned in chapter 9, the *specific absorption rate* (SAR), the rate at which RF power is absorbed per kilogram of tissue, increases roughly as the square of the field strength,

$$\mathrm{SAR}(B_0) \sim B_0^{\,2}, \qquad (9.6)$$

and the issue of tissue heating may become more problematic. Indeed, most of the MRI incidents reported to the FDA involve local tissue RF burns, often because a conductor has heated up dramatically. And cardiac electrophysiology (MRI pulse sequences that gate with the heart) may require new methodologies at very high fields (>3 T).

The FDA and others also recommend limits on how much RF power can be pumped into parts of the body, as quantified in terms of the *Specific Absorption Rate* (Table 15.2). The SAR is defined as the rate at which RF energy (measured in watts per kilogram of tissue) is deposited within the patient:

$$\mathrm{SAR} = \text{power/mass (W/kg)}, \qquad (15.2a)$$

SAR depends on, among other things, the densities and electrical conductivities of the affected tissues, patient size, and the strength of the electric field of the RF pulses [Wang et al. 2007]. It also increases with the main magnetic field strength roughly as $B_0^{\,2}$, so that while going from a 1.5 T field to 3.0 T may roughly double the signal-to-noise, it also quadruples the rate at which high-frequency power is pumped into tissues. A useful approximate expression for the SAR is a modification of Equation (9.6):

$$\mathrm{SAR} \sim B_0^{\,2}\alpha^2 \mathrm{TR}\,\Delta M, \qquad (15.2b)$$

where α is the flip angle, TR the repetition time (the RF duty cycle), and ΔM the mass of tissue being irradiated. Gradient-echo protocols may have flip angles considerably less than 90°, while spin-echo sequences, with their 90° and 180° RF pulses have correspondingly higher SAR. More accurate SAR calculations are complicated, and consider the patient's weight and positioning, the transmission coil characteristics, the pulse sequence employed, and other issues.

The amount of heat energy E deposited in tissue is a function, of course, of the duration of application of the RF power,

$$E = \mathrm{SAR} \times \Delta t. \qquad (15.2c)$$

That and the specific heat and cooling rate of the tissue material and other patient-specific factors will together determine the rise in local temperature and the potential for harm. Recent guidance about limitations on heating is discussed in a review article [Woods 2007].

There are additional biological interactions and effects known to occur above 3 T, including painful nerve stimulation, photophosphenes (patients may experience flashes in the eye), altered cardiac electrophysiology (pulse sequences that gate with the heart may require new methodologies), and hazards from stronger static field gradients near the ends of the magnet bore.

Much development and testing work is ongoing, however, and indicates that these and other problems may well be surmountable. A research area where the experimental complications are generally much less severe, and the biophysics possibilities unlimited, is in small animal studies. These are currently ongoing at fields typically from 7 T to near 20 T, with accompanying resolution of down to 10 microns. There are numerous reports of the imaging of amyloid plaques in transgenic mice that display Alzheimer's disease, for example, and of the movement of individual cells over time. This is stimulating the development of new contrast agents specific to particular cell-types and individual proteins involved in gene expression, and the behavior of intra-cellular structures. And that's just the beginning.

15.5 Contrast Agents and Injectors

A serious problem can arise from the use of chelated gadolinium contrast agent (Figure 10.6b). It has been found that associated with gadolinium-based contrast agents (GBCA), there is a small but serious risk of *nephrogenic systemic fibrosis* (NSF) in some patients—in particular those with renal disease, severe hepatic disease, history of hypertension or diabetes, or age over 60 (Table 15.4). In 2007 and after, the FDA called for prescreening of a patient for *glomerular filtration rate* (GFR) before contrast is used [FDA 2017]. It also recommended that pregnant patients *generally* avoid Gd-contrast. This is the responsibility of the attending physician, of course, but you should keep your eyes open for potential mistakes.

In addition, a bolus of the agent is commonly administered by an automated injector, which can itself produce too much pressure, rupturing veins. While not typically part of a physicist's testing responsibility, it is

Table 15.4 Some FDA recommendations concerning gadolinium-based contrast agents (GBCA) [FDA 2017]

MRI Safety – GBCA

possible Gd-contrast reaction

Small but serious risk of nephrogenic systemic fibrosis (NSF) in patients with:

- renal disease
- age >60
- history of hypertension
- history of diabetes
- severe hepatic disease

FDA advisories 2007, 2017:

- pre-screen for glomerular filtration rate (GFR)
- pregnant patients *generally* should not have GBCA

a good idea for physicists to understand how injectors work, are controlled, and inspected.

15.7 Siting, Installing, and Accepting an MRI Device

A good bad-news/good-news bad story:

Once upon a time at a clinic, not that long ago or far away, the head of radiology ordered a new MRI scanner and felt that there was no need to waste money on help in designing the site. He assumed that his own medical physicist could handle the job, even though she had never installed such a device before and was reluctant to jump into such a major undertaking without expert assistance.

The clinic prepared the MRI area with the needed magnetic and RF shielding, an adequate access control portal, an efficient operator's area, and everything else necessary. But as they began to roll their just-delivered machine inside, the staff discovered that the magnet was too broad to pass through the back door. The hospital's structural engineers managed to widened the door, which was actually an unfortunate achievement, because the floor of the imaging suite room itself turned out to be too weak to support the multi-ton superconducting magnet and, well, you can imagine the rest.

After they fixed this very expensive and time-consuming mess, everyone lived happily ever after, except for the heads of radiology and medical physics, who decided to move to Nome and Bismarck, respectively.

You may someday be responsible for procuring, installing, and acceptance testing an MRI machine, and let's say you've never undertaken this sort of task before. First thing to do, after you have settled upon a vendor and device and checked on their reputations, is go over *everything* in the specifications with your vendor's installation engineers, your own facility's architect and structural engineer, and the state regulator, preferably all together, until there is complete agreement on every point.

Check that sufficient electrical power, water, and HVAC are available, and that mechanical vibrations, dust, humidity, etc., at the site will not be excessive. A Faraday cage is needed to prevent RF interference, and passive and active magnetic shielding may be needed for the main magnet. There should be ample room outside the projected 0.5 mT safety zone for easily maneuvering of people and sensitive equipment.

Confirm that the magnet can pass readily into and through *all* the doors, corridors, elevators, etc., and temporary holes in walls, roofs, and floors to and at its final destination, and that *all* the floors can support its weight and that of its transport equipment. Yes, there have been little mistakes.... And do bear in mind that you may eventually need to remove the machine, or replace it with a larger magnet. (It may be desirable to install a 1.5 T magnet and upgrade it to 3 T at a later date.) What other changes might affect the facility in three or five years for any foreseeable reason?

Ensure that you have a plan and the necessary tools to carry out acceptance testing, and go over everything with the vendor's engineers and with AAPM colleagues. Check in with the MR specialists at the ACR—they can be, and actively want to be, very helpful; and you'll be dealing with them soon enough anyway for accreditation.

Confirm that what the contract agreement says about response time for repairs, penalties for downtime, etc., is acceptable, given your anticipated patient load.

In other words, get help from others who have done all this multiple times before. And enjoy the process. It can be a great learning experience, and even fun!

CHAPTER 16

On the Horizon: Imaging with a Crystal Ball

Earlier parts of this book went over the traditional MRI techniques, and some variations on those themes. Also mentioned were technologies being developed but not yet necessarily ready for widespread application in the clinic. This final chapter will mention several examples of the latter and provide a few paragraphs sketching how they operate. The selection here is just a sampler, and far from comprehensive.

In addition, it will comment on two of the directions computers are taking now—namely expanding artificial intelligence applications and harnessing neural and quantum computers. Hopefully the next edition will have more to say about them.

16.1 Evolution: Ongoing Changes

Quantitative imaging, including MR fingerprinting

Much of the evolution of clinical NMR/MRI has relied on the early observation that T1 and T2 relaxation rates could distinguish some neoplasia from healthy tissue [Damadian 1971]. With much refinement since then, basic T1-weighted and T2-*w* spin-echo and gradient-echo imaging have became the workhorses of clinical MRI. While such maps can often reveal only a spatial or textural irregularity on the monitor as a hyper- or hypo-intense area, they may lead the experienced radiologist to a specific diagnosis.

Often, however, this is not the case. Then *quantitative* tissue values that are only hinted at in an image—actual T1 or T2 times, for example—may be able to significantly enhance the utility to a study [Clarke et al. 2008; Cercignani et al. 2018]. Unlike a rapid T1-*w* image that shows gross qualitative differences, however, it generally takes several scans with multiple TR values to measure an actual T1 rate in tissue. That is, it takes more time to obtain absolute pixel values to serve as biomarkers for a disease.

Various initiatives—such as by the Quantitative Imaging Biomarkers Alliance (QIBA), which is sponsored by national and international organizations (e.g., RSNA, AAPM, NIH)—are underway to use parameter mapping of tissue for more selectivity and sensitivity. One interesting approach is a method called MR *fingerprinting* [Ma et al. 2013]. A coarse way to describe this technique is that the MR sequence plays out gradient pulses and readout in a particular randomized way, with variable TR and TE patterns acting like a mash-up of multiple sequences into one. Then the signal is compared with a 'dictionary' using pattern-recognition technology to infer properties such as T1 and T2 values from the data. The benefit is that the sequence effectively acquires the combination of several under-sampled datasets into one (via compressed sensing), allowing the user to pull out multiple parametric maps of the brain during the time it takes to execute a single sequence.

One difficulty with MRI becoming a fully quantitative imaging modality involves the development of new techniques that are fast enough to be implemented in the clinic practically, yet sensitive enough to provide a noticeable improvement. Another is the general professional resistance in the clinic to change, and a strong tendency to return to the tried-and-true imaging methods as the standard of care. Many of the quantitative imaging products have unfamiliar appearances. But without exposure to new methods during a medical residency or fellowship, and without publications showing dramatic improvements in healthcare using such techniques, adoption will be slow.

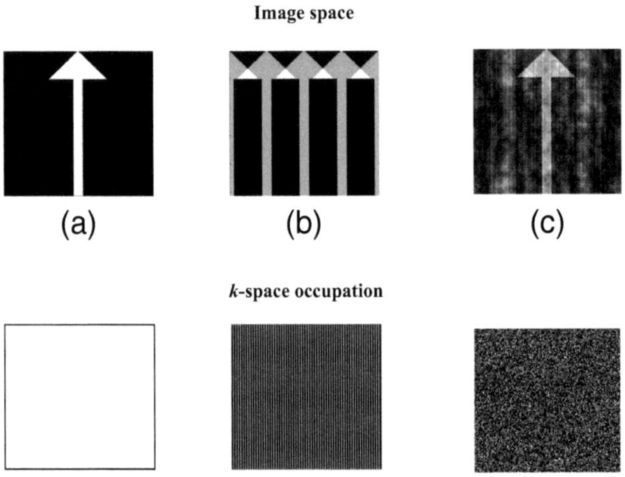

Figure 16.1 Images of an arrow produced with various degrees of under-sampling of data in **k**-space. The bottom row shows representations of **k**-space, where a white 'pixel' means that that region of **k**-space is filled with raw image data, and a black 'pixel' means that the corresponding datum in **k**-space has a value of 0 (i.e., it is not sampled). Column (a): Fully-filled **k**-space leads to a picture of an arrow after reconstruction. Column (b): One out of every four columns in **k**-space is sampled, leading to an acceleration of 4×. However, because of the periodic spacing of lines in **k**-space, the artifacts on the reconstructed image are prominent. Column (c): **k**-space is *randomly* under-sampled by the same amount as column (b) (1/4th of the data is present). Now, the image clearly shows an arrow masked with some amount of ghosting. The quality is not great, but the arrow is visible.

Compressed sensing

The objective of compressed sensing (as with a number of other technologies) is to reconstruct images from far fewer measurements than are traditionally thought necessary.

"Time is money," as they say (whoever *they* are), and that is certainly true for MRI exams. Techniques such as parallel imaging reduce acquisition time by not scanning parts of ***k***-space where the data would be redundant; many images contain a fair amount of air surrounding the patient, say, of no clinical interest. So also for an isointense bladder filled with fluid. To save time, it should possible to scan only the data representing the actual anatomy.

One way to reduce the amount of data to be dealt with is through compression (see Table 5.1). What if we combined data acquisition and data compression algorithms such that we scan compressed data directly? This question was posed by several groups in the mid 2000s, and the technique of *compressed sensing* (CS) was born; indeed, most new scanners have some form of compressed sensing on them [Candes et al. 2006; Donoho 2006; Lustig et al. 2007; Jaspan et al. 2015; Chen et al. 2014; Yao et al 2018]. This involves under-sampling the data in a manner that is time-efficient for the scanner. For one thing, gradient coils and other MR equipment must operate smoothly and not be driven hard by quick, frequent leaps across ***k***-space during sampling. Second, the under-sampling pattern must not create damaging artifacts. For example, Figure 16.1 shows images of an arrow that was constructed from data fully covering ***k***-space, and also from under-sampled ***k***-spaces. The image information is said to be *sparse* in the dark regions surrounding the arrow to its left and right. After under-sampling ***k***-space in a *periodic* fashion (Figure 16.1b), the reconstructed image has a dramatic aliased appearance, with the arrowhead region corrupted: the under-sampling method results in coherent artifacts due to the particular sparsity in the image. On the other hand, a *random* under-sampling (Figure 16.1c) leads to an image of shabby quality but with a recognizable image within.

If it can save so much time, why not use a dramatic amount of compressed sensing always? The reconstruction time for acceptable images is often slow. Also, general image quality suffers: detail blurring or ringing-type artifacts are common, requiring that acceleration rates be kept lower. Issues of image quality—including what compression is possible, why, and under what circumstances—are currently being explored.

Inexpensive low-field MRI

MRI started off at low fields, up to about 0.5 T with permanent magnets and electromagnets, but for nearly four decades these have been largely supplanted with superconducting devices. Superconducting scanners have significantly more homogeneous main fields and, since SNR is roughly linear in field strength, low-field units are more burdened with noise, requiring longer imaging time and larger NEX. Also, while several artifacts (chemical shift, susceptibility) are less apparent at low fields, some techniques (DTI, fMRI) cannot be performed there.

While it is often claimed that 1.5 T and 3 T machines lead to an optimal balance between image quality and cost, lower-field devices still perform adequately for several categories of tasks. Permanent and resistive magnets can be of open design, with wider examination space and greater access for interventional procedures, and no claustrophobia problems. Fringe fields are far lower, which reduces safety concerns.

Vendors have designed low-field MRI units expressly for the study of body extremities. These have bores that accommodate only hands, feet, and nothing larger than a knee over a region several inches in dimensions, and they can be an order of magnitude smaller than an ordinary superconducting-magnet system. Not only are they correspondingly less expensive, but also far more comfortable for patients. These have also been employed by veterinarians, such as on the legs of race horses.

Lower-field systems tend to have much lower purchase costs and, especially for permanent magnets, lower operating expenses. The ticket price for a standard 1.5 T whole-body scanner is typically above one million dollars, while a state-of-the-art 3 T device runs upward of $3M. (By comparison, CTs range from $1M for a 64-slice instrument to over $2M for a 256-slice machine). In addition, preparing the suite to house either can run a few hundred thousand more. A used, low-field scanner, on the other hand, can be had for perhaps a tenth of that and, with a permanent magnet, the operating and maintenance costs may be minimal. This would render it affordable to many small community hospitals and practices in third-world countries that would otherwise have to do completely without.

High-field imaging

The highest field at which clinical MRI is routinely practiced on humans is 3 T, although clinical research studies have been undertaken up to 7 T. The machines look much like their 3 T cousins (Figure 8.2), but are noticeably larger. They have demonstrated the benefits of high fields for modalities in which MR signals are especially weak, as in DTI and fMRI [Ugurbil et al. 2003].

Animal studies have been carried out above 12 T. As expected from the greater nuclear polarization, these units have displayed stronger signals and improved SNR. T1 is found to increase with field strength, and T2* gets shorter, which may contribute to enhanced image contrast. Some artifacts change in appearance, such as the dielectric effect (which becomes more pronounced and leads to image brightening in the middle of the image FOV). In MRS, the degree of chemical shift grows linearly with $\boldsymbol{B}_0$, and that can help to resolve NMR spectrographic lines, leading to closer examination of molecular proton environments.

The biophysical effects of applying fixed or varying high magnetic fields are not well understood on the molecular, cellular, or organism level. What *is* well known, on the other hand, is that instrument costs do tend to skyrocket—which is one reason we are unlikely to see many clinical MR devices that operate above 7 T, or even 3 T, anytime soon.

MRI on elements other than hydrogen; nuclear hyperpolarization

Hydrogen is the element most commonly examined in clinical MRI, by far, but several others display promise. MR studies of perfluorocarbons (PFC) containing ^{19}F have been ongoing for decades. ^{23}Na has been examined to assess the involvement of sodium in strokes and in the growth of neoplasms, and ^{13}C has revealed aspects of carbon metabolism. [Fain et al. 2010; Golman et al. 2006; Higuchi et al. 2005]. But of the elements indicated in the upper part of Table 2.2, it is phosphorous that has perhaps attracted the most attention, because of its critical roles in cell energetics and the relative strength of its resonance lines. Two ^{31}P spectra obtained *in-vivo* with MRS from live human muscle tissue are reproduced in Figure 16.2. The structure of each spectrum is determined by the chemical shifts for ^{31}P in different molecular environments, and by the relative concentrations of phosphorus-containing biomolecules of various types. The control spectrum was obtained under relaxed conditions, and the other follows 60 seconds of ischemic exercise. The peak designated P_i is attributable to inorganic phosphate; PCr belongs to phosphocreatine, and α, β, and γ to the three high-energy phosphate bonds of adenosine tri-phosphate (ATP).

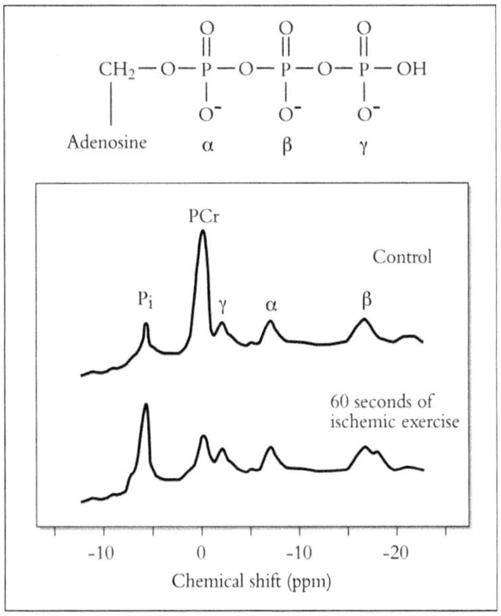

Figure 16.2 Typical MR spectra of ^{31}P muscle, one under normal metabolic conditions and the other after 60 seconds of ischemic exercise.

H-1
N-14
F-19
Na-23
P-31

At very low temperatures, near absolute-zero, electrons are polarized in a strong magnetic field in accord with the QM equivalent of the Boltzmann distribution, Equations (4.4). Because of their far greater masses, the relative number of nuclei in the lower state is lower by a factor of many thousands. *Hyperpolarization* is a process for enhancing the imaging capability of certain elements, in particular carbon and nitrogen, during their metabolism [Roos et al. 2015]. *Dynamic nuclear polarization* (DNP) is one process of transferring electron spin polarization, which can be considerable, to a population of nuclear spins, such as ^{13}C nuclei. This can be brought about through mixing among the electrons and carbon spin states, or by way of the several electron-nuclear cross relaxation mechanisms [Wolbarst 1972], or *spin-exchange optical pumping*. This can result in driving the carbons far from their normal equilibrium configuration, resulting in a hyperpolarization factor of as much as a of 10^3 to 10^5.

A sample of ^{13}C pyruvic acid, for example, may retain much of its hyperpolarization in a metastable state when brought to room temperature. It can be dissolved in an aqueous solution and injected in the study of the metabolism of cancer cells *in-vivo*. Likewise, molecules containing hyperpolarized ^{15}N and ^{19}F have shed light on general metabolic processes. And MRI on hyperpolarized ^{3}H and ^{129}Xe gases is effective in probing pulmonary structure and function.

Helium-free MRI

One of the significant expenses in running an MRI center used to be that of the liquid He coolant (currently on the order of $10 to $20 per liter). Regrettably, the global helium supply is dwindling as we speak, and consists of pockets that collect in natural gas wells as a by-product of the alpha decay of the members of the uranium and other natural decay chains within the Earth. Typical clinical scanners have a cryostat with a volume on the order of 1500 liters of liquid helium. During this decade, scanners have been developed with nominally 'zero boil-off' cryostats, for which few or no refills are needed during their clinical lifetimes. There are two other approaches to dealing with the problem. The first comprises a new family of magnet designs that seal the superconducting windings into a very small vessel (~20 liters) for a significant cost savings. Prototype magnets of this design have been manufactured, but they are not ready for clinical use yet. The second is the continued development of higher-temperature superconducting wire, with the hope that superconductivity above 77 °K can be achieved. Unfortunately, these materials are very brittle and expensive, and not yet ready for prime time.

Quiet imaging

Some MRI technology improvements are flashy, enriching the radiologists' diagnostic palate with increasingly specific and sensitive forms of physiological contrast. But others linger in the background, no fanfare save that of patients experiencing a less fearful exam. Consider audible noise.

The gradient coils are turned on and off rapidly and repeatedly to excite tissue, to spatially encode signal, and to spoil residual magnetization. The difficulty, of course, is that interactions between the gradient and the main fields try to intermittently force the gradient coils out of the scanner. In so doing, they send vibrations throughout the whole system, creating a cacophony of clicks, hums, buzzes, and booms that can rattle both the room and the patient.

In recent years, all MRI vendors have introduced imaging strategies that lower this noise. Most of these seek to activate and deactivate the gradient fields more gradually, leading to mean attenuation of the sound on the order of 10 to 15 dB or so, with higher attenuation at particular frequencies. Instead of sharp, square pulse-like gradients—like the gradient pulses in Figure 12.8a—a pulse with softer shoulders is used to reduce the mechanical vibrations (Figure 12.9). The drawback to these techniques is that they tend to use readout windows that are shorter or of higher bandwidth, leading to lower SNR and CNR.

Combining MRI with PET, NIR, US, etc.

Commercial Positron Emission Tomography (PET) instruments are almost always sold with an attached CT to provide clear anatomic points of reference. More recently, combined PET-MRI have been designed in which the MRI component not only delineates bodily structures, but also provides information on tissue physiology and pathology, like the PET itself (Figure 16.3). Such a system would clearly be advantageous in carrying out the planning of radiation, surgical, and other therapy procedures [Ouyang 2013]. Similar techniques arise from combining MRI and Near Infrared (NIR) spectroscopy [Mastanduno 2013].

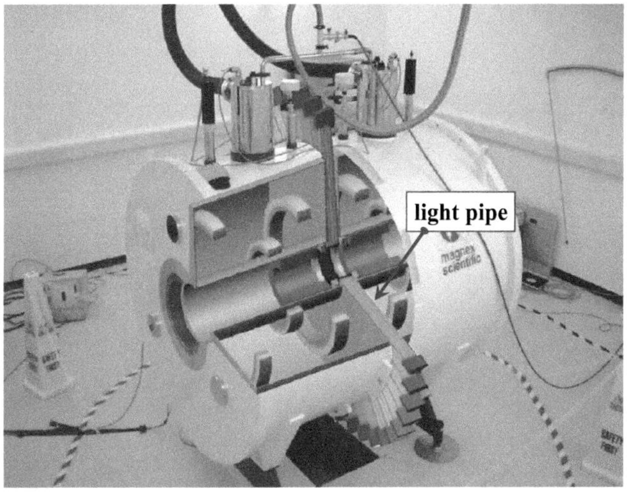

Figure 16.3 A combined MRI/PET device. The 511 keV photons from positron annihilation activate separate fluorescent detectors in a ring closely circling the bore, and the resulting light is brought out of the machine in light pipes.

Likewise, the measured biomechanical characteristics of a tissue may give diagnostic information about a malady. Pathological tissues are often more rigid than the surrounding normal tissue. Breast malignancies, for example, are generally much harder than healthy glandular tissue, and liver stiffness increases with the progression of hepatic fibrosis. *MR elastography* (MRE) is a noninvasive medical imaging technique for characterizing the elastic properties, such as stiffness and compressibility, of soft tissue *in-vivo* [Chen et al. 2013]. An external source of mechanical vibrations, such as an ultrasound transducer pressed against the skin surface, introduces shear waves into the tissues, and MR phase contrast techniques are utilized to acquire images of the wave propagation, which gives rise to an elastogram, a tissue-stiffness mapping.

Zero-quantum imaging

Protons separated by as much as a millimeter can, quite remarkably, sense one another by way of long-range magnetic dipolar interactions [Zhang et al. 2018]. While normally not detected in clinical MRI, signals

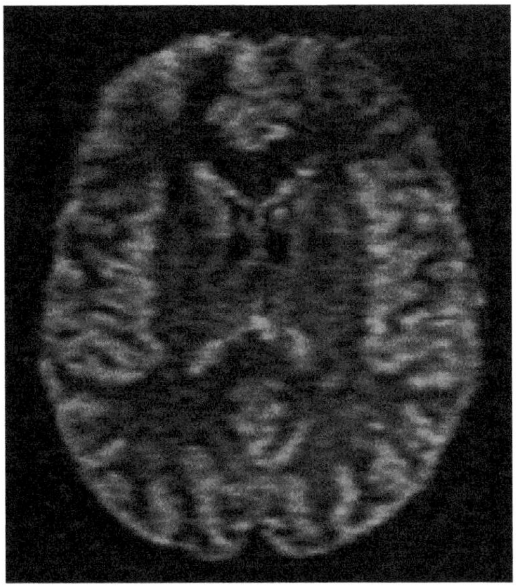

Figure 16.4 Zero-quantum coherence (ZQC) imaging, unlike standard MRI, is designed to be sensitive to the magnetic dipole interactions of protons as far distant as 1 mm.

from such pairings, or *intermolecular zero-quantum coherences* (ZQC), are rich enough to be brought out with an echo-planar imaging pulse sequence modified to incorporate an additional 45 degrees RF pulse and a correlation gradient [Davis et al. 2014]. ZQC images can generate a form of contrast that reflects variations in tissue susceptibility (Figure 16.4).

Radiation therapy treatment planning and follow-up

The three traditional methods of attacking solid cancers—with either curative or palliative intent—are surgery, chemotherapy, and radiotherapy, or some combination of them, with immunotherapy rapidly arising along the horizon. The objective of any of these is to subject the diseased tissues to tumoricidal or pain-relieving treatment, while at the same time not causing *unacceptable complications* (the term of art) in healthy essential organs nearby. It would be of questionable benefit to eradicate a tumor of the esophagus if the spinal cord were severely damaged in the process [Buck et al. 2016; Gore et al. 2010].

Roughly half of all cancer patients in the United States are treated with radiotherapy (RT), which makes use of high-intensity ionizing radiation produced by either an encapsulated radionuclide (*brachytherapy*) or a linear accelerator (*external beam*). A linac's photon x-radiation beam is significantly more energetic (megavoltage range) and penetrating than what is used in diagnostic imaging (kVp), and it delivers thousands of times more radiation dose during the procedure. Several such x-ray beams are directed into the patient from various angles, and they undergo Compton interactions (which predominate at high photon energies in soft tissue) as they pass through; the resulting Compton electrons ionize the tissues they traverse, depositing dose especially in the beams' crossfire region. (What about the Compton scatter photons?) The resulting production of free radicals and other molecular instabilities leads to damage to DNA and to the tumor microvasculature. That, in turn, is intended to kill the tumor cells while, ideally, managing to spare enough of the healthy tissues for their recovery.

This means that the treatment, and the planning of it, must be able to target the tumor with a great degree of accuracy and deliver the radiation nearly where it is supposed to go with as little as possible ending up elsewhere (Figure 16.5). The fairly recent introduction into the clinic of *intensity-modulated radiation therapy*

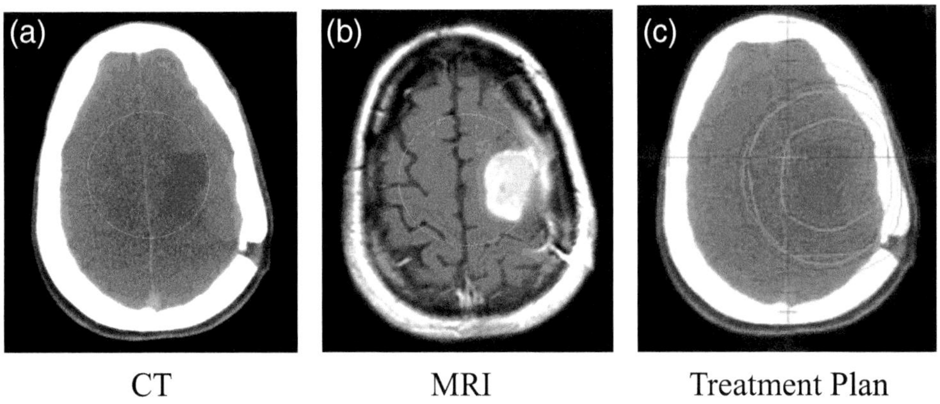

Figure 16.5 RT treatment planning. (a) As a first line of defense, CT may indicate the presence if a tumor. (b) MRI may be more effective in precisely delineating the target region, and perhaps its spatial characteristics, (c) which is critical in computerized radiotherapy or other treatment planning.

(IMRT) treatment, in which the beam direction, shape, intensity, etc., are varied continuously over time, has greatly enhanced the ability to deliver precisely sculpted three-dimensional dose distributions; the treatment planning process, however, has become correspondingly more complex.

Back in the day, the initial objective of the radiation oncologist and treatment planning team was to localize and delineate the tumor (as well as other anatomic structures), with several orthogonal *simulation* planar x-ray films. From this information, and with an assumed beam configuration, they would instruct the treatment planning computer to calculate the local distribution of ionizing radiation dose. It would then systematically modify beam angles, strengths, etc., so as to generate a good topographical dose distribution map that, ideally, will take the tumor to high dose, but little other nearby tissue.

This process was improved considerably with the development of digital planar and CT-based simulation, and the ability to localize tumors through PET input. The data from all of these are now routinely fed automatically into the treatment planning computer. The Digital Imaging and Communications in Medicine (DICOM) standard, moreover, provides compatibility among modalities in managing imaging data transfer, and it allows easy input to a RT treatment planning system from a wide variety of imaging modalities. A holy grail for all of this a system that optimizes treatment planning fully automatically. While this is an area of considerable research, what remains to be found is a biologically based optimization function that reaches a maximum at a 'best' plan [Wolbarst 1984; Wolbarst et al. 1982].

The point of this sub-section is that PET, MRI, and other technologies are playing increasingly central roles in defining the tumor region and critical tissues for both treatment planning and follow-up. As MRI, in particular, continues to evolve, new applications in planning and monitoring will accompany those changes.

16.2 Revolution: Artificial Intelligence and Quantum Computers

First appearing in the clinic in the 1970s, computers have enabled novel computation-intensive forms of imaging (CT, PET, MRI). Providing image processing, noise reduction, and analytical tools for all modalities, they and have also allowed the instantaneous storage, retrieval, display, and transmission of images and other information with Picture Archiving and Communications System (PACS), smart phones, electronic notepads, etc. In so doing, they have altogether changed how medicine is practiced.

Another advance, *artificial intelligence* (AI) is already having a profound impact in medicine and nearly everywhere else. AI algorithms for computer recognition of image features for *computer-aided detection/diagnosis* (CAD) and other applications are ushering in a whole new approach to all modalities of medical imaging. The only certainty is that what we see today is just the beginning. Tomorrow—perhaps the one-stop doc in a box? And after that?

Advanced programs play increasingly central roles in both software and hardware circuit design for future computers. Darwinian methods (i.e., 'genetic programming'), for example, occasionally but intentionally introduce small *random* 'errors' into code. These are nearly always deleterious, but on exceptionally rare occasions—much less frequently than one time out of a million—they may accidentally improve things, and such a change will be retained! (A major problem is to determine what is an 'improvement'. Nature has it easy in that department!) Such an approach can encourage some, and usually nonobvious, paths of positive 'evolution.' At some point, such machines and programs will (probably) evolve that incorporate something akin to the *curiosity* that drives cats and physicists—they may function in desirable ways, but the alterations that are responsible will most likely be far, far too subtle and complex for humans to understand. Similarly, while artificial intelligence is becoming ever more proficient at analyzing data patterns, it has made very little progress in the related area of understanding *causal relations*, which even children readily master early and on their own. A computer may see a stick strike a drum and simultaneously hear a loud noise—but was either responsible for the other? No clue!

Several branches of biotechnology are becoming involved in novel ways. Rather than sending currents of electrons along routes determined by the settings of silicon-based switches, a *DNA computer* would operate by snipping and joining segments of DNA in solution. Such a computer would consume a billion times less energy than its solid-state counterparts, require a trillionth of the space, and perform some kinds of computations that are otherwise impossible now.

The *neuron computer*, alternatively, might perhaps be grown as clusters of human nerve cells and be self-assembling according to building plans contained within a genetically modified DNA set, that itself might be allowed to improve through selective evolutionary forces. Already it can link directly with the cells of a human brain through attached electrodes or other means. It might interface with the outside world by way of a solid-state substrate—at the confluence of semiconductor, information, nano-, and cyber- biotechnologies. Conceivably, an electronic neural computer could be made to mimic the structure or function of some of our own neural complexes.

Artificial intelligence (AI)

It is widely believed among military leaders that the results of the next major armed conflict will be determined by which side can act most quickly. To that end, the Pentagon and the corresponding departments in China and Russia are all currently spending tens of *billions* of dollars (a number steadily growing) each year on developing *artificial intelligence* for warfare applications. Some of these, almost as afterthoughts, are finding implementations in medicine.

In particular, 'intelligent' computer programs are becoming ever more capable of detecting irregularities in all sorts of images and of suggesting explanations for them. The scientific underpinnings of artificial intelligence are still in an embryonic stage, yet computer-based *neural networks* and other AI systems are already augmenting certain roles of the physician as diagnostician. Computer-based programs are available that are highly effective at diagnosing skin cancers, and others already act as second readers in many mammography and other departments And they can do so for hours on end without being distracted by phone calls from angry teenage offspring, or growing the least bit bleary-eyed or grumpy.

Computers excel at analytic jobs that involve memory, logic, repetitively following directions, and applying simple if-then tests, as in playing chess or calculating, well, just about anything that involves logical decisions. They have made inroads, for example, in medical robotics [Christoforou et al. 2010], a field bound to expand. But even the most clever computers have a much harder time with the ordinary, elementary human activities of everyday life that require conscious intuition, ethical judgment, and simple common sense, where few algorithmic rules may apply. To a large extent, they still largely lack the kinds of learned visual skills that allow a radiologist to distinguish a liver with a small abscess from one that's normal, but just looks a little funny.

A great deal of progress has already been made in computer recognition of patterns in one dimension. Human speech, for example, may be viewed as a signal that is a function of only one variable, time, and highly accurate speech-processing programs are already available. Similarly, programs have been written that can analyze electrocardiograms (which also are one-dimensional images) with success rates comparable to those of skilled cardiologists; these devices routinely scan Halter signals for cardiac abnormalities.

The application of computer analysis to general two-dimensional images is orders of magnitude more difficult, however, and will be much slower in coming. It is not hard to imagine how a computer might locate a point in a two-dimensional mathematical area: it would sweep a vertical line horizontally until it makes contact, and then would repeat with a horizontal line moving upward. Likewise, it could find the two ends of a straight line—or even the shape, size, and orientation of an oval, rectangle, or other regular form—by comparison with shapes it has already 'learned.' But a search for significant spatial patterns in two or three dimensions is far more challenging and has to be sensitive to much more subtle and complex aspects of the image [Giger 2008; Wagner et al. 2008]

The intricacy of many of the clinically significant variations in detail and contrast, together with the degrading effects of statistical noise and overlapping tissues, along with the great range in normal and abnormal patterns occurring among different individuals, makes the general problem of automated diagnosis highly demanding. But while CAD is still just starting out, programs that search for spatial patterns characteristic of certain pathologic conditions in mammograms, chest films, virtual colonoscopy, ultrasonograms of the liver, etc., are already finding acceptance.

The designers of medical expert systems intend that they be able to accumulate the body of knowledge required of physicians, and then to some extent mimic their actual thought processes in decision-making. The algorithms are meant to consider the implications of the evidence at hand, weigh the probability of correctness of each possible explanation, perhaps demand more information to reduce the range of possibilities, and generally propose and test hypotheses. Deep learning neural networks—programs that use feedback information from humans to 'learn' decision rules from their own mistakes (without 'understanding' the reasons for them)—are highly promising for this kind of effort. We know next to nothing about how the intelligence of a human really works (or even that of a nematode, for that matter), and the 'intelligence' we develop for machines is likely to end up being quite different from our own—after all, an airplane doesn't flap its wings. In any case, computerized medical diagnosis, treatment decision-making, and medical informatics are highly promising fields, and it is to be expected that they will grow to reach higher-hanging fruit.

One area of great current activity is the correlation of the results of fMRI, PET, and other modalities with states of mind. Implementation of AI in such psychophysical studies is complicated by the difficulty in assessing what a subject (human or animal) is actually thinking or feeling, such as when exposed to a picture of a snarling dog or to a real angry pit bull—and how that reaction depends on past experience with animals, on any element of surprise or fear, etc. That problem is exacerbated by the effects of the study environment (clinic, laboratory), on the requirement to respond to a stimulus or to hold still, and so on. Still, the potential value of such investigations will be immense, once such problems are overcome.

In a related effort, the relatively young discipline of *biomedical informatics* is beginning to provide tools to search for, identify, retrieve, integrate, analyze, model, display, store, communicate, and manipulate all this information. Biomedical informatics is perhaps best known for its role in untangling the vast quantities of data coming from work on the genome. Another area of considerable interest is the need to find and display the few clinically essential visual items contained within an overabundance of imaging data from hundreds, perhaps more, of MRI, CT, and other sections per patient (with numerous such patients each day). A physician clearly cannot examine them all individually (despite the chance that a critical sign may lurk faintly in only one of them), so researchers have to continue to develop ways to search effectively through the various sorts of databases, which are growing in number, size, and complexity—expanding roles for AI.

Some experts expect that over the next few decades and centuries—assuming we're still around—general AI will be created in which computer systems are capable of replacing us in a wide range of human occupations. AI might not only drive cars and trade stocks, but also take proper care of babies and the elderly, manage all local and national security activities, accurately interpret countless forms of raw data, carry out original scientific research (including on AI), keep us alive indefinitely and, of course, fully control our lives. Or AI could even become terminally fed up with us. Keep an eye out for the next edition.

Quantum computers

The fundamental unit of information employed by a *digital computer* is the *bit*, or *binary digit*. It can assume only one of two possible values, commonly called *0* and *1*, and its state may be physically mimicked with a bipolar device such as a transistor switch that can be left only in an *on* or an *off* configuration. One such switch can provide the answer—*yes* or *no*—to a simple binary question like, "Did a heads come up in the coin toss?"

A digital computer operates by systematically operating on huge banks of such binary switches at lightening speed [Feynman 1996]. Functioning under the rules of the linear, "if-this-then-that" Boolean logic of everyday life, a digital computer can manipulate immense numbers of combinations of bits to resolve a broad range of complex numerical, logical, and communications problems. Not surprisingly, the potency of a conventional computer increases with the size of its *central processing unit* (CPU) and *memory*. The storage capacity of a personal computer is typically measured in *gigabytes*, or billions of bytes (a *byte* being a combination of eight bits), so increasing the memory by only one bit has little effect on its capabilities.

A *quantum computer* is another beast altogether, operating according to the totally counterintuitive rules of quantum mechanics [Feynman 1982; Milburn 1999; Rieffel 2014]. While digital computers operate with electronic switches that can be set in either the 'on' or the 'off' condition, quantum devices would manipulate *qubits*, subatomic quantum systems such as isolated electron spins or photons that can consist of a continuous range of *combinations* of the two. Just as a transistor switch will be either open or shut, or even an unfair coin will land either heads up or down, a proton that is disturbed while in a magnetic field often will eventually settle into a spin-up or spin-down state. But immediately *before* that happens, while it is just hovering there worrying about what to do next, it will exist in a mixed, *superimposed*, or *entangled* quantum state comprised of, say, both spin-up and spin-down states simultaneously. It's as if it were in both a 0 and a 1 configuration at the same time, or like our coin while it is still spinning in the air and displaying both faces. There are no simple classical analogs for so bizarre a situation, and it really doesn't make any sense to anyone, but such is the nature of a quantum system.

But one critically important aspect of all this can be readily affirmed: even though adding a single bit to the memory of a digital computer has negligible effect on its operations, an additional *quantum bit*, or *qubit*, to a quantum computer doubles its capacity to function. That is, while the power of a digital computer grows more or less linearly with the number of bits, that of a quantum computer increases exponentially with how many qubits it is working with. Expanding the set of qubits from 10 to 20, say, is equivalent to increasing its power from that of 1,024 classical bits to 1,048,576.

Quantum computers are not about to replace the digital kind anytime in the foreseeable future. There remain several major problems at present. It is extremely difficult to create physical devices that can store more than a few qubits for more than a fraction of a second. Such instruments must operate near absolute zero temperature and in a deep vacuum, and they are hypersensitive to even the slightest molecular disturbances.

Also, it takes a novel mind-set to program a quantum computer, and even if you can, it is likely to experience a high rate of error, an intolerable deficiency.

And while adept at separating monstrously humongous real numbers into prime factors, which is essential for code-breaking and the other tasks of cryptography, it is not clear what further applications are suited for it. Optimistic researchers suggest that these will become apparent in the not-too-near future and will include, in particular, transfiguring the field of artificial intelligence. Possibly the first real applications, relevant to imaging, will be quantum computing's ability to enhance AI.

16.3 Conclusion

Much of optical imaging, molecular imaging, thermography, electrocardiography, electroencephalography, magnetocardiography, magnetoencephalography, terahertz imaging, tissue impedance imaging, and electron-spin resonance imaging is still largely experimental, and some of the potential applications have achieved only limited clinical success so far. It is quite possible, however, that one or more of these technologies may burst without warning onto the clinical scene. High-technology companies are expending considerable resources to push MRI, CT, SPECT, PET, EEG, and other established modalities farther along. On the way, they may well find exactly what is needed to bring novel MRI applications to the fore. Just as likely, methods that combine modalities—such as what happened with PET plus fMRI imaging—will continue to break new ground. And then there are all those wild ideas not yet even imagined....

So maybe the clinic of 2050 will be as astonishing to us as MRI would be to Röentgen. You just never know!

"Everything that can be invented has been invented."

–Charles H. Duell, Commissioner
U.S. Patent Office, 1899

References

AAPM. American Association of Physicists in Medicine. *Acceptance Testing and Quality Assurance Procedures for Magnetic Resonance Imaging Facilities.* AAPM Report No. 100. (2010).

AAPM/RSNA. (2006). The AAPM/RSNA Physics Tutorial for Residents: MR Artifacts, Safety, and Quality Control." *Radiographics* 26(1):275–97. https://doi.org/10.1148/rg.261055134.

Abragam, A. *Principles of Nuclear Magnetism.* Oxford: Oxford University Press, 1961.

ACR. American College of Radiology. http:www.acraccreditation.org/modalities/mri. (2018).

ASTM. American Society for Testing and Materials International. *Standard Test Method for Measurement of Magnetically Induced Displacement Force on Medical Devices in the Magnetic Resonance Environment.* ASTM F2052 -06e1. West Conshohaken, PA: ASTM, 2006.

Baker, P. B. *Clinical Perfusion MRI.* New York: Cambridge University Press, 2013.

Balchandani, P. and T. P. Naidich. (2015). "Ultra-High-Field MR Neuroimaging." *AJNR Am. J. Neuroradiol.* 36(7):1204–15. doi:10.3174/ajnr.A4180.

Barrett, H. H. and K. J. Myers. *Foundations of Image Science.* Hoboken, NJ: John Wiley & Sons, 2004.

Bernstein, M. A., K. F. King, and X. J. Zhou. *Handbook of MRI Pulse Sequences.* Burlington, MA: Elsevier Academic Press, 2004.

Bloch, F. (1946) "Nuclear Induction." *Phys. Rev.* 70:460–74.

Block, K. T. and J. Frahm. (2005). "Spiral Imaging: A Critical Appraisal." *J. Magn. Reson. Imaging* 21:657–68.

Bloembergen, N. *Nuclear Magnetic Relaxation* (thesis reprint). New York: W. A. Benjamin, 1961.

Bloembergen, N., R. V. Pound, and E. M. Purcell. (1947). "Nuclear Magnetic Relaxation*."* *Nature* 160:475–6.

Breit, G. and I. I. Rabi. (1931). "The Measurement of Nuclear Spin." *Phys. Rev.* 38:2082.

Brown, R., Y. Wang, P. Spincemaille, and R. F. Lee. (2007). "On the Noise Correlation Matrix for Multiple Radio Frequency Coils." *Magn. Reson. Med.* 58:218–24.

Breuer, F. A., M. Blaimer, R. M. Heidemann, M. F. Mueller, M. A. Griswold, and P. M. Jakob. (2005). "Controlled Aliasing in Parallel Imaging Results in Higher Acceleration (CAIPIRINHA) for Multi-Slice Imaging." *Magn. Reson. Med.* 53:684–91.

Brown, R. W., Y-C. N. Cheng, E. M. Haacke, M. R. Thompson, and R. Venkatesan. *Magnetic Resonance Imaging: Physical Principles and Sequence Design,* 2nd Ed. Hoboken, NJ: Wiley Blackwell, 2014.

Bryan, R. N. *Introduction to the Science of Medical Imaging.* Cambridge: Cambridge University Press, 2010.

Buck, J. and J. Gore. "Frontiers of Biomedical Imaging Science (Highlights of the 2015 Vanderbilt Meeting)." In Godfrey, D. J., J. Van Dyk, S. K. Das, B. H. Curran, and A. B. Wolbarst, Eds. *Advances in Medical Physics 2016.* Madison, WI: Medical Physics Publishing, 2016.

Buck, R. C. *Advanced Calculus,* 3rd Ed. New York: McGraw-Hill, 2003.

Burgess, A. E. (1999). "The Rose Model, Revisited." *J. Opt. Soc. Am. A.* 16:633–46.

Bushberg J. T., J. A. Seibert, E. M. Leidholdt,Jr., and J. M. Boone. *The Essential Physics of Medical Imaging,* 3rd Ed. Philadelphia: Lippincott, Williams, and Wilkins, 2012.

Buxton, R. B. *An Introduction to Functional Magnetic Resonance Imaging: Principles and Techniques,* 2nd Ed. Cambridge: Cambridge University Press, 2009.

Carr, H. Y. Letter to the editor of *Physics Today.* Jan. 1993.

Carr, J. C. and T. J. Carroll. *Magnetic Resonance Angiography.* New York: Springer, 2012.

Carrington, A. and A. D. McLachlan. *Introduction to Magnetic Resonance.* New York: Chapman and Hall, 1967.

Chen, C-N and D. I. Hoult. *Biomedical Magnetic Resonance Technology,* 2nd Ed. Taylor & Francis, 2009.

Chen, J., M. Yin, K. J. Glaser, J. A. Talwalkar, and R. L. Ehman. (2013). "MR Elastography of Liver Disease: State of the Art." *Appl. Radiol.* 42:5–12.

Christoforou, E. G. and N. V. Tsekos. "Robotic Systems for MRI-Guided Interventions." In Wolbarst, A. B., A. Karellas, E. A. Krupinski, and W. R. Hendee, Eds. *Advances in Medical Physics 2010.* Madison, WI: Medical Physics Publishing, 2010.

Clarke, G. D. and L. Lee. "The Principles of Quantitative MRI." In Wolbarst, A. B., K. E. Mossman, and W. Hendee, Eds. *Advances in Medical Physics 2008.* Madison, WI: Medical Physics Publishing, 2008.

Collins, C. M., W. Liu, W. Schreiber, Q. X. Yang, and M. B. Smith. (2005). "Central Brightening Due to Constructive Interference With, Without, and Despite Dielectric Resonance." *J. Magn. Reson. Imaging* 21:192–96.

Collins, D. J. and A. R. Padhani. (2004). "Dynamic Magnetic Resonance Imaging of Tumor Perfusion." IEEWE Engineering in Medicine and Biology Magazine 23:65–83.

Damadian, R. (1971). "Tumor Detection by Nuclear Magnetic Resonance." *Science* 171:1151–53.

Damadian, R., M. Goldsmith, and L. Minkoff. (1977). "NMR in Cancer, XVI: FONAR Image of the Live Human Body." *Physiol. Chem. and Physics* 9:97–100.

Dawson, M. J. *Paul Lauterbur and the Invention of MRI.* Cambridge, MA: MIT Press, 2013.

Dennery, P. and A. Krzywicki. *Mathematics for Physicists.* Mineola, NY: Dover Publications, 1995.

Dietrich, T. J., E. J. Ulbrich, M. Zanetti, S. F. Fucentese, and C. W. A. Pfirrmann. (2011). *"PROPELLER Technique to Improve Image Quality of MRI of the Shoulder." Am. J. Roentgenol.* 197:W1093–1100.

Deshmane, A., V. Gulani, M. A. Grismold, and N. Seiberlich. (2012). "Parallel MR Imaging." *J. Magn. Reson. Imaging* 36:55–72.

Dixon, R. L. and K. W. Ekstrand. "Physical Foundations of Proton NMR: Part I." In *NMR in Medicine: The Instrumentation and Clinical Applications.* S. R. Thomas and R. L. Dixon, Eds. *AAPM Monograph No. 14.* New York, NY: American Institute of Physics, 1985.

Dixon, R. L., K. E. Ekstrand, and P. R. Moran. "Physical Foundations of Proton NMR: Part II—The Microscopic Description." In *NMR in Medicine: The Instrumentation and Clinical Applications.* S. R. Thomas and R. L. Dixon, Eds. *AAPM Monograph No. 14.* New York, NY: American Institute of Physics, 1985.

DOTmed.com. "FDA approves first MRI-compatible pacemaker." February 9, 2011. http://www.dotmed.com/news/story/15333?utm_campaigne=campaign&utmsource=2011-02-10&utm_medium=email.

Douek, P., R. Turner, J. Pekar, N. Patronas, and D. Le Bihan. (1991). "MR Color Mapping of Myelin Fiber Orientation." *J. Comput. Assist. Tomogr.* 15:923–29.

Duhamel, P. and M. Vetterli. (1990). "Fast Fourier Transforms: A Tutorial Review and a State of the Art." *Signal Processing* 19:259–99. doi:10.1016/0165-1684(90)90158-U.

Eisberg, R. and R. Resnick. *Quantum Physics of Atoms, Molecules, Solids, Nuclei, and Particles,* 2nd Ed. New York: John Wiley and Sons, 1985.

Elster, A. D. and J. H. Burdette. *Questions and Answers in Magnetic Resonance Imaging,* 2nd Ed. Philadelphia: Mosby, 2001. [On the web and continually updated since 2013 at http://mriquestions.com/index.html. Prepared for MRI radiology residents et al., but can be helpful for all!]

Ernst, R. R. and W. A. Anderson. (1966). "Application of Fourier Transform Spectroscopy to Magnetic Resonance." Rev. Sci. Instrum. 37:93.

Fain, S., M. L. Schiebler, D. G. McCormack, and G. Parraga. (2010). "Imaging of Lung Function Using Hyperpolarized Helium-3 Magnetic Resonance Imaging: Review of Current and Emerging Translational Methods and Applications." *J. Magn. Reson. Imaging* 32(6):1398–408. doi: 10.1002/jmri.22375.

FDA Drug Safety Communication. "FDA identifies no harmful effects to date with brain retention of gadolinium-based contrast agents (GBCA) for RIs; review to continue." www.fda.gov/Drugs/DrugSafety/ ucm559007.htm (Accessed 12-19-2017).

Feynman, R. P., R. B. Leighton, and M. Sands. *The Feynman Lectures,* Vols. I-III. Reading, MA: Addison Wesley Publishing Company, 1964.

Feynman, R. P. *The Character of Physical Law.* Cambridge, MA: The MIT Press, 1965.

Fleisch, D. *A Student's Guide to Maxwell's Equations.* New York: Cambridge University Press, 2008.

Garroway, A. N., P. K. Grannell, and P. Mansfield. (1974). "Image Formation in NMR by a Selective Irradiative Process." *J. Phys. C* 7:L457–62.

Giger, M. L. "Computer-Aided Detection/Computer-Aided Diagnosis." In Wolbarst, A. B., K. E. Mossman, and W. Hendee, Eds. *Advances in Medical Physics 2008.* Madison, WI: Medical Physics Publishing, 2008.

Golman, K. and S. J. Peterson. (2006). "Metabolic Imaging and Other Applications of Hyperpolarized C-13." *Acad. Radiol.* 13:932–42.

Gore, J. C., H. C. Manning, T. E. Peterson, C. C. Quarles, T. K. Sinha, and T. E. Yankeelov. "Quantitative Imaging Biomarkers of Cancer." In Wolbarst, A. B., A. Karellas, E. A. Krupinski, and W. R. Hendee, Eds. *Advances in Medical Physics 2010.* Madison, WI: Medical Physics Publishing, 2010.

Griswold, M. A., P. M. Jakob, R. M. Heidemannn, M. Nittka, V. Jellus, J. Wang, B. Kiefer, and A. Haasee. (2002). "Generalized Autocalibrating Partially Parallel Acquisition (GRAPPA)." *Magn. Reson. Med.* 47:1202–10.

Griswold, M. A. "Basic Reconstruction Algorithms for Parallel Imaging." In *Parallel Imaging in Clinical MR Applications.* Schoenberg, S. O., O. Dietrich, and M. F. Reiser, Eds. Berlin: Springer-Verlag, 2007.

Gudbjartsson, H. and S. Patz. (1995). "The Rician Distribution of Noisy MRI Data." *Magn. Reson. Med.* 34:910–14. https://www.ncbi.nlm.nih.gov/pmc/ articles/PMC2254141/.

Hahn, E. L. (1950). "Spin Echoes." Phys. Rev. 80:580–94.

Han, M., H. Jang, and J. Baek. (2018). "Evaluation of human observer performance on lesion detectability in single-slice and multi-slice dedicated breast cone beam CT images with breast anatomical background." *Med. Phys.* 45:5385–96.

Hand, J., H. Bosmans, C. Caruana, S. Keevil, D. G. Norris, R. Padovani, and O. Speck. (2012). "The European Federation of Organisations for Medical Physics Policy Statement No 14: The role of the Medical Physicist in the management of safety within the magnetic resonance imaging environment: EFOMP recommendations." *Phys. Med.* 29:122–25.

Harnsberger, H. R., A. G. Osborn, J. S. Ross, K. R. Moore, K. L. Salzman, C. R. Carrasco, B. E. Halmiton, H. C. Davidson, and R. H. Wiggins. *Diagnostic and Surgical Imaging Anatomy: Brain, Head and Neck, Spine.* 3rd ed. Salt Lake City: Amirsys, 2007.

Henkelman, R. M., G. J. Stanisz, and S. J. Graham. (2001). "Magnetization Transfer in MRI: A Review." *NMR Biomed.* 14(2):57–64.

Higuchi, M., N. Iwata, Y. Matsuba, K. Sato, K. Sasamoto, and T. C. Saido. (2005). "19F and 1H MRI Detection of Amyloid Beta Plaques in vivo." *Nat. Neurosci.* 8(4):527–33

Hindorean, C., D. P. Pfeifer, M. Chwialkowski, and R. Peshock. "Direct digitization of MR data." In *Proceedings of the Annual Conference on Engineering in Medicine and Biology* (pt 1 ed., Vol. 15, pp. 195). Piscataway, NJ: IEEE, 1993.

Hinshaw, W. S. (1976). "Image formation by Nuclear Magnetic Resonance: The sensitive-point method." *J. Appl. Phys.* 47:3709–21.

Hirokawa, Y., H. Isoda, Y. S. Maetani, et al. (2008). "MRI artifact reduction and quality improvement in the upper abdomen with PROPELLER and Prospective Acquisition Correction (PACE) technique." *Am. J. Roengenol.* 191(4):1154–58.

ICNIRP. (2009). Amendment to the ICNIRP. "Medical Magnetic Resonance (MR) Procedures: Protection of patients." HEALTH International Commission on Non-Ionising Radiation (ICNIRP). *Health Phys.* 97(3):259–61. doi: 10.1097/HP.0b013e3181aff9eb. http://www.icnirp.org/en/ publications/index.html.

IEC. International Electrotechnical Commission. Medical electrical equipment – particular requirements for the safety of magnetic resonance equipment for medical diagnosis. IEC 60601-2-33. Geneva, Switzerland: IEC, 2010.

IEC. International Electrotechnical Commission. ICE 60601-2-33, Third edition 2010-03, Medical Electrical Equipment – Part 2-33; Particular requirements for the basic safety and essential performance of magnetic resonance equipment for medical diagnosis. Geneva, Switzerland: IEC, 2010.

ISO. International Organization for Standardization. ISO/TS 10974 EDI. *Assessment of the safety of magnetic imaging for patients with active implantable medical device.* Geneva, Switzerland: ISO, 2012.

James, J. F. *A Student's Guide to Fourier Transforms.* Cambridge: Cambridge University Press, 1995.

Jezzard, P., P. M. Mathew, and S. M. Smith, Eds. *Functional Magnetic Resonance Imaging: An Introduction to Methods.* Oxford: Oxford University Press, 2003.

Jolesz, F. A. and N. McDannold. (2008). "Current Status and Future Potential of MRI-Guided Focused Ultrasound Surgery." *J. Magn. Reson. Imaging* 27(2):391–99. doi:10.1002/jmri.21261.

Jones, D. K. *Diffusion MRI.* New York: Oxford University Press, 2010.

Kanal, E., E. Barkovich, et al. (2013). "Expert Panel on MRS. ACR Guidance Document on MR Safe Practices." *J. Magn. Reson. Imaging* 37:501–30.

Katscher, U., T. Voigt, C. Findaklee, P. Vernickel, K. Nehrke, and O. Dossel. (2009). "Determination of Electrical Conductivity and local SAR via B1 Mapping." *IEEEW Trans. Med. Imaging* 28:1365–74.

Landau, L. D., E. M. Lifshitz, and L. P. Pitaevskii. *Electrodynamics of Continuous Media.* Oxford: Butterworth-Heinemann, 1984.

Larkman, D. J., J. V. Hajnal, A. H. Herlihy, G. A. Coutts, I. R. Young, and G. Ehnholm. (2001). "Use of Multicoil Arrays for Separation of Signal from Multiple Slices Simultaneously Excited." *J. Magn. Reson. Imaging* 13:313–17.

Lauderbur, P. C. (1973). "Image Formation by Induced Local Interactions: Examples Employing Nuclear Magnetic Resonance." *Nature* 242:190-91.

Le Bihan, D. and P. Van Zijl. (2002). "From the Diffusion Coefficient to the Diffusion Tensor." *NMR Biomed.* 15:431–34.

Liang, Z-P. and P. C. Lauterbur. *Principles of Magnetic Resonance Imaging: A Signal Processing Perspective.* New York: IEEE Press, 2000.

Ma, D., V. Gulani, N. Seiberlich, L. Kecheng, J. L. Sunshine, J. L. Duerk, and M. S. Griswold. (2013). "Magnetic Resonance Fingerprinting." *Nature* 495:187–92.

Mansfield, P. *MRI in Medicine: The Nottingham Conference.* Chapman and Hall, 1995.

Mariappan, Y. K., K. J. Glaser, and R. L. Ehman. (2010). "Magnetic Resonance Elastography: A Review." *Clin Anat.* 23(5):497–511. doi: 10.1002/ca.21006.

Markl, M., M. T. Draney, M. D. Hope, J. M. Levin, F. P. Chan, M. T. Alley, N. J. Pelc, and R. J. Herfkens. (2004). "Time-Resolved 3-Dimensional Velocity Mapping in the Thoracic Aorta: Visualization of 3-Directional Blood Flow Patterns in Healthy Volunteers and Patients." *J. Comput. Assist. Tomogr.* 28:459–68.

Mattson, J. and M. Simon. *The Pioneers of NMR and Magnetic Resonance in Medicine: The Story of MRI.* Jericho, New York: Bar-Ilan University Press, 1996.

Mastanduno, M. A., K. E. Michaelsen, S. C. Davis, S. Jiang, and B. W. Pogue. "Combined Magnetic Resonance Imaging and Near-Infrared Spectral Imaging." In Anastasio, M. A. and P. L. Rivieere. *Emerging Imaging Technologies in Medicine.* Boca Raton, FL: CRC Press, 2013.

Maxwell, J. C. "A Dynamical Theory of the Electromagnetic Field." (1865). *Phil. Trans. R. Soc. London* 155:459–512.

MHRA. (2015) *Safety Guidelines for Magnetic Resonance Imaging Equipment in Clinical Use.* Medicines and Healthcare Products Regulatory Agency (GB), v4.2.

Milburn, G. J. *The Feynman Processor: Quantum Entanglement and the Computing Revolution.* Cambridge, MA: Helix Books, 1999.

Muthupillai, R., D. J. Lomas, P. J. Rossman, J. F. Greenleaf, A. Manduca, and R. L. Ehman. (1995). "Magnetic Resonance Elastography by Direct Visualization of Propagating Strain Waves." *Science* 269:1854–57.

Noll, D. C, S. J. Peltier, and F. E. Boada. (1998). "Simultaneous Multislice Acquisition Using Rosette Trajectories (SMART): A New Imaging Method for Functional MRI." *Magn. Reson. Med.* 39:709–16.

Obuchowski, N. A. and J. A. Bullen. (2018). "Receiver Operating Characteristic (ROC) Curves: Review of Methods with Applications in Diagnostic Medicine." *Phys. Med. Biol.* 63(7):07TR01. doi:10.1088/1361-6560/aab4b1.

Oesterle, C., R. Strohschein, M. Kohler, M. Schnell, and J. Henning. (2000). "Benefits and Pitfalls of Keyhole Imaging, Especially in First-Pass Perfusion Studies." *J. Magn. Reson. Imaging* 11:312–23.

Oh, C. H., Y. C. Ryu, J. H. Hyun, S. H. Bae, S. T. Chung, H. W. Park, and Y. G. Kim. (2010). "Dynamic Range Expansion of Receiver by using Optimized Gain Adjustment for High-Field MRI." *Concepts Magn. Reson. Part A* 36A:243–54.

Oppelt, A. *Imaging Systems for Medical Diagnostics: Fundamentals, Technical Solutions and Applications for Systems Applying Ionizing Radiation, Nuclear Magnetic Resonance and Ultrasound.* Erlangen, Germany: Publicis KommunikationsAgentur GmbH, 2006.

Ouyang, J., Q. Li, and G. El Fakhri. "*Multimodality: Positron Emission Tomography—Magnetic Resonance Imaging.*" In Anastasio, M. A. and P. L. Rivieere. *Emerging Imaging Technologies in Medicine.* Boca Raton, FL: CRC Press, 2013.

Parker, J. A. *Image Reconstruction in Radiology.* Boca Raton, FL: CRC Press, 2018.

Parkes, L. M. and P. S. Tofts. (2002). "Improved Accuracy of Human Cerebral Perfusion Measurements Using Arterial Spin Labeling; Accounting for Capillary Water Permeability." *Magn. Reson. Med.* 48:27–41.

Paschal, C. B. and H. D. Morris. "K-Space in the Clinic." (2004). *J. Magn. Reson. Imaging* 19(2)145–59.

Pipe, J. G. (1999). "Motion correction with PROPELLER MRI: application to head motion and free-breathing cardiac imaging." *Magn. Reson. Med.* 42:963–69.

Pipe, J. G. and N. Zwart. (2006). "Turboprop: Improved PROPELLER Imaging." *Magn. Reson. Med.* 55(2):380–85.

Placidi, G. *MRI: Essentials for Innovative Technologies*. Boca Raton, FL: CRC Press, 2012.

Poole, M. and R. Bowtell. (2007). "Novel Gradient Coils Designed Using a Boundary Element Method." *Concepts Magn. Reson. Part B*. 31B(3):161–75.

Powell, D. and C. Smith. "*Diffusion Tensor Imaging: Neuroscience Applications.*" In Wolbarst, A. B., P. Capasso, D. J. Godfrey, R. R. Price, B. R. Whiting, and W. R. Hendee, Eds. *Advances in Medical Physics 2012.* Madison, WI: Medical Physics Publishing, 2012.

Press, W. H, S. A. Teukolsky, W. T. Vetterling, and B. P. Flannery. *Numerical Recipes: The Art of Scientific Computing*, 3rd Ed. New York: Cambridge University Press, 2007.

Price, R. R. and A. B. Wolbarst. "MRI Sequences Above and Beyond Spin-Echo." In Godfrey, D. J., S. K. Das, and A. B. Wolbarst, Eds. *Advances in Medical Physics 2014.* Madison, WI: Medical Physics Publishing, 2014.

Pruessmann, K. P., M. Weigner, M. B. Scheidegger, and P. Boesiger. (1999). "SENSE: Sensitivity Encoding for Fast MRI." *Mag. Reson. Med.* 42:952–62.

Purcell, E. M., R. V. Pound, and H. C. Torrey. (1946). "Measurement of Magnetic Resonance Absorption by Nuclear Moments in a Solid." *Phys. Rev.* 69:681.

Purcell, E. M. and D. J. Morin. *Electricity and Magnetism*, 3rd Ed. Cambridge, UK: Cambridge University Press, 2013.

Rabi, H., H. F. Ramsey, and J. Schwinger. (1954). "Use of Rotating Coordinates in Magnetic Resonance Problems." *Rev. Mod. Phys.* 26:167–71. (Three Nobel laureates!)

Robson, M. D. and G. M. Bydder. (2006). "Clinical Ultrashort Echo Time Imaging of Bone and Other Connective Tissues." *NMR Biomed.* 19(7):765–80.

Rockmore, D. N. (2000). "The FFT: An Algorithm the Whole Family Can Use." *Comput. Sci. Eng.* 2:60–64. doi:10.1109/5992.814659.

Roos, J., H. McAdams, S. Kaushik, and B. Driehuys. (2015). "*Hyperpolarized Gas MR Imaging: Technique and Application.*" *Magn. Reson. Imaging Clin. N Am.* 23(2):217–29.

Scarlett, J. (2010). "Advancements in MRI Architectures Reduce Design Complexity while Improving Image Quality." *Analog Devices*. Technical Article MS-2058.

Sears, F. W., M. W. Zemansky, H. D. Young, and R. A. Freedman. *University Physics with Modern Physics*, 14th Ed. Harlow, England: Pearson Higher Education, 2015.

Seo, J. K., E. J. Woo, U. Katscher, and Y. Wang. *Electro-Magnetic Tissue Properties MRI.* London: Imperial College Press, 2014.

Servoss, T. G. and J. P. Hornak. (2011). "Converting the Chemical Shift Artifact to a Spectral Image." *Concepts Magn. Reson. Part A* 38:107–16.

Shellock, F. G. and J. V. Crues. (2004). "*MR Procedures: Biologic Effects, Safety, and Patient Care.*" *Radiology* 232:635–52.

Shellock, F. G. *Reference Manual for Magnetic Resonance Safety, Implants, and Devices,* 2016 Ed. Biomedical Research Publishing Group, 2016.

Slichter, C. P. *Principles of Magnetic Resonance,* 3rd Ed. Berlin: Springer-Verlag, 1996.

Smith, H-J. and F. N. Ranallo. *A Non-Mathematical Approach to Basic MRI.* Madison, WI: Medical Physics Publishing, 1989.

Smith, S. W. *The Scientist and Engineer's Guide to Digital Signal Processing*. San Diego, CA: California Technical Publishing, 1997.

Sodickson, D. K., M. A. Griswold, P. M. Jakob, R. R. Edelman, and W. J. Manning. (1999). "Signal-to-Noise Ratio and Signal-to-Noise Efficiency in SMASH Imaging." *Mag. Reson. Med.* 41:1009–22.

Spratt, K. S., K. M. Lee, and P. S. Wilson. (2018). "Champagne Acoustics." *Phys. Today* 71:66–67.

Stanisz, G. J., E. E. Odrobina, J. Pun, M. Escaravage, S. J. Graham, M. J. Bronskill, and R. M. Henkelman. (2005). "T1, T2 Relaxation and Magnetization Transfer in Tissue at 3T." *Magn. Reson. Med.* 54:507–12.

Steckner, M. C. "MRI Update." In Wolbarst, A. B., K. E. Mossman, and W. Hendee, Eds. *Advances in Medical Physics 2008.* Madison, WI: Medical Physics Publishing, 2008.

Steckner, M. C. "*Current Issues in MRI Safety*." In Wolbarst, A. B., P. Capasso, D. J. Godfrey, R. R. Price, B. R. Whiting, and W. R. Hendee, Eds. *Advances in Medical Physics 2012.* Madison, WI: Medical Physics Publishing 2012

Strang, G. *Introduction to Applied Mathematics.* Wellesley: Wellesley-Cambridge Press, 1986.

Suetens, P. *Fundamentals of Medical Imaging.* New York: Cambridge University Press, 2002.

Symms, M., H. R. Jäger, K. Schmierer, and T. A. Yousry. (2004). "A Review of Structural Magnetic Resonance Neuroimaging." *J. Neurol. Neurosurg. Psychiatry* 75:1235–44.

Tan, E. T., J. Huston, N. G. Campeau, and S. J. Reiderer. (2010). "Fast Inversion Recovery Magnetic Resonance Angiography of the Intracranial Arteries." *Magn. Reson. Med.* 63:1648–58.

Titchmarsh, E. C. *Introduction to the Theory of Fourier Integrals.* Oxford: Clarendon Press, 1962.

Twieg, D. B. (1983). "The k-Trajectory Formulation of the NMR Imaging Process with Applications in Analysis and Synthesis of Imaging Methods." *Med. Phys.* 10(5):610–21.

Uğurbil, K., G. Adriani, P. Anderson, W. Chen, M. Garwood, R. Gruetter, P. G. Henry, S. G. Kin, H. Lieu, L. Tkac, T. Vaughn, P. F. Van De Moortele, E. Yacoub, and X. H. Zhu. (2003). "Ultrahigh Field Magnetic Resonance Imaging and Spectroscopy." *Magn. Reson. Imaging* 21(10):1263–81.

van Wijk, K. and S. Hitchman. (2017). "Apple Seismology." *Phys. Today* 70:94–95.

Wagner, R. F., W. A. Yousef, and W. Chen. "*Finite Training of Radiologists and Statistical Learning Machines: Parallel Lessons.*" In Wolbarst, A. B., K. E. Mossman, and W. Hendee, Eds. *Advances in Medical Physics 2008.* Madison, WI: Medical Physics Publishing, 2008.

Wang, W., J. C. Lin, W. Mao, W. Liu, M. B. Smith, and C. M. Collins. (2007). "SAR and Temperature: Simulations and Comparison to Regulatory Limits for MRI." *J. Magn Reson. Imaging* 26:437–41.

Wang, Y. *Principles of Magnetic Resonance Imaging.* 2016.

Wang, Y. and T. Liu. (2015). "*Quantitative Susceptibility Mapping (QSM): Decoding MRI Data for a Tssue Magnetic Biomarker.*" *Magn. Reson. Med.* 73(1):82–101.

Webb, S. *Webb's Physics of Medical Imaging,* 2nd Ed. Boca Raton, FL: CRC Press, 2012.

Wiesinger, F., P. Boesiger, and K. P. Pruessmann. (2004). "Electrodynamics and Ultimate SNR in Parallel MR Imaging." *Magn. Reson. Med.* 52(2):376–90.

Wolbarst, A. B. (1983). "On Competitive Failure Modes and the Usefulness of a "Survival Curve Point of View." *Med. Phys.* 10(2):232–36.

Wolbarst, A. B., P. Capasso, and A. Wyant. *Medical Imaging—Essentials for Physicians.* Hoboken, NJ: Wiley-Blackwell Medicine, 2013.

Wolbarst, A. B. *Physics of Radiology,* 2nd Ed. Madison, WI: Medical Physics Publishing, 2005.

Woods, T. O. (2007). "Standards for Medical Devices in MRI: Present and Future." *JMRI* 26(5):1186–89.

Yanasak, N. E. and A. B. Wolbarst. "Parallel Imaging: Moving Beyond the Use of Gradients for Spatial Encoding." In Godfrey, D. J., S. K. Das, and A. B. Wolbarst, Eds. *Advances in Medical Physics 2014.* Madison, WI: Medical Physics Publishing, 2014.

Young, H., et al. *University Physics with Modern Physics,* 14th Ed. Summerfield, NC: eBookAir.com, 2015. [Original authors F. Sears and M. Zemanski.]

Zaremba, L. A. *FDA Guidelines for Magnetic Resonance Equipment Safety.* Center for Devices and Radiological Health. undated.

Zhang, L., A. McCallister, K. M. Koshlap, and R. T. Branca. (2018). "Correlation distance dependence of the resonance frequency of intermolecular zero quantum coherences and its implication for MR thermometry." *Magn. Reson. Med.* 79(3):1429–38.

Zhuo, J. and R. P. Gullapalli. "Artifacts in MRI." In Wolbarst, A. B., A. Karellas, E. A. Krupinski, and W. R. Hendee, Eds. *Advances in Medical Physics 2010.* Madison, WI: Medical Physics Publishing, 2010.

Symbols

Figures and tables are indicated by an italic *f* or *t* following the page number.

Symbol	Definition/Meaning	Page
$\cdot$	scalar product of vectors	5, 78-80, 79, 80, 236
$\times$	vector product	24, 110
$\|\uparrow>$	quasi-quantum spin-up state	32
42.58 MHz/T	Larmor frequency for proton	6, 35
∇	del or nabla: in grad, div, curl	25
$\mathrm{III}(t)$	Dirac comb, sha function	100
A	amplitude	2.5, 5.12
A(v), dE/dt(v)	resonance amplitude, lineshape	45, 109, 109f
$A(t)$	activity of radioactive sample	76
A	coil area	31
$\|\boldsymbol{\alpha}\|$, alpha	absolute value of vector $\boldsymbol{\alpha}$	3
α_E	Ernst angle	169, 227
$\alpha°$	tip angle	167–169, 168*f*, 226*f*, 227, 227*f*
B	magnetic field	25
$\boldsymbol{B}_0$	main magnetic field	27*t*, 36, 40, 135, 136*f*, 137*t*
$\boldsymbol{B}_z(x)$	*z*-field in voxel at *x*	4
$B_1(t)$	RF magnetic field	27, 40, 111, 135, 136*f*, 137*t*
BW	bandwidth	43, 147, 204, 204*f*, 205
b	diffusion *b*-value	249
c	speed of EM radiation	26
$[C]$, C(x,t)	concentration	248
C	contrast	47, 193
CD	contrast-detail diagram	40, 52, 55–56, 55*f*, 103
$C_{min}\ A^{1/2}$ SNR	Rose criterion	103
$\gamma_n/2\pi$, gamma	gyromagnetic ratio, type-n nucleus	3, 5, 31, 126
Γ, gamma	universal resonance function	109, 109f
Δ, δ, delta	small change or difference	5
Δ, δ, G_{diff}, b	diffusion sequence parameters	249
$\delta(x-x')$	Dirac delta function	83–85, 84f
δ_{ij}	Kronicker delta function	78
D, D_x, D_{ij}	diffusivity: isotropic, anisotropic, tensor	247*f*, 248, 251, 251*f*
DQE(k)	Detective Quantum Efficiency	55
$E(x, t)$	electric field	23, 26
E, ΔE	energy	4, 26, 27, 262
E	inter-proton dipole-dipole	32
ΔE_{Zeeman}	Zeeman energy splitting	5, 34
ε_0, epsilon	electric permittivity, free space	25, 26
FID	free induction decay	123
FOV	field of view	42, 131, 148

Symbol	Definition/Meaning	Page
$FOV_{\boldsymbol{k}}$	FOV in ***k***-space	102
FT, FT^{-1}	Fourier (and inverse) Transform	82*f*, 87, 125*f*, 126, 146*f*, 207*f*
F_x	magnetic force on object	140, 260
[Gd]	gadolinium agent concentration	11f, 29, 167*f*, 179, 180, 263*t*
GE, GRE	gradient (recall) echo sequence	226, 226*f*, 230*f*
$\boldsymbol{G_x}(t) = \partial \boldsymbol{B_z} / \partial x$	*x*-gradient	6, 27, 40, 41, 45f, 121, 123f, 126, 136*f*, 141, 142*f*, 183*f*, 197*f*, 198*f*
h	Planck constant	4, 33, 35
θ, theta	general angle, phase	33, 79, 79*f*
i	current	25, 31
I	proton spin angular momentum	3
IR	inversion recovery sequence	200–202
$\boldsymbol{J}$	classical angular momentum	107
$J(v)$	noise power spectrum	160*f*, 162
$\boldsymbol{k}$	wave-number, -vector	77, 89, 206, 206*f*
Δk_x	step in ***k***-space	129*f*
k_{max}	max spatial wavelength	101–102
$k_{\boldsymbol{x}}(t) = (\gamma/2\pi)G_{\boldsymbol{x}}\, t$	conversion to $k_{\boldsymbol{x}}$-space	128, 128*f*, 129*f*, 195, 196*f*
k_B	Boltzmann constant	67
κ, kappa	Hooke's law constant	76, 80, 176
L	light intensity	50
ln(x)	natural logarithm	75
LSF(x)	line-spread function	87
λ, lambda	decay constant	75–76
λ	wavelength	25, 25*f*, 26*t*, 50, 77
$\boldsymbol{m}(x,t)$	magnetization of voxel	7, 33, 63, 67, 155
$\boldsymbol{m_z}(x,t)$	component of $\boldsymbol{m}(x,t)$ along ***z***-axis	64, 184, 189
$\boldsymbol{m_{xy}}(x,t)$	component of $\boldsymbol{m}(x,t)$ in ***x-y*** plane	117, 123*f*, 187
$\lvert\boldsymbol{m_z}(t_+)\rvert = \lvert\boldsymbol{m_{xy}}(x,t_-)\rvert$	dynamics of 90° pulse	112, 116, 163, 164*f*, 184, 188, 190
$\boldsymbol{m}_0(x)$	$\boldsymbol{m}(x)$ at thermal equilibrium	64, 67
$d\boldsymbol{m}(t)/dt$	Bloch equations	110, 110*t*, 158, 175, 248
$\boldsymbol{m}(x,k_x)$	$\boldsymbol{m}$ expressed in terms of phase	129
$\boldsymbol{M}(t)$	magnetization from all voxels together	63
M	electrons' magnetization field	29
$\mathcal{M}(k)$	modulation	50
MTF(v)	modulation transfer function	49, 52, 87
μ, mu	statistical mean	95
μ	magnetic moment of loop	31
$\boldsymbol{\mu}$	nuclear magnetic moment	3, 5, 31–32
$\boldsymbol{\mu_z}$	*z*-component of $\boldsymbol{\mu}$	4
$\mu = \lvert\boldsymbol{\mu}\rvert$	magnitude of magnetic moment	3
μ_0	magnetic permeability of vacuum	24*t*, 25, 26
n	integer number of echoes	186
N_-	number of spins in low-energy state	70
$N \equiv [N_- + N_+]$	total number of spins in voxel	64
$n \equiv [N_- - N_+]$	excess spins in low-energy state	64, 156
n_0	equilibrium value of n	67

Symbol	Definition/Meaning	Page
N_+/N_-	Boltzmann	67
$dn(t)/dt$	population change rate	70, 75
N_n	number of neutrons in nth nucleus	3
N_{TP}	number of true positives	57
$(N_{TP} + N_{TN}) / N_{total}$	accuracy	58
$n \equiv Z_n + N_n$	atomic weight	3
$N(t)$	noise signal	55, 94, 96
$N_{batches}, N_{slices}$	multiplexed multiple slice	220–223
NEX	number of repetitions, averaged	98, 182*t*, 219, 224, 231
ν, nu	frequency of signal, wave	6, 26*t*, 27, 77, 107
ν_1	fundamental frequency	81
ν_{beat}	beat frequency	123*f*, 124
ν_{IF}	intermediate frequency	144–145
ν_{res}	natural resonant frequency	106f
ν_{Larmor}	Larmor frequency	5–6, 35–36, 44, 106, 106*f*, 107–109, 108f
	Bloch equations	109–111
	lab. vs. rotating frame	113–114
	nutation of $\boldsymbol{m}(t)$ about $\boldsymbol{B}_1$	111, 112*f*
	selecting z-slice	120
	spin-state transition	158–160
$\nu_{nutation}$	nutation frequency	112f, 114
$\nu_{rotate}(\boldsymbol{x})$	offset frequency in rotating frame	121, 160–161, 160*f*
ν_{samp}	sampling rate	99–100, 147
[O]	oxygen concentration in blood	48t, 251–253, 252*f*
$P(x)$	fluid pressure at x	244, 244I
$P_{\mu,\sigma}(n)$	Normal/Gaussian distribution	96, 96*t*
$P_{\mu}(n)$	Poisson distribution	97, 98
$P_{q,\sigma}(n)$	Rician distribution	99
$P(X\|Y)$	conditional probability of X given Y	59
$P_{RF}(t)$	RF power	123*f*
$PD(x)$	proton density at x	38, 43, 154*t*
$PSF(x)$	point spread function	52–53, 87–88, 88*f*
r	radius of circle	31
$\boldsymbol{r}$	3D position vector	40
R	Reynolds number	244
R, Δx	resolution	49, 101, 102, 130
Re, Im	real, imaginary parts of complex number	79
Rect(t)	rectangle function = **FT**{Sinc}	84*f*, 87, 122, 205
ROC	receiver operating characteristics	56–59, 102
RF	radio-frequency	143*f*
ρ, rho	charge density, density	25, 244
$s(x,t)$	MR signal from voxel at x	9, 38, 67, 116, 190
$S(t)$	signal from multiple voxels	10, 126*f*, 127*f*, 128*f*, 195, 238
$S(\nu)$	frequency-spectrum of $S(t)$	125, 126, 126*f*, 127*t*, 146*f*
$S(k)$	$\boldsymbol{k}$-space representation of $S(t)$	126*f*, 127*t*, 128*f*
SAR(B_0)	specific absorption rate	163, 257, 257*t*, 262
SE	spin-echo	181–200

Symbol	Definition/Meaning	Page
SE/SW	SE spin-warp	212, 212*f*, 213*f*
Sinc(t)	Sinc function = **FT**{Rect}	84*f*, 87, 122
SNR	signal-to-noise ratio	53, 54*f*, 55, 96, 98, 103, 150, 193*t*, 234
σ, sigma	statistical standard deviation	95
σ	chemical shift	12, 12f, 29, 61
t, Δt	time, interval	–
Δt_{phase}	phase-encoding time	213
$t_{1/2}$	half-life	75
$t_{dwell} = 1/\nu_{samp}$	1/sampling frequency	147
T	tesla	3, 24
T	general relaxation time	8, 106, 117–118
T	period of wave	77
T_{acq}	acquisition time	218*f*, 219, 224
$1/T_{SFd\text{-}P}$	rate of Static Field de-Phasing	176, 177, 187, 190
T1	longitudinal spin relaxation time	8, 9, 11, 33, 153–157, 154*t*, 160*f*, 162, 164*f*
T1-*w*	T1-weighting in SE image	191, 194, 194*t*
T1(B_0, [Gd])	dependence on field, contrast agent	162, 180
T2	transverse spin relaxation time	9, 11, 28, 33, 125*t*, 154*t*, 171–177, 172*f*
T2*	T2 star	177, 186
TE	echo time	183*f*, 183–188, 193*t*
TI	inversion time	200, 201
TR	repetition time	163–169, 193*t*
TNf	true negative fraction (specificity)	57*t*
TPf	true positive fraction (sensitivity)	57*t*
τ, tau	torque	107, 107*f*
τ_c	correlation time	1, 71, 161
$V(t)$	voltage	148
V_S, V_N	signal, noise voltages	149, 150
$v(x)$	bulk fluid velocity, flow	242, 244
W_+, W_-, W_{RF}	spin transition rates	70
φ	phase	77
$\varphi(x,t)$	phase wave	131
φ_x, φ_y	x- and y-phases of $\boldsymbol{m}_{xy}(x,t)$	76, 197f, 198f, 212
$\Delta\varphi_y$	phase encoding step along k_y-axis	212
$\{e^{-2\pi i k_p x}\}$	basis vectors	83–85
Φ, phi	magnetic flux	116, 117, 149
Δx	step, width in x-space	131
χ, chi	magnetic susceptibility	29t, 30
ψ, psi	quantum state	33
$\omega = 2\pi\nu$	angular frequency	35, 77
Z_n	atomic number	3
$z = \lvert z\rvert e^{i\theta}$	complex number $z = x + iy$	79
Δz	slice thickness	204–205, 204f

Units, Constants, Relationships, Principles, and the Periodic Table

THE SEVEN BASIC SI UNITS

Quantity (symbol)	SI Unit
length* (x)	meter (m)
mass (m)	kilogram (kg)
time (t)	second (s)
electric current (I)	ampere (A)
temperature (T)	kelvin (K)
amount of substance	mole (mol)
luminous intensity	candela (cd)

* 1 inch = 2.54 cm; 1 angstrom (Å) = 10^{-10} m

SOME SI DERIVED UNITS

Quantity (symbol)	SI Unit
force (F)	newton ($N = kg\ m/s^2$)
work, energy (E)	joule ($J = N \times m$)
power (P)	watt ($W = J/s$)
electric charge (Q,q)	coulomb ($C = A \times s$)
electric potential (V)	volt ($V = J/C$)
electric field (E)	volt per meter (V/m)
resistance (R)	ohm ($\Omega = V/A$)
magnetic flux density (B)	tesla ($T = Wb/m^2$)
temperature (T)	degree Celsius (°C)
frequency (v)	hertz ($Hz = s^{-1}$)
wavelength (λ)	(m^{-1})
velocity (v,c)	(m/s)

SOME SI PREFIXES

10^{10}	tera (T)
10^{9}	giga (G)
10^{6}	mega (M)
10^{3}	kilo (k)
10^{-2}	centi (c)
10^{-3}	milli (m)
10^{-6}	micro (μ)
10^{-9}	nano (n)
10^{-12}	pico (p)

SOME PHYSICAL CONSTANTS

speed of light (c)	2.9979×10^{8} m/s
Planck's constant (h)	6.626×10^{-34} J s
charge on electron (e)	1.9602×10^{-19} C
mass of electron (m_e)	9.1094×10^{-31} kg
mass of proton (m_p)	1.6726×10^{-27} kg
mass of neutron (m_n)	1.6749×10^{-27} kg
Boltzman's constant (k)	1.3806×10^{-23} J-K^{-1}
gravitational constant (G)	6.6726×10^{-11} N-m^{2}-kg^{-2}
Avagadro's number (N_a)	6.0221×10^{23} mol^{-1}
absolute zero (0 K)	–273.15 °C (–459.67 °F)

SOME BASIC PHYSICAL RELATIONSHIPS

electric power	power = $I \times V$
Ohm's law	$V = I \times R$
propagation of waves	$\lambda \times v = v$ or c
Einstein Relation	$E = hv$

SOME USEFUL MATHEMATICAL RELATIONSHIPS

Exponentials and Logarithms

e = 2.71828....

$e^{-a} = 1/e^{a}$

$e^{b} e^{c} = e^{(b+c)}$

$\ln w = z$ means $w = e^{z}$

$\log_{10} w = z$ means $w = 10^{z}$

Probability

$P(A) = N(A) / N(\text{total})$	where there are N(A) different ways that event A can occur
mutually exclusive events	P (A or B) = P(A) + P(B)
independent events	P (A and B) = P(A) $\times$ P(B)

Poisson Statistics

If μ Poisson events occur per unit of time, distance, etc., on average, then:

Measurement will yield a result in the range	With a probability, (P)
$-\sqrt{\mu}$ to $\mu+\sqrt{\mu}$	0.683
$-2\sqrt{\mu}$ to $\mu+2\sqrt{\mu}$	0.95
$-3\sqrt{\mu}$ to $\mu+3\sqrt{\mu}$	0.995

IUPAC Periodic Table of the Elements

Key:

atomic number
Symbol
name
conventional atomic weight
standard atomic weight

1	2	3	4	5	6	7	8	9	10	11	12	13	14	15	16	17	18
1 **H** hydrogen 1.008 [1.0078, 1.0082]																	2 **He** helium 4.0026
3 **Li** lithium 6.94 [6.938, 6.997]	4 **Be** beryllium 9.0122											5 **B** boron 10.81 [10.806, 10.821]	6 **C** carbon 12.011 [12.009, 12.012]	7 **N** nitrogen 14.007 [14.006, 14.008]	8 **O** oxygen 15.999 [15.999, 16.000]	9 **F** fluorine 18.998	10 **Ne** neon 20.180
11 **Na** sodium 22.990	12 **Mg** magnesium 24.305 [24.304, 24.307]											13 **Al** aluminium 26.982	14 **Si** silicon 28.085 [28.084, 28.086]	15 **P** phosphorus 30.974	16 **S** sulfur 32.06 [32.059, 32.076]	17 **Cl** chlorine 35.45 [35.446, 35.457]	18 **Ar** argon 39.95 [39.792, 39.963]
19 **K** potassium 39.098	20 **Ca** calcium 40.078(4)	21 **Sc** scandium 44.956	22 **Ti** titanium 47.867	23 **V** vanadium 50.942	24 **Cr** chromium 51.996	25 **Mn** manganese 54.938	26 **Fe** iron 55.845(2)	27 **Co** cobalt 58.933	28 **Ni** nickel 58.693	29 **Cu** copper 63.546(3)	30 **Zn** zinc 65.38(2)	31 **Ga** gallium 69.723	32 **Ge** germanium 72.630(8)	33 **As** arsenic 74.922	34 **Se** selenium 78.971(8)	35 **Br** bromine 79.904 [79.901, 79.907]	36 **Kr** krypton 83.798(2)
37 **Rb** rubidium 85.468	38 **Sr** strontium 87.62	39 **Y** yttrium 88.906	40 **Zr** zirconium 91.224(2)	41 **Nb** niobium 92.906	42 **Mo** molybdenum 95.95	43 **Tc** technetium	44 **Ru** ruthenium 101.07(2)	45 **Rh** rhodium 102.91	46 **Pd** palladium 106.42	47 **Ag** silver 107.87	48 **Cd** cadmium 112.41	49 **In** indium 114.82	50 **Sn** tin 118.71	51 **Sb** antimony 121.76	52 **Te** tellurium 127.60(3)	53 **I** iodine 126.90	54 **Xe** xenon 131.29
55 **Cs** caesium 132.91	56 **Ba** barium 137.33	57-71 lanthanoids	72 **Hf** hafnium 178.49(2)	73 **Ta** tantalum 180.95	74 **W** tungsten 183.84	75 **Re** rhenium 186.21	76 **Os** osmium 190.23(3)	77 **Ir** iridium 192.22	78 **Pt** platinum 195.08	79 **Au** gold 196.97	80 **Hg** mercury 200.59	81 **Tl** thallium 204.38 [204.38, 204.39]	82 **Pb** lead 207.2	83 **Bi** bismuth 208.98	84 **Po** polonium	85 **At** astatine	86 **Rn** radon
87 **Fr** francium	88 **Ra** radium	89-103 actinoids	104 **Rf** rutherfordium	105 **Db** dubnium	106 **Sg** seaborgium	107 **Bh** bohrium	108 **Hs** hassium	109 **Mt** meitnerium	110 **Ds** darmstadtium	111 **Rg** roentgenium	112 **Cn** copernicium	113 **Nh** nihonium	114 **Fl** flerovium	115 **Mc** moscovium	116 **Lv** livermorium	117 **Ts** tennessine	118 **Og** oganesson

57 **La** lanthanum 138.91	58 **Ce** cerium 140.12	59 **Pr** praseodymium 140.91	60 **Nd** neodymium 144.24	61 **Pm** promethium	62 **Sm** samarium 150.36(2)	63 **Eu** europium 151.96	64 **Gd** gadolinium 157.25(3)	65 **Tb** terbium 158.93	66 **Dy** dysprosium 162.50	67 **Ho** holmium 164.93	68 **Er** erbium 167.26	69 **Tm** thulium 168.93	70 **Yb** ytterbium 173.05	71 **Lu** lutetium 174.97
89 **Ac** actinium	90 **Th** thorium 232.04	91 **Pa** protactinium 231.04	92 **U** uranium 238.03	93 **Np** neptunium	94 **Pu** plutonium	95 **Am** americium	96 **Cm** curium	97 **Bk** berkelium	98 **Cf** californium	99 **Es** einsteinium	100 **Fm** fermium	101 **Md** mendelevium	102 **No** nobelium	103 **Lr** lawrencium

For notes and updates to this table, see www.iupac.org. This version is dated 1 December 2018.

United Nations Educational, Scientific and Cultural Organization

2019 IYPT International Year of the Periodic Table of Chemical Elements

IUPAC

Index

Figures and tables are indicated by an italic *f* or *t* following the page number.

G

M

Q

R

T